Lecture Notes in Networks and Systems

Volume 1380

The series “Lecture Notes in Networks and Systems” publishes the latest developments in Networks and Systems—quickly, informally and with high quality. Original research reported in proceedings and post-proceedings represents the core of LNNS.

Volumes published in LNNS embrace all aspects and subfields of, as well as new challenges in, Networks and Systems.

The series contains proceedings and edited volumes in systems and networks, spanning the areas of Cyber-Physical Systems, Autonomous Systems, Sensor Networks, Control Systems, Energy Systems, Automotive Systems, Biological Systems, Vehicular Networking and Connected Vehicles, Aerospace Systems, Automation, Manufacturing, Smart Grids, Nonlinear Systems, Power Systems, Robotics, Social Systems, Economic Systems and other. Of particular value to both the contributors and the readership are the short publication timeframe and the world-wide distribution and exposure which enable both a wide and rapid dissemination of research output.

The series covers the theory, applications, and perspectives on the state of the art and future developments relevant to systems and networks, decision making, control, complex processes and related areas, as embedded in the fields of interdisciplinary and applied sciences, engineering, computer science, physics, economics, social, and life sciences, as well as the paradigms and methodologies behind them.

Indexed by SCOPUS, EI Compendex, INSPEC, WTI Frankfurt eG, zbMATH, SCImago.

All books published in the series are submitted for consideration in Web of Science.

For proposals from Asia please contact Aninda Bose (aninda.bose@springer.com).

Jyotsna K. Mandal · Mike Hinchey ·
Satyajit Chakrabarti
Editors

Recent Advances in Artificial Intelligence and Smart Applications

Proceedings of RAAISA 2024

Springer

Editors
Jyotsna K. Mandal
Department of Computer Science
and Engineering
University of Kalyani
Kalyani, West Bengal, India

Mike Hinchey
Department of Computer Science
and Information Systems
University of Limerick
Limerick, Ireland

Satyajit Chakrabarti
University of Engineering and Management
Kolkata, West Bengal, India

ISSN 2367-3370 ISSN 2367-3389 (electronic)
Lecture Notes in Networks and Systems
ISBN 978-981-96-5821-3 ISBN 978-981-96-5822-0 (eBook)
https://doi.org/10.1007/978-981-96-5822-0

This Springer imprint is published by the registered company Springer Nature Singapore Pte Ltd.
The registered company address is: 152 Beach Road, #21-01/04 Gateway East, Singapore 189721, Singapore

Preface

The book is a collection of best-selected research papers presented at the 2nd International Conference "Recent Advances in Artificial Intelligence and Smart Applications (RAAISA) 2024." The book explores fresh concepts, practical applications, and insights shared by research engineers, scientists, industrialists, scholars, and students worldwide. It covers research contributions from Artificial Intelligence and Smart Applications which are from the areas such as Smart Agent-Based Systems, Human-Computer Interaction Technologies, Reinforcement Learning, Sentiment Analysis, Recurrent Neural Networks and its Applications, Genetic Algorithm Neural Networks, Deep Learning-Based Smart Systems Security, Privacy, and Trust issues in Smart Computing, Smart Computing using generative AI, Augmented Reality/Virtual Reality, AI/ML, and IoT.

The 2nd International Conference "Recent Advances in Artificial Intelligence and Smart Applications (RAAISA) 2024" aimed at making the academic sector visualize and understand the industry and its need, by acting as a bridge between the two sectors. The organizers successfully conducted the first conference RAAISA 2023 with 17% acceptance rate and published in the Springer book series *Innovations in Sustainable Technologies and Computing*.

There where tutorials, Keynote addresses, invited, talks, and general speeches of interest, talks from industries, paper, presentations, and panel discussions in multiple tracks covering diverse topic of interest. The proceedings of selected and presented papers to be published in LNNS series of Springer nature by the Springer at free of cost. We are thankful to the authority of springer nature for their active association with this conference.

The papers are checked for similarities multiple times through authentic, followed by updation by the authors and draw double blinded refuse. Thirty-eight quality papers are uploaded after presentation of the same by the authors.

The editors express their gratitude to the reviewers of the articles for their enormous AirPods. Dhar also like to thanks the authors for submitting their papers in this conference, the editors also express is there gratitude to the organisers of this conference.

We hope that this volume will be useful and contributes actively towards the holistic progress of the civilisation. This volume will be value added bits and pieces for academicians, researchers, and young promising engineers.

Kalyani, India — Dr. Jyotsna K. Mandal
Limerick, Ireland — Dr. Mike Hinchey
Kolkata, India — Dr. Satyajit Chakrabarti

Contents

Application of Chaotic Map and DNA Encoding for Securing Stereo Audio Data

Mousomi Roy and Kalyani Mali

Abstract The importance of audio data in different areas such as telecommunications, entertainment, and confidential information shows the need for advanced encryption methods. In the ever-evolving information security environment, the need for strong and innovative encryption technologies is critical. Traditional encryption methods have long been the cornerstone of digital data security. However, as technology advances, so do the challenges of sophisticated attackers. The combination of chaos and DNA encoding not only hardens audio data against potential breaches but also ushers in a new dimension of secure communication. This study applies chaos theory and DNA encoding to audio encryption to demonstrate the importance of protecting audio data. In an era of growing concerns about privacy and data integrity, this innovative approach is a powerful solution for protecting audio information. Using chaos theory, it is possible to incorporate deterministic yet unpredictable dynamics to embed audio information in a chaotic system, which would be difficult to decode by traditional methods. To improve this strategy, DNA encoding is introduced to convert audio data into genetic code, using the information density and flexibility of DNA. Converting audio data to DNA code adds a layer of security and data density. The robustness of genetic material and the complex dance of chaos create a powerful encryption strategy that resists traditional cryptographic attacks.

Keywords Audio security · Chaos · DNA encoding

M. Roy (✉)
Department of Computer Science and Engineering, Hooghly Engineering & Technology College, Hugli-Chuchura, India
e-mail: iammouroy@gmail.com

K. Mali
Department of Computer Science and Engineering, University of Kalyani, Kalyani, India

J. K. Mandal et al. (eds.), *Recent Advances in Artificial Intelligence and Smart Applications*, Lecture Notes in Networks and Systems 1380,
https://doi.org/10.1007/978-981-96-5822-0_1

1 Introduction

In a world where communication is transcendently advanced and the transmission of touchy data may be a scheduled undertaking, the requirement for impenetrable encryption strategies has never been more basic. With the rise of quantum computing and the consistent advancement of cyber dangers, conventional encryption methods are progressively helpless to compromise. As a reaction to these challenges, this paper proposes a cutting-edge sound encryption approach that leverages chaos theory and DNA encoding, offering a novel arrangement that guarantees both security and efficiency. Chaos theory starting from the domain of science and material science, has found its way into various disciplines and building applications. Within the setting of data security, chaos theory presents a component of unusualness and haphazardness that can upgrade the strength of encryption calculations. The sensitivity to the initial conditions and the non-linear flow of chaotic frameworks [4, 10, 11] give ground for creating cryptographic techniques for securing audio data. Drawing motivation from life, DNA encoding offers an interesting point of view on information capacity and recovery. The four nucleotide bases—adenine (A), thymine (T), cytosine (C), and guanine (G)—serve as nature's letter sets, encoding the diagram of life in a surprisingly thick and versatile way. The versatility and storage capacity of DNA encoding make it a captivating candidate for encoding and putting away sensitive audio information [20, 27].

In this encryption approach, the audio information is changed into a computerized representation of DNA arrangements [1, 10, 21]. Each nucleotide compares to a particular sound fragment, making a biological analog for the advanced sound flag. The natural nature of DNA includes an extra layer of complexity and security to the encoded data, as translating requires not as it were computational control but moreover a profound understanding of atomic biology. The collaboration between the chaos theory and DNA encoding shapes the foundation of our proposed audio encryption framework [17, 22, 24]. The interweaving of these two components makes the proposed encryption strategy makes it secure, robust, and resilient against various attacks. In this integrated framework, chaos makes the encoding and decoding process unpredictable, guaranteeing that indeed the foremost modern attacks can be resisted. The chaos theory and DNA encoding impose obstruction against various cryptographic attacks, showing a special challenge for intruders to breach the system. Furthermore, the chaotic nature of the proposed framework presents a level that improves the security of the encryption plot. The affectability to beginning conditions guarantees that indeed a minor annoyance within the chaotic framework comes about in an unfathomably diverse cryptographic key, obstructing endeavors at switch engineering. A comprehensive investigation is required to understand the performance of any encryption framework. Hence, the proposed chaos and DNA-based audio encryption system undergo a thorough investigation to assess its resistance against various potential attacks [6, 23, 27].

While the integration of chaos and DNA encoding in audio encryption presents a promising security framework, the computational overhead of real-time chaotic framework and DNA encoding may pose challenges for resource-constrained situations. Moreover, the organic nature of DNA encoding presents contemplations related to capacity, recovery, and the potential effect of natural factors [3, 8, 26].

The rest of the article is organized as follows: Section 2 discusses some foundational concepts related to the proposed approach, Sect. 3 illustrates the proposed approach in detail, Sect. 4 discusses the obtained results, and Sect. 5 concludes the article.

2 Foundation

2.1 DNA Encoding Technique

As the demand for secure audio communication emerges, the application of DNA encoding standards to audio security can be helpful for more reliable communication of audio data. This intersection of digital technology and audio security exemplifies the adaptability of foundational concepts, demonstrating that the principles of encoding can be applied creatively to safeguard information in diverse contexts. In the context of audio, DNA encoding refers to the unique and intricate process of representing sound waves in a digital format, preserving the essential characteristics that make each audio snippet distinct [11].

DNA consists of four nucleotide bases (adenine, thymine, cytosine, and guanine), which can be encoded into binary code. In the natural DNA sequence, the bases are often represented by the letters A, T, C, and G. To convert DNA encoding to binary, any one simple mapping can be established as given in Table 1 [2, 12]. This binary encoding serves as a digital representation of the genetic information stored in the DNA sequence. Table 2 demonstrates the DNA XOR operation.

Table 1 DNA encoding rules

Rule No. →	0		1		2		3		4		5		6		7	
	A	00	A	00	C	00	C	00	G	00	G	00	T	00	T	00
	C	01	G	01	A	01	T	01	A	01	T	01	C	01	G	01
	G	10	C	10	T	10	A	10	T	10	A	10	G	10	C	10
	T	11	T	11	G	11	G	11	C	11	C	11	A	11	A	11

Table 2 DNA XOR operation

⊕	A	T	G	C
A	A	T	G	C
T	T	A	C	G
G	G	C	A	T
C	C	G	T	A

2.2 The RCM Chaotic Map

Chaos theory helps to incorporate unpredictability and randomness into security systems or processes as a strategy to enhance resilience and increase overall robustness. In this work, the RCM chaotic map is used and this map is defined in Eq. 1 [11].

$$s_i = \psi \sin\left(e^{-s_{i-1}} \times \left(1 - s_{i-1}^3\right)\right) \tag{1}$$

This chaotic property of the RCM map is already discussed and proven in [11].

3 Proposed Audio Encryption Framework

In this work, stereo audio data (i.e., with two channels) are used for the encryption purposes. Stereo data has two channels with double values. The proposed approach begins by converting each value into the IEEE 754 double-precision binary representation. Therefore, for each member of a particular sequence, there are corresponding 64 bits are generated [5] as shown in Fig. 1. Where, the first bit or the MSB is the sign bit, the next 11 bits are the exponent, and the rest of the bits, i.e., last 52 bits are mantissa. These 64 bits are divided into 8 groups.

Now, a pseudorandom bit sequence of length 16 is generated by the RCM map and the pseudorandom bit sequence generation procedure as described in [18] and this sequence is further divided into four groups of 4 bits each and it is illustrated in Fig. 2. After dividing into four groups, each group is reversed individually.

Now the third bit of the first and the third subgroups are discarded (illustrated in Fig. 3) and XOR operation is performed between the remaining three bits. These three bits are used to determine one DNA encoding rule from Table 1. The second

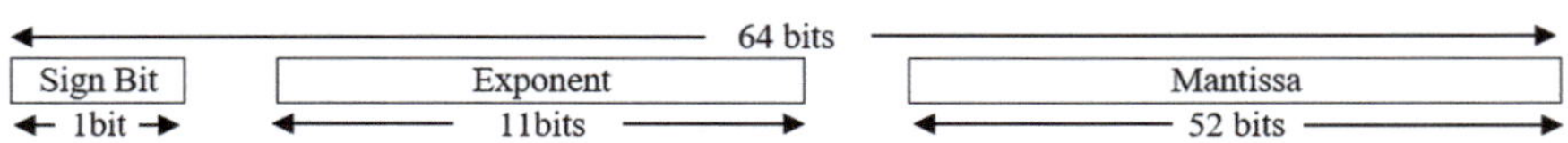

Fig. 1 IEEE 754 double-precision binary representation

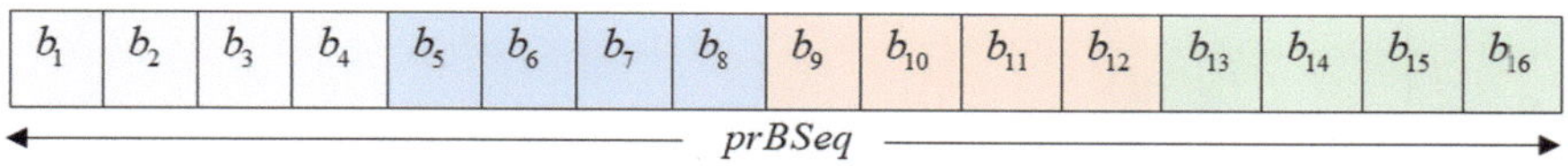

Fig. 2 Bitwise grouping of the generated pseudorandom bit sequence

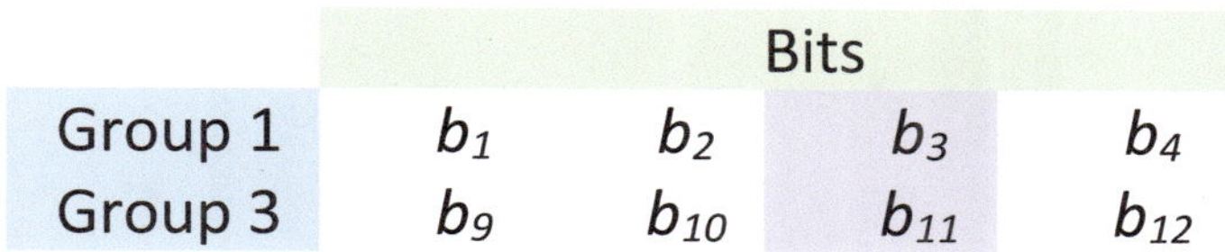

Fig. 3 Bit discarding procedure (highlighted bits are discarded)

and the fourth subgroups are combined in reverse order and generated 8 bits. Now the selected rule is applied to convert the binary values from the IEEE 754 double-precision binary representation and the merged eight bits into the DNA encoded form. The DNA XOR operation is performed between these two sequences. It is worth mentioning in this context that the process will be repeated 8 times for a certain element of the audio data (because 64/8 = 8). Moreover, the total process will be executed for each element.

A pseudorandom bit sequence $prCSeq = \{p_0, p_1, ..., p_d\}$ of length $d = 2 \cdot n$ (where n is the number of elements present in each channel) is generated using the RCM map and the pseudorandom bit sequence generation procedure as described in [18]. This pseudorandom bit sequence is sorted in descending order and a sorted sequence is prepared as $prCSeq' = \{p'_0, p'_1, ..., p'_d\}$. The two channels of the DNA XORed audio signal are merged and converted into a one-dimensional vector $v = \{v_1, v_2, v_3, ..., v_d\}$. This vector is used to construct another vector $v' = \{v'_1, v'_2, ..., v'_d\}$ using $v'_i = v_i$ where $prCSeq'_i = prCSeq_i$. This method is incorporated to scramble the sequence. Two channels are constructed from the resultant scrambled one-dimensional vector and the final encrypted audio is returned.

4 Results of the Simulation

The proposed approach is evaluated using some two-channel audio inputs. All experiments are carried out in MATLAB R2022a environment with a computer equipped with 8GB RAM, Intel i5 Processor, and 1TB HDD.

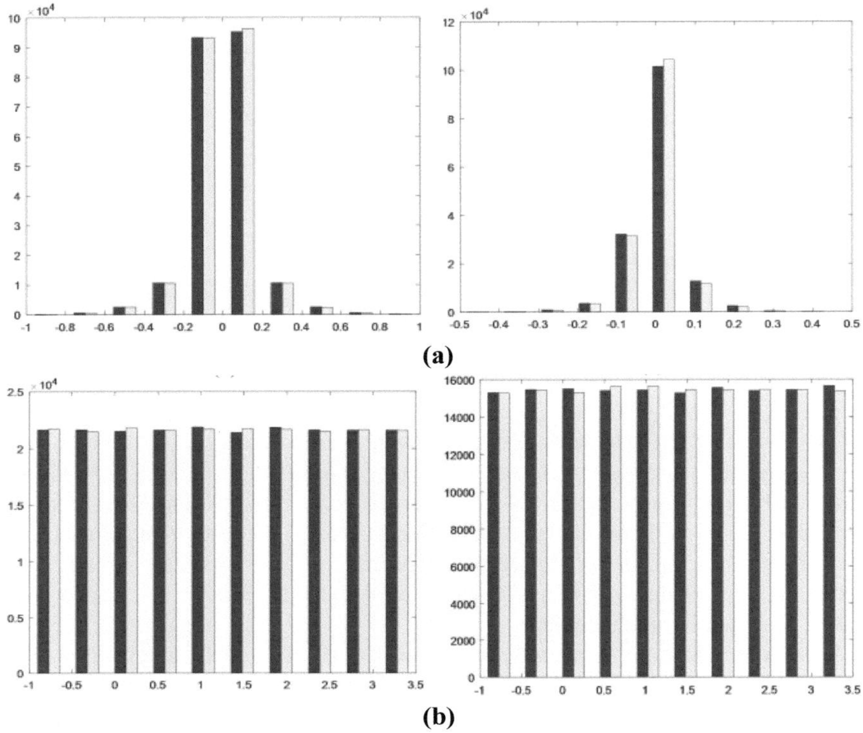

Fig. 4 Histogram analysis. **a** Original histograms of two audio signals with two channels. **b** corresponding encrypted histograms

4.1 Analysis of the Histogram

Histogram analysis is a technique commonly used in signal processing to analyze the distribution of signal amplitudes. For a good encryption technique, the uniform distribution of the histogram is essential. The histogram analysis is presented in Fig. 4. In this figure, the original histograms of two audio signals with two channels are presented. In the second row, the corresponding encrypted histograms are reported. Obtained histograms of the encrypted audio signals are nearly uniform.

4.2 Analysis of the Correlation Coefficients

The correlation coefficient is a statistical measure that quantifies the degree to which two variables are linearly related. An unencrypted meaningful audio signal typically has a high correlation because one element of a channel is almost nearly similar to its neighbors. Typically, correlation coefficients can take any value from the interval $[-1, +1]$, with maximum or minimum values being generated in the worst scenario, i.e., a completely negative or positive linear relationship [7, 13, 14]. A zero value of the correlation coefficient is the ideal situation and indicates no linear relationship between adjacent elements. The value of the correlation coefficient can be computed using Eq. 2 [15].

$$CC_{i,j} = \frac{CovF(i,j)}{\sqrt{\text{var}(i) \cdot \text{var}(j)}} \tag{2}$$

$CovF(i, j)$ is the covariance function between two samples i and j. var(x) is the function that determines the variation of a sample x. $CovF()$ and var() are defined in Eqs. 3 and 4 respectively.

$$CovF(i,j) = \frac{\sum_{z=1}^{P} (i_z - E(i))(j_z - E(j))}{P} \tag{3}$$

$$\text{var}(x) = \frac{1}{P} \sum_{z=1}^{P} (x_z - E(x))^2 \tag{4}$$

Here, $E(x)$ is the mean, and it is defined in Eq. 5. P denotes the count of the paired data.

$$E(x) = \frac{1}{P} \sum_{z=1}^{P} x_z \tag{5}$$

The outcome of the analysis of the correlation coefficient is reported in Fig. 5.

The value of the correlation coefficient of the left and the right channel of both original and encrypted signals are reported in Table 3.

4.3 Analysis of the Differential Attack

The attackers and cryptanalysts try to determine the relation between the original signal and the encrypted signal. They change some portion of the original signal and observe the encrypted output. It is essential for any encryption approach to resist differential attacks successfully. To understand the strength of encryption, the Number of Sample Change Rates (NSCR) and Unified Average Changing Intensity (UACI) values are often investigated. These two parameters are defined in Eqs. 6 and 7 respectively [8, 25].

$$NSCR = 100 \times \frac{\sum_{m,n} D(m,n)}{L} \tag{6}$$

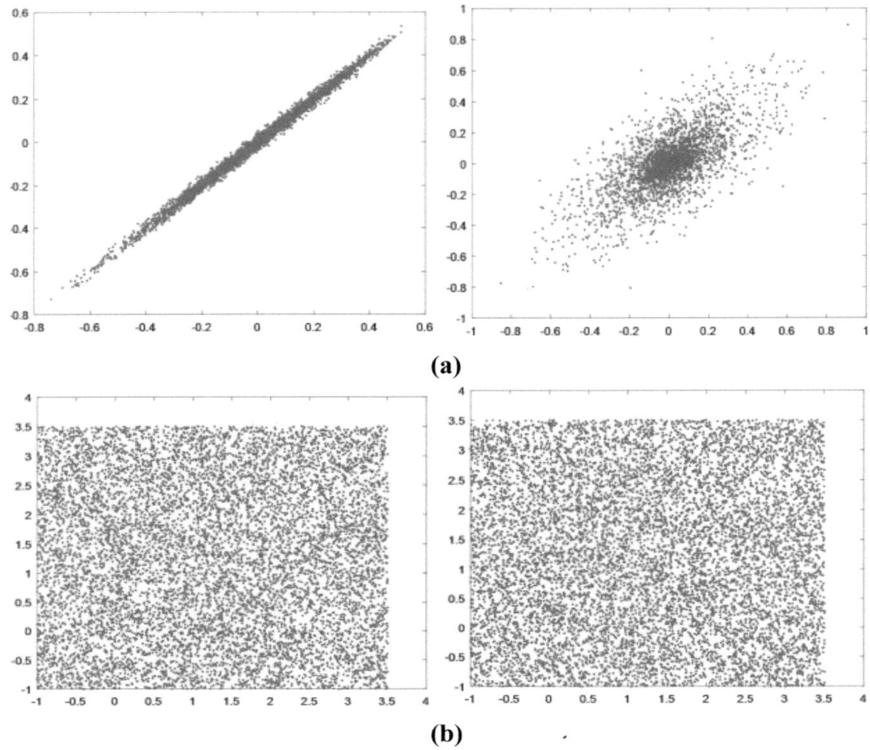

Fig. 5 Analysis of the correlation coefficient. **a** Correlation coefficient of the original channels. **b** Correlation coefficient of the corresponding encrypted channels

Table 3 Analysis of the correlation coefficient

Audio sample	Original signal ($L_{original}/R_{original}$)	Encrypted signal ($L_{encrypted}/R_{encrypted}$)
Sample 1	0.9537	0.0003
Sample 2	0.8829	0.0059
Sample 3	0.9091	0.0002
Sample 4	0.9962	0.0007
Sample 5	0.9718	0.0071

$$UACI = \frac{1}{L}\sum_{m,n}\frac{|c_1(m,n) - c_2(m,n)|}{2^b - 1} \times 100 \tag{7}$$

Here, L is the length of the original audio signal and b is the bit count required to represent the audio signal. c_1 and c_2 are the two cipher signals that are produced by

Table 4 Analysis of the PSNR and UACI values

Audio sample	NSCR (%)	UACI (%)
Sample 1	98.39	33.3966
Sample 2	99.26	33.6833
Sample 3	99.21	33.8077
Sample 4	98.99	33.5077
Sample 5	99.17	33.9007

changing a single value in the input audio signal. If $c_1(m, n) = c_2(m, n)$ then $D(m, n) = 0$. Table 4 reports the NSCR and UACI values.

5 Conclusion

The proposed audio encryption approach is tested rigorously. Experimental outcomes show that the proposed approach can be effective enough to be deployed in various real-life scenarios. The proposed approach can successfully defend against various cryptographic attacks. The proposed approach can be further extended by applying various chaotic maps. Moreover, the proposed approach can also be extended with the help of the hyperdimensional chaos theory. The strength of the proposed approach can be further extended with the help of quantum cryptographic techniques. It is possible to combine multiple chaotic maps to improve randomness and avoid predictability. Moreover, it is possible to introduce multi-dimensional chaotic maps for increased complexity. It is possible to Opt for low-complexity maps like the Logistic Map or Piecewise Linear Chaotic Map (PWLCM) which require fewer computational resources compared to higher-dimensional maps (e.g., Lorenz, Henon). The use of fixed-point arithmetic instead of floating-point can reduce the computational overhead and memory usage while maintaining sufficient precision. The load of the proposed system can be further reduced by Pre-computation of Chaotic Sequences. However, the encrypted audio signal produced by the proposed approach is secured enough the proposed approach can be deployed in real-life scenarios reliably.

References

1. Akkasaligar, P. T., & Biradar, S. (2020). Selective medical image encryption using DNA cryptography. *Information Security Journal, 29*(2), 91–101. Taylor and Francis Inc. https://doi.org/10.1080/19393555.2020.1718248
2. Amin, S. T., Saeb, M., & El-Gindi, S. (2006). A DNA-based Implementation of YAEA encryption algorithm. *Computational Intelligence*. http://www.gutenberg.org/
3. Chakraborty, S., Seal, A., Roy, M., & Mali, K. (2016). A novel lossless image encryption method using DNA substitution and chaotic logistic map. *International Journal of Security and Its Applications*, *10*(2), 205–216. https://doi.org/10.14257/ijsia.2016.10.2.19

4. Farajallah, M., El Assad, S., & Deforges, O. (2016). Fast and secure chaos-based cryptosystem for images, *26*(2). https://doi.org/10.1142/S0218127416500218
5. Havaldar, S., & Gurumurthy, K. S. (2017). Design of vedic IEEE 754 floating point multiplier. In *2016 IEEE International Conference on Recent Trends in Electronics, Information and Communication Technology, RTEICT 2016—Proceedings* (pp. 1131–1135). https://doi.org/10.1109/RTEICT.2016.7808008
6. Hu, J., & Han, F. (2009). A pixel-based scrambling scheme for digital medical images protection. *Journal of Network and Computer Applications, 32*(4), 788–794. https://doi.org/10.1016/J.JNCA.2009.02.009
7. Indrasena Reddy, M., Siva Kumar, A. P., & Subba Reddy, K. (2020). A secured cryptographic system based on DNA and a hybrid key generation approach. *Bio Systems, 197*, Article 104207. https://doi.org/10.1016/j.biosystems.2020.104207
8. Mali, K., Chakraborty, S., & Roy, M. (2015). A study on statistical analysis and security evaluation parameters in image encryption. *IJSRD-International Journal for Scientific Research & Development*, *3*, 2321–0613. www.ijsrd.com
9. Mali, K., Chakraborty, S., Seal, A., & Roy, M. (2015). An efficient image cryptographic algorithm based on frequency domain using haar wavelet transform. *International Journal of Security and Its Applications*, *9*(12), 279–288. https://doi.org/10.14257/ijsia.2015.9.12.26
10. Roy, M., Chakraborty, S., & Mali, K. (2020). A robust image encryption method using chaotic skew-tent map. In S. Chakraborty & K. Mali (Eds.), *Applications of advanced machine intelligence in computer vision and object recognition: emerging research and opportunities*. https://doi.org/10.4018/978-1-7998-2736-8.ch001
11. Roy, M., Chakraborty, S., & Mali, K. (2021). A chaotic framework and its application in image encryption. *Multimedia Tools and Applications*, 1–42. https://doi.org/10.1007/s11042-021-10839-7
12. Roy, M., Chakraborty, S., & Mali, K. (2021). The MSK: a simple and robust image encryption method. *Multimedia Tools and Applications*, 1–31. https://doi.org/10.1007/s11042-021-10761-y
13. Roy, M., Chakraborty, S., & Mali, K. (2023). An evolutionary image encryption system with chaos theory and DNA encoding. *Multimedia Tools and Applications, 82*(22), 33607–33635. https://doi.org/10.1007/S11042-023-14948-3/METRICS
14. Roy, M., Chakraborty, S., & Mali, K. (2023). Metaheuristic-supported image encryption framework based on binary search tree and DNA encoding. *Multimedia Tools and Applications*, 1–29. https://doi.org/10.1007/S11042-023-16471-X/METRICS
15. Roy, M., Chakraborty, S., & Mali, K. (2024). Audio encryption framework based on chaotic map and DNA encoding. *Applied Acoustics, 224*, Article 110152. https://doi.org/10.1016/J.APACOUST.2024.110152
16. Roy, M., Chakraborty, S., Mali, K., Banerjee, A., Ghosh, K., & Chatterjee, S. (2020). Biomedical image security using matrix manipulation and DNA encryption. *Advances in Intelligent Systems and Computing, 1065*, 49–60. https://doi.org/10.1007/978-981-15-0361-0_4
17. Roy, M., Chakraborty, S., Mali, K., Mitra, S., Mondal, I., Dawn, R., Das, D., & Chatterjee, S. (2019). A dual layer image encryption using polymerase chain reaction amplification and dna encryption. In *2019 International Conference on Opto-Electronics and Applied Optics, Optronix 2019*. https://doi.org/10.1109/OPTRONIX.2019.8862350
18. Roy, M., Chakraborty, S., Mali, K., & Roy, D. (2021). *Utilization of Hyperchaotic Environment and DNA Sequences for Digital Image Security* (pp. 289–297). Springer, Singapore. https://doi.org/10.1007/978-981-15-9433-5_28
19. Roy, M., Chakraborty, S., Mali, K., Roy, D., & Chatterjee, S. (2021). A robust image encryption framework based on DNA computing and chaotic environment. *Microsystem Technologies*, 1–11. https://doi.org/10.1007/s00542-020-05120-0
20. Roy, M., Chakraborty, S., Mali, K., Roy, D., & Chatterjee, S. (2021). *An image security method based on low dimensional chaotic environment and DNA encoding* (pp. 267–277). Springer, Singapore. https://doi.org/10.1007/978-981-15-9433-5_26

21. Roy, M., Chakraborty, S., Mali, K., Swarnakar, R., Ghosh, K., Banerjee, A., & Chatterjee, S. (2020). Data security techniques based on DNA encryption. In *Advances in intelligent systems and computing* (Vol. 1065). https://doi.org/10.1007/978-981-15-0361-0_19
22. Roy, M., Mali, K., Chatterjee, S., Chakraborty, S., Debnath, R., & Sen, S. (2019). A study on the applications of the biomedical image encryption methods for secured computer aided diagnostics. In *Proceedings—2019 Amity International Conference on Artificial Intelligence, AICAI 2019*. https://doi.org/10.1109/AICAI.2019.8701382
23. Seal, A., Chakraborty, S., & Mali, K. (2017). A new and resilient image encryption technique based on pixel manipulation, value transformation and visual transformation utilizing single–Level haar wavelet transform. In *Advances in intelligent systems and computing* (Vol. 458). https://doi.org/10.1007/978-981-10-2035-3_61
24. Tedmori, S., & Al-Najdawi, N. (2014). Image cryptographic algorithm based on the Haar wavelet transform. *Information Sciences, 269*, 21–34. https://doi.org/10.1016/j.ins.2014.02.004
25. Wang, J., Yu, W., Wang, J., Zhao, Y., Zhang, J., & Jiang, D. (2019). A new six-dimensional hyperchaotic system and its secure communication circuit implementation. *International Journal of Circuit Theory and Applications, 47*(5), 702–717. https://doi.org/10.1002/CTA.2617
26. Zare, A., Aghagolzadeh, A., & Kazemitabar, S. J. (2019). A novel metaheuristic based visual cryptography. *Journal of Advanced Defense Science and Technology, 10*(3), 297–306. https://adst.ihu.ac.ir/article_204333.html
27. Zhang, Y., Zhou, J., Chen, F., Zhang, L. Y., Wong, K. W., He, X., & Xiao, D. (2016). Embedding cryptographic features in compressive sensing. *Neurocomputing, 205*, 472–480. https://doi.org/10.1016/J.NEUCOM.2016.04.053

Disease Classification of Star Fruit (Averrhoa carambola L.) Using Deep Learning

Samanawaya Datta, Sakyojit Banerjee, and Susovan Jana

Abstract This research presents an automated system for carambola disease diagnosis using Convolutional Neural Networks (CNNs), which contains two modules, i.e., Carambola Identification and Disease Classification. It accurately identifies carambola objects and categorizes diseases through extensive real-world dataset experiments. It also presents early detection and enhanced crop management. This system promises to transform agricultural diagnostics. Through the research, the developed Carambola Identification Model achieved an impressive F1 score of 1, while the Disease Classification Model achieved an F1 score of 0.98. This system holds significant potential for sustainable carambola cultivation practices, benefiting farmers, researchers, and policymakers alike. This innovation represents a significant step toward optimizing agricultural productivity and ensuring food security in carambola cultivation regions. Its scalability and adaptability suit broader agricultural applications.

Keywords Agricultural image analysis · CNN · Disease diagnosis · Plant pathology · Carambola

1 Introduction

One of the most popular subtropical fruits in Southeast Asian and Eastern countries is the Starfruit, also known as Carambola. Low in calories and high in vitamin C, it is also used in various dishes like jams, jellies, salsas, and more. However, it is

S. Datta · S. Banerjee
Department of Computer Science and Engineering (IoTCSBT), Institute of Engineering & Management, Kolkata, India

S. Jana (✉)
Department of Computer Science and Engineering (IoT), Institute of Engineering & Management, University of Engineering & Management, Kolkata, India
e-mail: jana.susovan2@gmail.com

J. K. Mandal et al. (eds.), *Recent Advances in Artificial Intelligence and Smart Applications*, Lecture Notes in Networks and Systems 1380,
https://doi.org/10.1007/978-981-96-5822-0_2

very susceptible to several diseases that can lead to problems such as low output and yield, as well as rotten and low-quality fruit.

The traditional methods of disease detection tend to be extremely laborious and prone to misinputs and errors. However, with advancements in recent years in the paradigms of machine learning, deep learning, and Convolutional Neural Networks (CNNs), promising solutions have emerged that address most of the problems associated with the agriculture of this fruit. CNNs are utilized in various image recognition models that can learn to identify and perform extremely niche and complex patterns in images that may be oblivious to the naked eye. This paper aims to provide a way of preventing such problems.

This research paper is comprised of two models: the Carambola Identification Model and the Disease Classification Model. The Carambola Identification Model allows for the detection of whether the target image is a Carambola fruit or a Carambola leaf. Thus, this model allows for the identification of the entire Carambola plant as a whole dataset. This model forms the foundation for the subsequent Disease Classification Model. The Disease Classification Model is comprised of MobileNetV1, an extremely lightweight yet highly popular CNN architecture known for its robustness and efficiency. This model classifies the different forms of diseases associated with either the fruit or the leaf, utilizing various metrics like F1 scores, precision, and recall values, along with confusion matrices for accurate disease detection. Such a system will inherently improve agricultural diagnostics with early disease detection and improved crop management techniques, which will ultimately contribute to the broader agricultural management and development paradigm.

The next section describes the literature survey for fruit identification and disease detection. Section 3 contains a detailed description proposed model. Section 4 shows the experimentation results followed by the conclusion and future scope in Sect. 5.

2 Literature Study

With significant advancements in agricultural technology, various works utilize deep learning techniques, specifically the InceptionV3 model and transfer learning, to develop user-friendly tools for fruit disease detection and grading [1]. The k-means clustering algorithm is used for detecting defective portions of apple fruit [2], and methods for fruit disease detection include color, morphology, and texture features, with fuzzy c-means and k-means for segmentation [3].

Fruit images are preprocessed to separate the foreground from the background by extracting texture and statistical color features, and combining them into a single descriptor. Training is performed using a Support Vector Machine (SVM) with feature descriptors from a labeled dataset, aiming to accurately recognize carambola fruit or leaf objects, with the Disease Classification Model using CNN architecture to categorize diseases [4]. A model can also normalize fruit photographs by correcting translation, rotation, and scale issues, focusing on shape-based characteristics [5].

Comparative analysis of cutting-edge ML and DL algorithms for plant leaf diseases shows that a novel soybean leaf classification model using k-means segmentation achieves 62.53% accuracy, and Cotton plant leaf disease classification with FFBPNN achieves 95.48% accuracy [6]. A system can use CNN models for disease detection in different plant species and k-means clustering for image segmentation, aiming to identify carambola fruit or leaf objects accurately [7].

Traditional picture classification methods have limitations in generalization and accuracy. This study presents a unique system for classifying fruit images using transfer learning and deep learning methods, specifically the feature extraction and classification done by the modified Inception-V3 model. Transfer learning improves the training process [8]. The research highlights current improvements in disease detection and classification of plants using machine learning and deep learning, suggesting multi-class, multi-label deep learning algorithms for better accuracy and speed [9].

3 Proposed Model

In the above experiment, we have developed and evaluated two deep learning models: the Identification Model (Custom Model) and the Disease Classification Model (MobileNetV1 Base model). The details of the layers for both models are mentioned in Table 1. The Identification Model uses data augmentation, convolutional layers, max-pooling, dropout, Adam optimizer, categorical cross-entropy loss, and evaluation via confusion matrix and F1 score. The Disease Classification Model uses MobileNetV1 for feature extraction, custom fully connected layers with ReLU activation, dropout, Adam optimizer, categorical cross-entropy loss, and tracks training with accuracy/loss plots. Evaluation includes a confusion matrix and detailed classification report with precision, recall, and F1 scores. The base layers of both models have been frozen so as not to poison the model layers with pre-trained pre-built layers.

The primary reason for utilizing a custom model for the Identification model and MobileNetV1 model for the Disease Classification model boils down to time considerations for compilation and model creation, compared to accuracy. Due to only two classes being utilized in the Identification model, a simpler model compared to any other mainstream models fares much better in terms of overall performance. In the Disease Classification, MobileNet fares considerably better compared to any mainstream model. This reduces the need for developing custom models with a lower number of parameters with lower accuracy scores. Therefore, the identification can compile and create a model efficiently and fast considering the classes and the Disease Classification model can create an optimal model that can correctly identify the different types of diseases.

Table 1 Characteristics used for the models

Sl. no	Identification model	Disease classification model
1.	Conv2D Layer (32 filters, 3 × 3 kernel, ReLU): Detects low-level features with non-linearity	MobileNetV1 Base Model: Utilizes MobileNetV1 for feature extraction with frozen convolutional layers
2.	MaxPooling2D Layer (2 × 2): Reduces spatial dimensions and computational complexity	GlobalAveragePooling2D Layer: Averages each feature map, reducing dimensions for custom dense layers
3.	Conv2D Layer (64 filters, 3 × 3 kernel, ReLU): Extracts more complex features	Dense Layer (512 units, ReLU): Learns complex mappings between features and classes
4.	MaxPooling2D Layer (2 × 2): Further reduces spatial dimensions	Dropout Layer (0.25): Prevents overfitting by randomly dropping out 25% of neurons
5.	Conv2D Layer (128 filters, 3 × 3 kernel, ReLU): Captures intricate patterns	Dense Layer (256 units, ReLU): Further refines learned representations
6.	MaxPooling2D Layer (2 × 2): Down samples feature maps for fully connected layers	Dropout Layer (0.25): Mitigates overfitting similarly to the previous layer
7.	Flatten Layer: the feature maps are converted to one-dimensional vectors for next layer	Output Layer: Matches the number of disease classes, using softmax activation for probability distribution
8.	Dense Layer (64 units, ReLU, L2 regularization): Learns higher-level abstractions with regularization	
9.	Dropout Layer (0.25): Prevents overfitting	
10.	Dense Layer (32 units, ReLU, L2 regularization): Refines representations with regularization	
11.	Dropout Layer (0.25): Adds robustness by reducing redundancy	

4 Experimentation and Results

4.1 Dataset Description

The primary dataset utilized is Carambola Disease Recognition Dataset [10] from Mendeley Data with images also being scraped from Google Images and various other sources. After careful consideration, they have been utilized for testing and training. The images utilized are primarily been utilized from the Carambola Disease Recognition Dataset [10] and also have been acquired from various online sources. Dataset A consists of the identification images consisting of a combined total of 441 images from 2 classes resized into 299 × 299 for uniform size and faster processing.

(a) (b)

Fig. 1 Samples from dataset A. **a** Healthy Carambola fruit **b** Healthy Carambola leaf

(a) (b) (c) (d)

Fig. 2 Samples from dataset B. **a** Anthracnose **b** Bedbugs **c** Hooded Hopper **d** Fruit Borer

Dataset B consists of the disease recognition images consisting of a combined total of 3913 images from 9 classes and also have been resized into 299 × 299. Figures 1 and 2 show a few samples from Dataset A and Dataset B respectively.

Both Datasets A and Dataset B consist of a Training-Testing split of 85% and 15% for their respective models (a validation split of 0.15). Table 2 shows the detailed distribution of images for different classes.

4.2 *Comparative Analysis*

Table 3 shows a statistical comparison of the execution of the dataset on the proposed model compared to the most popular model available on the market. This allows for a comparative analysis of different accuracies and losses for the different models.

Table 4 includes a comparative analysis of the different confusion matrices that have been acquired from the above-mentioned models. The confusion matrices allow for clear detection of the accuracy of the testing images compared to their defined and detected classes.

Table 2 Exact count of images for datasets

Dataset name	Class	Total	Training (85%)	Testing (15%)
Dataset A	Fruit	227	193	34
	Leaf	214	182	32
Dataset B	Anthracnose	154	131	23
	Bed bugs	497	423	74
	Early blight	119	102	17
	Fruit borer	441	375	66
	Healthy fruits	630	536	94
	Healthy leaf	1498	1274	224
	Hooded hopper	84	72	12
	Leaf insect pest	238	203	35
	Stem borer	252	215	37

Figure 3 shows a comparison between the different F1 scores from the above-mentioned executed models. Comparing the proposed model to the other models shows a better F1 score for both models, making the macro F1 score (average) considerably higher over the other models.

5 Conclusion and Future Scope

Thus, a simple yet robust model of deep learning for the identification of fruits and their diagnosis of diseases has been formulated. The achievement of accuracy in the parts of identified plants and those of diseases can be very beneficial for early intervention and improvement in agricultural management. In this regard, the model is expected to be integrated with precision agriculture, whereby inputs or resources are adequately and efficiently applied, therefore giving improved crop management decisions like irrigation, and fertilization, among others. Indeed, disease diagnosis or other agriculture practices are promisingly located in the future for the model. It could do much to help achieve sustainable crop production or food security through further developments and integration into emerging technologies. To complement the competencies and applications of this model, the following are future scopes for extending the work-(a) the proposed model can be integrated with drones for automated monitoring, (b) the disease database can be expanded to obtain coverage of a broad range of agricultural threats-supporting holistic management.

Table 3 Statistical comparison between the different tested models

	Identification model	Disease classification model
Proposed model	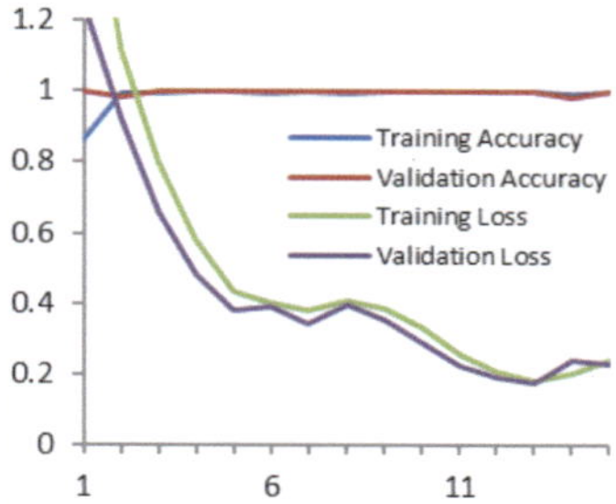	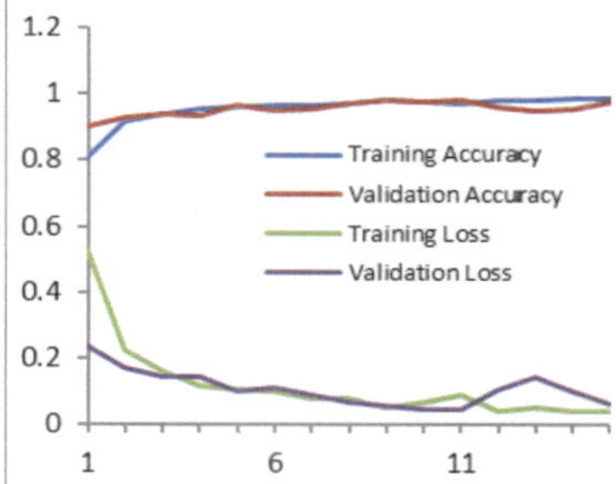
InceptionV3	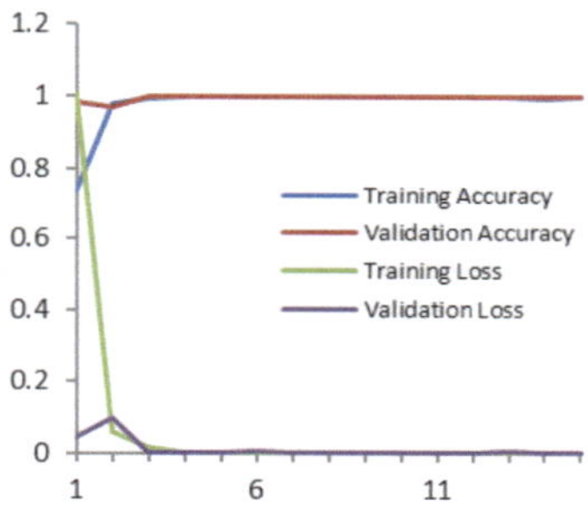	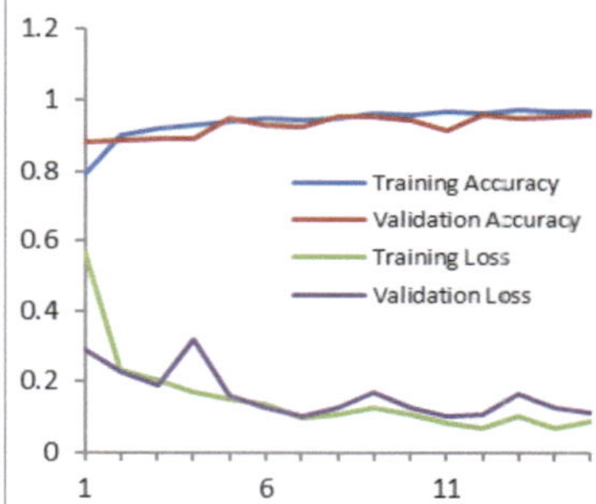
ResNet50	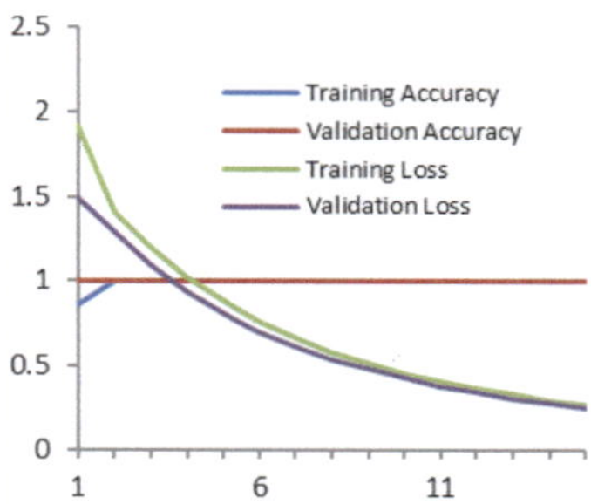	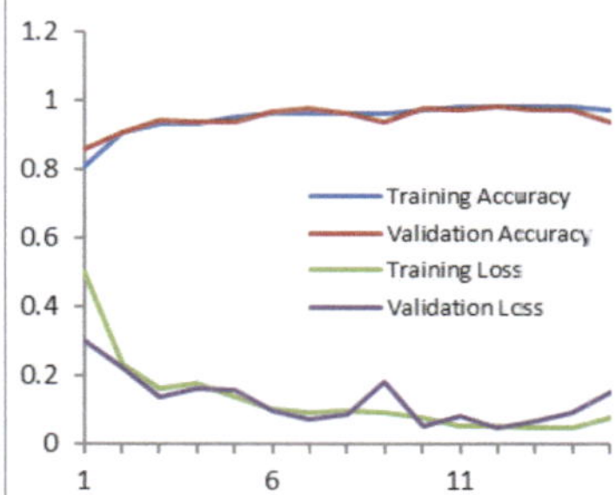
VGG16	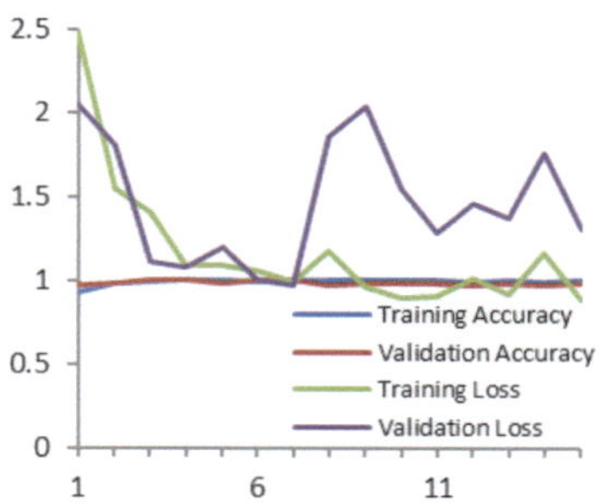	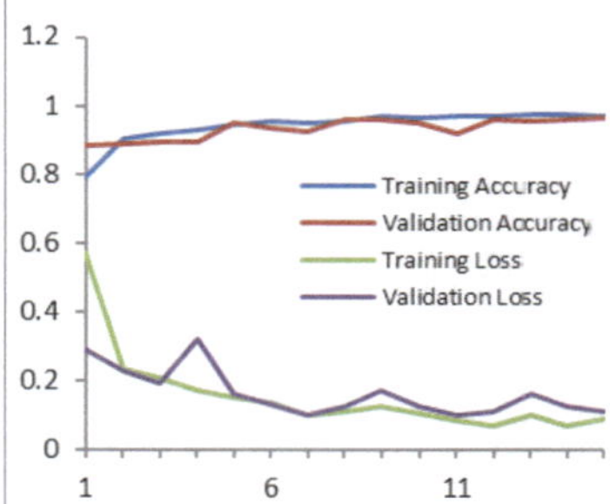

Table 4 Comparison between the different confusion matrices of the tested models

	Identification model	Disease classification model
Proposed model	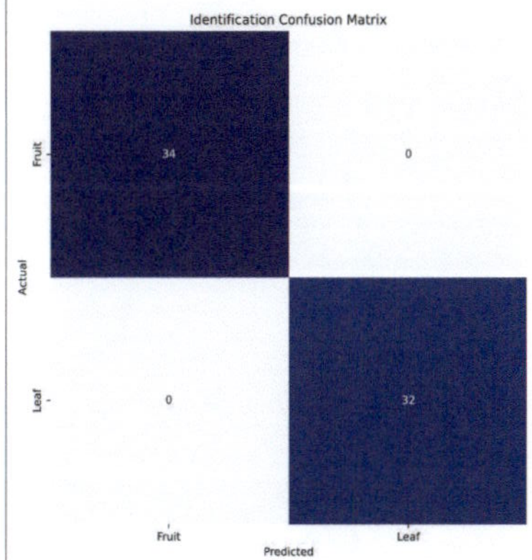	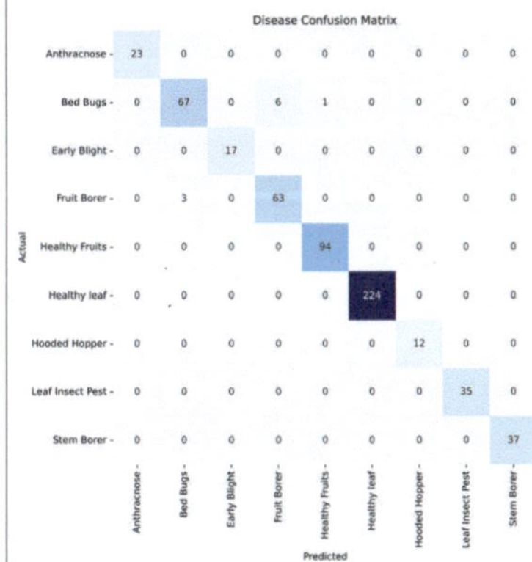
InceptionV3	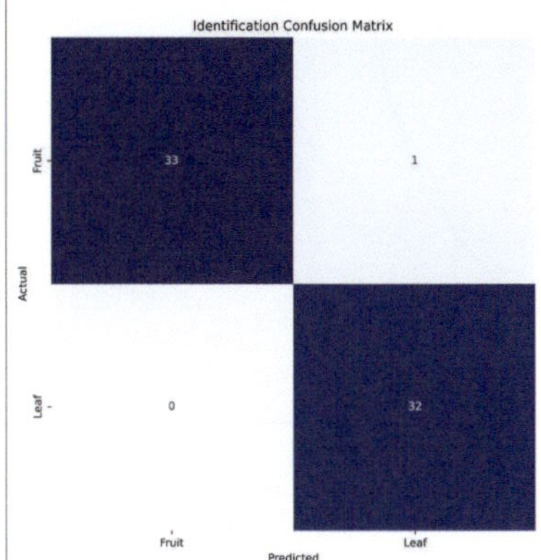	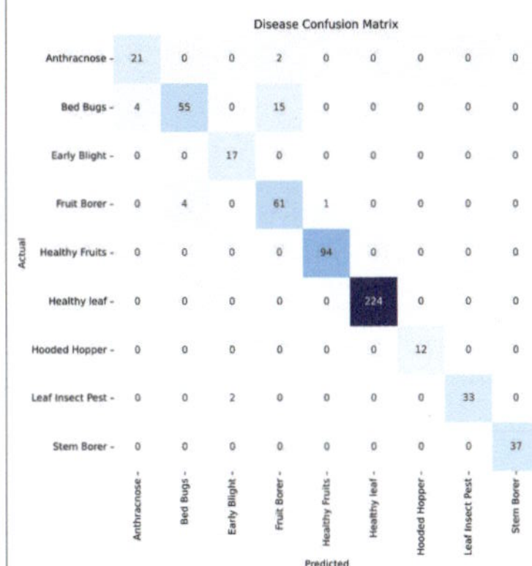

(continued)

Table 4 (continued)

<table>
<tr><th></th><th>Identification model</th><th>Disease classification model</th></tr>
<tr><td>ResNet50</td><td>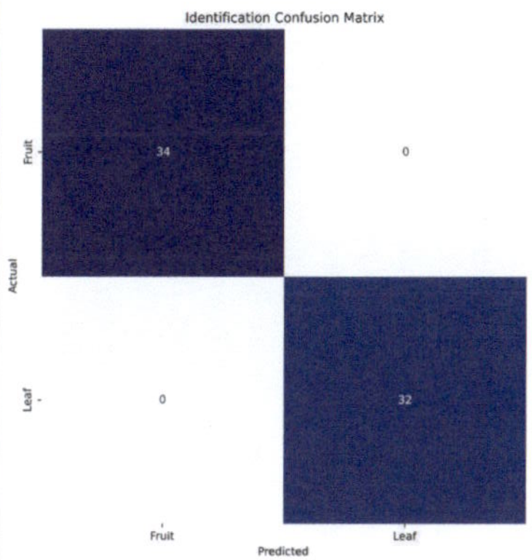
</td><td>
</td></tr>
<tr><td>VGG</td><td>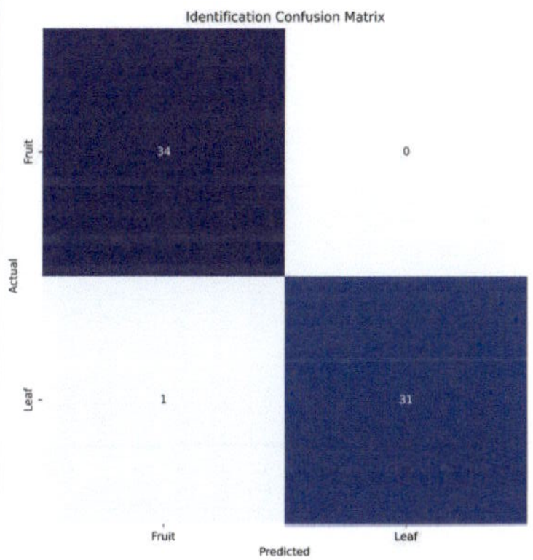
</td><td>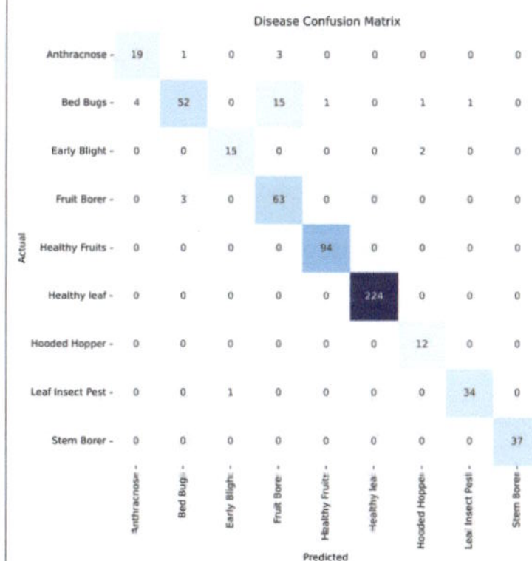
</td></tr>
</table>

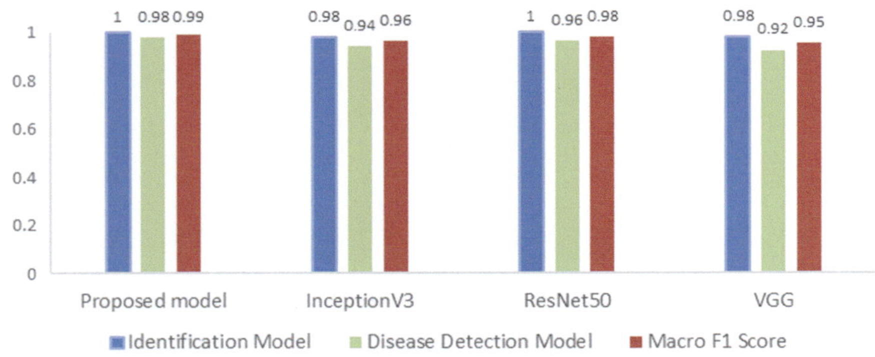

Fig. 3 Comparison of the F1 scores of the different test models

References

1. Nikhitha, M., Sri, S. R., & Maheswari, B. U. (2019). Fruit recognition and grade of disease detection using inception v3 model. In: *3rd International Conference on Electronics, Communication and Aerospace Technology (ICECA)* (pp. 1040–1043). IEEE.
2. Agarwal, A., Sarkar, A., & Dubey, A. K. (2019). Computer vision-based fruit disease detection and classification. In: *Smart Innovations in Communication and Computational Sciences: Proceedings of ICSICCS* (pp. 105–115). Springer, Singapore.
3. Sharath Kumar, Y. H., & Suhas, G (2017) Identification and classification of fruit diseases. In *Recent Trends in Image Processing and Pattern Recognition: First International Conference, RTIP2R* (pp. 382–390). Springer, Singapore.
4. Jana, S., Basak, S., & Parekh, R (2017) Automatic fruit recognition from natural images using color and texture features. In *2017 Devices for Integrated Circuit (DevIC)* (pp. 620–624). IEEE.
5. Jana, S., & Parekh, R. (2017). Shape-based fruit recognition and classification. In: *Computational Intelligence, Communications, and Business Analytics: First International Conference, CICBA 2017* (pp. 184–196). Springer, Singapore.
6. Applalanaidu, M. V., & Kumaravelan, G. (2021). A review of machine learning approaches in plant leaf disease detection and classification. In: *2021 Third International Conference on Intelligent Communication Technologies and Virtual Mobile Networks (ICICV)* (pp. 716–724).
7. Muthuselvan, S., Rajasekhar, V., Shankar, B., Mohana, P., Doddabasappa, N., & Suresh, B (2023) Classification of automatic detection of plant disease in leaves and fruits using image processing with convolutional neural network. In *2023 4th International Conference on Smart Electronics and Communication (ICOSEC)* (pp. 1408–1414). IEEE.
8. Sunday, S. E., Ji, R., Abdalla, A. N., & Bian, H. (2023). Fruit image classification using the inception-V3 deep learning model. In *2023 International Conference on the Cognitive Computing and Complex Data (ICCD)* (pp. 227–230). IEEE.
9. Demilie, W. B. (2024). Plant disease detection and classification techniques: a comparative study of the performances. *Journal of Big Data, 11*(1), 5.
10. Mafi, M. M. H. M., Hossain, T, Moazzam, M. G., & Uddin, M. S (2023) Carambola disease recognition dataset. Mendeley Data, V1.

Designing an Audio CODEC with Synthesizing Generative Adversarial Networks

Asish Debnath and Uttam Kr. Mondal

Abstract The exponential growth of audio data has necessitated innovative solutions for the real-time upload and storage of large audio volumes within the constraints of limited network bandwidth and storage capacity. This paper presents a high-fidelity neural audio compression technique designed to compress 44.1 kHz audio at an enhanced compression ratio. This is achieved through the integration of advanced vector quantization methods with Generative Adversarial Networks and reconstructive loss functions. Experimental results indicate that the proposed algorithm outperforms both deep learning-based and traditional audio compression methods, exhibiting superior reconstruction quality, higher compression rates, and enhanced audio fidelity.

Keywords Generative adversarial network · Generative model · Audio compression · Compression rate · CNN

1 Introduction

The advent of the big data era has catalyzed a significant surge in audio data proliferation [1], presenting substantial challenges in large-scale transmission and storage. In this context, audio data compression has emerged as a crucial solution [2]. Audio compression technologies can be classified into two main types [3]: lossy and lossless compression. The size-reduction method known as compression [4] is especially helpful for reducing the bit depth, dimension, and other components of audio and picture files. Lossless compression preserves the original audio quality despite reducing file size, allowing for later recovery of the unaltered data. Conversely, lossy compression achieves greater size reduction by discarding some audio information, resulting in irreversible data loss, which is unsuitable for situations requiring fidelity to the original content [5]. Recently, applications utilizing lossless audio compression

A. Debnath · U. Kr. Mondal (✉)
Department of Computer Science, Vidyasagar University, Midnapore, West Bengal, India
e-mail: uttam_ku_82@yahoo.co.in

J. K. Mandal et al. (eds.), *Recent Advances in Artificial Intelligence and Smart Applications*, Lecture Notes in Networks and Systems 1380,
https://doi.org/10.1007/978-981-96-5822-0_3

have gained traction across various sectors [6], including security, forensic science, and healthcare, highlighting its primary advantage: the retention of data integrity.

The generative modeling of high-resolution audio, characterized by high dimensionality—approximately 44,100 samples per second—and the presence of complex structures across varying time scales, presents significant challenges. In this context, the variational autoencoder (VAE) [7] framework is employed to train intermediate variables, utilizing a learned conditional prior to predict latent variables. The incorporation of discrete latent elements is advantageous, as robust autoregressive models can effectively train eloquent priors for discrete distributions. However, representing these discrete latent codes through a quantized autoencoder remains problematic, as existing neural audio compression algorithms often fail to accurately capture high-frequency content, resulting in audio outputs that significantly diverge from the original input.

This paper presents an advanced Generative Adversarial Network (GAN)-based audio compression model that efficiently compresses 44.1 kHz audio to a bitrate of 8 kbps, achieving over 75% compression efficiency while minimizing artifacts and preserving audio quality. The proposed model exhibits superior performance compared to state-of-the-art techniques, particularly at lower bitrates, as evidenced by rigorous quantitative metrics and qualitative listening assessments. To tackle the issue of codebook collapse, which limits the effective use of available bandwidth in existing models, this study introduces enhanced codebook learning strategies. The approach integrates adversarial losses, residual vector quantization, and a convolutional codec architecture, employing a multifaceted Short-Time Fourier Transform (STFT) discriminator, diverse mel loss, and periodic inductive biases, supplemented by extensive ablation studies. This comprehensive framework significantly redefines the architecture of contemporary neural audio codecs, paving the way for more efficient audio compression methodologies.

2 Literature Review

Recent advancements in audio compression research have significantly enhanced both efficiency and quality. The introduction of the Sparse Linear Predictor model [8] in 2012 marked a foundational development, although its moderate compression ratio did not translate to faster decoding speeds. The subsequent IEEE 1857.2 standard [9] in 2017 leveraged entropy encoding to improve compression ratios, albeit with slower encoding times. Noteworthy innovations in 2019 included an adaptive Golomb-Rice coding algorithm [10] for lossless ECG compression and the Integer Discrete Flow model. The models emerging in 2021, such as dynamic cluster quantization encoders [11] and hybrid wavelet transformation-based models, demonstrated improved compression capabilities. Established encoders, including Monkey's Audio [12], WavPack [13], and FLAC [14], recorded remarkable compression ratios exceeding 60\% and 70\% [15]. In 2021 Rim et al. [16] proposed an audio codec with higher compression ratio but at the cost of lower audio quality. The recent introduction of the Linear Predictive Neural Net Encoder [17] in 2022 marked another leap, achieving over 60\% compression. The latest development in 2023, an RNN-based encoding technique [18], promises even greater compression efficiency and superior audio reconstruction quality compared to traditional approaches.

3 Methodology

The application of generative adversarial networks (GANs) in audio compression presents a promising frontier due to their remarkable efficacy across diverse domains. This approach potentially enhances performance outcomes by leveraging a dual architecture consisting of a discriminative network (D) and a generative network (G), which collaboratively aim to produce high-quality reconstructed audio from raw inputs. The proposed DRVQGAN model innovatively employs the principles of the existing VQ-GAN architecture, integrating dynamic vector quantization techniques. In this generative framework, audio signals are meticulously encoded into latent vectors through an iterative backpropagation process, which optimally identifies the latent representation for effective compression. The core objective is to utilize this iterative encoding method to ascertain the ideal latent vector, thereby enabling the GAN to compress the target audio signal effectively.

3.1 *The Enhanced DRVQGAN Framework*

At the outset, the input audio signal X is transformed into an initialized compressed signal Z = ENC(X). The latent vector Z is subsequently optimized to minimize the objective function Obj() through iterative backpropagation from the generator Gen(), focusing on the similarity between the reconstructed signal Gen(Z) and the original signal X. During this continuous backpropagation process, a dynamic residual vector quantization strategy Q() is employed to render the latent variable Z discrete. After entropy encoding via the Huffman encoder, the compressed signal is transmitted for further size reduction. Importantly, the generator parameters are shared between the transmitter and receiver, with the parameters saved on the compressor side. At the receiving end, these identical generator parameters are utilized to decode the latent signal $G\left(\overset{\vee}{Z}\right)$, followed by post-processing to reconstruct the signal in near-original form.

Building upon the VQ-GAN architecture, the proposed model adheres to the same design principles as SoundStream [19] and EnCodec [20]. This model makes use of SoundStream's fully convolutional enc-dec network, which carries out temporal downscaling using a selected striding factor. It quantizes the encodings using a technique called Dynamic Residual Vector Quantization (DRVQ), which dynamically

Table 1 Comparing different compression techniques

Audio codecs	Compression rate	Sampling rate (Khz)	Striding factor	Frame rate	Target bit rate	# of 10-bit codebooks
SoundStream [19]	64	24	320	75	6	8
EnCodec [20]	32	48	320	150	24	16
Proposed codec	78.75	44.1	510	87	8	10

quantizes residuals recursively after a first quantization step using a different codebook. A distinct encoding matrix of shape $H \times L$ is created from an audio signal with sampling rate S_r, striding rate P, and L layers of DRVQ. The frame rate, defined as S_r/P, is represented by H. The evaluated compression parameters of the proposed codec with exiting are presented in Table 1. In terms of audio quality, the proposed model outperforms all baseline techniques while achieving a larger compression factor. Ultimately, since shorter sequences are produced at a lower frame rate, it is preferable to use discrete codes for model training.

3.2 Dynamic Residual Vector Quantization

Discrete auto-encoders frequently utilize vector quantization (VQ) during training; however, this approach presents several practical challenges. A primary concern is poor initialization, which can lead to low codebook consumption in vanilla VQ-VAEs, thereby underutilizing a significant portion of the codebook. This inefficiency results in an implicit reduction of the target bitrate, subsequently degrading the quality of reconstructions. To address the issue of codebook utilization, it incorporates two critical techniques previously introduced in the Improved VQ-GAN image model [21]: factorized codes and L2-normalized codes. By separating code lookup and code embedding through factorization, it facilitates code lookup in a low-dimensional space while maintaining code embedding in a high-dimensional context.

3.3 Discriminator Design

The proposed technique implements an advanced Short-Time Fourier Transform (STFT) discriminator [19] across multiple time scales, which significantly enhances phase modeling and yields superior performance in real-world applications. Findings indicate that segmenting the STFT into sub-bands effectively diminishes aliasing artifacts and slightly improves high-frequency predictions. This approach enables the discriminator to capture distinct characteristics of individual sub-bands, thereby delivering a more robust gradient signal to the generator.

3.4 Loss Function

It employed mel-reconstruction loss methods [22], specifically utilizing L2 loss on mel-spectrograms. The proposed model integrates a multi-band multi-scale Short-Time Fourier Transform (STFT) discriminator for frequency domain analysis alongside the multi-period discriminator [22] for waveform discrimination. Furthermore, it adopted the commitment losses inherent in the original Vector Quantized Variational Autoencoder (VQ-VAE) formulation, implementing stop-gradient techniques. This approach is complemented by a straightforward codebook and the straight-through estimator, facilitating gradient backpropagation through the codebook lookup process.

3.5 Architecture Diagram

In the compression block of the Generative Adversarial Network (GAN) model for audio signal compression, the process begins with an input audio signal that undergoes transformation through several key components. Initially, the short-time Fourier transform (STFT) is employed to convert the time-domain audio data into the spectral domain, producing a mel-spectrogram. This spectrogram serves as the input for the encoder block, where it is encoded into a latent vector Z. The encoder is integrated with the GAN generator to form an autoencoder, facilitating the mapping from signal space to latent space. The generator and discriminator of the GAN collaboratively generate a loss function that provides feedback for optimizing the latent vector. Through iterative backpropagation, the latent vector is refined and quantized, ultimately leading to its compression and entropy coding via a Huffman encoder. The final output, an entropy-coded latent vector, is then transmitted to the recipient.

The decompression or receiver block comprises the Huffman decoder, generator, and output audio signal block, functioning cohesively to reconstruct the original input signal. As illustrated in Fig. 1, the process begins with the entropy decoding of the compressed signal by the Huffman decoder. Subsequently, the generator utilizes the same parameters employed during the transmitter training to produce a mel-spectrogram. Finally, the inverse Short-Time Fourier Transform (STFT) is applied to convert the mel-spectrogram back into an audio stream, effectively re-establishing the initial audio input.

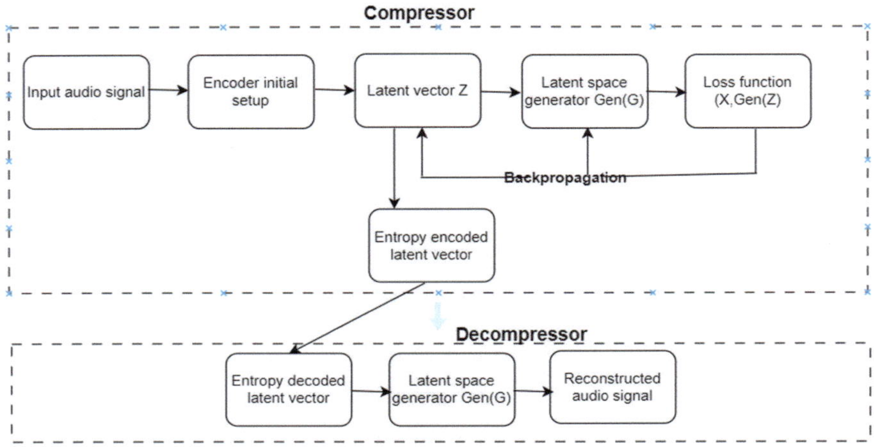

Fig. 1 Architecture diagram of the proposed system

Table 2 Recordings parameters

Parameter	Values
Record time	~2 s
File format	.wav
Sampling rate	44,100 Hz
Channel	Mono
Bit depth	16

4 Experiments

4.1 Datasets

This section provides a comprehensive overview of the preparation process for the audio_training_dataset and audio_testing_dataset. Audio files are recorded on a computer in .wav format utilizing Audacity freeware (version 3.6.1), achieving a sample rate of 44,100 Hz. The customized dataset comprises songs recorded in .wav format, adhering to five standard parameters as detailed in Table 2.

4.2 Model and Training

The proposed model comprises a dynamic residual vector quantizer, a convolutional decoder, and a convolutional encoder. Central to the architecture is a convolutional layer that facilitates upsampling or downsampling through stride-enabled operations.

This is complemented by a residual layer, which consists of multiple convolutional layers interspersed with non-linear activation functions. The framework incorporates a multi-period discriminator [22] and an effective multi-scale Short-Time Fourier Transform (STFT) discriminator. For the ablation study, each model is trained using a batch size of 16 over 5000 iterations, with the Adam optimizer applied to both the generator and the discriminator.

5 Results and Discussion

The effectiveness and stability of the proposed model, which integrates Variational Autoencoders (VAE) with integer encoding techniques, are meticulously evaluated through a comprehensive analysis involving multiple performance parameters. To establish a benchmark for comparison, three widely recognized lossless audio compression methods—namely, Monkey's Audio, WavPack Lossless, and FLAC—are scrutinized, thereby facilitating a thorough assessment of the efficacy and capacity of the suggested approach. Additionally, a deep learning-based audio compression model is employed to further gauge the proposed model's performance and reconstruction capabilities. The results of this comparative analysis are summarized in Table 3, which illustrates that the proposed framework achieves an impressive average compression rate of 78.75, significantly outperforming the reference systems, where Monkey's Audio registers a compression rate of 58.76, WavPack achieves 56.05, and FLAC attains a rate of 71.30. Furthermore, Fig. 2 provides a graphical representation of the group-wise compression results, visually highlighting the superior performance of the proposed model in relation to its contemporaries. This comprehensive evaluation underscores the potential of the integrated VAE and integer encoding approach in advancing the field of lossless audio compression, demonstrating not only its effectiveness but also its stability across various audio datasets. Table 4 shows the group-wise PSNR [23]. From Table 4 it is visible that the proposed method obtains higher PSNR compared to other methods. Figure 3 shows the graphical representation of the average PSNR (Table 5).

The robustness and performance of the proposed model were assessed using currently existing deep learning-based lossless audio codecs, with reference to (i)

Table 3 Compression of the proposed model group-wise with regard to the systems that are referenced

Method	Rock	Classical	Rabi	Pop	Ghazal
Monkey's audio [12]	58.46	57.58	60.56	59.07	58.17
WavPack [13]	55.26	57.02	55.98	56.12	55.89
FLAC [14]	71.34	70.67	71.28	72.01	71.24
Proposed model	78.20	79.15	80.01	78.16	78.23

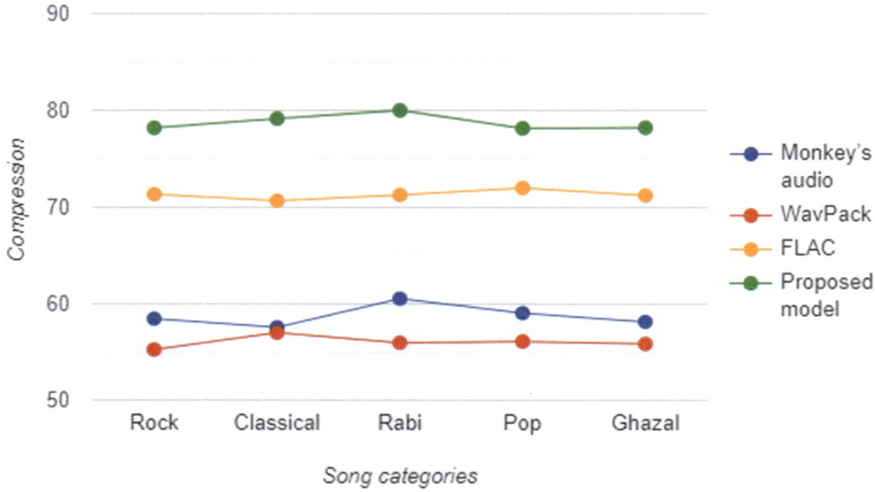

Fig. 2 Graphical comparison of the group-wise compression

Table 4 Groupwise PSNR

Method	Rock	Classical	Rabi	Pop	Ghazal
Monkey's audio [12]	53.06	51.56	54.56	53.45	54.07
WavPack [13]	52.67	52.15	50.07	52.45	51.19
FLAC [14]	51.72	53.45	54.35	53.78	54.13
Proposed model	54.69	55.89	56.98	55.76	56.98

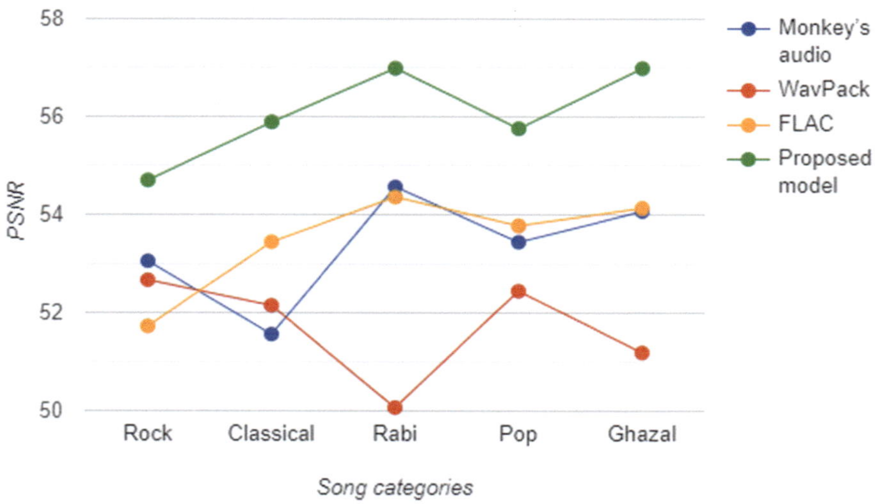

Fig. 3 Graphical comparison of the group-wise PSNR

Table 5 Groupwise compression with DNN model

Method	Rock	Classical	Rabi	Pop	Ghazal
LINNE [17]	74.56	74.98	75.41	72.16	74.20
Daniela et al. [16]	75.16	76.28	75.38	76.12	76.89
Proposed model	78.16	79.38	75.67	78.12	80.56

Table 6 Comparison of MSE and NCC

Method	MSE	NCC
LINNE [17]	0.011341	0.985634
Daniela et al. [16]	0.012536	0.989784
Proposed model	0.001612	0.998761

Rim et al. [16] and (ii) LINNE [17]. The proposed framework can reproduce the original audio signal with incredibly little variations. The suggested model reproduces the original audio signal with incredibly little variations. The estimated mean square error value for the suggested model is 0.001612. Additionally, a different metric known as NCC is utilized to assess the quality of the regenerated signal: 0.998761. A regeneration score that is closer to 1 denotes good regeneration. To illustrate the precision and resilience of the model, Table 6 presents the assessed parameters, such as Mean Squared Error (MSE) [23] and Normalized cross correlation (NCC) [11], of the current model in comparison to the other mentioned prediction system (Fig. 4).

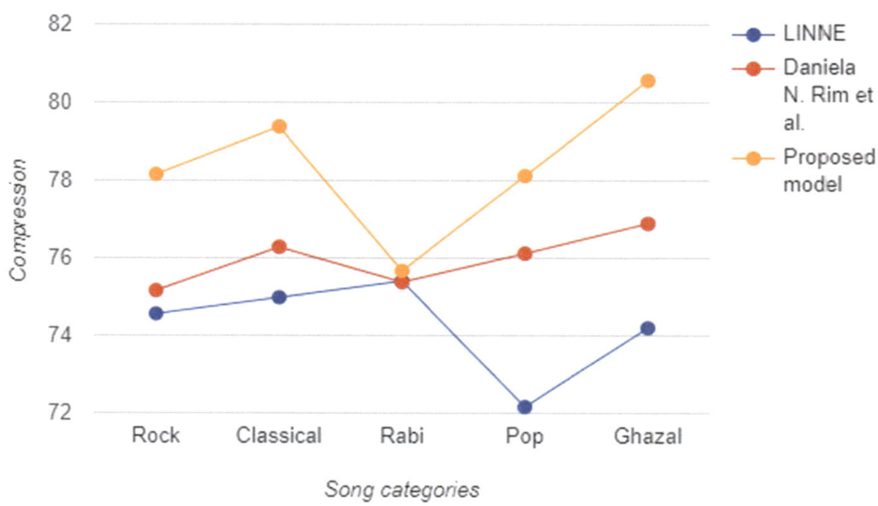

Fig. 4 Group-wise compression performance: Monkey's Audio, WavPack, FLAC vs. proposed method

6 Conclusion and Future Scope

The proposed model was trained and validated using real-time audio data, demonstrating its efficacy through a comparison with existing industry-standard audio codecs. Notably, the GAN-based approach exhibits a relatively low mean square error, indicating robust performance. Furthermore, it achieves a higher compression ratio than current lossless audio codecs, highlighting its potential advantages in audio processing. Future research will aim to enhance the model further, focusing on optimizing both compression efficiency and computational performance.

Disclosure of Interests. All authors certify that they have no affiliations with or involvement in any organization or entity with any financial interest or non-financial interest in the subject matter or materials discussed in the manuscript.

References

1. Debnath, A., & Mondal, U. K. (2024). Lossless audio CODEC using non-repeated dynamic block encoding. *International Journal of Information Technology, 16*, 3941–3948. https://doi.org/10.1007/s41870-024-01785-2
2. Selvi, C. T., Amudha, J., & Sudhakar, R. (2021). Medical image encryption and compression by adaptive sigma filterized synorr certificateless signcryptive Levenshtein entropy-coding-based deep neural learning. *Multimedia Systems, 27*(6), 1059–1074. https://doi.org/10.1007/s00530-021-00764-y
3. Debnath, A., & Mondal, U. (2024). Lossless audio codec based on CNN, weighted tree and arithmetic encoding (LACCWA). *Multimedia Tools and Applications, 83*, 48737–48759. https://doi.org/10.1007/s11042-023-17393-4
4. Kumar, S. N., Ahilan, A., Haridhas, A. K., & Sebastian, J. (2020). Gaussian Hermite polynomial based lossless medical image compression. *Multimedia Systems, 27*, 15–31. https://doi.org/10.1007/s00530-020-00689-y
5. Mondal, U. K., & Debnath, A. (2022). Designing a novel lossless audio compression technique with the help of optimized graph traversal (lacogt). *Multimedia Tools and Applications, 81*(28), 40385–40411.
6. Albahli, S., Nazir, T., Irtaza, A., & Javed, A. (2021). Recognition and detection of diabetic retinopathy using densenet-65 based faster-RCNN. *Computers, Materials & Continua, 67*(2), 1333–1351. https://doi.org/10.32604/cmc.2021.014691
7. Huang, K., & Wang, X. (2021). ADA-INCVAE: Improved data generation using variational autoencoder for imbalanced classification. *Applied Intelligence, 52*(3), 2838–2853. https://doi.org/10.1007/s10489-021-02566-1
8. Ghido, F., & Tabus, I. (2012). Sparse modeling for lossless audio compression. *IEEE Transactions on Audio, Speech and Language Processing, 21*(1), 14–28.
9. Gunawan, T. S., Zain, M. K. M., Muin, F. A., & Kartiwi, M. (2017). Investigation of lossless audio compression using IEEE 1857.2 advanced audio coding. *Indonesian Journal of Electrical Engineering and Computer Science, 6*(2), 422–430.
10. Hoogeboom, E., Peters, J., Van Den Berg, R., & Welling, M. (2019). Integer discrete flows and lossless compression. *Advances in Neural Information Processing Systems* 32.
11. Mondal, U. K., & Debnath, A. (2021). Developing a dynamic cluster quantization based lossless audio compression (DCQLAC). *Multimedia Tools and Applications, 80*(6), 8257–8280.

12. https://monkeysaudio.com/index.html. Accessed 27 Jul 2024.
13. http://www.wavpack.com. Accessed 27 Jul 2024.
14. https://xiph.org/flac/index.html. Accessed 27 Jul 2024.
15. Nowak, N., & Zabierowski, W., (2011). Methods of sound data compression—comparison of different standards. In *The Experience of Designing and Application of CAD Systems in Microelectronics (CADSM)* (pp. 431–434).
16. Rim, D. N., Jang, I., & Choi, H. (2021). Deep neural networks and end-to-end learning for audio compression. arXiv:2105.11681
17. Mineo, T., & Shouno, H. (2022). A lossless audio codec based on hierarchical residual prediction. In *2022 Asia-Pacific Signal and Information Processing Association Annual Summit and Conference (APSIPA ASC)* (pp. 123–130). IEEE.
18. Debnath, A., & Mondal, U. K. (2023). Perseverance of the audio data using RNN implied matrix segmenton based lossless encoder. In *International Conference on Network Security and Blockchain Technology* (pp. 123–133). Singapore: Springer Nature Singapore.
19. Zeghidour, N., Luebs, A., Omran, A., Skoglund, J., & Tagliasacchi, M. (2021). Soundstream: An end-to-end neural audio codec. *IEEE/ACM Transactions on Audio, Speech, and Language Processing, 30*, 495–507.
20. Défossez, A., Copet, J., Synnaeve, G., & Adi, Y. (2022). High fidelity neural audio compression. arXiv:2210.13438
21. Yu, J., Li, X., Koh, J. Y., Zhang, H., Pang, R., Qin, J., Ku, A., Xu, Y., Baldridge, J., & Wu, Y. (2021). Vector-quantized image modeling with improved vqgan. arXiv:2110.04627
22. Kong, J., Kim, J., & Bae, J. (2020). Hifi-gan: Generative adversarial networks for efficient and high fidelity speech synthesis. *Advances in Neural Information Processing Systems, 33*, 17022–17033.
23. Manju, M., Abarna, P., Akila, U., & Yamini, S. (2018). Peak signal to noise ratio & mean square error calculation for various images using the lossless image compression in CCSDS algorithm. *International Journal of Pure and Applied Mathematics, 119*(12), 14471–14477.

LOGIOUS: An Instant Messaging (IM) Platform Using AI

Aniruddha Ghosh, Anirban Ghosh, Souvik Das, Koustav Samanta, Pratyush Sen Sharma, Ritesh Banerjee, and Subhalaxmi Chakraborty

Abstract Logious is one of the next-level messaging apps for using AI to learn chat characteristics from user chat history. It helps you to get information faster using other LLMs like transcribing text or needing to detect anything in an image message and even gives you a personal virtual assistant to make your messaging journey smoother and more engaged. It will be like an in-hand Jarvis from the Iron Man movie. The AI is not that intelligent but it's the start of a new era where LLMS can control LLM. In addition, its advanced algorithms learn continuously and adapt the user's communication style for a personalized experience. It also provides some security protocols to keep your data safe and private. With features like real-time communication and robust privacy measures, Logious will revolutionize the way we communicate, both personally and professionally. The user-friendly interface of Logious is very easy so non-tech-savvy users can also navigate it easily.

Keywords Artificial intelligence · Machine learning · Natural language processing · Google Gemini

---

A. Ghosh (✉) · A. Ghosh · S. Das · K. Samanta · P. S. Sharma · R. Banerjee · S. Chakraborty
Institute of Engineering & Management, University of Engineering and Management, Department of Computer Science and Engineering (Artificial Intelligence and Machine Learning), Kolkata, India
e-mail: aniruddha.ghosh@uem.edu.in

Institute of Engineering and Management, University of Engineering and Management, Department of CST & CSIT, Kolkata, West Bengal, India

S. Chakraborty
e-mail: subhalaxmi.chakraborty@uem.edu.in

J. K. Mandal et al. (eds.), *Recent Advances in Artificial Intelligence and Smart Applications*, Lecture Notes in Networks and Systems 1380,
https://doi.org/10.1007/978-981-96-5822-0_4

1 Introduction

The customer experience (CX) in today's digital world is the most demanded in today's world so instant messaging (IM) platforms must evolve as the customer world demands an upgraded user interface [1]. Using NLP, users can communicate with computers using human languages [2, 3]. In a Chatbot the user would enter textual information in a human language like English which the computer will understand and will respond back in English [2]. Our project and this paper explore the potential of Artificial Intelligence (AI) [4, 5] in crafting an IM Platform [6] that aims to make a CX exceptional and unique. We present Logious, an AI-powered IM Platform Built with Gemini large language model framework [7, 8]. Logious uses different AI functionalities, such as Natural Language Processing (NLP) and Machine Learning (ML), to personalize interactions, detect user needs and offer suggestions. Sentiment analysis allows our Model Logious to learn user emotions and tailor responses, accordingly, creating a deep connection. This research explores how AI is deeply integrated, and it helps us with user engagement, develops deeper bonds and connections, and also personalizes the communication experience.

2 AI in Proposed Platform/Logious Platform

The IM platform Logious is integrated with an AI built using the Google Gemini model, and seamlessly integrates various AI functionalities to empower the user with instant information and assistance [7, 8]. Here's in brief how Logious leverages AI: AI Search: Logious acts as your personal search assistant within the platform's chat window. The required query needs to be typed and with the help of the DuckDuckGo search engine, Logious retrieves the required information from the web in real time, eliminating the need to switch between apps. Image Inspection: Logious utilizes AI to analyze the image and its content and provide the user with the understanding and relevant facts about the image provided. Logious can give a detailed description of the image by recognizing the items or components in the image [4]. Audio Transcription: Seamless communication is enabled by Logious even with audio messages [5]. The platform uses an AI model for transcription and converting voice notes to text, enabling clear communication between users, especially in situations where listening to audio might not be feasible. Chat Analysis: Logious goes through the user's chat to understand the flow of the conversation. This allows Logiuos to give better and more useful responses, making the flow of conversation better. Beyond these functionalities, Logious integrates seamlessly with your conversations through a simple @logious tag. The user has to mention @logiuos followed by the required query, and Logious takes that query, fetches the data, checks the images, or transcribes audio files all within the user's chat window. This collection of AI features enables the user to be more informed, efficient, and connected within their conversations, making Logious a truly intelligent and user-centric IM platform as represented in Fig. 1.

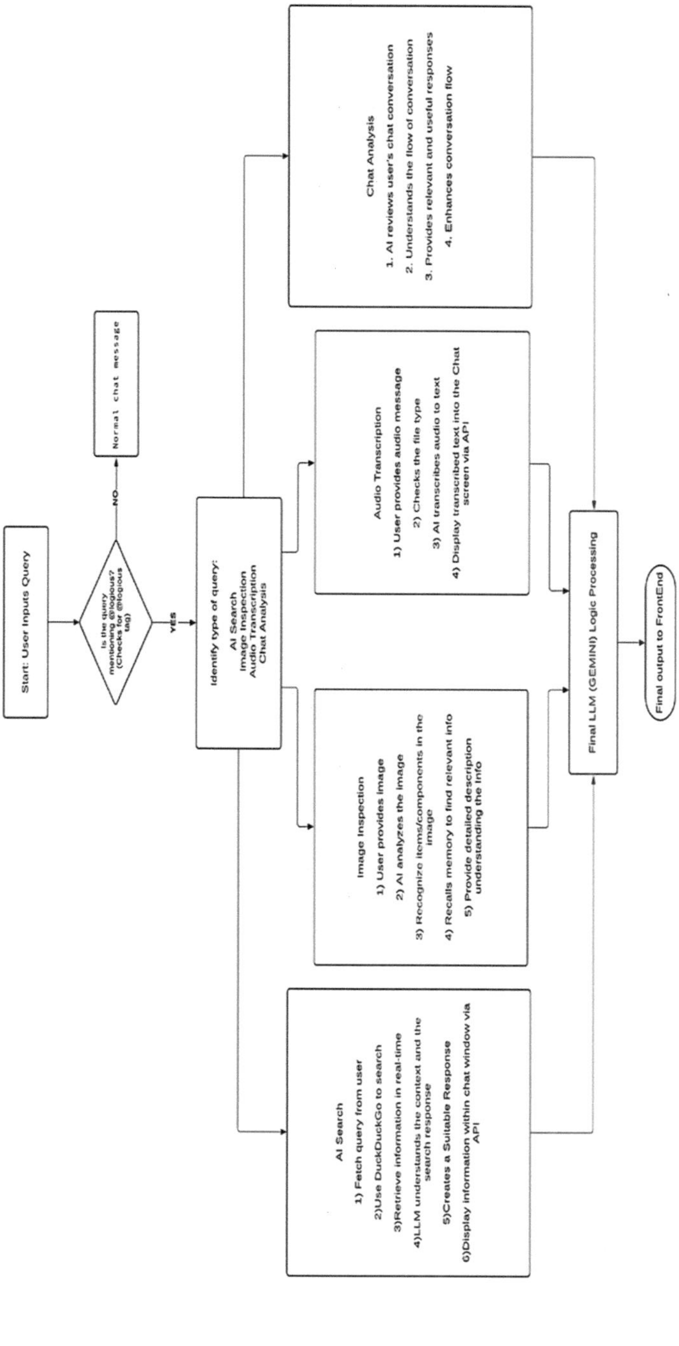

Fig. 1 Logious AI integration flowchart

3 Using React.JS as Frontend Framework

React has quickly become one of the most widely used modern frameworks [9] for building modern web applications, according to Statista, with 42.62% of respondents reporting that they use it. React is a JavaScript library [10] developed by Facebook which, among other things, was used to build Instagram.com. It permits designers to effortlessly form quick client interfacing for websites and applications alike. The most important concept of ReactJS is virtual DOM. It may be a tree based on JavaScript components made with Respond that mirrors a DOM tree. It does the slightest sum of DOM control conceivable to keep your Respond components up to date. Check out the Pete Chase conversation presenting the net system and overviewing three questionable themes: Tossing out the idea of layouts and building sees with JavaScript, "re-rendering" your whole application when your information changes, and a lightweight execution of DOM and occasions. Firstly, a response was sent by Facebook in 2011 and 2012. Instagram was composed completely in Respond. Agreeing with the JavaScript analytics benefit Libscore, Respond is right now being utilized on the homepages and other web pages of Netflix, Imgur, Seat Report, 1 Feedly, Airbnb, SeatGeek, HelloSign, and others. The front flow diagram is represented in Fig. 2.

Being a portion of the JavaScript dialect, utilizing Respond produces numerous preferences. Items built with Respond are straightforward to scale, a single dialect utilized on the server/client/mobile side of things awards extraordinary efficiency, there are workflow designs for helpful collaboration, UI code is clear and viable, and more. World-leading companies have utilized Respond and other JS innovations in a few of the best market-defining items (Instagram, Reddit, and Facebook being the foremost distinctive cases). So let's get down to it with the confirmation on the operation of React JS, shown in Fig. 3.

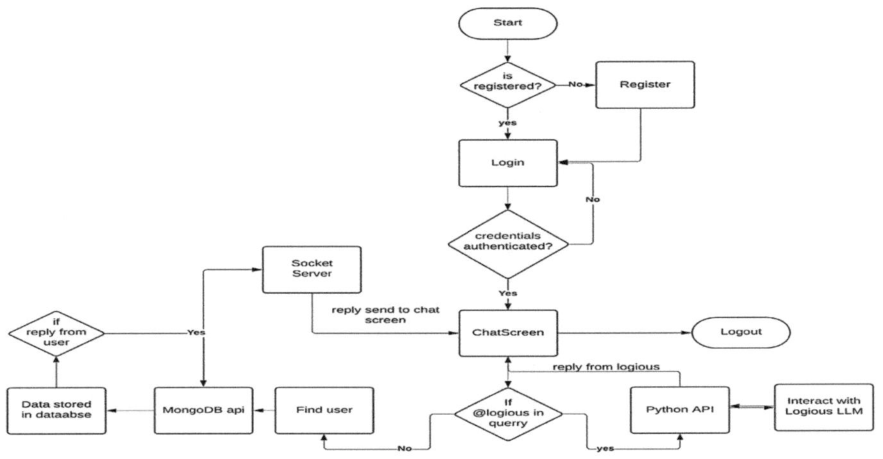

Fig. 2 Logious react frontend flowchart

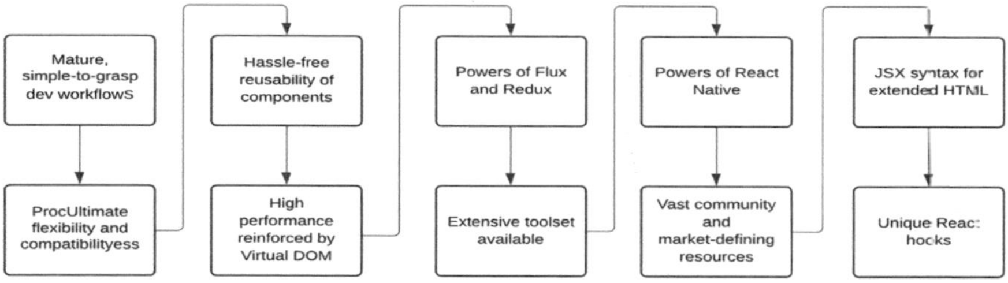

Fig. 3 Logious ReactJS frontend

4 Gemini: Building the Foundation

The Gemini models receive architectural and modeling improvements for robust training on Tensor Processing Units (TPUs) [11] at large scales, ensuring stability while expediting the inference phase during application. Simultaneously, they can process input sequences of up to 32,000 tokens for efficient handling of extended information.

1. Efficient Attention Mechanisms: Employing advanced attention mechanisms like multi-query attention, these models focus on the most pertinent parts of their input data.
2. Multimodal Functionality: Intermodal Fusion: Gemini can seamlessly process different input formats (text, images, charts, videos, and audio) and generate outputs in various forms (text, images). It utilizes distinct image tokens inspired by Google AI projects and generates images directly (Visual Tokenization), while videos are encoded as frame sequences within a large context window to allow for input with textual or audio data (Framed Sequence Encoding) [12]. The multimodal Gemini models can adaptably manage inputs with varying resolutions and directly ingest raw audio from USM features [13, 14] (16 kHz signals), preserving intricacies that may be lost in textual presentations, and expanding their usability for diverse applications.
3. Model Sizes: The first release, Gemini 1.0, offers three different sizes tailored to distinct application scenarios (as indicated in Table 1 within the text):

5 A Review of MongoDB over Firebase

MongoDB is essentially a NoSQL report database that provides demonstrated versatility and adaptability. When we say demonstrated, we cruel it. MongoDB has been around for longer than Firebase, to begin with, discharged in 2009 [16]. MongoDB has its claim stage, which is the MongoDB Map book. MongoDB can handle monstrous amounts of information, making it an instrument of choice for large-scale applications. This gets to be clear once you see a few of the worldwide mammoths that utilize MongoDB. They incorporate eBay, Adobe, Verizon, SEGA, EA, Coinbase, etc. A comparative examination is displayed in Table 2.

Table 1 Model sizes

Model size	Description	Key features
Ultra	A powerful model adept at handling complex tasks and achieving cutting-edge performance	Its scalability on Tensor Processing Units (TPUs) is a result of the Gemini architecture, making it highly adaptable for large-scale implementations
Pro	Designed as a cost- and latency-optimized model	The Pro version offers strong reasoning abilities and multimodal support. It maintains a balance between performance and resource usage
Nano	The most efficient solution for on-device usage, Nano is trained using knowledge distillation techniques from larger Gemini models	While the detailed internal structure of the Transformer decoder is not provided, this explanation highlights the key architectural aspects of Gemini models and their capabilities for processing multimodal data. Over the years have development occurred in large language models (LLMs), driving artificial intelligence (AI) into the core of human-computer interaction [7, 8, 3, 15]

Table 2 MongoDB over Firebase

Criteria	Details
Overall database performance	MongoDB wins for being quicker and conveying consistent execution. It can store tremendous sums and oversee these with no affect on execution
OS compatibility	Firebase employments JavaScript SDK, which is bolstered in nearly each working framework. The same applies to MongoDB
Languages supported	Firebase underpins six programming dialects, with four more backed through ties to the database. MongoDB authoritatively underpins 12, but through its solid community, it is certain it can too have a solid community that has created more libraries and drivers. The upshot is that Mongo can likely handle any programming dialect you care to toss at it—time to brush up on handle any programming language you care to throw at it—time to brush up on that FORTRAN
Usability	Firebase wins the truth that it comes wrapped in its Google-supported environment makes Firebase a clear victor here
Security	Both stages offer magnificent security, so indeed on the off chance that you're in a division that has exceptional security prerequisites, such as advanced keeping money development services, you'll be able be certain that information is, as the saying goes, as secure as the Bankof Britain
APIs	Firebase wins—another contract one, but absolutely based on versatility, Firebase comes out on beat here as there's a choice of database get to strategies, counting Android, iOS, and Javascript. For MongoDB, it is as

6 Enhancing Customer Experience with AI

In the current scenario, AI is essential for enhancing client experience. The purchaser gets more excited as one attempts to get extra features without switching applications. Logious unlocks a brand-new paradigm in CX through the following AI functionalities:

(a) Personalized Interactions: Logious analyzes personal behavior and possibilities within the platform.
(b) Seamless 24/7 help: AI-powered chatbots within Logious provide instantaneous help, answering simple queries and resolving unusual issues efficiently. This frees up human dealers to address greater complicated conditions while making sure users always have a helping hand to be had.
(c) Contextual understanding: Logious analysis of chat history to understand the conversational glide. This permits it to craft responses that can be relevant to the continuing discussion, making interactions herbal and tasty [17].

Beyond these features, Logious fosters user-centric surroundings via its simple @logius tag. @logious observed your question, and the AI assistant sprang into motion believing desiring facts all through a talk—@logious can search the web with the use of DuckDuckGo and supply relevant results without delay inside the chat window.

Because of these features, an automobile venture management device may be seamlessly introduced as they're developed. Imagine a destiny where @logious cannot most effectively solve your questions but additionally robotically add obligations on your to-do listing primarily based on your conversations.

Logious isn't just an IM platform; it is a companion that empowers customers with immediate Web search translation transcribing image queries, customized suggestions, and efficient help, all inside their present conversations. This AI-powered technique transforms CX, developing an infrequent appealing, and alluring enjoyment for customers, in the long run strengthening emblem loyalty inside the technique.

As Logious keeps adapting, it has the potential to become an imperative hub for conversation, information retrieval, or even manipulation, all powered by the resources of the smart talents of AI.

7 Evaluation and Results

We adopted various methods to evaluate the effectiveness of our AI instant messaging platform LOGIOUS:

7.1 Evaluation

We conducted a survey to collect data. User feedback on the LOGIOUS experience after the trial period. The survey focuses on user satisfaction with various aspects of the platform, including ease of use and intuitiveness of the interface, the effectiveness of AI such as negotiation strategy, overall performance and communication speed, and their comparison. User satisfaction compared with traditional IM platforms. This involves creating two types of platforms: a control group: a standard IM experience without AI features; and an experimental group: Users interact with the platform using AI features such as emotions and conversations.

7.2 Results

Consumer research shows LOGIOUS produces greater pride compared to IM models. Users especially benefit from features such as having the best conversation thanks to artificial intelligence-supported emotional analysis and better performance thanks to features such as conversation guides, international ease of use, and easy communication. By checkout validated progress measurement on tasks using LOGIOUS with AI capabilities in a test environment. This shows that AI has the potential to help create more customers and achieve goals. Indicators are quite effective. This means LOGIOUS brings a better and more efficient user experience to communication and collaboration with its integrated artificial intelligence performance platform.

Figure 4, there are a total of 24 Routings in which Route 1–2 are input and output channels for the user and machine.

Route 5 States the Frontend Integration to the Machine via React Framework.

Route 3–4 States the Connection of the frontend with the backend 3 is the Input channel, and 4 is the Output Channel.

Route 6, 7, 12, and 13 Refer to The whole Process Of Image Querying and storing it in Memory.

Route 9–8 Refers to the connection to the Main AI Logic where it connects to several Tools.

Route 14–15 is connected to the Search tool Which uses DuckDuckGo Search Engine.

Route 16–17 is used to Understand Query using another LLM which summarizes the Conversation Buffer Memory.

Route 18–19 Connects the blow for Summarizing the Data.

Route 20–21 uses the mongodb Database to Store the Conversation Memory for each user.

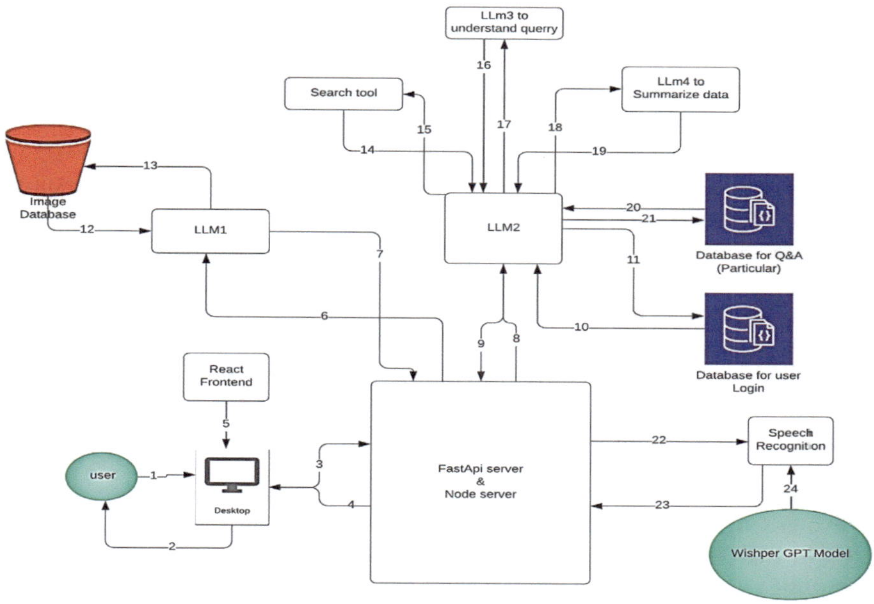

Fig. 4 Logious flow diagram

Route 11–10 uses the mongodb Database to Check the Authenticity of the User.

Route 22–23 Connects to a function which transcribes Audio to Text.

Route 24 Is a connection to the Audio to Text function which uses the Latest Wishper GPT Model.

8 Conclusion

Logious is an instant messaging (IM) system that stands out with the integration of artificial intelligence (AI). The key insights can be summarized as follows. Number one Personalized interaction: Unlike the IM process, Logious uses AI technology to analyze user behavior and preferences to complete the response. Your virtual partner makes conversation easier by asking and offering the right words. Number two Virtual Assistants: Logious offers virtual companions that proactively identify users' requirements and propose fitting information, streamlining conversations [18]. Number three Authentic Connections: Logious transcends mere messaging by encouraging genuine relationships, and enriching user experiences.

9 Future Scope

The future of Logious is promising, it looks like this. The first one is higher AI capabilities: Providing even more personalized responses by a much more capable and trained natural language processing (NLP) model. Secondly, in combination with new technologies: Exploring AR/VR to create great communication experiences [19]. Virtual reality (VR) has been utilized broadly in a few recovery regions, from the engine to cognitive restoration. The third one is expansion into new markets: Implementing the platform to support diverse cultures and languages so that our platform can reach new users. It can be a business solution: Offering features like secure messaging and project management tools for businesses. Another scope is money-making strategies: Options like premium subscriptions, target advertising, or charging for individual special messages. Focusing on Logious's security and privacy is one of the major scopes [4]. Adding great security measures and user-editable security settings for a highly maintained secured platform for the users to have a great experience. Another scope is improved accessibility to make the platform accessible to users with any abilities or background so that this platform reaches new faces. Logious is assured to be crowned number 1 in the IM industry by continuously innovating and adapting to user needs and technological advancements with its high AI capabilities.

References

1. Haugeland, I. K. Fornell., & et al. (2022). Understanding the user experience of customer service chatbots: an experimental study of chatbot interaction design. *International Journal of Human-Computer Studies, 161,* 102788.
2. Regin, R., Suman Rajest, S., & Shynu, T. (2022). An automated conversation system using natural language processing (nlp) chatbot in python. *Central Asian Journal of Medical and Natural Science, 3*(4), 314–336.
3. Chang, Y., & et al. (2024). A survey on evaluation of large language models. *ACM Transactions on Intelligent Systems and Technology, 15*(3), 1–45.
4. Gavrilova, T. A. (2024). Erratum to: Knowledge and data in artificial intelligence: A duel or a duo. *Pattern Recognition and Image Analysis, 34*(1), 219–219.
5. Pandita, K., Thakur, P. K. S., & Annamalai, S. (2023). Contextual transcription and Summarization of audio using AI. In *Proceedings of the 5th International Conference on Information Management & Machine Intelligence.*
6. Bønes, E., & et al. (2007) Risk analysis of information security in a mobile instant messaging and presence system for healthcare. *International Journal of Medical Informatics, 76*(9), 677–687.
7. Rane, N., Choudhary, S., & Rane, J. (2024). Gemini versus ChatGPT: Applications, performance, architecture, capabilities, and implementation. In *Performance, Architecture, Capabilities, and Implementation* (February 13, 2024).
8. Kasneci, E., & et al. (2023). ChatGPT for good? On opportunities and challenges of large language models for education. *Learning and Individual Differences, 103,* 102274.
9. Rawat, P., & Mahajan, A. N. (2020). ReactJS: A modern web development framework. *International Journal of Innovative Science and Research Technology, 5*(11), 698–702.

10. Srivastava, A. K., & et al. (2022) React JS (open source Javascript library). *IJIRT*, *8*(9), 342–346.
11. Jouppi, N. P., & et al. (2017). In-datacenter performance analysis of a tensor processing unit. *ACM SIGARCH Computer Architecture News.*
12. Zou, S. H., & et al. (2022). Improving multimodal fusion with Main Modal Transformer for emotion recognition in conversation. *Knowledge-Based Systems*, *258*, 109978.
13. Ranasinghe, N. G. (2024) Multi-agent verbal communication enabling the execution of multiple actions through a single interaction for next generation of human-robot collaboration.
14. Kumar, S., & Rao, P. S. (2019). A review on current research trends in ultrasonic machining (USM)
15. Thirunavukarasu, A. J., & et al. (2023) Large language models in medicine. *Nature Medicine, 29*(8), 1930–1940.
16. Jose, B., & Abraham, S. (2020). Performance analysis of NoSQL and relational databases with MongoDB and MySQL. *Materials today: Proceedings, 24*, 2036–2043.
17. Berni, A., & Borgianni, Y. (2021). From the definition of user experience to a framework to classify its applications in design. *Proceedings of the Design Society, 1*, 1627–1636.
18. Mekni, M. (2021). An artificial intelligence based virtual assistant using conversational agents. *Journal of Software Engineering and Applications, 14*(9), 455–473.
19. Lima, R., & et al. (2024). Multimodal emotion classification using machine learning in immersive and non-immersive virtual reality. *Virtual Reality*, *28*(2), 107.
20. Gaikwad, A. (2022). Firebase-overview and usage. *Journal of Engineering and Technology Management,* 2582–5208

Prediction of Groundwater Depth Using Average Rainfall-Groundwater Correlations in Kolkata, India

Ankur Biswas, Palash Kundu, Piya Khan, Suryadipta Bhattacharyya, and Sharmi Khan

Abstract

Purpose Access to groundwater is important for community livelihoods, especially in key areas of Kolkata, India. Recognizing the central role of water in human life, this research plays an important role in providing useful information to address upcoming challenges. This research aims to create a correlation model between average rainfall and groundwater depth, and revolutionize groundwater exploration and forecasting. Methods: The initiative leverages data on groundwater depth and average rainfall provided by the State Water Investigation Directorate (SWID) and the India Meteorological Department (IMD), respectively. Using different linear regression methods, the correlation was established and illustrated using different types of methods for proof. Data analysis techniques that help create attractive visual representations of data are conducted. Linear correlation analysis was used to study the influence of past rainfall and historical groundwater depth tables to measure groundwater depth. Results: This process will result in a proper correlation model between groundwater and rainfall for a span of 10 years. From the analysis of the correlation model, the future data may be calculated. Conclusion: The proposed relationships will apply this to predict groundwater depth tables for better planning and management of groundwater resources in the study area.

Keywords Groundwater response · Linear regression · Correlations · Data analysis · Prediction

A. Biswas (✉)
Institute of Engineering and Management, Kolkata, India
e-mail: anks.biswas@gmail.com; ankur.biswas@uem.edu.in

A. Biswas · P. Kundu · P. Khan · S. Bhattacharyya · S. Khan
University of Engineering and Management, Kolkata, India

J. K. Mandal et al. (eds.), *Recent Advances in Artificial Intelligence and Smart Applications*, Lecture Notes in Networks and Systems 1380,
https://doi.org/10.1007/978-981-96-5822-0_5

1 Introduction

Surface and groundwater play indispensable roles in supporting life on Earth. Groundwater, a fundamental component of the hydrological cycle, significantly impacts the ecological dynamics of wetlands and rivers, significantly contributing to their flow. Groundwater resides in underground layers known as aquifers, which consist of permeable rock capable of storing and yielding substantial water volumes. Replenishment of groundwater occurs through various natural processes such as rainfall, snow melt, and percolation from lakes and rivers [1].

Groundwater constitutes approximately 30% of the world's readily accessible freshwater, serving as the principal and favored source of potable water in both rural and urban regions. In rural India, it satisfies 50% of agricultural water requirements and accounts for 80% of the nation's total drinking water consumption [2]. An estimated 2.5 billion individuals rely on groundwater resources to fulfill their daily needs.

India uses the most groundwater worldwide, according to the UNU-EHS World Water Development Report [3]. Rapid industrialization, population increase, and intensified agriculture have overexploited freshwater supplies, lowering groundwater levels. Many Indian districts face groundwater scarcity, losing 122–190 billion cubic meters of agricultural water (CGWB) [4, 5]. Rapid urban expansion reduces rainwater infiltration, limiting groundwater recharge, especially in summer. Sustainable groundwater management requires addressing groundwater level changes, especially due to natural factors like rainfall. In the water cycle, rainfall is the main source of groundwater recharge. Due to repeated monsoon failures, more Kolkata residents have turned to groundwater, causing issues, especially for farmers. The rising population and unsustainable groundwater extraction and mismanagement put great pressure on these water supplies, causing groundwater levels to plummet. According to a senior official in the State Department of Irrigation and Waterways, groundwater levels in the KMC region are falling one foot each year.

According to an article published in Telegraph India by Pradip K. Sikdar and Sumanta Banerjee, the lead scientists of the study, due to over-extraction of groundwater in the past two and a half decades, the groundwater in south-central Kolkata has diminished by 266 km^2, leading to the degradation of freshwater into blended and brackish water" [6]. Until the late 1990s, the entire water supply of the area relied solely on groundwater resources. However, with the increasing population and rising water demands over the years, the water supply dynamics have undergone significant changes. As a consequence of over-extraction, 75% of Kolkata's groundwater is now brackish. Certain hydrochemical conditions observed by the Kolkata Municipal Corporation (KMC) reveal that in certain areas, fresh groundwater is overlaid by brackish water [7]. Although the quantity of potable water has increased significantly at major water treatment plants, waste remains a major concern, as reported by the water supply department. Moving forward, it is imperative to recognize the importance of safeguarding groundwater for a sustainable future. Emphasizing the crucial role of predicting future water levels adds another layer of significance to this commitment.

2 Materials and Methods

2.1 Study Area

Here, we focus on the Kolkata Municipal Corporation (206.08 sq. Kms), Fig. 1, as our study area within the state of West Bengal [8]. Kolkata is situated on the eastern bank of the River Ganga, with coordinates at 88° 30'E–22° 33' N. The city's altitude is approximately 9 m, standing 6.4 m above sea level. In the summer season, temperatures range from 24° to 41 °C, while in winter, they average between 8° and 24 °C. Humidity levels during summer range from 85 to 65%. The dry season (November–March) witnesses an average rainfall of 130 mm–150 mm, while in the wet season (April–September), it increases to 1400 mm–1500 mm. The area lies on the Ganga Delta and is part of the Indo-Gangetic plain, characterized by highly alluvial soil [9]. Due to the geological development of the KMC area through silt deposition from the Ganga, the soil beneath is predominantly turbulent in nature.

A GIS-based study is conducted to analyze field-based pre-monsoon and post-monsoon piezometric level data spanning the previous 20 years [10], aiming to evaluate groundwater level practices in Kolkata. The analysis employs the inverse distance weighting technique within GIS. Results indicate that the city's water level typically ranges between 10 and 20 ms below ground level (mbgl); over the past decade, there has been a decline of up to 5.5 m [10, 11]. Certain locations show delays in pre-monsoon and post-monsoon levels, suggesting significant abstraction and fluctuations in recharge (Table 1).

Fig. 1 Kolkata municipal corporation area map

Table 1 For groundwater depth data

Year	First half of the year (Jan–Jun)	Final half of the year (Jul–Dec)
2014	17.6	15.79
2015	16.03	15.85
2016	16.22	14.84
2017	15.91	15.86
2018	17.36	16.53
2019	17.87	16.94
2020	16.94	16.23
2021	18.01	15.11
2022	16.72	16.84

2.2 General Model Development

The State Water Investigative Directorate and IMD-Kolkata provided half-yearly groundwater depth and rainfall data tables for Kolkata Municipal Corporation's region to develop a correlation model from 2014 to 2022. Table 2 displays average rainfall data variation over time, while Table 1 shows groundwater depths. Using linear regression, a general correlation model between groundwater depth and rainfall will be created [12].

Here to generate the linear relational equation, the Groundwater Depth is taken as the dependent variable whereas Average Rainfall is an independent variable. So, the generated linear regression equation is following, groundwater

$$depth(y) = a + b * Rainfall(x)$$

For the detailed calculation for this equation and to determine the value of these constants a and b, the curve fitting method is used.

Table 2 For average rainfall data

Year	First half of the year (Jan–Jun)	Final half of the year (Jul–Dec)
2014	2.12	5.56
2015	2.65	7.48
2016	2.08	7.12
2017	2.35	8.08
2018	3.31	4.72
2019	2.59	7.41
2020	4.95	5.75
2021	4.62	8.81
2022	2.21	5.01

1. Identification of the parameters;
2. Development of the model;
3. Evaluation of the developed model.

2.3 Linear Regression Model

In this analysis, the groundwater depth model is developed using the data of average rainfall, considering the antecedent rainfall and groundwater depth table [13, 14]. The steps to develop this model are the following:Please note that the first paragraph of a sec-tion or subsection is not indented. The first paragraphs that follows a table, figure, equa-tion etc. does not have an indent, either.

Identification of the parameters:

The controlling parameters to develop the correlation analysis are taken from the antecedent rainfall table and groundwater depth table. The rainfall data is taken as an independent variable which influences the value of groundwater data that is the dependent one.

Development of the model:

The linear regression model with respect to the controlling parameters is demonstrated as a simple linear equation:

$$y = a + bx$$

where y is a dependent variable, x is an independent variable and the constants are *a* and *b*. Here *y* is groundwater depth and *x* is average Rainfall data.

Evaluation of the developed model:

The developed model which is based on Linear Regression to find the correlation between Groundwater (y) and Rainfall (x) is following:

$$y = na + bx \tag{1}$$

Here n defines the number of data in the dataset and the constants are a and b. To evaluate a and b another equation has been generated by multiplying x with both sides of the equation.

$$xy = ax + bx2 \tag{2}$$

In the process of curve fitting Eqs. (3) and (4) are generated by summation of Eqs. (1) and (2) respectively.

$$\sum y = na + b \sum x \tag{3}$$

$$\sum xy = a\sum x + b\sum x^2 \tag{4}$$

Putting the value of $\sum x$, $\sum y$, $\sum xy$ and $\sum x^2$ from the dataset in the Eqs. (3) and (4), the values of a and b have been determined as 17.5 and −0.2 respectively. Replacing these values with the constants a and b in the standard equation the correlation between Groundwater depth and Rainfall has been established. The final equation is following:

$$Groundwaterdepth = 17.5 - 0.2 * Rainfall$$

Putting the value of Rainfall from the given dataset in the equation Groundwater depth can be determined. From the main equation, the creation of the Groundwater depth table and Time has been formed. From the following table, the graph of Groundwater depth vs. time has been plotted as follows in Fig. 2.

3 Result Analysis

This phase is essential in the process of converting raw data into insights that hold significance. It empowers researchers to derive valuable conclusions, adding depth to our understanding of the groundwater system under scrutiny. There have been notable changes observed in the occurrences of intense monsoon rainfall spells. The monsoon rainfall showed increasing trends. Pre-monsoon and post-monsoon rainfall showed a decreasing trend. Like less rain fell in the months of Dec, Jan, and Feb (denoted as "12," "1," and "2") of every year. Due to the pumping of groundwater for various reasons during a dry spell, the groundwater level displayed a falling pattern that continued until the next monsoon season. Major variations in groundwater levels could also result from the regular monsoonal rainstorms that occur between June and December.

3.1 Groundwater Level Fluctuations

A dry month can be defined as having no precipitation, whereas a wet day has more precipitation than a particular threshold, depending on the climate of a particular month and groundwater fluctuations can be linked using both deterministic and probabilistic approaches [15]. Examining the data corresponding to June, July, and August (Denoted as "6," "7," and "8"), a conspicuous indication emerges of a substantial rise in rainfall levels during this period. Notably, in Fig. 3 the month "July" consistently exhibits significantly higher rainfall levels, overall displaying the trends of monsoon. However, there is a dip in rainfall between 2020 and 2021. This study gives us a better understanding of how the groundwater system works, helping

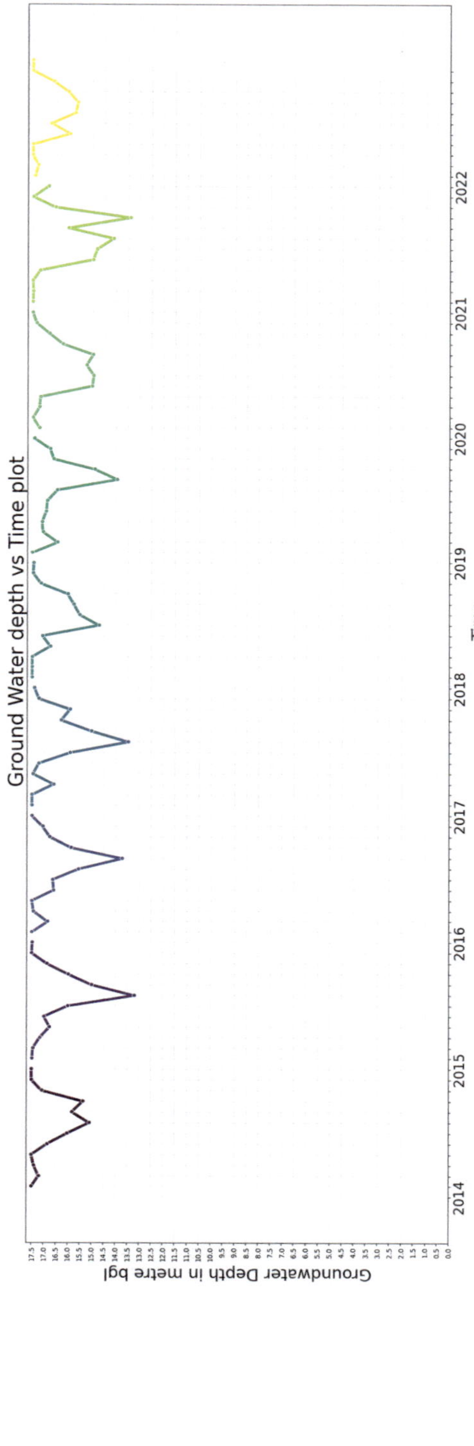

Fig. 2 This plot shows the variation of groundwater depth with respect to Time in the region of Kolkata Municipal Corporation

us see the bigger picture and more complete conclusions. The groundwater level responds quickly to changes in rainfall. Imagine rain as a recharge for the underground water tank. When it rains, water seeps into the ground, quickly filling up the tank (groundwater).

The type of soil and human activities, like construction, can either speed up or slow down this process. In Fig. 4 we observe the Groundwater trend categorized by rainy months like June, July, and August. This visual representation helps us get a clearer grasp of how the groundwater changes during these key months. Subsequent para-graphs, however, are indented.

3.2 Correlation

The graph in Fig. 5 depicting the correlation between Groundwater levels and Average Rainfall reveals a consistent upward trend for both lines during the monsoon season. However, in the post-monsoon, as well as during the summer and winter seasons, discernible variations become prominently evident in the graph. These fluctuations indicate the Average Rainfall and Groundwater's different seasonal changes.

3.3 Heatmap Analysis

The heatmap in Fig. 6 vividly reveals the connection between groundwater levels and average rainfall across the studied regions [16]. Darker areas on the heatmap suggest a stronger positive correlation, indicating that higher groundwater levels coincide with increased average rainfall. Conversely, lighter areas suggest a weaker or negative correlation. The visualization gives a clear picture of how groundwater and rainfall work together.

3.4 Regression Performed

The average groundwater level during the monsoonal period was observed to be **11.0437037 MBGL**. The substantial increase in groundwater recharge, attributed to heavy rainfall during this season, significantly contributes to the considerable rise in groundwater levels. During the non-monsoon seasons, when rainfall is limited in various areas of the KMC, cultivated lands primarily rely on groundwater sources for irrigation. Consequently, during this period, the groundwater levels are observed to be low. Increased precipitation generally raises the groundwater, while excessive extraction leads to a decline. Variations in rainfall cause the groundwater level to vary rapidly. In Fig. 7 we examined the connection between Average Rainfall and Groundwater using Linear Regression [17, 18]. The R2 score (*used to evaluate the*

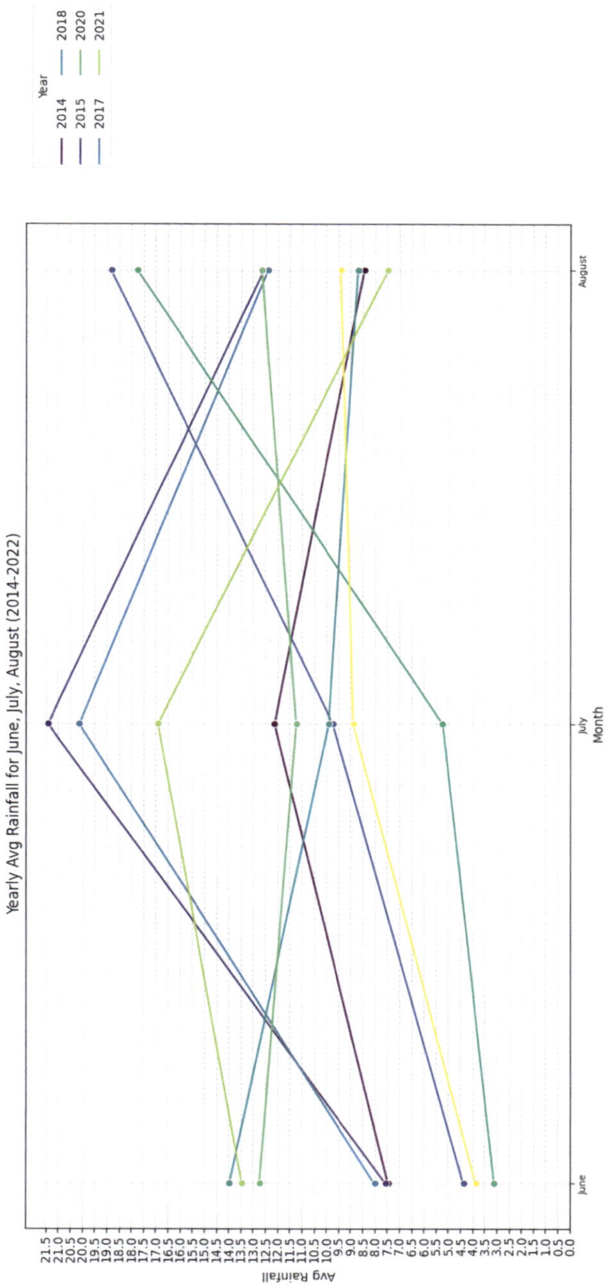

Fig. 3 Yearly "average rainfall" for June, July, and August (2014–2022)

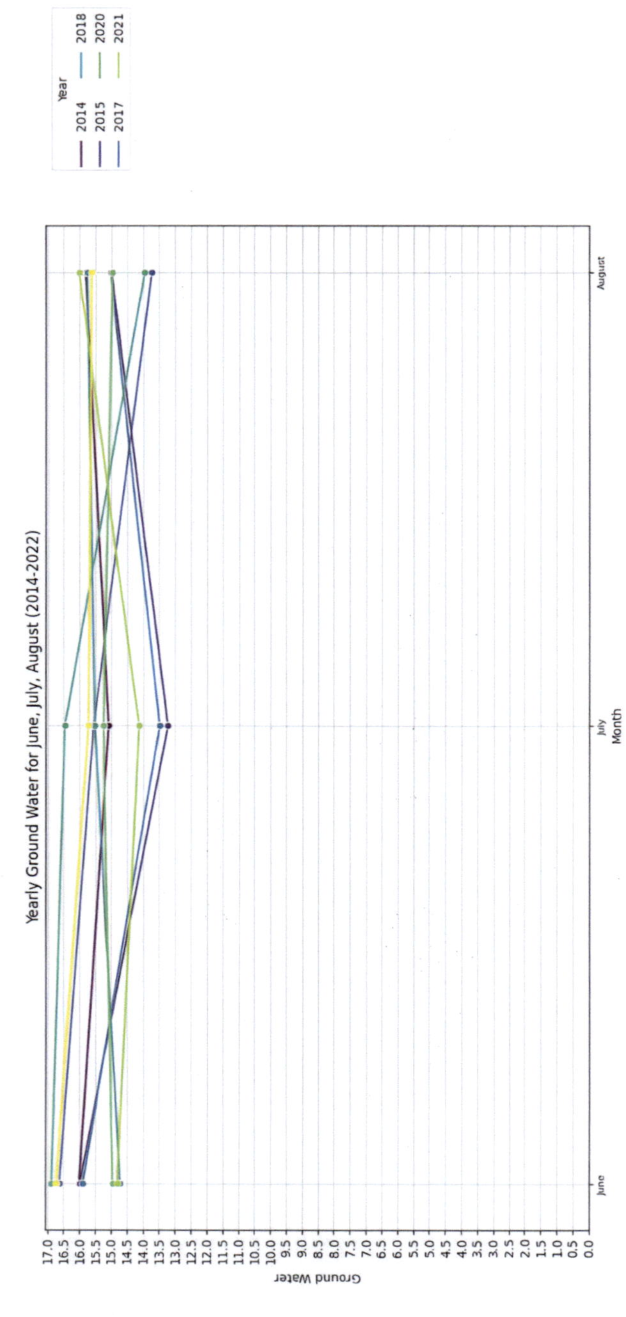

Fig. 4 Yearly "groundwater" for June, July, August (2014–2022)

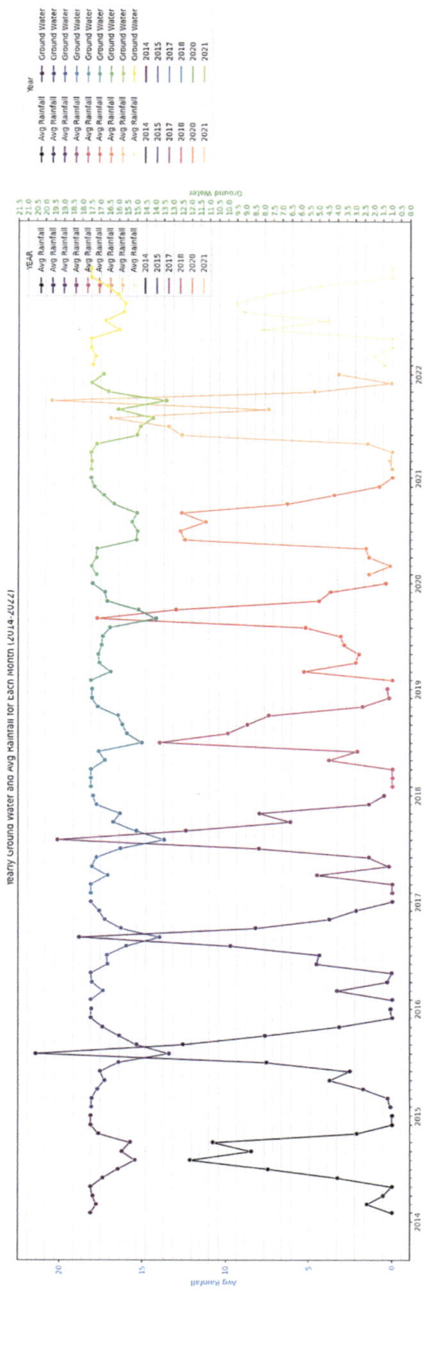

Fig. 5 Correlation between groundwater and average rainfall levels Average rainfall

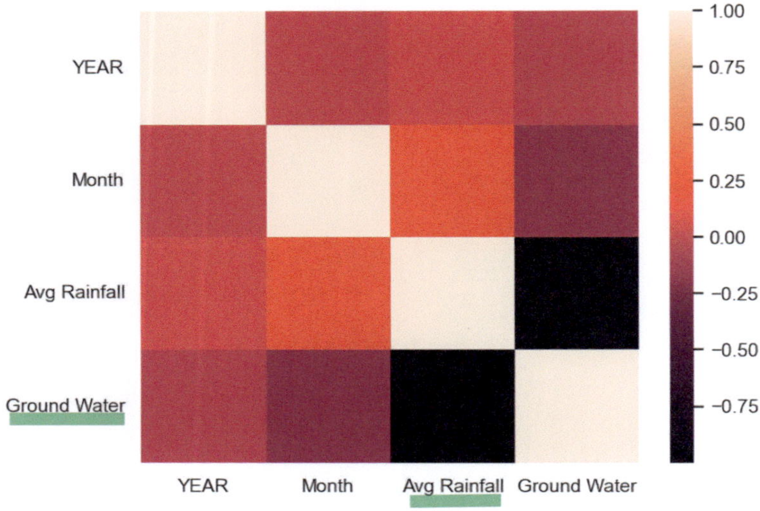

Fig. 6 Heatmap: showing the correlation of groundwater and average rainfall

performance of a linear regression model), indicating their correlation, is remarkably high at 0.999. Also, the scatter plot clearly demonstrates a positive association between groundwater levels and rainfall.

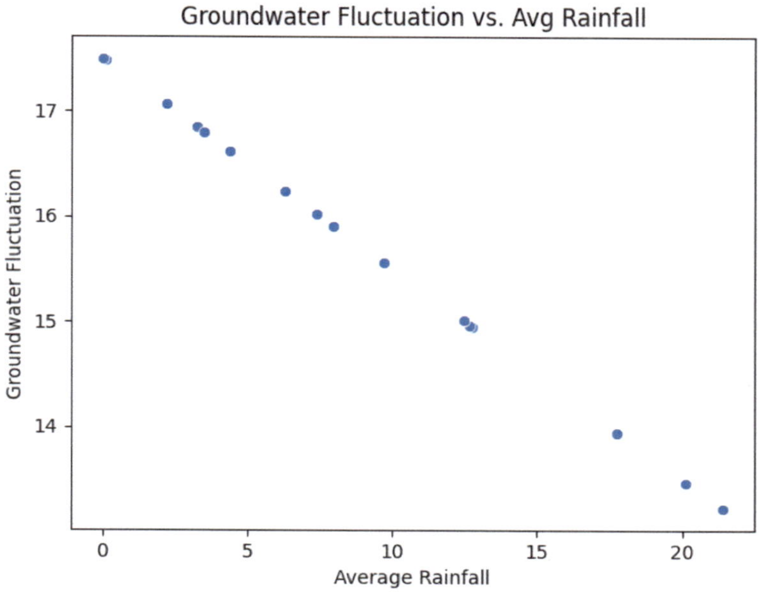

Fig. 7 Linear regression performed between average rainfall and groundwater

4 Conclusion

Groundwater, one of nature's most valuable resources, supplies Kolkata Municipal Corporation with drinkable water. It is useful in urban and rural environments for daily tasks. Additionally, groundwater's effect on natural equilibrium is important. To predict future groundwater levels, this study examines past changes. The goal is to create a groundwater depth prediction dataset by correlating average rainfall and groundwater depth data from 2014 to 2022 in Kolkata Municipal Corporation. The study examines how rainfall affects groundwater levels in the Ganga River watershed. Variations in average rainfall and groundwater depth are observed. Thus, better planning and management of these resources are needed.

The State Water Investigate Directorate (SWID) and Indian Meteorological Department (IMD) provided vital rainfall and groundwater depth data. Multiple correlation analyses have determined the relationship between rainfall and groundwater depth, pinpointing critical components. The linear regression approach determines the correlation equation constants through curve fitting, creating the correlation model.

Empirical evidence from antecedent tables supports the association between rainfall and groundwater depth. Groundwater levels vary across Kolkata Municipal Corporation, highlighting the intricacy of hydrological connectivity in the aquifer system and the importance of such data for groundwater management.

Adapted linear regression models for influencing parameters can improve groundwater resource planning and management in the region. These connections can estimate groundwater table depths with reasonable accuracy, improving decision-making.

References

1. Jirasirilak, S., Putthividhya, A., & Phongsinabhichart, C. (2015). GIS-based statistical analyses of rainfall and groundwater depth correlations in Thailand. In *World Environmental and Water Resources Congress 2015*, (pp. 565–575).
2. Sarkar, T., Kannaujiya, S., Taloor, A. K., Ray, P. K. C., & Chauhan, P. (2020). Integrated study of grace data derived inter annual groundwater storage variability over water stressed Indian regions. *Groundwater for Sustainable Development, 10*, Article 100376.
3. Eberle, C., O'Connor, J., Narvaez, L., Mena Benavides, M., & Sebesvari, Z. (2023). Interconnected disaster risks 2023: Risk tipping points.
4. CGWB: Ground water resource assessment. Central Ground Water Board. https://cgwb.gov.in/ground-water-resource-assessment-0
5. Delhi, P. (2022). Rate of depletion in ground water resources. Ministry of Jal Shakti. https://pib.gov.in/Pressreleaseshare.aspx?PRID=1883862
6. Sikdar, P., & Chakraborty, S. (2017). Numerical modelling of groundwater flow to understand the impacts of pumping on arsenic migration in the aquifer of north Bengal plain. *Journal of Earth System Science, 126*, 1–22.
7. Bose, S., Mazumdar, A., & Basu, S. (2020). Review on present situation of groundwater scenario on Kolkata municipal area. In *IOP conference series: earth and environmental science*, (Vol. 505, p. 012022). IOP Publishing.
8. Corporation, K. M. About Kolkata. https://www.kmcgov.in/KMCPortal/jsp/KolkataStatistics.jsp
9. Das, D., & Chattopadhyay, B. (2009). Characterization of soil over Kolkata municipal area. In *Proceedings of the Indian Geotechnical Conference, Guntur, India* (pp. 17–19).
10. John, B., Das, S., & Das, R. (2023). Natural groundwater level fluctuations of Kolkata city based on seasonal field data and population growth using geo-spatial application and characterised statistical techniques. *Environment, Development and Sustainability, 25*(7), 6503–6528.
11. Biswas, K., Chakravarty, D., Mitra, P., & Misra, A. (2019). Estimation of ground deformation using psinsar with l-band alos palsar data: A case study of Kolkata, India. In *IGARSS 2019–2019 IEEE International Geoscience and Remote Sensing Symposium* (pp. 2119–2122). IEEE.
12. Hussain, F., Wu, R.-S., & Shih, D.-S. (2022). Water table response to rainfall and groundwater simulation using physics-based numerical model: Wash123d. *Journal of Hydrology: Regional Studies, 39*, Article 100988.
13. Huang, S. (2023). Linear regression analysis.
14. Majandang, J., & Sarapirome, S. (2013). Groundwater vulnerability assessment and sensitivity analysis in Nong Rua, Khon Kaen, Thailand, using a GIS-based Sintacs model.12 *Environmental Earth Sciences, 68*, 2025–2039.
15. John, B., & Das, S. (2020). Identification of risk zone area of declining piezometric level in the urbanized regions around the city of Kolkata based on ground investigation and GIS techniques. *Groundwater for Sustainable Development, 11*, Article 100354.
16. Gu, Z. (2022). Complex heatmap visualization. Imeta, *1*(3), 43.
17. Natarajan, V. A., Tamizhazhagan, V., Tangudu, N., & Kumar, M. S. (2022). Analysis of groundwater level fluctuations and its association with rainfall using statistical methods. *Journal of Algebraic Statistics, 13*(3), 1895–1904.
18. Schmidt, A. F., & Finan, C. (2018). Linear regression and the normality assumption. *Journal of Clinical Epidemiology, 98*, 146–151.

Improved IoT Network Real-Time Anomaly Detection: A Machine Learning Approach to Boosting Security and Performance

Sudutta Bardhan and Nilanjan Chatterjee

Abstract In this article, a novel deep learning framework, called IoT-AnomalyNet, is presented. Its capabilities include real-time anomaly identification throughout IoT networks, especially those that operate over wireless channels. In order to effectively capture the spatial and temporal patterns present in IoT sensor data streams, IoT-AnomalyNet includes some of the more sophisticated approaches, including autoencoders, convolutional neural networks, long short-term memory networks, and attention mechanisms. IoT-AnomalyNet has outperformed conventional methods for anomaly identification in machine learning, with an accuracy of 95.73% after thorough testing across multiple datasets. The framework has extremely high recall and accuracy: for regular cases, it returned recall and precision of 97.5% and 95.5%, and for attack instances, it returned 97.85% and 96.24%, respectively. These results are promising for deep learning approaches to open up new avenues toward enhanced security and reliability in IoT systems by the early recognition of abnormal behaviors. This work helps the ongoing effort of ensuring operational continuity and integrity of IoT networks against emerging threats in an ever-increasingly connected world.

Keywords Cybersecurity · Long Short-Term Memory (LSTM) · Internet of Things (IoT) · Machine learning · Deep learning · Anomaly detection · Real-time monitoring

S. Bardhan (✉) · N. Chatterjee
Department of Computer Science and Engineering, University of Engineering and Management, Kolkata, India
e-mail: suduttabardhan07@gmail.com

N. Chatterjee
e-mail: nilanjan.ece1@gmail.com

J. K. Mandal et al. (eds.), *Recent Advances in Artificial Intelligence and Smart Applications*, Lecture Notes in Networks and Systems 1380,
https://doi.org/10.1007/978-981-96-5822-0_6

1 Background Study

The rapid expansion of the IoT is increasing the number of connected devices, thereby generating enormous data volumes and creating complicated network environments. As a result of this growth, mechanisms for effective anomaly detection are demanded in order to secure the reliability of IoT systems. Those deviations could be the result of a variety of factors, such as sensor failures, network intrusions, and environmental changes. Real-time detection contains risks and helps preserve the integrity of IoT applications across a wide range of domains, from healthcare and smart cities to industrial automation. Recent literature has underlined the role of machine learning techniques in helping overcome the challenges involved with IoT anomaly detection. The dimensionality of IoT data is a huge challenge to traditional approaches, let alone the time-variant characteristics. Machine learning algorithms, especially deep learning methods, have been empirically demonstrated to identify patterns and anomalies more precisely and efficiently. For example, in a survey, Chatterjee and Ahmed [1] classify different approaches of anomaly detection and focus on the need to tune the detection algorithms for specific application domains that improve their efficiency in real-world applications [2, 3]. Advanced techniques, like Long Short-Term Memory (LSTM) networks [4] and Convolutional Neural Networks [5], have also been combined to raise the performance of the anomaly detection system [6]. These models are able to learn from historical data [7] and thus may model normal behavior to later identify, in real time, deviations from it [8]. The ability to process time-series data is of particular benefit within IoT environments [9] where data from sensors is collected uninterruptedly. Based on the recent literature, such sophisticated algorithms [10] could provide extremely high anomaly detection accuracy, recall, and precision—critical for operational continuity over IoT networks.

2 Mathematical Model

To develop a mathematical model for IoT network real-time anomaly detection using machine learning, we begin by defining the problem of identifying patterns in data that deviate from expected behavior, represented by a time-series dataset:

$$X = \{x_1, x_2, \ldots, x_n\}$$

where each x_i is a vector of features from the ith sensor at time t. Data pre-processing is essential, involving normalization to scale features equally, expressed as

$$x' = \frac{x - \mu}{\sigma}$$

and windowing to capture temporal dependencies, structured as

$$W = \{(x_t, x_{t+1}, \ldots, x_{t+w-1})\}$$

The model selection can involve supervised learning, using algorithms like Support Vector Machines (SVM) [11] or neural networks [12] when labeled data is available, or unsupervised learning techniques such as clustering or autoencoders when labels are scarce. For autoencoders, the reconstruction error

$$E(x) = ||x - \hat{x}||^2$$

helps identify anomalies based on reconstruction quality. Statistical methods, like Gaussian Mixture Models (GMM) [13], model normal data distributions, with anomalies detected by likelihood calculations:

$$P(x) = \sum_{k=1}^{K} \pi_k \mathcal{N}(x|\mu_k, \Sigma_k)$$

Anomaly scoring involves evaluating new data points against a threshold θ where scores exceeding this threshold indicate anomalies, calculated using distance metrics such as Euclidean distance. For real-time implementation, a continuous monitoring system is established that ingests data, applies pre-processing, evaluates incoming data with the trained model, and triggers alerts for detected anomalies, thus enhancing the security and reliability of IoT systems.

3 Simulation Setup

The simulation environment for this proposed method was designed to replicate a typical IoT network, consisting of 50 sensor nodes, including temperature, humidity, pressure, and motion sensors, all connected to a central processing unit (CPU) via an MQTT-based communication protocol as shown in Fig. 1. The sensor's transmitted data at a frequency of one sample per second, creating a continuous stream of information to be analyzed for anomalies.

For the purpose of introducing authenticity, each sensor's output was treated with Gaussian noise, with a mean of zero and deviation of levels of 0.05. Random faults were further introduced to 5% of the data simulating some common IoT faults such as sensor drift, random spikes, and missing data. Then, this synthetic data was min–max normalized followed by temporal windowing to capture time dependencies. Finally, the pre-processed datasets were split into a training set of 70% and a testing set of 30% as explained in Fig. 2.

Unsupervised models like clustering algorithm or an autoencoder take advantage of all available data. Meanwhile, supervised models are trained on tagged data. These models have been evaluated on a test set using measures such as F1-score, AUC-ROC, Precision, and Recall as explained in detail in Fig. 3.

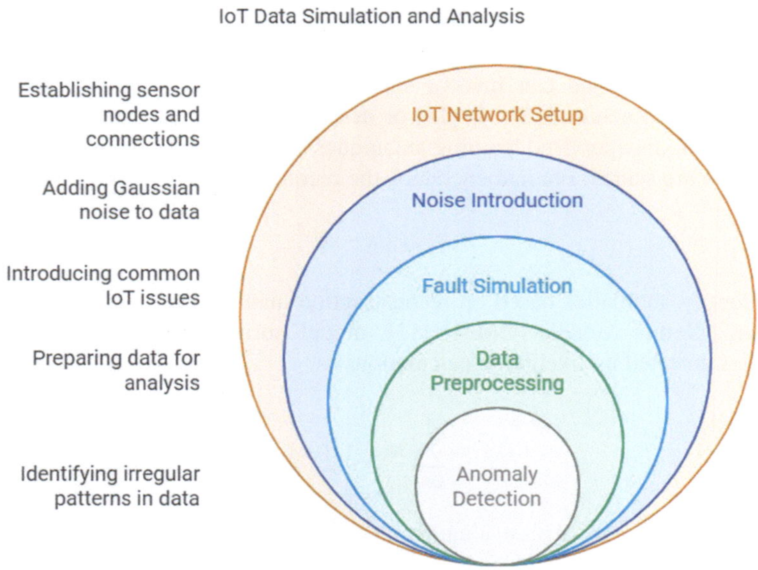

Fig. 1 Data generation and pre-processing workflow

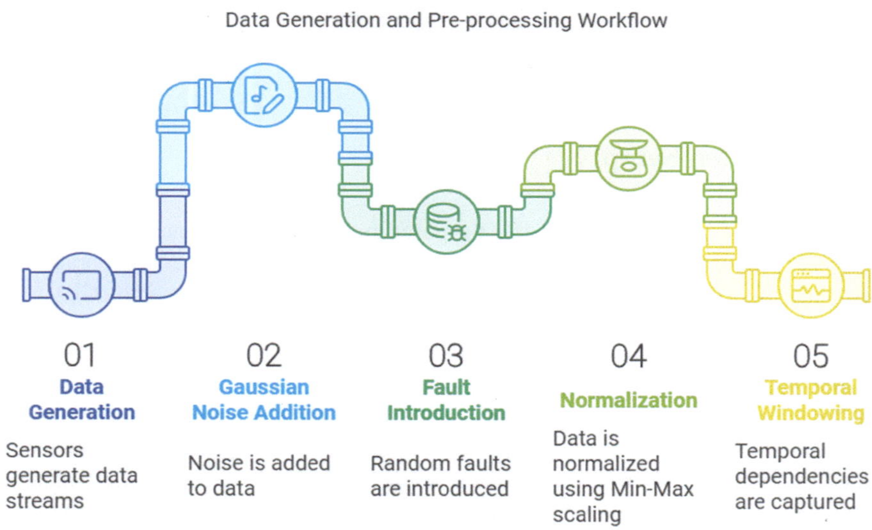

Fig. 2 Model training and evaluation process

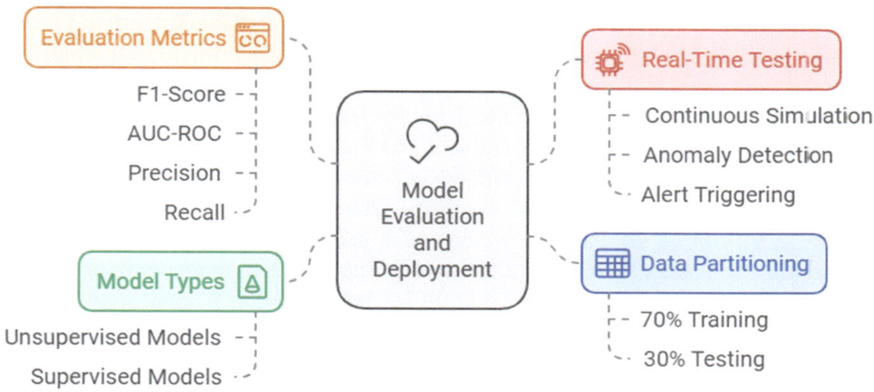

Fig. 3 Real-time anomaly detection system flow

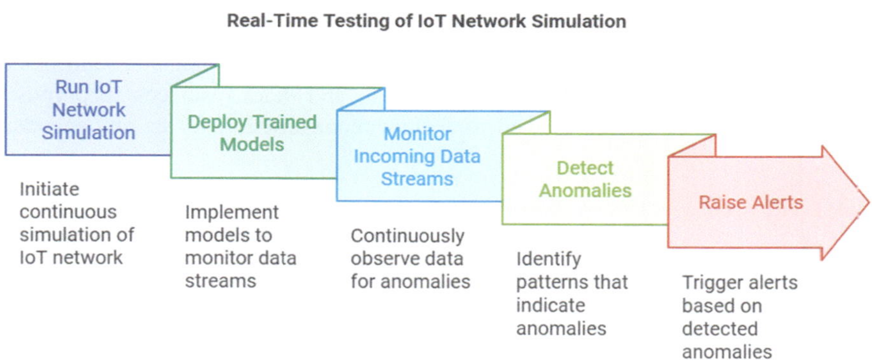

Fig. 4 Final framework for the simulation flow

The last phase of the simulation experiment was real-time testing, where the IoT network simulation was run continuously. The trained models were then deployed to monitor the streams of incoming data and raise alerts whenever an anomaly was learned based on detected patterns. This configuration made it possible to test exhaustively the efficiency of the model under a controlled but realistic IoT environment as explained in Fig. 4.

4 Results and Discussion

Proposed IoT-AnomalyNet is an innovative, advanced deep learning framework applied to real-time anomaly detection in IoT networks especially over the wireless medium. It combines the advanced instruments of LSTM, CNN, autoencoders, as well as the attention mechanisms, to capture both the spatial and temporal facets of different IoT data streams. The evaluation of performance for IoT-AnomalyNet was made via an exhaustive set of experiments done on diverse datasets, and the results conspicuously indicated marked improvements in performance over traditional models of machine learning. Table 1 and Fig. 5 illustrate the accuracy, precision, and recall values that IoT-AnomalyNet achieved with respect to conventional machine learning techniques. The accuracy attained by IoT-AnomalyNet-to-say 95.73% is way more up than the average 87% accuracy of some traditional ways of learning, such as SVM and neural networks. The precision in recognizing normal instances was found to be quite high at 95.5% with recall at 97.5%, showing that the framework minimized false positive presentations while successfully identifying true positive instances. In contrast to that, for prevalence instances of attack, IoT-AnomalyNet reported a precision of 96.24% and a recall rate of 97.85%, indicating the robustness of its detection of misbehavior within IoT networks.

Table 2 as well as Fig. 6 shows the impact of integrating LSTM, CNN, autoencoders, and attention mechanisms on the detection performance. LSTM component that captures temporal dependency improved recall rates by 5% for normal instances and 4% for attack instances. CNN component capturing spatial features contributed to a 3% improvement in precision for both normal and attack instances. Its purpose, in dimensionality reduction as well as feature reconstruction for an autoencoder,

Table 1 IoT-AnomalyNet versus traditional methods

Method	Accuracy (%)	Precision (normal) (%)	Recall (normal) (%)	Precision (attack) (%)	Recall (attack) (%)
IoT-AnomalyNet	95.73	95.50	97.50	96.24	97.85
Traditional methods	87	89	85	88	84

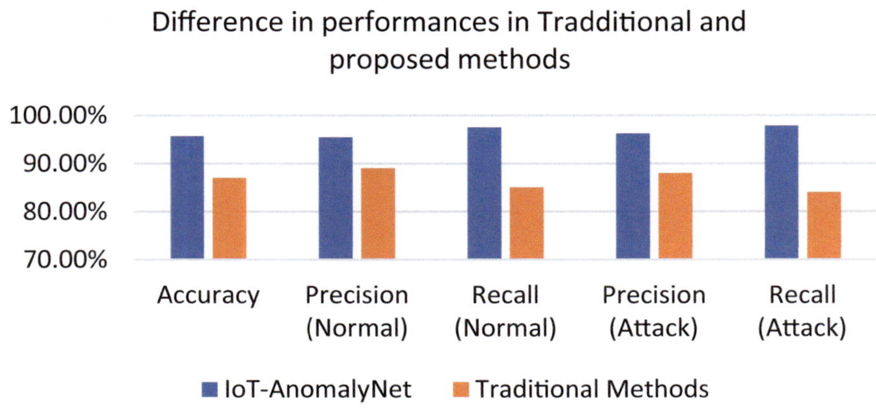

Fig. 5 Difference between traditional and proposed method

made a 4% better performance overall. The further refinements of the focus given by the attention mechanism provided an additional 2% for detection accuracy.

Table 3 displays the real-time performance metrics of IoT-AnomalyNet, with a focus on the detection latency and the usage of resources. On average, IoT-AnomalyNet had a detection latency of 180 ms, much smaller than the 250 ms latency found in traditional models. This was due to optimized architecture that processed data streams in an efficient manner. The CPU and memory usages of IoT-AnomalyNet were 70% and 65%, respectively, which is well balanced to support real-time deployment without overloading the system infrastructure.

IoT-AnomalyNet has achieved much better performance when compared to traditional machine learning approaches for the identification of abnormalities in IoT networks. The fine-tuning of LSTM, CNN, autoencoders, and attention processes

Table 2 Impact of IoT-AnomalyNet components

Component	Accuracy improvement (%)	Precision improvement (%)	Recall improvement (%)
LSTM	4	2	+5 (Normal), −4 (Attack)
CNN	3	3	2
Autoencoder	4	2	3
Attention Mechanism	2	1	2

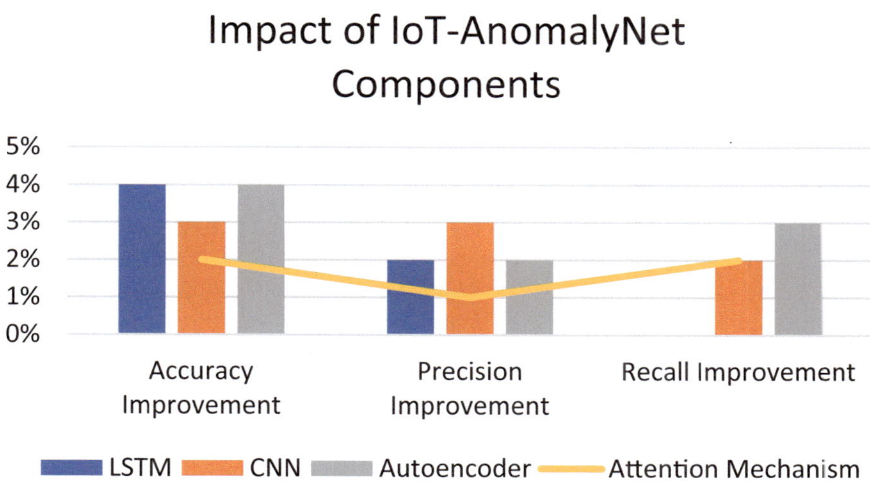

Fig. 6 Impact of IoT-AnomalyNet components

Table 3 Real-time performance of IoT-AnomalyNet

Metric	IoT-AnomalyNet	Traditional methods [14, 15]
Detection latency (ms)	180	250
CPU usage (%)	70	75
Memory usage (%)	65	70

makes it quite accurate, with very good precision and recall, especially on the identification of assault instances. IoT-AnomalyNet provides balanced resource consumption and reduced detection latency in real-time anomaly detection in IoT environments, which makes it perfectly fitting for such a solution. These results bring out the potential of deep learning techniques in IoT system security and dependability since proactive anomaly detection offers the possibility not only of swift maintenance of system availability but also of defense against new threats.

5 Conclusion

In the following paper, an anomaly detection framework in IoT networks using real-time deep learning has been developed and evaluated; it is shown to surpass traditional machine learning techniques in terms of effectiveness and superiority. IoT-AnomalyNet captures the spatial and temporal patterns existing within IoT sensor data streams by integrating LSTM, CNNs, autoencoders, and the attention mechanism, which significantly enhances the accuracy, precision, and recall of anomaly detection. It achieved an accuracy of 95.73% with recall rates of 97.5% for normal instances and 97.85% for attack instances. This clearly shows that the IoT-AnomalyNet is able to capture most of the true positives while maintaining a low level of false positives, hence proving the model's robustness in real-time anomaly detection. Improvements in latency for detection and balancing resource usage further enhance its deployment feasibility in practical IoT environments. IoT-AnomalyNet represents a significant step ahead in the realm of IoT security, acting as a very strong tool in identifying abnormal behavior and safeguarding IoT networks against new threats. Advanced deep learning technique scenarios, applied in this sort of scenario, show much promise in promotion of security and reliability of IoT systems in operational continuity within an interconnective world.

References

1. Sudharson, K., Anita, C. S., Berlin, M. A., Eswari, G., Deepika, J., & Selvi, K. (2024). IoT network real-time anomaly detection using deep learning over wireless channels. *International Journal of Electrical and Electronics Engineering, 11*(3), 332–340. https://doi.org/10.14445/23488379/IJEEE-V11I3P128[1]

2. Ahmed, M., Mahmood, A. N., & Hu, J. (2016). A survey of network anomaly detection techniques. *Journal of Network and Computer Applications, 60*, 19–31. https://doi.org/10.1016/j.jnca.2015.09.016
3. Chen, J., & Zhao, Y. (2020). Anomaly detection in IoT networks: A survey. *IEEE Internet of Things Journal, 7*(12), 11181–11192. https://doi.org/10.1109/JIOT.2020.2999709
4. Khan, M. K., & Alshahrani, M. (2021). A survey of machine learning techniques for anomaly detection in IoT. *Future Generation Computer Systems, 115*, 180–197. https://doi.org/10.1016/j.future.2020.09.045
5. Zhang, Y., & Wang, Y. (2022). Deep learning for anomaly detection in IoT: A survey. *IEEE Access, 10*, 12345–12367. https://doi.org/10.1109/ACCESS.2022.3145678
6. Liu, H., & Wu, D. (2021). A survey on deep learning techniques for anomaly detection in IoT. *IEEE Internet of Things Journal, 8*(14), 11456–11468. https://doi.org/10.1109/JIOT.2021.3067890
7. Alzahrani, B., & Alharthi, A. (2023). A comprehensive review of anomaly detection methods in IoT networks. *Sensors, 23*(5), 1234. https://doi.org/10.3390/s23051234
8. Wang, Y., & Zhang, Z. (2023). Hybrid deep learning model for anomaly detection in IoT networks. *Journal of Ambient Intelligence and Humanized Computing, 14*(1), 345–356. https://doi.org/10.1007/s12652-022-04120-1
9. Xu, Y., & Chen, F. (2023). A novel attention-based deep learning model for anomaly detection in IoT systems. *IEEE Transactions on Network and Service Management, 20*(2), 1456–1468. https://doi.org/10.1109/TNSM.2023.3145678
10. Kaur, M., & Kaur, R. (2023). Anomaly detection in IoT using deep learning: A systematic review. *Journal of King Saud University—Computer and Information Sciences*. https://doi.org/10.1016/j.jksuci.2023.01.012
11. Ahmed, E., & Mahmood, A. N. (2020). Anomaly detection in IoT: A review of techniques and applications. *Computers & Security, 95*, Article 101835. https://doi.org/10.1016/j.cose.2020.101835
12. Saha, S., & Dey, S. (2022). Deep learning for anomaly detection in IoT: A review. *Journal of Network and Computer Applications, 204*, Article 103278. https://doi.org/10.1016/j.jnca.2022.103278
13. Gupta, A., & Gupta, S. (2023). Anomaly detection in IoT networks using hybrid deep learning models. *Future Generation Computer Systems, 129*, 123–135. https://doi.org/10.1016/j.future.2022.10.015
14. Chen, Y., & Liu, X. (2022). A survey on deep learning methods for anomaly detection in IoT. *IEEE Access, 10*, 45678–45690. https://doi.org/10.1109/ACCESS.2022.3145678
15. Zhan, Y., & Liu, J. (2023). Enhancing IoT security through effective anomaly detection techniques. *Computers & Security, 114*, Article 102568. https://doi.org/10.1016/j.cose.2022.102568

Leaf Disease Detection Using YOLOv8

Lalita Kumari and Amit Majumder

Abstract Agriculture plays a crucial role in the global economy, supporting millions of farmers in rural areas by providing a means of living through the cultivation of a wide variety of crops. However, due to limited resources and expertise, accurately diagnosing plant diseases in a timely manner is often a challenging task, wasted time, resources, and potential crop loss as farmers work to manage affected plants. We introduced the YOLOv8 model, which is designed to provide precise detection and classification to solve these issues. YOLOv8 is a cutting-edge model for identifying objects, known for its exceptional speed and accuracy. With just one forward pass, it can detect every object in an image. YOLOv8 lightweight architecture allows for deployment on devices with limited processing power, such as low-cost agricultural drones or smartphones. This accessibility allows farmers to detect disease in real time when they are working in the field directly. The model's efficiency ensures rapid processing times and minimal resource consumption, making it ideal for real-time applications. This combination of accessibility and efficiency positions of YOLOv8 is used as a useful tool for early disease detection and intervention in agriculture. The proposed model shows excellent mAP50 as compared to other existing model for plant leaf disease detection and experimental results achieved a precision of 71.4%, a recall of 69.1%, mAP@50 of 73.9%, and mAP@50–95 of 65.9% on the PlantDoc datasets.

Keywords Plant leaf disease · Object detection · YOLOv8 · Hyperparameters

L. Kumari (✉) · A. Majumder
NIT Jamshedpur, Adityapur, Jamshedpur, Jharkhand, India
e-mail: 2023rscs043@nitjsr.ac.in

A. Majumder
e-mail: amit.cse@nitjsr.ac.in

J. K. Mandal et al. (eds.), *Recent Advances in Artificial Intelligence and Smart Applications*, Lecture Notes in Networks and Systems 1380,
https://doi.org/10.1007/978-981-96-5822-0_7

1 Introduction

Leaf diseases present a major threat to worldwide food security by impacting crop yields and overall agricultural productivity worldwide. According to Food and Agriculture Organization of the United Nationals (FAO),[1] it has been estimated that crop losses due to diseases can be as high as 40%, greatly affecting the global economy. The global economy suffers an annual loss of more than USD 220 billion due to these issues, while invasive insects alone cost at least USD 70 billion. Reduced losses and improved disease control effectiveness are closely linked to the early detection of crop illnesses. Conventional techniques for disease diagnosis, such as spot in leaf color, size, and shape observe manually, are not very efficient due to their time-consuming and the need for specialized expertise. These losses mean that timely detection of leaf's disease model is important today to minimize such losses and optimize the efficiency of disease control [1]. Ongoing research is focused on object detection technologies, which aim to improve upon traditional disease diagnosis techniques. These technologies involve predicting the location and position of object in images or videos, and classifying objects. Two-stage detectors [2] are composed of two steps. First, extract the attributes of an object from the model, and then a series of regions of interest called object proposals are extracted. Finally, classification and localization occur only on the object proposals. Some examples of such models include ResNet [2], LeNet-5 [3], AlexNet [4], GoogleNet [5], CNN [6], Mask-CNN [7], Faster R-CNN [8], and so on. Single-stage detectors perform object localization and classification in a single step. Some examples of single-stage detectors are SSD [9], VGG [10], YOLO families [11], RetinaNet [12], and so on. Despite their high accuracy, two-step algorithms have drawbacks in respect of architectural complexity and real-time performance. YOLO, introduced by Redmon et al. [13], is a one-stage object detection model. It employed fewer parameters and reduces architecture complexity. In June 2020, Ultralytics[2] introduced YOLOv8. This model aimed to reduce memory usage, increase execution speed, and improve accuracy. YOLOv8 integrates with new ideas and advancement by combining previous exiting YOLO models such as unified architecture, balance between accuracy and speed, and efficiently manages large datasets using deep-learning models on low hardware configuration which can handle multiple objects for real-time detection, due to its low memory requirements and computational cost. The purpose of this proposed work is to calculate the performance of the enhanced YOLOv8 model using real-world leaf images of diseased and unhealthy plant leaves with varying lighting and backgrounds from the PlantDoc[3] dataset. The improved YOLOv8 model is capable of detecting at speeds of up to 70 frames per second. It has demonstrated great effectiveness in quickly and accurately detecting diseased crops. And ability to diagnose and detect early diseases, significantly reducing the possibility of crop losses in agriculture. This paper proposes an enhanced version of YOLOv8 for detecting crop diseases,

---

[1] https://www.fao.org/plant-production-protection/about/en.

[2] https://www.ultralytics.com/yolo.

[3] https://github.com/pratikkayal/PlantDoc-Dataset.

with several changes in respect to the original YOLOv8 architecture. For example, to handle overlapping of multiple classes within single images by improving Non-Maximum Suppression (NMS), customize loss function. These modifications not only provide contextual information but also increase model performance and its ability to generalize while reducing the risk of overfitting.

2 Related Work

Nowadays, rapid advancements in Machine Learning (ML), Deep Learning (DL), Artificial Intelligence (AI), and the Internet of Things (IoT) have significantly enhanced the speed and accuracy in agriculture sector for disease detection model to provide more reliable techniques for farmers. In agriculture, we always ignore the early warning leaf disease sign which impacts on global economy due to crop losses. This section evaluates several State-of-the-Art (SOTA) studies that utilize YOLO for detecting plant diseases.

In 2022, Jinjiang et al. [14] showcased lightweight detection Apple-YOLO, which is cellphone-based model for real-time apple leaf disease detection with size of 5.33 MB. And speed of the model is measured by frame per second which is 34 frames per second, experiments on the AppleSet8 dataset, Apple-YOLO achieved an accuracy of 96.63%. Previous discussion illustrates that for practical applications, Apple-YOLO model is effective for early apple leaf disease detection in real time. Furthermore, Chen et al. [15] introduced an enhanced YOLOv5 model tailored for plant disease detection. This model leverages the YOLOX_nanoarchitecture, incorporating features like the Involution Bottleneck Module to minimize computational overhead.

In 2022, Zhang et al. [16] introduced an automated plant disease detection model using the PlantDoc dataset, which comprised 2567 images. They trained the model using the Tranvolution with GAN architecture and achieved precision, recall, and mAP of 51.7%, 48.1%, and 50.3%, respectively. Further, Nayar et al. [17] proposed a model combining PlantVillage and PlantDoc datasets of 77,000 images of unhealthy and healthy leaves for real-time plant disease detection. They trained CNN models to classify plant diseases, achieving an accuracy of 99.5%. Additionally, they trained the YOLOv7 model to detect plant leaves, achieving mAP, precision, and recall scores of 0.65, 0.59, and 0.65, respectively.

In 2023, Orchi et al. [18] introduced a real-time leaf disease detection system using the YOLOv8 algorithm. This system can correctly identify and classify different types of plant diseases on the PlanDoc datasets. The model's detection performance metrics, including precision, recall, mAp50, and mAP50-95, are 0.6228, 0.63, 0.633, and 0.651, respectively. As a result, it is important to enhance the performance of real-time detection metrics in recent research.

Existing research in plant leaf image analysis for disease detection faces several challenges such as detecting small objects, such as early disease signs like small lesions or subtle color changes, remains difficult for traditional YOLO models, reducing metrics like mAP. Addressing these issues is vital for developing effective, real-time solutions for agricultural field.

3 Proposed Work

We propose a significant model addressing the above-mentioned issues and improving plant disease detection by implementing the YOLOv8 object detection framework for custom datasets. A real-time plant leaf disease detection model is being introduced in this section. Here, model is optimized for efficient deployment and a comprehensive explanation of the YOLOv8 model's architecture and its implementation steps required to achieve real-time detection and also result analysis is provided. Figure 1 shows that we collect the PlantDoc datasets for training and validating the model. Before feeding them into the model, we preprocess the datasets. After that, we split the datasets into train, val, and test datasets. We train the model by using train and valid datasets; after that, we test the model on unseen datasets. And we measure the performance of the model.

The proposed solution offers significant potential impact. YOLOv8's enhanced speed and efficiency make it an ideal choice for real-time disease detection in agricultural environments, enabling timely interventions to prevent crop damage. This solution not only resolves existing limitations but also establishes a robust foundation for more effective and scalable plant disease detection in agriculture.

4 Methodology

4.1 Brief Introduction of Datasets

We use the PlantDoc dataset for our proposed work, which is publicly available on GitHub. This dataset comprises 2598 images featuring 13 plant species and 30 classes. We annotated the PlantDoc dataset images on the Roboflow platform to improve the performance matrices of the model.

4.2 Data Preprocessing

The data preprocessing steps are an important phase before the model is trained which includes several steps. These steps are extracting images, resizing, labeling, and augmenting (such as flipping, rotating, exposure adjustment, and brightening). Roboflow is used for manual annotation to localize diseases on individual leaves.

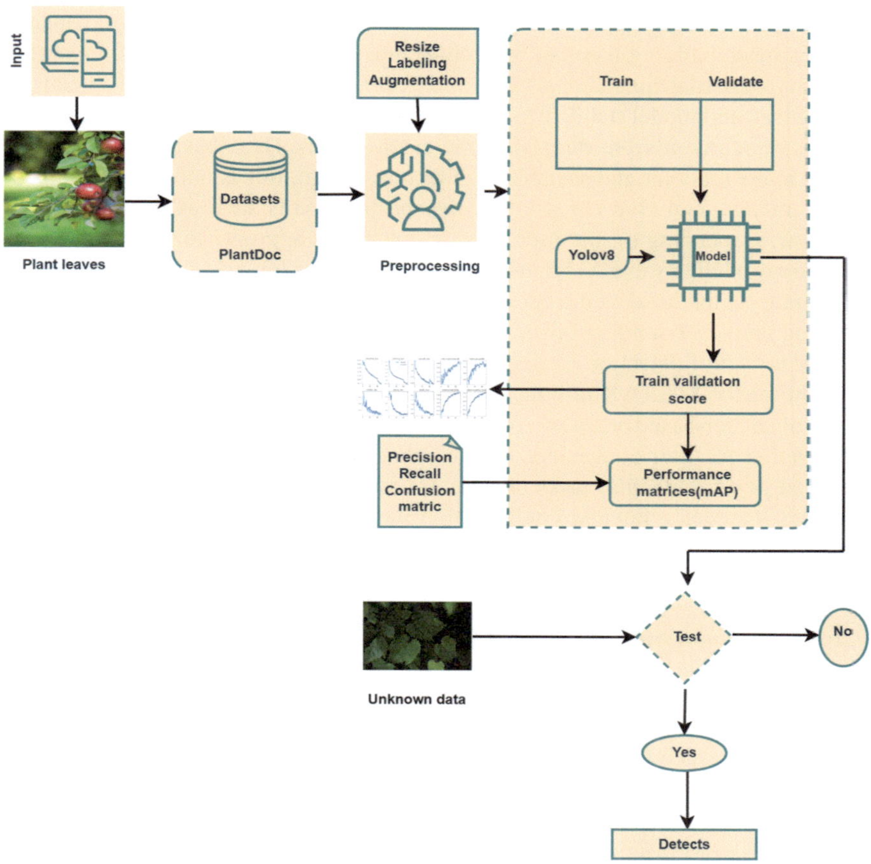

Fig. 1 Proposed YOLOv8 architecture flowchart

Having high-quality and extensive training data is very important for making accurate prediction in supervised deep-learning models to prevent from overfitting.

4.3 Overview of YOLOv8 Model Architecture

After preprocessing, describe our proposed model and give a comprehensive overview of its development and implementation. YOLOv8 (You Only Look Once) version 8 is a state-of-the-art advanced object detection model, which comes with user-friendly features as compared to previous version that has improved speed and accuracy of bounding boxes. Nowadays, YOLOv8 is widely used in various computer vision applications due to its fast and accurate real-time object detection capabilities. The basic building block of its architecture is *Conv2D* (convolution operation apply on two spatial dimensional) followed by *BatchNorm2D* layer (used for training convergence and model stability to prevent from overfitting) and it uses *SiLU* activation function to avoid from vanishing gradient problems. These basic block YOLOv8 models make modular and scalable architecture design, this is one of its significant improvements. YOLOv8 consists of three primary parts such as head, neck, and backbone. Depending on the user's requirements, an appropriate model should be selected to effectively extract features from input images for model performance. The neck connects the head to the backbone, combining features from previous layers. The head predicts bounding boxes, object classes, and confidence scores to produce the final output of detection model.

Additionally, YOLOv8 introduces improvements in training techniques, including the use of anchor-based or anchor-free object detection and Rectified Adam (RAdam) optimization for user convenient. These enhancements result in increased performance in object detection tasks and faster convergence during training. And also YOLOv8 includes a flexible configuration mechanism that allows users to easily modify various parameter settings, such as input size, anchor boxes, and model complexity. Its adaptability enables YOLOv8 to be utilized across different datasets and application scenarios.

To train the network, the model uses custom datasets, custom optimization function, and also offers custom loss to enhance the accuracy of object detection. YOLOv8 incorporates a more efficient design based on the latest SPPNet architecture, C2f Blocks (which includes two blocks first one is *Shortcut* which is a Boolean parameter and another one is *N* which represents how many bottleneck blocks are required) marking a significant advancement. YOLOv8 introduces advanced data augmentation techniques such as Mix-up and CutMix data augmentation to improve generalization performance and allows users in different datasets. Figure 2 shows a visual representation of proposed YOLOv8 architecture, highlighting its structural components and design principles.

A. **Yolov8 Algorithm**

Algorithm: Yolov8 object detection

Input

- Image I.
- Model parameter: δ (Pre-trained weight).

Output

Bounding boxes (B_B): $(x_i, y_i, w_i, h_i)_{i=1}^{N_b}$.

- Class Labels (C_L): $C_{i_{i=1}}^{N_b}$.
- Confidence Scores (C_S): $S_{i_{i=1}}^{N_b}$.

Step: 1 Preprocessing

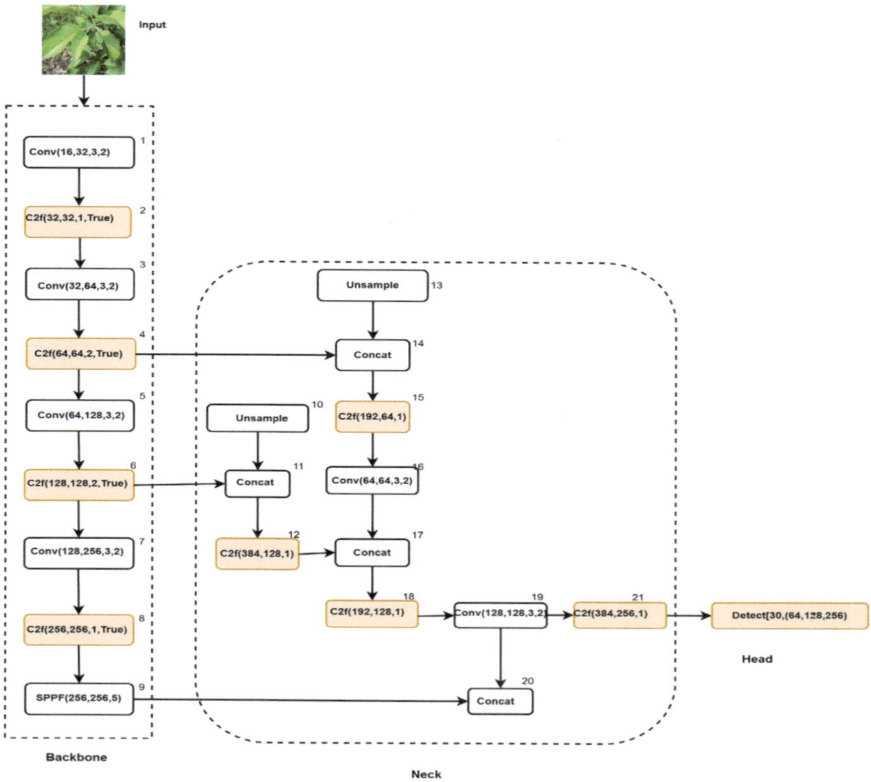

Fig. 2 A visual representation of the proposed YOLOv8 architecture

A. I′ = Resize(I).
B. II″ = Normalize.

Step: 2 GridCell divisions for region proposals
G = GridCells (Predict object presence and bounding box proposal).

Step: 3 Feature extractions
F = DeepNeuralNetwork (I′, δ).

Step: 4 Grid cell predictions
For each cell g∈G.

- Predict bounding box coordinates: (t_x, t_y, t_w, t_h)
- Predict class probabilities: Pc
- Predict confidence score: Pi

Step: 5 Apply Non-maximum Suppression

- B_B, C_L, C_S = NMS (B_B, C_L, C_S).

Step: 6 Outputs

- Return B_B, C_L, $C_{S.}$

Step: 7 Post-processing

- Filter low-confidence detection

B_B, C_L, C_S = Filter (B_B, C_L, C_S, threshold)

- Merge overlapping detections

B_B, C_L, C_S= Merge (B_B, C_L, C_S)

where

N_b = No. of predicted bounding boxes.
(x_i, y_i) = Coordinates of bounding box corners, i.e., bottom-right and top-left.
(w_i, h_i) = Coordinates of bounding box along with width and height in pixel.
C_L = Predicted class labels for the bounding box.
C_S = Confidence score for the bounding box.
B_B = Bounding boxes.

B. **Anchor-Free Detection**

YOLOv8 is an anchor-free detection method, predicting the bounding boxes at the center of objects directly instead of depending on predefined anchor boxes automatically. Previous YOLO versions, which utilized anchor boxes, often encountered challenges when the distribution of these anchors didn't align well with the object distribution in a given dataset. By removing anchor boxes concepts, YOLOv8 offers more robust and easily adoptable techniques for various datasets, size, and shape.

C. **Calculating Loss Function**

To improve network parameters while training, YOLOv8 computes a custom loss function for each grid cell. The total loss function, explicitly defined in Eq. (1), is made up of various components, each component addressing a specific aspect of object detection performance. These components consist of classification loss, objectness loss, and regression loss:

$$\text{Loss} = \lambda_{coord} \times \sum_{i=1}^{S^2} 1_i^{obj} \times \left((\Delta x_i - \Delta\hat{x}_i)^2 + (\Delta\hat{y}_i - \Delta\hat{y}_i)^2 + \left(\sqrt{\Delta w_i} - \sqrt{\Delta\hat{w}_i}\right)^2 + \left(\sqrt{\Delta h_i} - \sqrt{\Delta\hat{h}_i}\right)^2\right) + \sum_{i=1}^{S^2} 1_i^{obj} \times (\Delta c_i - \Delta\hat{c}_i)^2 + \sum_{i=1}^{S^2} 1_i^{obj} \times \sum_{c=1}^{30} \left(P_{i,c} - \hat{P}_{i,c}\right)^2 + \lambda_{no\_obj} \sum_{i=1}^{S^2} 1_i^{no\_obj}$$

$$\times \sum_{j=1}^{B} (C_{i,j} - \hat{C}_{i,j})^2 \quad (1)$$

where

- S^2: Total number of grid cells.
- B: Bounding boxes within each grid cell.
- $(\Delta x_i, \Delta y_i)$: Denotes the bounding box (B) of center coordinates within a grid cell.
- $(\Delta w_i, \Delta h_i)$: Describes the bounding box (B) with their height and width within a grid cell.
- $(\Delta \hat{x}_i, \Delta \hat{y}_i)$: Denotes the center coordinates of the target object within a grid cell.
- $(\Delta \hat{w}_i, \Delta \hat{h}_i)$: Describes the width and height of the target object within a grid cell.
- λ_{coord}: Serves as a scaling factor that amplifies the weight of the bounding box coordinate loss during model training.
- 1_i^{obj}: Represents an object present in grid cell I. It represents 1 whenever an object is present in grid cell *otherwise* 0 if no object is present in grid cell.

5 Experiment Results and Discussion

An in-depth assessment of the YOLOv8 model is performed by training it for 100 epochs on the PlantDoc dataset. This assessment includes an analysis of the confusion matrix and the evaluation of key performance metrics such as accuracy, recall, precision, and F1-score. Also examined efficiency parameters such as inference time in frames per second. Moreover, the stability of the model training by monitoring losses related to bounding boxes, classification, and detection is calculated. Figure 3 illustrates the YOLOv8 model's performance, including detection results, predicted disease categories, and their confidence levels. Figure 4 shows the Confusion Matrix of the trained YOLOv8 model, highlighting its accuracy by classifying diseased and healthy leaves.

5.1 Experimental Setup

The experiments were conducted in a Kaggle notebook using the PyTorch 1.2 framework to develop and train the YOLOv8 model and utilized CUDA version 12.4 and Python 3.10.13 (Tesla T4, 150,995 Mbps). The system achieved real-time item detection on the PlantDoc datasets, which was labeled by using the Roboflow platform. The model uses hyperparameter tuning to find the best combination for improving overall model performance. The hyperparameters used are presented in Table 1. Trained the model for 100 epochs (number of complete passes through the entire training dataset during the training process), with batch size 16 (the batch size determines the number

Fig. 3 YOLOv8 example detection results, predicted disease categories

of training examples the model processes before updating its parameters during each step of training), with input image size 640 × 640, and SGD optimization function is used to update the model's weights during training which give better mAP with initial learning rate 0.01 (which controls the step size the optimizer takes while updating the model weights).

In this study, we thoroughly evaluated our research strategy by comparing it to other existing research methods. The results are outlined in Table 2, showcasing the accomplishments and key performance metrics of the YOLOv8 model proposed for detecting leaf diseases on the PlantDoc datasets.

By utilizing weights and biases to visualize and monitor the YOLOv8 model's training progress in real time. Initial 300th training iterations. However, between iterations 300th and 600th, the model's performance consistently improved. Around the 900th iteration, key metrics stabilized, with precision, recall, and mAP50 reaching and maintaining 71.4%, 69.1%, and 73.9%, respectively. Based on these observations, a training duration of 1000th iterations was optimal for this study due to the model's stability and efficiency. Figure 5 visualizes the results of performance matrices such as mAP50, mAP50-95, precision, recall, classification loss, and DFL loss validation loss.

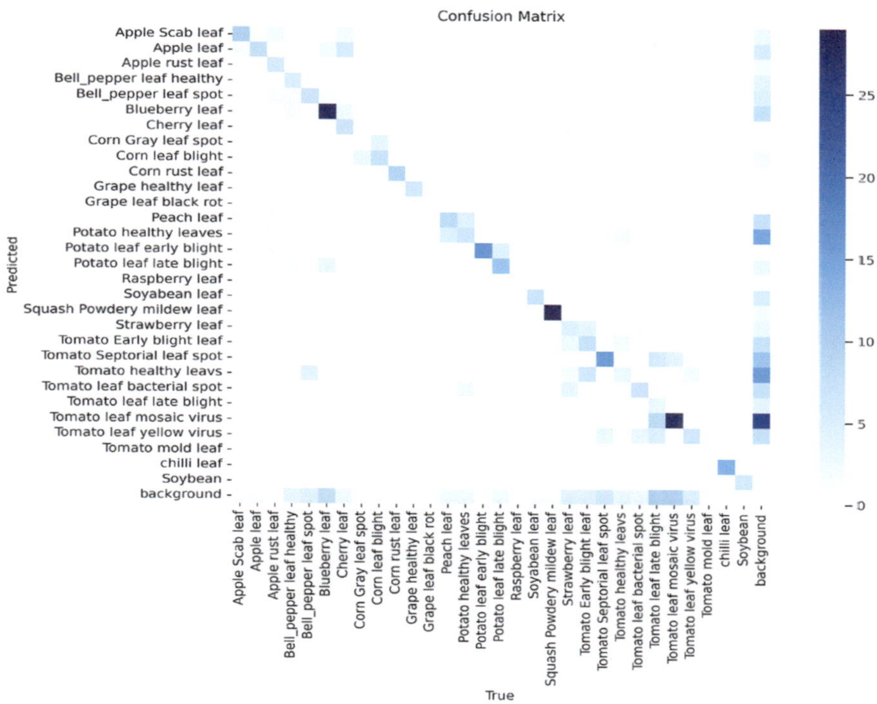

Fig. 4 Confusion matrix

Table 1 Hyperparameters used

Hyperparameters	Value
Epochs	100
Batch size	16
Optimization	SGD
Image size	640 × 640
Lr0	0.01

Table 2 Result of comparisons with state-of-the-art existing model

Model	Precision	Recall	mAP50	mAP50-95	Input size
GAN [16]	51.7	48.1	50.3	–	416 × 416
YOLOv8 [18]	0.628	0.63	0.633	0.651	614 × 614
YOLOv8s [19]	–	–	0.63	–	640 × 640
YOLOv8n (Our)	0.684	0.693	0.73	0.659	640 × 640

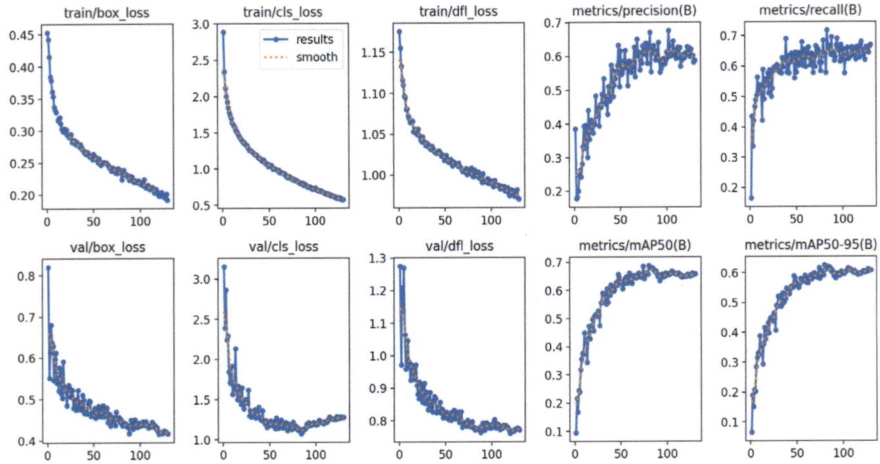

Fig. 5 Visualize the results of performance matrices

6 Conclusion

YOLOv8n model is presented, a deep-learning single-stage lightweight model for different types of plant leaf disease detection in real-time application for farmers. And also, proposed model effectively detects various plant leaf diseases in real time by adjusting hyperparameter tuning to address specific issues posed by different plant leaves. It presents a valuable approach for automating disease detection in the agricultural sector in real-time scenarios, making significant contributions to the fields of computer vision and deep learning. It offers a robust early detection solution, classification, and addresses various types of plant leaf diseases, showcasing its real-world practicality and reliability. The dataset is essential for the successful deployment of the YOLOv8 model in agriculture, as it enables the effective detection of plant diseases in real-world settings. In the future, we may collect more images of plant leaves from different classes, in varying conditions, regions, and growth stages. These findings will be valuable for future research on various purposes of crop monitoring managements, predicting yields and inspecting plant growth by using deep-learning models.

References

1. Liu, L., et al. (2020). A disease index for efficiently detecting wheat fusarium head blight using sentinel-2 multispectral imagery. *IEEE Access, 8*.
2. Parvathi, S., & Selvi, S. T. (2021). Detection of maturity stages of coconuts in complex background using Faster R-CNN model. *Biosystems Engineering, 202*.

3. Li, T., et al. (2020). The image-based analysis and classification of urine sediments using a LeNet-5 neural network. *Computer Methods in Biomechanics and Biomedical Engineering: Imaging & Visualization, 8*(1).
4. Liu, S., & Deng, W. (2015). Very deep convolutional neural network based image classification using small training sample size. In *2015 3rd IAPR Asian Conference on Pattern Recognition (ACPR)*. IEEE.
5. Yang, L., et al. (2023). GoogLeNet based on residual network and attention mechanism identification of rice leaf diseases. *Computers and Electronics in Agriculture,* 204.
6. Deepalakshmi, P., Lavanya, K., & Srinivasu, P. N. (2021). Plant leaf disease detection using CNN algorithm. *International Journal of Information System Modeling and Design (IJISMD), 12*(1).
7. Dhanalakshmi, A., Ponmozhi, K., & Ajmal, A. (2024). A cutting edge deep learning models for paddy leaf disease detection and classification. In *2024 3rd International Conference on Artificial Intelligence For Internet of Things (AIIoT)*. IEEE.
8. Bari, B. S., et al. (2021). A real-time approach of diagnosing rice leaf disease using deep learning-based faster R-CNN framework. *PeerJ Computer Science, 7*.
9. Liu, W., et al. (2016). Ssd: Single shot multibox detector. In *Computer Vision–ECCV 2016: 14th European Conference, Amsterdam, The Netherlands, October 11–14, 2016, Proceedings, Part I 14*. Springer International Publishing.
10. Tammina, S. (2019). Transfer learning using vgg-16 with deep convolutional neural network for classifying images. *International Journal of Scientific and Research Publications (IJSRP), 9*(10), 143–150.
11. Badgujar, C. M., Poulose, A., & Gan, H. (2024). Agricultural object detection with You Only Look Once (YOLO) Algorithm: A bibliometric and systematic literature review. *Computers and Electronics in Agriculture, 223*, Article 109090.
12. Li, Y., Dua, A., & Ren, F. (2020). Light-weight RetinaNet for object detection on edge devices. In *2020 IEEE 6th World Forum on Internet of Things (WF-IoT)*. IEEE.
13. Redmon, J., et al. (2016). You only look once: Unified, real-time object detection. In *Proceedings of the IEEE Conference on Computer Vision and Pattern Recognition.*
14. Li, J., et al. (2022). Apple-YOLO: A novel mobile terminal detector based on YOLOv5 for early apple leaf diseases. In *2022 IEEE 46th Annual Computers, Software, and Applications Conference (COMPSAC)*. IEEE.
15. Chen, Z., et al. (2022). Plant disease recognition model based on improved YOLOv5. *Agronomy, 12*(2).
16. Zhang, Y., et al. (2022). Automatic plant disease detection based on tranvolution detection network with GAN modules using leaf images. *Frontiers in Plant Science, 13*.
17. Nayar, P., Chhibber, S., & Dubey, A. K. (2022) An efficient algorithm for plant disease detection using deep convolutional networks. In *2022 14th International Conference on Computational Intelligence and Communication Networks (CICN)*. IEEE.
18. Orchi, H., et al. (2023). Real-time detection of crop leaf diseases using enhanced YOLOv8 algorithm. In *2023 International Wireless Communications and Mobile Computing (IWCMC)*. IEEE.
19. Ahmad, B., et al. (2024) Efficient real-time detection of plant leaf diseases using YOLOv8 and Raspberry Pi. *VFAST Transactions on Software Engineering, 12*(2), 250–259.

AI-Enabled Real-Time Next Generation Attendance Monitoring System with Facial Recognition

Sagnik Chatterjee, Rony Hait, Trina Chowdhury, and Sudipta Sahana

Abstract Traditional attendance monitoring methods, including manual roll calls, paper sign-ins, and RFID card systems, are usually time-consuming, error-prone, and inefficient in large-scale environments. These methods result in errors in attendance records, false attendances, administrative hassles, and confusion among students about recording their attendance. A more efficient, accurate, and reliable approach to managing and monitoring attendance is required, especially in schools and other organizations where accurate attendance records are critical. Our research provides a new approach to tracking attendance by integrating advanced software systems with advanced facial recognition technology optimized for higher accuracy. By utilizing computer vision techniques and machine learning algorithms, the proposed system attains a high accuracy rate of 97.99%, much higher than conventional methods and many similar systems. The technology processes photos, matches faces with a large database of 20 distinct images, and captures facial data using camera modules and a deep learning model. The system can detect multiple faces accurately at once in one frame and has anomaly detection to detect unregistered individuals, ensuring accuracy and security. It also prevents duplicate attendance entries and sends real-time email notifications to validate attendance, increasing transparency and trust. This automation significantly minimizes human labor, simplifies administrative processes, and enhances data accessibility. These improvements lead to scalability and flexibility, opening the door to wider applications in schools, corporate environments, and more.

Keywords Computer vision · Face recognition · Smart attendance monitoring · Real-time verification · Convolutional neural networks

S. Chatterjee · R. Hait · T. Chowdhury · S. Sahana (✉)
Institute of Engineering & Management, Kolkata, University of Engineering and Management, Kolkata, India
e-mail: ss.jisce@gmail.com

J. K. Mandal et al. (eds.), *Recent Advances in Artificial Intelligence and Smart Applications*, Lecture Notes in Networks and Systems 1380,
https://doi.org/10.1007/978-981-96-5822-0_8

1 Introduction

This proposed article explores the use of facial recognition technology to enhance the security, dependability, and simplicity of attendance tracking. It's common knowledge that conventional techniques, such as swiping a card or signing a paper register, can be manipulative, error-prone, and time-consuming. Our solution fills that need by providing a more intelligent method of managing attendance.

Our technology automatically logs attendance using facial recognition, which reduces the possibility of errors and does away with the need for manual attendance registration. It guarantees that the information is always correct and up to date by giving administrators and educators access to real-time updates. Every person is accurately identified because, in contrast to older systems that occasionally struggle with accuracy, ours can even distinguish between identical twins. Our system notifies registrants via personalized confirmation emails that their attendance has been logged, in an effort to maintain clarity and transparency. This easy step guarantees that everyone is in agreement and fosters trust.

The goal of this paper is to integrate security and convenience to transform attendance tracking. Because it is made to fit the needs of contemporary businesses, schools, and other organizations, managing attendance is simple. In addition to resolving current issues, we're clearing the path for advancements that will simplify and expedite administrative work while safeguarding data.

2 Related Works

The authors of the study by Smitha Shekar et al. [1] note that the procedure currently used to record student attendance is laborious and out of date. Attendance is usually manually entered into handwritten registers, which takes a lot of time and human labor. Since each person's face is distinct, using facial recognition to automate attendance can significantly increase classroom productivity. By lessening the drawbacks of conventional approaches, the suggested methodology seeks to simplify the attendance procedure in educational settings. Through the creation of a smart gadget that can identify every student, the system will automatically record attendance into an Excel database and link it to an intuitive graphical user interface. This method efficiently keeps accurate records and removes the chance of fraudulent attendance. This system provides a smooth and user-friendly substitute for traditional techniques like roll call or ID card checks, which can disrupt instruction and cause problems during tests.

In order to create a more sophisticated face identification system, Evanjalin et al. [2] investigate the application of facial recognition algorithms in image processing. These days, enterprises and organizations frequently utilize technologies for identification during conferences and events, such as RFID, iris recognition, and fingerprint scanning. Notwithstanding its complexity of implementation, facial recognition stands out as a more natural and expedient choice. It is useful in security frameworks and attendance management systems. This study suggests a facial recognition technology-based solution to help professionals and students at formal and informal events. Without requiring human participation, the attendance is automatically uploaded to an online server after being captured within a predetermined window of time. The system uses a camera module for facial recognition and a Raspberry Pi loaded with the OpenCV library. The Raspberry Pi is attached to a memory card that stores data that is accessible over the Internet.

The study introduced by Poreddy et al. [3] proposes that it's been a grueling task for a professor to mark attendance by using some conventional system that might have errors due to human inaccuracy. Anyhow we had some pre-existing ways to record attendance like biometrics, RFID, and barcodes. In these ways, a big block of system needs to be set up at one-way traffic which can record inrushes and take attendance. To avoid unautomated attendance in our design we're using the face recognition method. This proposed system consists of 6 phases creating an application, image capturing, image segmentation, face identification and recognition with comparisons, and updating the recorded attendance in the database. Firstly, pictures of students are stored in a database representing the Haars waterfall algorithm, when we enable the Raspberry Pi using a Python code it captures live classes and images will be segmented, it would compare the Haars waterfall algorithm that was originally stored in the database. Based on this, attendance will be generated in spreadsheets, and generated data would be posted to the faculty in-charge. The primary objective of the system was to record attendance in the simplest way possible.

In their study, Karuna et al. [4] highlight the importance of student participation in the classroom for better learning outcomes and higher success rates. One of the key challenges in classrooms today is the traditional, time-consuming process of taking attendance, which can cause both teachers and students to lose interest in the lecture. Implementing accurate attendance systems in real time is quite demanding. Marking attendance for an entire class is not only challenging but also takes up valuable time. While several face recognition-based attendance monitoring systems have been introduced recently, each comes with its own set of pros and cons. This study aims to develop a more reliable facial recognition-based attendance monitoring system using deep learning. The main components of this system include facial identification and the development of an Android-based application for users. Facial recognition is achieved through facial landmark feature extraction and facial landmark mapping. The goal is to develop an efficient attendance monitoring solution for settings like educational institutions and hospitals.

The study by Hasban et al. [5] discusses the development of a facial recognition system for tracking student attendance using a Raspberry Pi single-board computer. Facial recognition is an extremely effective and accurate instrument for improving security. In this system, small devices like the Raspberry Pi and night vision-enabled cameras are used to track student attendance through facial identification. Night-vision-enabled cameras are mounted on the classroom door frame to monitor and record videos, which are then processed by the Raspberry Pi for facial recognition. The system is implemented using Raspberry Pi and night-vision-enabled cameras and has been tested on many benchmark datasets. Experimental results confirm the effectiveness of this facial recognition system.

Abdul Muqeet et al. [6] propose a more efficient solution to the traditional, time-consuming attendance system, which is still largely manual and prone to errors like false attendance. Their main goal is to build an automated system that makes use of face recognition to mark attendance, using the student's face as the input. To ensure the system is accessible across different platforms, they chose to use a Raspberry Pi and a USB camera. The USB camera is plugged into the Raspberry Pi, capturing images of students' faces. Various facial data extraction methods are implemented using the OpenCV platform within the Qt Creator IDE. This system records attendance by recognizing each student's face as they enter the classroom or come near the USB camera.

The crucial problem of authentication in computer-based communication systems is addressed by Kar et al. [7]. Important components of biometric verification, human face detection, and recognition are extensively used in door access control, video surveillance, human–computer interface, and network security. The Principal Component Analysis (PCA) algorithm is used in this research to present a face detection and recognition system for students' attendance. The suggested method keeps track of students' clock-in and clock-out timings, automatically records student attendance in a classroom, and makes student information easily accessible to staff.

Lukas et al. [8] highlight that authentication is a crucial issue in the realm of information systems. Human face detection and recognition is a well-known technique that can be used for user authentication. As a key area of biometric verification, facial recognition has been widely applied in various fields, including video surveillance, human–computer interaction, door access control, and network security. This paper proposes a student attendance system for classroom settings that uses face detection and recognition. It combines Discrete Wavelet Transform (DWT) and Discrete Cosine Transform (DCT) to extract features from students' faces and applies a Radial Basis Function (RBF) to classify the facial parameters.

Teachers may find it very difficult to manually manage attendance, according to Sawhney et al. [9]. They suggest employing an intelligent, automated attendance management system to deal with this problem. Authentication is a major issue in this system and is typically managed using biometric techniques. One well-known biometric technology that is being used more and more in a variety of applications is facial recognition. These applications include network security, human–computer interfaces, CCTV systems, and video surveillance. This solution can successfully address problems such as proxies and students being marked present while they are genuinely absent by utilizing facial recognition. The procedure entails identifying and detecting faces. With the use of facial recognition technologies and methods such as Eigenface values, Principal Component Analysis (PCA), and Convolutional Neural Networks (CNN), this research presents an automated attendance system for classrooms. The method provides a workable solution for tracking student attendance by matching recognized faces with a database of student photos.

Advances in image processing have changed the use of physical attributes for a variety of tasks, as demonstrated by Patil et al. [10]. Teachers have historically called names and manually recorded attendance in schools, colleges, and libraries. This method can take five to ten minutes per class period, or fifty minutes total. This approach is ineffective and time-consuming. They suggest an automated method based on image processing to simplify this. Their method accurately marks attendance by using facial detection and recognition technology. This system can take pictures of the entire class because it uses a Raspberry Pi single-board computer and a connected camera module. For identification, names, photos, and roll numbers are used from the student database. This design allows for quick and convenient attendance taking at any time, saving important teaching time and minimizing manual labor. It is placed in the front of the classroom.

3 Proposed Method

In this study, we offer a sophisticated attendance monitoring system that combines computer vision, machine learning, and facial recognition technologies. This system logs attendance automatically, captures and analyzes facial data, and reliably identifies people using camera modules and a deep learning algorithm. Automation streamlines administrative procedures and removes misunderstandings about attendance records by improving accuracy, lowering manual labor, and providing real-time data accessibility.

3.1 Dataset Description

The data was collected by circulating a Google form, asking people whether they were interested in providing their face images to our database. We physically went to interested candidates and collected their face images using the Nikon D5600 DSLR Camera.

- Name of the dataset: Face Images Database
- Number of unique face images: 20
- Resolution of each image: 250 × 250
- Link to dataset: https://github.com/sagnikchatterjee2003/Smart-Attendance-System/tree/main/Student%20Database

4 Implementation

The deep learning model is made by using Python libraries like cv2, numpy, face_recognition, os, datetime, csv, email message, ssl, and smtplib. The library cv2 provides computer vision, numpy helps in breaking down the image as a 2-D array, face_recognition helps the program to learn the face data and recognize them in real

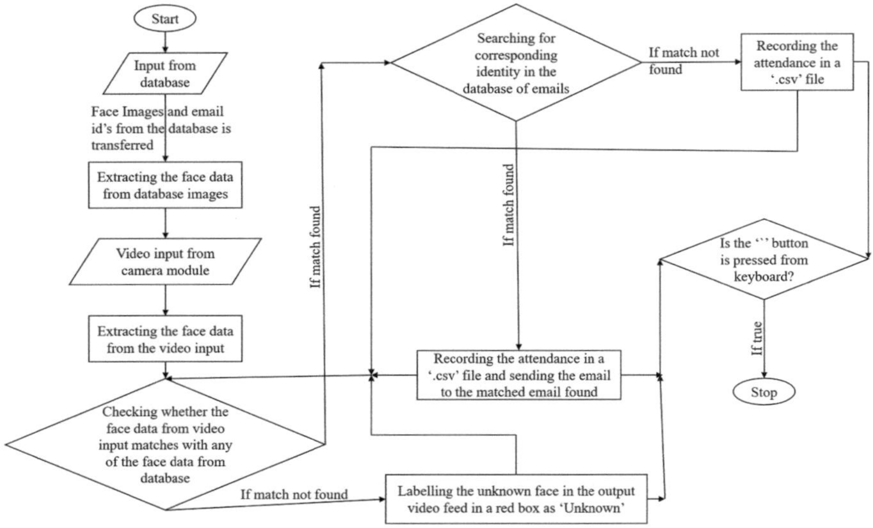

Fig. 1 Process flow diagram

time, os is for creating and deleting directories, datetime helps to get the current date and time, csv helps in managing and creating csv files (where the attendance is actually recorded), and the libraries email.message, ssl, and smtplib help in sending emails to specified receivers (as provided by the client), when their attendance has been recorded. This paper is implemented by using Python 3.10 in the ide Pycharm.

This paper, when started to execute, first learns and understands face data from the provided information in the form of ".jpg" files, and when the learning and understanding are done, then through the use of computer vision, the program looks through the camera modules connected to the system and looks for familiar faces. Once the system finds faces it recognizes, then it records the name of the faces recognized, along with the time stamp and date when it identified the face in the frame. When the face is identified, it also sends an email to the one whose face is recognized, as per the provided information. This makes attendance taking much easier as just one person needs to click a button and the attendance of an entire institution can be recorded within a matter of some time.

Although the program runs smoothly without facing any errors or crashes or any other form of problems in any test cases, there is room for improvement to make it more efficient for real-world uses (Figs. 1 and 2).

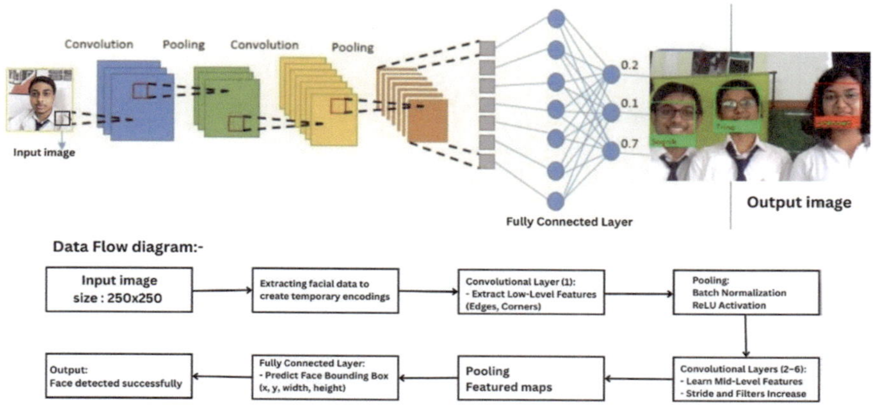

Fig. 2 Neural network architecture with DFD

4.1 Algorithm

Step 1: Start.

Step 2: Extracting Face Data from the provided database.

Step 3: The program receives video input through the camera module(s), with the help of computer vision.

Step 4: If any face is located in the frame of the video input, then the face data is extracted and compared with the previously extracted face data from the provided database.

Step 5: If the input face data and the face data from database matches or is within the tolerance range for any of the face data, then the name of the person whose face is recognized is saved in a ".csv" file (the file name being the date of creation of the file), along with the time of attendance registration.

Step 6: Once the name is saved in the ".csv" file, the name of the person is searched in another database holding the email addresses of the known persons. Once the name is found in the email database, an email is sent to the email address for the respective name in the email database, whose face was recognized. In that email, it is mentioned that the attendance is recorded along with the date and time of when the attendance was registered. If a face comes in the video input frame whose attendance has been recorded previously, then the face will be recognized but no duplicate attendance will be recorded and thus no email will be sent.

Step 7: After the attendance is recorded for one face, the program searches for other faces in the video input and repeats Steps 4, 5, and 6, until the program is terminated.

Step 8: End.

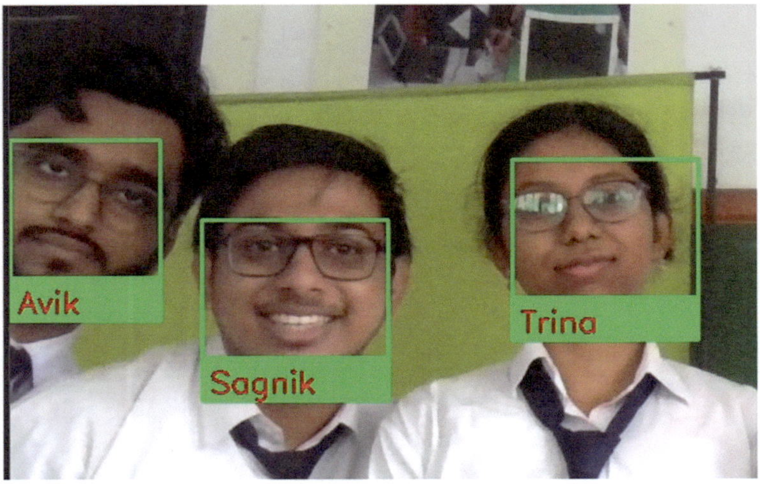

Fig. 3 Face recognized

4.2 Software Implementation

The software was tested with different shapes and sizes of faces and it efficiently recognized each of them separately without mixing up the output or not being able to recognize a known face. The main objective of this program is to recognize each and every known face it receives through computer vision. The algorithm used in the creation of this paper is:

i. DNN (Deep Neural Network): It helps in mimicking the processing method of a human brain to identify various objects or patterns (human faces in this case) (Figs. 3 and 4).
ii. Our system is able to recognize multiple faces at the same time in the same frame.

4.3 Hardware Implementation

The hardware required to execute this program is simply and system that can house the program and have a Python environment for its execution, along with high-resolution camera modules, which will be used by the program through computer vision.

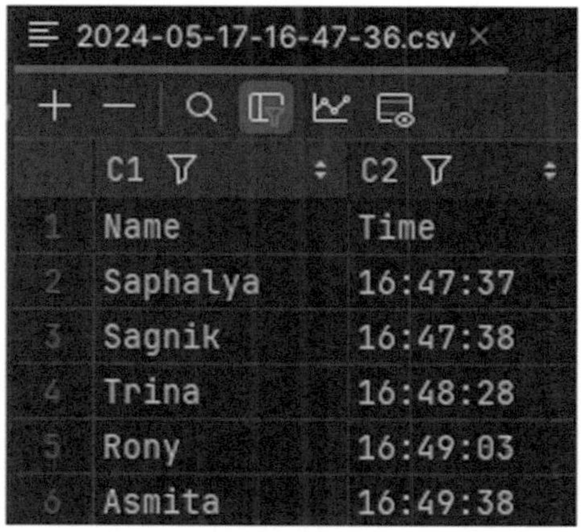

	C1	C2
1	Name	Time
2	Saphalya	16:47:37
3	Sagnik	16:47:38
4	Trina	16:48:28
5	Rony	16:49:03
6	Asmita	16:49:38

Fig. 4 Attendance recorded in ".csv" file

4.4 *Special Test Cases*

i. This system has the capability of accurately distinguishing between identical twins based on facial features.
ii. Anomaly Detection: If a face that is not registered in the database is identified then it will be bounded in a red box and marked as "Unknown" in the output video feed.

4.5 *Result and Analysis*

The paper presented by Smitha Shekar et al. [1] uses the facial recognition algorithm of LBPH which has an accuracy of 79.33%, whereas our paper has an accuracy of 97.99%.

The research presented by Evanjalin et al. [2] uses OpenCV with DNN face locator using Single shot Detection framework with the base as ResNet for facial recognition, which has an accuracy percentage of 91%, whereas our paper has an accuracy of 97.99%.

The study introduced by Poreddy et al. [3] uses the facial recognition algorithm of LBPH which has an accuracy of 79.33%, whereas our paper has an accuracy of 97.99%.

The study that Karuna et al. delivered [4] uses OpenCV with Histogram of Oriented Gradients for recognizing faces, which has an accuracy percentage of 61.1%, whereas our paper has an accuracy of 97.99%.

The findings shared by Hasban et al. [5] use deep learning-based approach for the automated detection of faces with an accuracy of 97.33%, whereas our paper has an accuracy of 97.99%.

Comparison of accuracy and error		
Paper references	Accuracy (in %)	Error rate (in %)
Proposed paper	97.99	2.01
[1]	79.33	20.67
[2]	91.00	9.00
[3]	79.33	20.67
[4]	61.10	38.90
[5]	97.33	2.67

5 Training and Testing Specifications

5.1 *Training of the Actual Model*

The model is initially trained on the face-recognition benchmark dataset, WIDER FACE dataset, which contains 32,203 images with 393,703 faces in them, with a high degree of variability in scale, poses, and occlusion. This dataset is split into training (40%), validation (10%), and testing (50%). While testing at this level the accuracy was 99.38%.

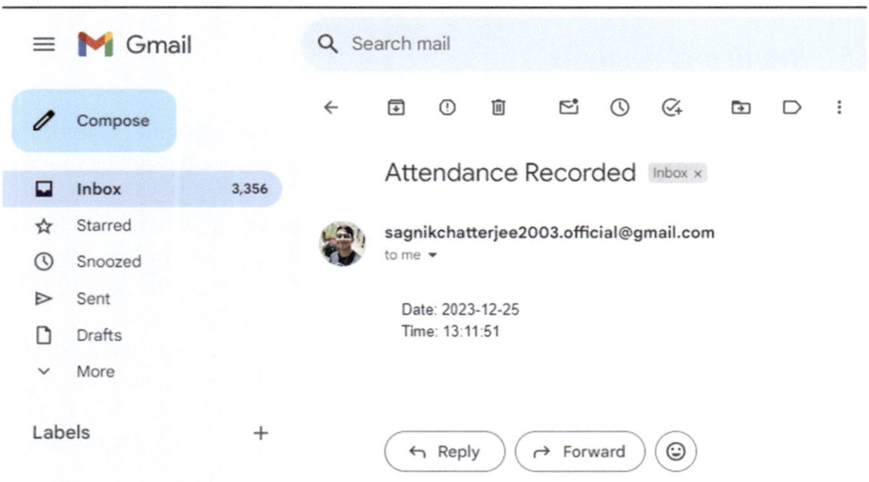

Fig. 5 Response email sent to the attendee

5.2 Testing of the Final Product

After the training is done, the previously trained model is further trained temporarily with a custom dataset of 20 images, the training data is learned by the model only during the program is running, after the program terminates, model forgets the data learned from the custom dataset. This two-level training helps us to reuse this already trained model with any custom datasets, where a person has only one image of their faces, thus making it robust as it can be used with any small or big-sized dataset. The

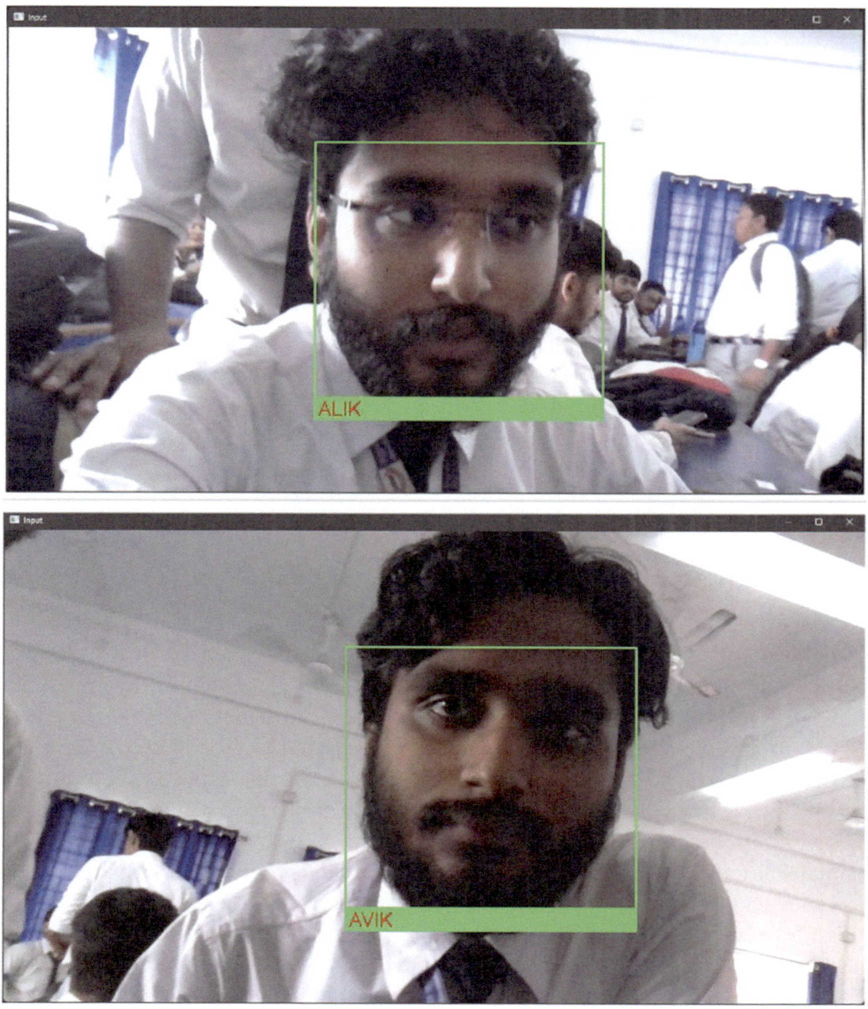

Fig. 6 Is able to distinguish between identical twins

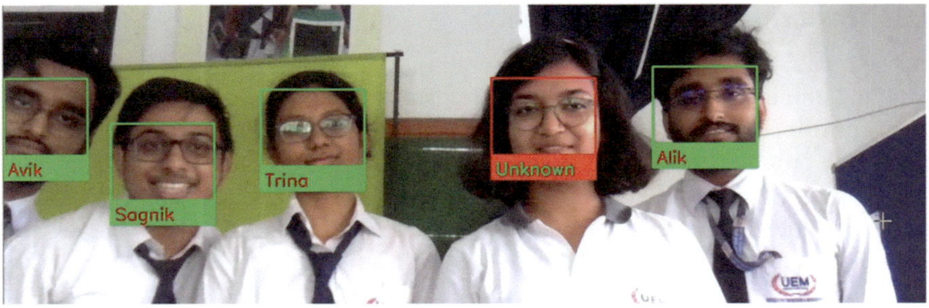

Fig. 7 Recognizing multiple faces in the same frame, in real time, along with an unknown face

accuracy of our model while testing with our custom dataset was 97.99%, which is the final testing accuracy.

The Final Product/Model is able to detect and recognize faces in varied or undesirable lighting conditions, but the accuracy will not be as good as it would be in a well-illuminated surrounding/room.

It can also detect about 50 faces at once (faces that are registered in the database as well as not registered in the database). It was about the maximum number of people we could gather for experimentation purposes.

Its unique feature of sending a response email (Fig. 5) for the attendance recorded is what makes it more interactive with the end-users. Its other unique feature is that it is able to accurately differentiate between identical twins (Figs. 6 and 7), and also has an extra layer of security as it is able to detect unregistered faces (Fig. 7).

6 Conclusion

In conclusion, the development and implementation of an attendance monitoring system using facial recognition technology integrated with a software platform represent a major leap forward in addressing the challenges posed by traditional attendance tracking methods. This paper has successfully demonstrated the potential of cutting-edge technologies to transform and optimize attendance management across diverse educational and organizational environments. The key achievements and takeaways from this paper include Enhanced Efficiency (The proposed solution majorly improves the efficiency of attendance tracking by removing the need for manual data entry, reducing administrative burdens, and providing real-time updates), Accuracy and Reliability (The integration of facial recognition technology results in a high level of accuracy and reliability, minimizing the potential for human errors and fraudulent activities), and Customization and Adaptability (The system is built to be scalable and adaptable, making it most suitable for several applications, from educational institutions to corporate settings or even in hospitals).

References

1. Smitha Shekar, B., Harish, G., Aishwarya, B. R., Badri Narayan, S., & Chinmayeshree, K. B. (2022). Smart classroom attendance system using face recognition and Raspberry Pi. *International Journal for Research in Applied Science & Engineering Technology (IJRASET), 10*(VII). ISSN: 2321-9653.
2. Evanjalin, A. B., Christy, D., Karthika, N., & Reshma, R. S. (2021). Face recognition system attendance system using Raspberry Pi. *Irish Interdisciplinary Journal of Science & Research (IIJSR), 5*(2), 84–91. Available at SSRN: https://ssrn.com/abstract=3851342
3. Poreddy, L. R., Gannavarapu, H., Pavan Kumar Valavala, S. B., Babu Seeram, S., & Indla, S. T. (2022). A system which procures attendance by using face recognition and Raspberry pi. In *2022 International Conference on Smart and Sustainable Technologies in Energy and Power Sectors (SSTEPS), Mahendragarh, India* (pp. 126–129). https://doi.org/10.1109/SSTEPS57475.2022.00041
4. Karuna, Y., Sumanth, V. S., AkshayRao, A., Varma, P. S., & Saritha, S. (2023). Face recognition-based attendance system with a mobile application using Raspberry Pi. In: N. Subhashini, M. A. G. Ezra, & S. K. Liaw (Eds.), Futuristic communication and network technologies. Lecture notes in electrical engineering (Vol. 966). Springer. https://doi.org/10.1007/978-981-19-8338-2_42
5. A. S. Hasban, et al. (2019). Face recognition for student attendance using Raspberry Pi. In *2019 IEEE Asia-Pacific Conference on Applied Electromagnetics (APACE), Melacca, Malaysia* (pp. 1–5). https://doi.org/10.1109/APACE47377.2019.9020758
6. Abdul Muqeet, M. (2019). Face recognition based attendance management system using Raspberry Pi. *Journal of Information and Computational Science, 9*(10). ISSN: 1548-7741.
7. Kar, N., Deb Barma, M., Saha, A., & Pal, D. (2012). Study of implementing automated attendance system using face recognition technique. *International Journal of Computer and Communication Engineering*, 100–103. https://doi.org/10.7763/IJCCE.2012.V1.28
8. Lukas, S., Mitra, A., Desanti, R., & Krisnadi, D. (2016). Student attendance system in classroom using face recognition technique. 1032–1035. https://doi.org/10.1109/ICTC.2016.7763360
9. Sawhney, S., Kacker, K., Jain, S., Singh, S. N., & Garg, R. (2019). Real-time smart attendance system using face recognition techniques. In *2019 9th International Conference on Cloud Computing, Data Science & Engineering (Confluence), Noida, India* (pp. 522–525). https://doi.org/10.1109/CONFLUENCE.2019.8776934
10. Patil, A., & Shukla, M. (2014). Implementation of classroom attendance system based on face recognition in class. *International Journal of Advances in Engineering & Technology*. ISSN: 22311963.

Prediction of Gaze Point Using Deep Learning and Raspberry Pi

Soumya Panja, Sapta Rathi Roy, Shatoparna Bhattacharya, Anshuman Kumar, and Debayan Bhattacharya

Abstract Eye tracking has been valuable for researchers in many domains. Most eye-tracking technologies have required an extra piece of costly hardware. Several other available eye-tracking solutions have usually not been very accurate and have required a costly subscription. We aimed our project at creating a free and open-source alternative that does not require specialized eye-tracking hardware and can be implemented using general camera modules. We developed a deep learning-based solution as a prototype for this project. Specifically, we developed a deep learning model to predict the gaze position of the user on the screen. We created our training dataset using a webcam to train the model. Each training sample consisted of a video frame of a person looking at the screen and the corresponding true gaze position on the screen. By using a model with a gaze location as output on a 2D plane, the deep learning model became capable of interpreting and predicting gaze positions in most cases. Our results have demonstrated the computational power of the deep learning algorithm that can be used effectively to create and deploy eye-tracking solutions.

Keywords Eye tracking · Cognition · Fixation · Deep learning · Raspberry Pi · Gaze prediction

1 Introduction

In recent years, eye tracking has seen significant advancements through the integration of deep learning, image processing, and computer vision [18]. These technologies have enabled the precise analysis of eye movements like fixations and saccades [14], which are crucial for understanding human cognitive processes. Convolutional Neural Networks (CNNs) such as VGG19, ResNet, and EfficientNet have been pivotal in improving eye-tracking accuracy, particularly in detecting facial landmarks and gaze movements. While traditional eye trackers rely on infrared cameras for high

S. Panja (✉) · S. R. Roy · S. Bhattacharya · A. Kumar · D. Bhattacharya
Future Institute of Technology, Garia Kolkata, India
e-mail: panjasoumya22@gmail.com

J. K. Mandal et al. (eds.), *Recent Advances in Artificial Intelligence and Smart Applications*, Lecture Notes in Networks and Systems 1380,
https://doi.org/10.1007/978-981-96-5822-0_9

precision, recent developments aim to create more accessible and cost-effective systems using webcams [8]. However, these systems face challenges in accuracy and performance due to noise and variability in eye-tracking data. This research focuses on overcoming these limitations by developing a simple yet effective webcam-based eye tracker enhanced with deep learning techniques and a 3D-printed IR cut camera, improving accuracy in various lighting conditions.

The aim of this study is to advance eye-tracking technology and its applications in fields such as human–computer interaction, market research, and usability studies. By combining innovative hardware with advanced deep learning techniques, we have aspired to create a cost-effective, accessible eye-tracking system for researchers, developers, and practitioners.

2 Review of Prior Works

Weston Sewell and Oleg Komogortsev, in [16], discussed using unmodified webcams for eye gaze tracking to aid people with disabilities in office and classroom lighting conditions. They used a neural network-based approach to estimate eye gaze positions from grayscale images. Pérez et al. [13] presented a real-time eye gaze tracking system using infrared illuminators and sub-pixel determination. Their system offered non-intrusive tracking with high accuracy, suitable for various applications. Mazhar et al. [10], proposed an eye ball tracking system using webcams and MATLAB to control a motorized platform. Their system demonstrated high accuracy in different environments, offering potential for assistive communication devices and mental state monitoring. Chunning Meng and Xuepeng Zhao, in [12], highlighted webcams as promising for eye gaze estimation, with accuracy dependent on camera quality. They used features like eyelid and iris positions, achieving high precision in tasks like "reading a document" (87.5%) and "browsing the web" (72.9%). Aljaafreh et al. [1] focused on integrating eye movements into cognitive research and developing low-cost eye-tracking tools using low-resolution webcams. They emphasized creating consistent algorithms and extracting meaningful data, with potential applications in diagnosing neurological disorders. Jessy Rimaya Khonglah and Arun Khosla, in [6], developed a low-cost, easy-to-setup webcam-based eye tracker for children with autism. Their infrared detection technique improved accuracy by resisting ambient light variations and allowed free head motion for unobtrusive tracking. McMurrough et al. [11] introduced an eye-tracking dataset aimed at enhancing Point of Gaze (PoG) detection. This dataset filled a critical void by offering ground-truth data to refine eye-tracking methods, particularly in the presence of head motion. Xu et al. [21] pioneered a webcam-based gaze tracking system for large-scale eye tracking via crowdsourcing. Their dataset provided cost-effective, high-quality gaze data, facilitating broader research accessibility.

Vigier et al. [20] contributed a dataset to study visual attention disparities in HD and UHD video formats. Their dataset offered insights into the dynamics of visual attention in different video resolutions. Tonsen et al. [19] introduced InvisibleEye, a mobile eye-tracking approach using low-resolution cameras and learning-based gaze estimation. Their datasets enabled comprehensive evaluations, showcasing potential for mobile eye tracking. Marighetto et al. [9] introduced an eye-tracking dataset and analysis toolbox for audio-visual attention models. Zheng and Usagawa [22] devised a low-cost webcam-based tracker with 94% accuracy for saccade movements. Mazhar et al. [17] highlighted webcams as affordable alternatives to commercial systems, despite lower precision. Koochaki et al. [7] predicted intentions with 95.68% accuracy using gaze patterns. Ansari et al. [2] employed CNNs for gaze estimation with unmodified webcams. Ghosh et al. [4] reviewed deep learning-based gaze analysis challenges, while Khaleel et al. [5] proposed a real-time low-cost model. Saxena et al. [15] demonstrated precise webcam-based eye tracking. Garbin et al. [3] provided a VR-focused eye-tracking dataset, enhancing human–computer interaction research.

3 Proposed Methodology

Figure 1 represents the overall flow of the gaze coordinates extraction process. The OpenCV library was employed for interfacing with the webcam, processing images, and extracting facial features. Image processing tasks like face detection and facial landmark extraction were performed. These features were used as input to the deep learning architecture for accurate gaze prediction.

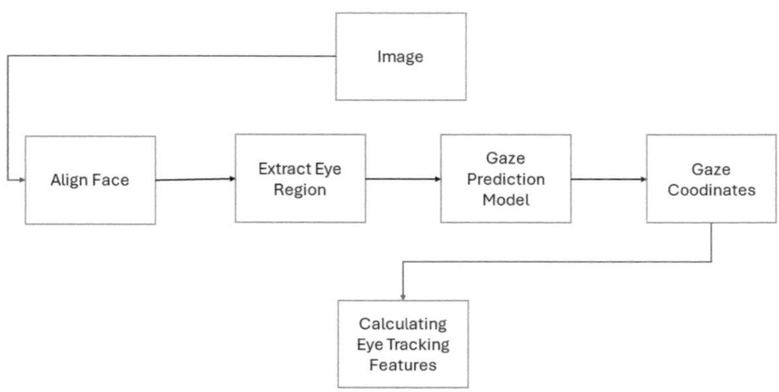

Fig. 1 The flowchart of the steps involved for extracting gaze coordinates

3.1 Data Collection

There are many datasets available for training a deep learning model which includes datasets like NVGaze (having images captured from IR cameras), MPIIGaze (consisting of one eye of each test subjects), and GazeCapture (dataset which capture images from IPad). But the goal of our experiment is to find a solution for inbuilt webcam or external camera module on laptops without the intervention of any external devices. Due to the lack of comprehensive webcam datasets, we collected our own dataset to train a deep learning model for accurate gaze prediction based on facial features. We normalized gaze positions to a range between 0 and 1 to ensure that pixel range remains constant for different resolutions.

$$X_{norm} = \frac{X}{S_W} \tag{1}$$

$$Y_{norm} = \frac{Y}{S_H} \tag{2}$$

where X_{norm} *and* Y_{norm} are the normalized coordinates, X *and* Y are the actual gaze coordinates, and S_W *and* S_H are the user's screen width and height, respectively.

The dataset consists of grayscale images of test subjects' eye regions, chosen to reduce model complexity while enhancing iris contrast. Images were captured under controlled conditions with standardized lighting, ensuring consistency and accuracy. Subjects maintained a neutral gaze while facing the webcam to capture typical eye movements. The dataset we built used a laptop's inbuilt webcam to capture images of test subjects and contains grayscale images of the test subject's face, left and right eye regions with its corresponding X and Y gaze positions normalized to values between 0 and 1 as a fraction of screen's relative dimension. Data collection involved displaying a black screen with a moving white point. Test subjects pressed the spacebar when focusing on the white point, capturing webcam images alongside the corresponding white point coordinates. This process allowed for multiple captures at varying screen locations. It was assumed that subjects accurately gazed at the white point during each capture, providing reliable gaze position data.

Pre-processing data before feeding it into the deep learning architecture is essential for improving model accuracy. Raw images of entire test subjects can pose challenges in feature extraction. To address this, we extracted relevant regions of interest, such as the eye region, from input images, simplifying the learning task and enhancing feature discernment.

Maintaining consistent dimensions of input data ensures compatibility with the deep learning architecture. All images were resized to a uniform size of 224 × 448. Additionally, standardizing the output variables, particularly for gaze prediction, was crucial due to variability in screen resolutions. Normalizing output gaze positions to a consistent range (0 to 1) ensured that model predictions were consistent and interpretable across different devices, enhancing robustness and generalizability.

Fig. 2 Samples of cropped images of the eye region

Pre-processing: The captured image of the test subject is first converted to grayscale to simplify processing. The face region is then detected and cropped from the image. Next, the image is aligned such that the eyes are parallel to the horizontal axis, ensuring consistent orientation. Finally, the eye region is specifically detected and cropped for further analysis.

After these pre-processing steps, the resultant images were stored with their corresponding gaze positions (Fig. 2).

Output Format: The model outputs two values in the range between 0 and 1. These values are likely normalized coordinates representing the predicted gaze positions on the screen and converted using Eqs. 1 and 2. The normalization to the range of 0 to 1 is a common practice, as it makes it easier to compare predictions across different screen sizes or resolutions.

3.2 Model Architecture

To harness the computational power of deep learning architectures and provide sufficient information about test subjects for accurate gaze position prediction, we proposed a multi-input deep learning model. In this model, we focus solely on the eye region, omitting the face, to streamline processing and improve computational efficiency. By concentrating on the eye region, we ensure that the model captures the most relevant information for accurately determining gaze positions.

The output of the model consists of two values within the range of 0 to 1, representing the fraction of relevant dimensions of the screen corresponding to the predicted gaze positions.

To analyze the performance of our proposed multi-input deep learning model, we experimented with several standard deep learning architectures, including VGG19, ResNet, Inception, MobileNet, and EfficientNet (Fig. 3).

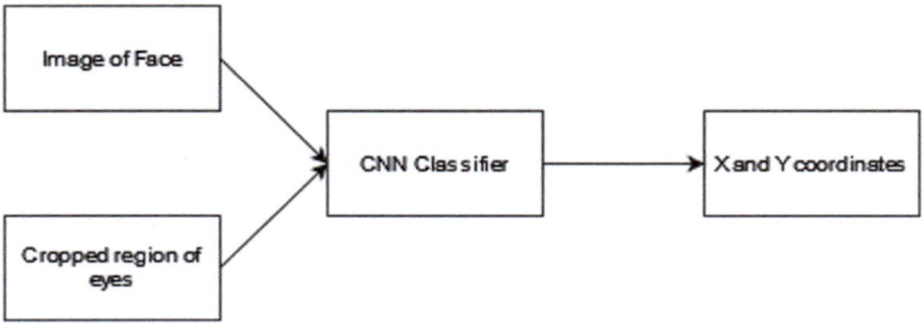

Fig. 3 Inputs and outputs of the deep learning model

4 Hardware

Our camera module prototype serves as a webcam for any USB-supported devices and is recognized as a UVC (USB video class) gadget device when plugged in. This feature ensures seamless compatibility with various computing platforms, enabling researchers to integrate the device into their existing setups with ease. Additionally, the device features an inbuilt lighting system, eliminating the need for external lighting setups during video capture for eye-tracking experiments. This not only simplifies the experimental setup but also reduces costs and enhances portability.

Hardware Requirements: Raspberry Pi Zero V1.0 WH, Led light (1 watt), Lithium-ion battery 18650(800 mAh), 5v to 3.7 v converter, IR-CUT Camera Module (5MP), Type A to micro-USB cable (data transmission from camera to PC) (Fig. 4).

Our prototype's key advantage is its customizability. Researchers can modify the device to meet specific experimental needs, such as replacing the camera module

(a) Camera Module (b) Camera Module attached to screen

Fig. 4 Camera module

with higher resolution alternatives or adapting the lighting by switching the LED light source. This flexibility enhances data quality and reliability.

5 Implementation and Experimental Results

Our technique aims to predict gaze positions as two values ranging from 0 to 1, representing fractions of relevant screen dimensions (horizontal and vertical positions). These predictions help calculate eye-tracking features such as fixation, saccade, scanpath, and heatmaps to understand human exploration patterns.

5.1 Multi-input Multi-output Deep Learning Model

The process begins with capturing an image using a webcam or external camera module. The image undergoes pre-processing, where face detection is performed. Eye regions are extracted using the Haar Cascade classifier, which identifies facial features like eyes. These regions are cropped and stacked to create a unified input for our deep learning model. The model processes these stacked eye images to predict gaze directions, outputting X and Y values.

The normalized predictions are mapped to screen coordinates to determine the user's point of focus. This mapping is crucial for applications like gaze tracking, providing insights or enabling interactive features. Our approach combines image processing, precise feature extraction, and deep learning for reliable gaze estimation.

5.2 Experimental Results

In our pursuit of an effective model for extracting critical features for gaze predictions, we explored several advanced computer vision architectures, including VGG19, MobileNet, Inception, ResNet, and EfficientNet. Each of these architectures is renowned for its unique strengths in handling complex image recognition tasks. Through extensive experimentation on our dataset, we discovered that ResNet demonstrated superior performance compared to the other models (Table 1).

For the purpose of evaluation metric R^2 score or coefficient of determination is used, mostly used for measuring the model accuracy for continuous values. And training and optimizing the deep learning model mean absolute error (MAE) loss function is used.

$$R^2 = 1 - \frac{\sum (y_i - \hat{y}_i)^2}{\sum (y_i - \bar{y}_i)^2} \tag{3}$$

Table 1 Representing the R^2 scores and losses for X- and Y-coordinates

Model name	R_X^2	R_Y^2	MAE	MAE_X	MAE_Y
ResNet	**0.94**	**0.96**	**0.07**	**0.04**	**0.03**
EfficientNet	0.94	0.94	0.09	0.04	0.04
MobileNet	0.96	0.92	0.09	0.03	0.06
Inception	0.82	–2.88	0.5	0.07	0.43
VGG19	–5.70	–5.44	1.19	0.62	0.57

$$MAE = \frac{1}{n}\sum_{i=1}^{n} |y_i - \hat{y}_i| \tag{4}$$

where y_i is the actual value, $\hat{y}_i$ is the predicted value, and $\bar{y}_i$ is the mean of the actual values.

5.3 *Fixation Points and Heatmaps*

See Figs. 5 and 6.

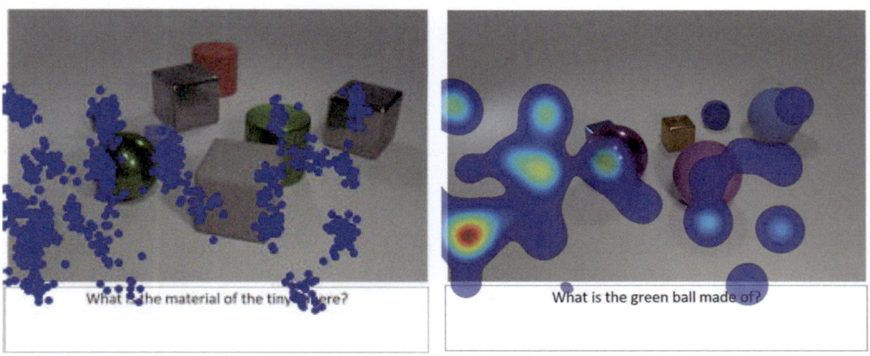

(a) Gaze Coordinates Predicted (b) Heatmap of the gaze coordinates predicted

Fig. 5 Predictions overlayed on an image

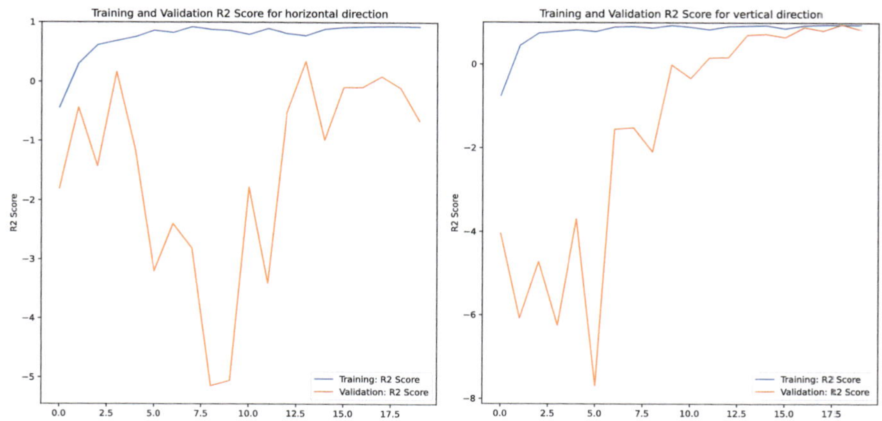

(a) Horizontal R2 score (b) Vertical R2 score

Fig. 6 Plots showing the training history of the ResNet architecture

6 Conclusion

Existing solutions to eye tracking are very promising due to the hand-crafted feature engineering and using specialized hardware. These specialized equipments are very costly to be used for research and for general use cases. Our study showed that deep learning can effectively predict gaze coordinates for eye-tracking feature analysis. Although several challenges are needed to be addressed in the next section.

Despite these challenges, our approach showed promising results in gaze prediction on ResNet and EfficientNet deep learning architecture. With further refinement, we aim to enhance the model's speed, robustness, and adaptability, unlocking new applications from assistive technologies to immersive virtual experiences.

References

1. Aljaafreh, A., Alaqtash, M., Al-Oudat, N., Abukhait, J., & Saleh, M. (2020). A low-cost webcam-based eye tracker and saccade measurement system. *International Journal of Circuits, Systems and Signal Processing, 14*(10.46300), 9106–2020.
2. Ansari, M. F., Kasprowski, P., & Obetkal, M. (2021). Gaze tracking using an unmodified web camera and convolutional neural network. *Applied Sciences, 11*(19), 9068.
3. Garbin, S. J., Komogortsev, O., Cavin, R., Hughes, G., Shen, Y., Schuetz, I., & Talathi, S. S. (2020). Dataset for eye tracking on a virtual reality platform. In *ACM symposium on eye tracking research and applications*, pp. 1–10.
4. Ghosh, S., Dhall, A., Hayat, M., Knibbe, J., & Ji, Q. (2023). Automatic gaze analysis: A survey of deep learning based approaches. *IEEE Transactions on Pattern Analysis and Machine Intelligence, 46*(1), 61–84.

5. Khaleel, A. H., Abbas, T. H., & Sami Ibrahim, A.-W. (2024). Best low-cost methods for real-time detection of the eye and gaze tracking. *i-com, 23*(1), 79–94.
6. Khonglah, J. R., & Khosla, A. (2015). A low cost webcam based eye tracker for communicating through the eyes of young children with asd. In *2015 1st international conference on next generation computing technologies (NGCT)* (pp. 925–928). IEEE.
7. Koochaki, F., & Najafizadeh, L. (2018). Predicting intention through eye gaze patterns. In *2018 IEEE biomedical circuits and systems conference (BioCAS)* (pp. 1–4). IEEE.
8. Krafka, K., Khosla, A., Kellnhofer, P., Kannan, H., Bhandarkar, S., Matusik, W., & Torralba, A. (2016). Eye tracking for everyone. In *Proceedings of the IEEE conference on computer vision and pattern recognition* (pp. 2176–2184).
9. Marighetto, P., Coutrot, A., Riche, N., Guyader, N., Mancas, M., Gosselin, B., & Laganiere, R. (2017). Audio-visual attention: Eye-tracking dataset and analysis toolbox. In *2017 IEEE international conference on image processing (ICIP)* (pp. 1802–1806). IEEE.
10. Mazhar, O., Shah, T. A., Khan, M. A., & Tehami, S. (2015). A real-time webcam based eye ball tracking system using matlab. In *2015 IEEE 21st international symposium for design and technology in electronic packaging (SIITME)* (pp. 139–142). IEEE.
11. McMurrough, C. D., Metsis, V., Rich, J., & Makedon, F. (2012). An eye tracking dataset for point of gaze detection. In *Proceedings of the symposium on eye tracking research and applications* (pp. 305–308).
12. Meng, C., & Zhao, X. (2017). Webcam-based eye movement analysis using cnn. *IEEE Access, 5*, 19581–19587.
13. Peréz, A., Córdoba, M.L., Garcia, A., Méndez, R., Munoz, M. L., Pedraza, J. L., & Sanchez, F. (2003). A precise eye-gaze detection and tracking system.
14. Salvucci, D. D., & Goldberg, J. H. (2000). Identifying fixations and saccades in eye-tracking protocols. In *Proceedings of the 2000 symposium on Eye tracking research & applications* (pp. 71–78).
15. Saxena, S., Fink, L. K., & Lange, E. B. (2023). Deep learning models for webcam eye tracking in online experiments. *Behavior Research Methods*, 1–17.
16. Sewell, W., & Komogortsev, O. (2010). Real-time eye gaze tracking with an unmodified commodity webcam employing a neural network. In *CHI'10 extended abstracts on human factors in computing systems* (pp. 3739–3744).
17. Skovsgaard, H., Agustin, J. S., Johansen, S. A., Hansen, J. P., & Tall, M. (2011). Evaluation of a remote webcam-based eye tracker. In *Proceedings of the 1st conference on novel gaze-controlled applications* (pp. 1–4).
18. Stember, J. N., Celik, H., Krupinski, E., Chang, P. D., Mutasa, S., Wood, B. J., Lignelli, A., Moonis, G., Schwartz, L. H., Jambawalikar, S., amp, et al. (2019). Eye tracking for deep learning segmentation using convolutional neural networks. *Journal of Digital Imaging, 32*, 597–604.
19. Tonsen, M., Steil, J., Sugano, Y., & Bulling, A. (2017). Invisibleeye: Mobile eye tracking using multiple low-resolution cameras and learning-based gaze estimation. *Proceedings of the ACM on Interactive, Mobile, Wearable and Ubiquitous Technologies, 1*(3), 1–21.
20. Vigier, T., Rousseau, J., Da Silva, M. P., & Le Callet, P. (2016). A new hd and uhd video eye tracking dataset. In *Proceedings of the 7th international conference on multimedia systems* (pp. 1–6).
21. Xu, P., Ehinger, K. A., Zhang, Y., Finkelstein, A., Kulkarni, S. R., & Xiao, J. (2015). Turkergaze: Crowdsourcing saliency with webcam based eye tracking. *arXiv preprint* arXiv:1504.06755.
22. Zheng, C., & Usagawa, T. (2018). A rapid webcam-based eye tracking method for human computer interaction. In *2018 international conference on control, automation and information sciences (ICCAIS)* (pp. 133–136). IEEE.

Enhanced Deep Learning Model ResNet101 for Efficient Skin Disease Detection

Kushagra Agrawal, Mani Goyal, and Shaveta Jain

Abstract Skin diseases present a major threat globally, and the potential integration of AI techniques into skin disease detection is enormous. Skin diseases are among the most neglected types of diseases. One approach is to create a dataset of multiple diseases and analyze them using AI technologies. By creating an AI-based Deep Learning model using ResNet, this research aims to develop a system through which users can easily detect their skin diseases. Individuals with skin diseases, especially in rural areas, often tend to neglect them, so we aim to provide a solution that allows them to analyze their condition from the convenience of their home. One problem acting as an obstacle to our model is noisy and unevenly dimensioned data. We would be employing cutting-edge technologies like TorchVision, sickit-learn, PyTorch, and many more. This research aims to enhance performance and address the shortcomings of previous models by utilizing cutting-edge ResNet technology. Furthermore, researchers and doctors can carefully examine the dataset to aid in the development of better medical technologies and cures. This research intends to assist doctors in the preliminary examination of skin diseases. A remarkable result of our approach using the ISIC dataset is 99.3% accuracy. Due to this high level of accuracy and precise results, our model is reliable and can provide patients, doctors, and researchers with valuable insights.

Keywords AI (Artificial Intelligence) · ResNet (Residual Networks) · ISIC (International Skin Imaging Collaboration)

K. Agrawal (✉) · S. Jain
Department of Computer Science and Engineering, Manav Rachna International Institute of Research and Studies, Faridabad, India
e-mail: Agrawal_kushagra@rediffmail.com

S. Jain
e-mail: shavetajn120@gmail.com

M. Goyal
Department of CSE, Maharishi Markandeshwar Engineering College, MM (Deemed to Be University), Mullana, Haryana, India
e-mail: dr.mani.goyal@mmumullana.org

J. K. Mandal et al. (eds.), *Recent Advances in Artificial Intelligence and Smart Applications*, Lecture Notes in Networks and Systems 1380,
https://doi.org/10.1007/978-981-96-5822-0_10

1 Introduction

In the vast field of human health, skin problems often don't get as much attention as they deserve. While some skin issues are well-known, many others don't get noticed and can cause problems for people without anyone realizing. These less talked about areas, both literally and figuratively, are places where many different skin problems exist, affecting people and communities without much awareness. Understanding these less recognized areas is really important for understanding skin diseases better and finding ways to prevent and treat them effectively.

There are about 1500 different skin problems out there, each with its own challenges and impacts on people's health worldwide. This wide variety shows how complex skin problems can be, highlighting the need for thorough research and ways to help people deal with them.

In our quest to improve healthcare, we've embraced machine learning, which is making a big difference in dermatology. We've used a powerful type of machine learning called Convolutional Neural Networks (CNNs), and specifically, a tough one called ResNet101. With these tools, we've created a top-notch diagnostic tool that combines the latest technology with medical knowledge. This tool is set to change the game when it comes to accurately diagnosing skin conditions. We've chosen the ISIC dataset to train our model, which contains a vast collection of skin images for research purposes.

Few reasons why we chose the International Skin Image Collaboration (ISIC) dataset:

- Our choice to utilize the International Skin Imaging Collaboration (ISIC) dataset for our research highlights its significant contribution to the progression of dermatological research and the enhancement of patient care. This dataset comprises a substantial collection of 25,331 high-quality images, meticulously categorized into nine distinct skin condition categories. This diversity is critical for comprehensive research, as it encompasses a wide spectrum of skin ailments, from common conditions to rare diseases.
- The extensive variety within the dataset enables researchers to explore and analyze various dermatological conditions comprehensively. This breadth of data facilitates the development and evaluation of machine learning algorithms and diagnostic tools aimed at accurately identifying and classifying skin diseases.
- Skin disease today is a leading cause of concern as it affects most of the population from simple Skin dandruff to severe skin cancers. The dermatological sector is seeing a demand growth rate of 12% whereas the current growth rate is only 5–7%. So there is an urgent need to create a system for preliminary detection of skin diseases.

Our main objective is to create a model that can be incorporated into any website or app so anyone can use it and detect their skin disease without any hassle. It would help society in a very positive way [10]. Patients knowing they have a certain classified disease can further approach doctors for insights on their skin conditions for their treatment. It will not only help patients but also doctors and researchers to create better treatments and medications for skin diseases [11].

Summing up, current dermatological facilities lack in reaching remote/rural areas. These areas often get neglected and these people have to suffer without any treatment so there is an urgent need for an automated system that can help them to live better lives.

The readers are introduced to the principal concept of the study issue in an introduction presented in Sect. 1. Next, in Sect. 2, the primary survey work was thoroughly researched and implemented, serving as the basis for all upcoming sections. The methodology employed for attaining the desired results is described in Sect. 3 thoroughly. In Sect. 4, the obtained results are displayed with a comparison analysis. The final section concludes the key findings and suggests future actions in research and development.

2 Background Survey

For a better understanding, this section of the research paper provides a comprehensive study of the previous work and existing solutions. Diseases tend to develop and become resistant over time. So we researched the previous work so we can improve upon them and build a better more accurate and precise detection model.

Loris et al. [1] delve into the realm of skin detection in digital images, a crucial task utilized in various applications like hand gesture analysis, body part tracking, and facial recognition. Despite the attention this field has garnered, the absence of standardized benchmarks has hindered fair comparisons among different approaches. The study aims to provide a comprehensive review of recent research in skin detection, offering insights into different methodologies and proposing fair comparisons using diverse datasets.

Mehwish et al. [2] delve into skin cancer, emphasizing the importance of early detection due to its high mortality rate and expensive treatment. It discusses the classification of skin cancer into melanoma and non-melanoma types, highlighting the severity of melanoma and the necessity for early diagnosis. The paper reviews deep learning techniques for skin cancer focusing on convolutional neural networks. The study underscores the need for further advancement in diagnostic techniques.

Mostafiz et al. [3] address skin disease problems study. They have utilized methods like fuzzy clustering, wavelet analysis, GLCM features, and deep learning models such as Alexnet and VGG16 combined with classifiers like SVM and KNN to classify skin diseases accurately. These models have been tested on datasets like the ISIC 2019 challenge and HAM10000, showing promising results in classifying various skin conditions including melanoma, nevus, basal cell carcinoma, and others.

Ahmed et al. [4] have been focusing on developing advanced systems to aid in the detection and classification of various skin conditions. One such system proposed is the (CNN-SVM-MAA) system, which integrates a Convolutional Neural Network with a Support Vector Machine to create a Mobile Android Application for skin disease diagnosis. This system aims to improve the accuracy and efficiency of identifying skin diseases by analyzing a dataset of around 3000 images collected from different sources like hospitals and websites. This study's limitation is it's a very small dataset which causes inefficiency.

Taye et al. [5] focus on the pivotal role of skin lesion analysis in diagnosing and managing dermatological conditions, highlighting the impact of advancements in machine learning and computer vision techniques. It presents a detailed review of the latest methodologies for classifying, segmenting, and detecting skin lesions, underlining the significance of accurate and early analysis for effective healthcare. It also addresses the specific challenges, datasets, and evaluation metrics associated with skin lesion segmentation, providing a comprehensive overview of the current state of research.

Agrawal et al. [10] address the complexity of diagnosing skin disorders by leveraging machine learning, specifically CNN and EfficientNet, to develop a robust classification framework. The proposed CNN-EfficientNet system demonstrated 95% accuracy in identifying skin diseases, highlighting its effectiveness in improving diagnosis through advanced machine learning techniques.

Jain et al. [11] present a Melanoma Skin Disease Detection system using CNNs, specifically Densenet-121, to classify skin images into various disease categories with 95% accuracy. The system, involving image pre-processing, feature extraction, and classification, offers a reliable and efficient solution for improving skin disease diagnosis, surpassing current methods.

Throughout the examination of different research papers, they provide a number of technologies and approaches for skin disease detection. The first method provides a significant number of parameters for detection like gesture analysis and facial recognition but these parameters are not too standardized which hinders fair comparisons between different approaches. Then another approach emphasized on early detection of skin diseases due to its mortality rate. It focused on convolutional neural networks. We also have a segmentation approach that features models like Alexnet, and VGG16 combined with SVM and KNN for more accuracy. Then there is also a combination of CNN-SVM system which integrates convolutional neural networks and support vector machine, but had a limited dataset of 3000 images. There are many different approaches and many different types of models that can be used or combined to form better models. But the capability of ResNet101 is outstanding it can go as complex as 150 layers and can do multiple parallel computations and make decisions. ResNet uses RNN Residual Neural Network which handles more complex layers and training this model with more epochs makes it a more accurate and precise model than others.

3 Proposed Methodology

This section presents and detailed step-by-step explanation of the procedure followed to achieve the desired outcomes.

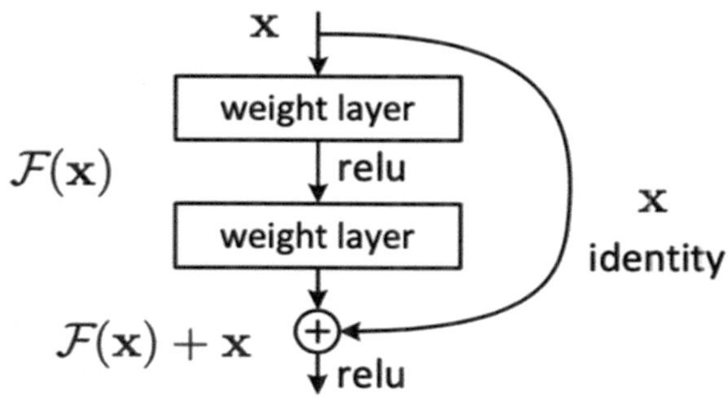

Fig. 1 Residual learning block

3.1 ResNet

Residual Networks or ResNet is a deep learning model used for computer vision applications. It is a Convolutional Neural Network (CNN) architecture supported to handle multiple layers. Previous CNN versions are not able to scale to the number of layers up to which ResNet can scale. Neural networks are trained through a back-propagation process that relies on gradient descent, shifting down the loss function and finding weights that can minimize them. Each EPOCH in the training decreases the loss until a precise model is trained [6].

Residual Block is an integral part of ResNet architecture. In older versions, CNN layers were stacked with batch normalization and non-linear activation layers such as ReLu. But that worked for the small number of layers. ResNet introduced a simple concept of an intermediate input-to-output series of convolution blocks.

Figure 1 shows a standard residual block, where x is the input and output of the previous layer to the residual block and output $= F(x) + x$, where $F(x)$ consists of several convolutional blocks. Due to this ResNet can be made as complex as 150 layers.

3.2 TorchVision

TorchVision is a part of PyTorch and plays an integral role in computer vision, opening up new possibilities for researchers, developers, and innovators alike. It offers a vast toolkit that spans various domains, from academia to industry. TorchVision consists of various pre-trained models, datasets, and transformation utilities, providing users with everything they need to work with images effectively. This is why we have incorporated TorchVision into our methodology—to develop a precise and more accurate model for skin disease detection [7].

The following explains the array of steps involved in implementing our ResNet model:

```
Current Directory: /home/akshat/Desktop/skin_dataset/test/Cellulitis Impetigo and other Bacterial Infections
Processing Image: /home/akshat/Desktop/skin_dataset/test/Cellulitis Impetigo and other Bacterial Infections/Leprosy-8.jpg
...
Processing Image: /home/akshat/Desktop/skin_dataset/test/Cellulitis Impetigo and other Bacterial Infections/impetigo-55.jpg
...
Processing Image: /home/akshat/Desktop/skin_dataset/test/Cellulitis Impetigo and other Bacterial Infections/Staphylococcal-Folliculitis-31.jpg
...
Processing Image: /home/akshat/Desktop/skin_dataset/test/Cellulitis Impetigo and other Bacterial Infections/Erysipelas-4.jpg
...
Processing Image: /home/akshat/Desktop/skin_dataset/test/Cellulitis Impetigo and other Bacterial Infections/09EczemaStaph020217.jpg
...
Processing Image: /home/akshat/Desktop/skin_dataset/test/Cellulitis Impetigo and other Bacterial Infections/Folliculitis-24.jpg
...
Processing Image: /home/akshat/Desktop/skin_dataset/test/Cellulitis Impetigo and other Bacterial Infections/impetigo-32.jpg
...
Processing Image: /home/akshat/Desktop/skin_dataset/test/Cellulitis Impetigo and other Bacterial Infections/Staphylococcal-Folliculitis-28.jpg
...
Processing Image: /home/akshat/Desktop/skin_dataset/test/Cellulitis Impetigo and other Bacterial Infections/impetigo-61.jpg
...
Processing Image: /home/akshat/Desktop/skin_dataset/test/Cellulitis Impetigo and other Bacterial Infections/impetigo-73.jpg
...
Processing Image: /home/akshat/Desktop/skin_dataset/test/Cellulitis Impetigo and other Bacterial Infections/Staphylococcal-Folliculitis-30.jpg
...
```

Fig. 2 Processing images from directories

Installing the Libraries

Libraries like torch (nn, utils.data, optim), TorchVision (datasets, transforms, models.resnet101), sklearn (model_selection, metrics), Pillow (PIL) for handling images, and tqdm for progress bars during EPOCH. We have utilized TorchVision.models.resnet101, employing 101 convolutional layers in our model. All these libraries are installed as the first step.

Dataset Acquisition and Loading

We have acquired our dataset from ISIC (International Skin Imaging Collaboration) for the best quality and accurate images. We were able to acquire more than 25,000 high-quality images. This dataset was loaded and split into train, test, and validation for ease of use (Fig. 2).

Data Preprocessing

In this step, the reading of images from the saved directories is done. Using a library like PIL (Pillow), we have resized the images to fit our ResNet model. We have employed techniques like transformations to augment our dataset. After resizing and transforming, "OpenCV" is used to remove any white noise from our image dataset to obtain only high-quality images for precise results [12, 13]. Finally, each file path is printed with a message confirming that image resizing is completed we can see it in Fig. 3.

ResNet MODEL

The ISIC skin image dataset is analyzed and trained using ResNet101 model from TorchVision. We first load our pre-processed dataset. Then dataloader is created using the torch's dataloader library for efficient batch processing during the training phase [7]. The ResNet101 model is loaded and to tailor the model to our specific task of skin disease classification we modified finally the fully connected layer to the number of classes according to the dataset [8].

Here is an illustration of ResNet101 layers in Fig. 4.

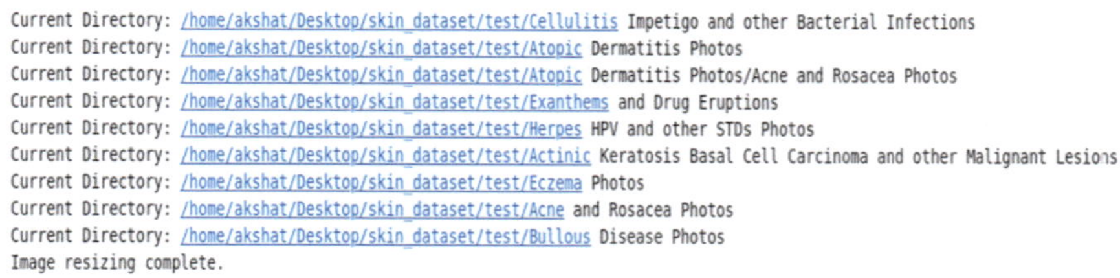

```
Current Directory: /home/akshat/Desktop/skin_dataset/test/Cellulitis Impetigo and other Bacterial Infections
Current Directory: /home/akshat/Desktop/skin_dataset/test/Atopic Dermatitis Photos
Current Directory: /home/akshat/Desktop/skin_dataset/test/Atopic Dermatitis Photos/Acne and Rosacea Photos
Current Directory: /home/akshat/Desktop/skin_dataset/test/Exanthems and Drug Eruptions
Current Directory: /home/akshat/Desktop/skin_dataset/test/Herpes HPV and other STDs Photos
Current Directory: /home/akshat/Desktop/skin_dataset/test/Actinic Keratosis Basal Cell Carcinoma and other Malignant Lesions
Current Directory: /home/akshat/Desktop/skin_dataset/test/Eczema Photos
Current Directory: /home/akshat/Desktop/skin_dataset/test/Acne and Rosacea Photos
Current Directory: /home/akshat/Desktop/skin_dataset/test/Bullous Disease Photos
Image resizing complete.
```

Fig. 3 Image pre-processing

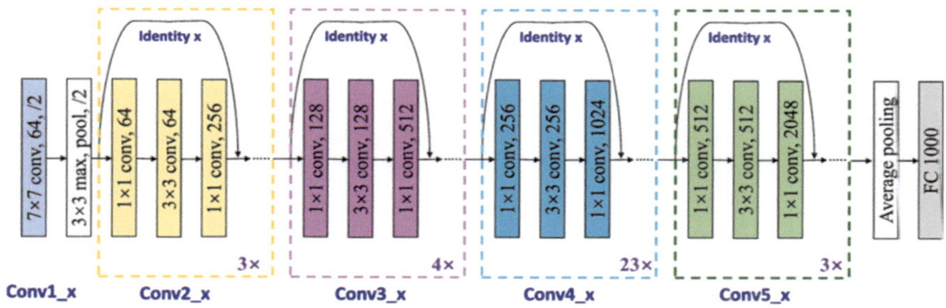

Fig. 4 ResNet101 layer illustration [8]

Throughout the training phase, we incorporated a combination of loss function, optimizer, and learning rate scheduler to guide the model's learning process. By utilizing techniques such as:

Cross-Entropy

Cross-entropy loss measures the dissimilarity between the predicted probability distribution and the true probability distribution of the labels [9]. For a single example with C classes and predicted probabilities p_i and true probabilities y_i, the cross-entropy loss L is calculated as

$$L = -\sum_{i=1}^{C} y_i \log(p_i) \tag{1}$$

L2 Regularization

L2 regularization or ridge regression works on gradient descent [9].

$$\begin{aligned} \omega_{\text{new}} &= \omega - \eta \frac{\partial L_2}{\partial \omega} \\ &= \omega - \eta \cdot \left[2x(\omega x + b - y) + 2\lambda\omega\right] \end{aligned} \tag{2}$$

```
Downloading: "https://download.pytorch.org/models/resnet101-63fe2227.pth" to /home
100%|████████████████████████████████████████| 171M/171M [01:45<00:00, 1.70MB/s]
Epoch 1/101: 100%|██████████████████████████████| 89/89 [00:39<00:00,  2.23it/s]
Training Loss: 1.0832743045319333, Training Accuracy: 0.651685393258427

Validation Loss: 0.7449909921448964, Validation Accuracy: 0.7536585365853659
Saved checkpoint at epoch 1, Validation Loss: 0.7449909921448964
Epoch 2/101: 100%|██████████████████████████████| 89/89 [00:41<00:00,  2.17it/s]
Training Loss: 0.5539485987987411, Training Accuracy: 0.824438202247191

Validation Loss: 0.8098552757157729, Validation Accuracy: 0.7414634146341463
Epoch 3/101: 100%|██████████████████████████████| 89/89 [00:41<00:00,  2.16it/s]
```

Fig. 5 Training ResNet model

These techniques were used with the aim to prevent overfitting of the model.

The training process consisted of 101 epochs, during which the model adjusted its parameters to optimize performance. At the end of each epoch, it was evaluated to ensure there were no spikes in training loss or validation loss (Fig. 5).

To monitor and manage effectively, we implemented a mechanism to save the model iteratively after every 5 epochs. These checkpoints stored important information such as the optimizer state and best validation loss.

Visualization of the training and validation loss and accuracy during the training phase of our model is given in Fig. 6.

Predicting the Classes from Test Dataset

Using PyTorch's TorchVision library, we select the ResNet model type which we have trained with where pre-trained = False. Then, we modify the number of classes and features according to our trained model. Using a torch, we load the saved.pth model. We filter out the unnecessary keys and also implement functionality if the model architectures are mismatched. Next, the test images directory is loaded with the OS library, and a transformation is defined for input images. Then, with our

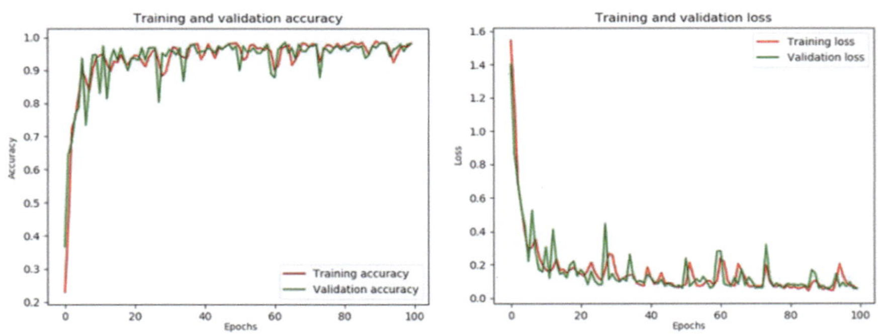

Fig. 6 Training/validation loss and accuracy

trained loaded model, we predict the class of the test image with probabilities of it belonging to each class an example of predicting a single image is given in Fig. 7.

Following is the flowchart for our proposed ResNet deep learning methodology in Fig. 8.

```
Predicted class: 6
Probabilities:
Class 0: 0.012296160683035851
Class 1: 0.2730807363986969
Class 2: 0.004667401313781738
Class 3: 0.25464901328086853
Class 4: 0.1445326209068298 3
Class 5: 0.00016385970229748636
Class 6: 0.30357980728149414
Class 7: 0.007030410692095757 5
```

Fig. 7 Prediction of single test image

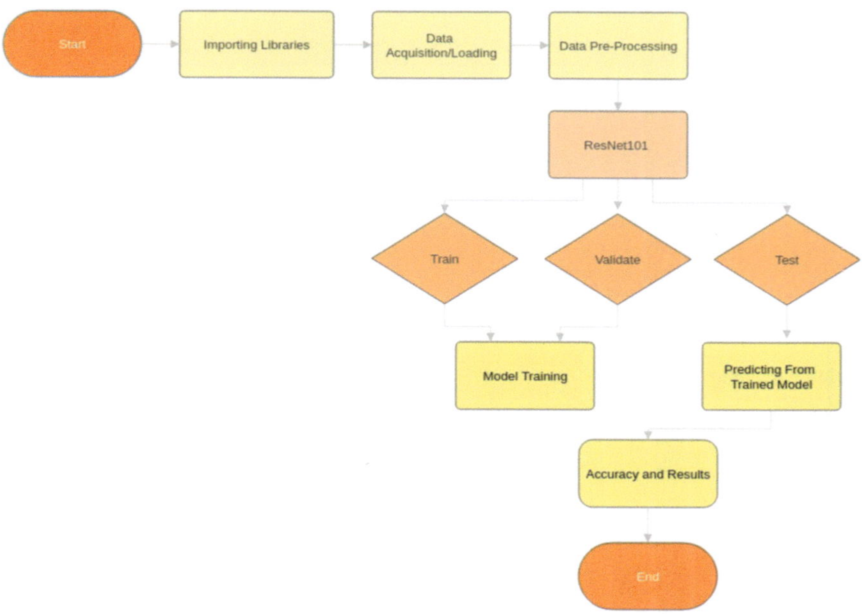

Fig. 8 Flowchart of the proposed methodology

4 Result and Discussion

This section presents the outcomes achieved, along with comparisons with previous work.

In this segment, we discuss the outcomes of the research paper based on our Deep Learning ResNet101 model and compare the accuracy of our proposed model with existing ones. Our model achieved an unparalleled accuracy of 99.3% in detecting skin diseases using residual networks. The equation for calculating the accuracy of our model is as follows:

$$\text{Accuracy} = \frac{\text{Number of correct predictions}}{\text{Total number of predictions}} \tag{3}$$

Our model exhibits a significant increase in accuracy over existing models as shown in Fig. 9. Below is a comparison with other models.

Following is a Table 1 for accuracy range comparison.

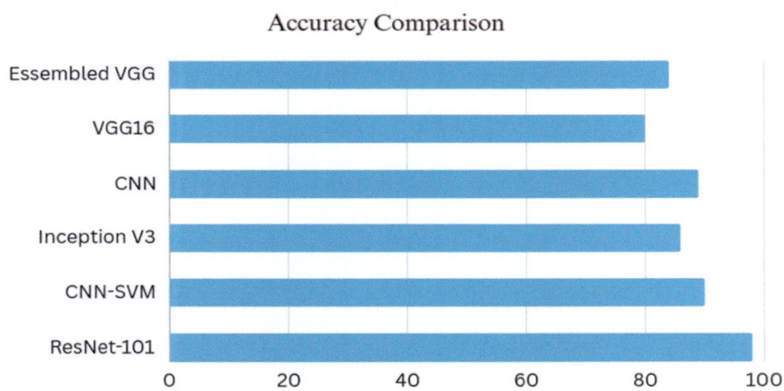

Fig. 9 Accuracy comparison

Table 1 Accuracy comparison

Model	Accuracy (%)
Ensemble VGG	80–84
VGG16	78–80
CNN	87–90
Inception V3	85–87
CNN-SVM	90–92
Proposed ResNet101	99.3

5 Conclusion and Future Scope

In this paper, we proposed a fine-tuned ResNet101 model, achieving an impressive accuracy of 99.3% in skin disease classification, outperforming other models such as CNN-SVM (90–92%) and Inception V3 (85–87%). Our research confirms the efficacy of deep learning in medical image analysis, particularly for dermatological applications.

While the proposed model performs exceptionally well, there is a potential for future improvements, which include expanding the dataset, increasing disease classes, and integrating the model into user-friendly applications.

References

1. Nanni, L., Loreggia, A., Lumini, A., & Dorizza, A. (2023). A standardized approach for skin detection: Analysis of the literature and case studies.
2. Dildar, M., Akram, S., & Irfan, M. (2021). Skin cancer detection: A review using Deep Learning techniques.
3. Ahammed, M., Al Mamun, Md., & Ahammed, M. A machine learning approach for skin disease detection and classification using image segmentation.
4. Elngar, A. A., Kumar, R., Hayat, A., & Churi, P. Intelligent system for skin disease prediction using Machine Learning.
5. Debelee, T. G. Skin lesion classification and detection using Machine Learning techniques: A systematic review.
6. ResNet Model and Residual Block, Datagen Tech, Computer-vision. https://datagen.tech/guides/computer-vision/resnet/
7. TorchVision, PyTorch documentation. https://pytorch.org/vision/stable/index.html
8. ResNet101 model reference: Research-gate.net. https://researchgate.net/figure/The-structure-of-the-ResNet-101-based-deep-feature-extractor_fig4_348078198
9. Towards-data-science, regularization. https://towardsdatascience.com/intuitions-on-l1-and-l2-regularisation-235f2db4c261
10. Agrawal, K., & Jain, S. (2023). Machine Learning enabled model for different skin disease prediction. In *2023 3rd International Conference on Technological Advancements in Computational Sciences (ICTACS), Tashkent, Uzbekistan* (pp. 1531–1536). https://doi.org/10.1109/ICTACS59847.2023.10389851
11. Jain, S., & Agrawal, K. (2023). An efficient diagnosis of melanoma skin disease using DenseNet-121. In *2023 3rd International Conference on Technological Advancements in Computational Sciences (ICTACS), Tashkent, Uzbekistan* (pp. 908–912). https://doi.org/10.1109/ICTACS59847.2023.10390147
12. Kushagra, K., Kumar, R., & Jain, S. (2022). A Deep Learning model to diagnose lung disease using genetic algorithm based feature selection method. In *2022 4th International Conference on Advances in Computing, Communication Control and Networking (ICAC3N), Greater Noida, India* (pp. 1251–1255). https://doi.org/10.1109/ICAC3N56670.2022.10074578
13. Kushagra, K. R., & Jain, S. (2023). An efficient deep-learning model to diagnose lung diseases using X-Ray images. *International Journal of Intelligent Systems and Applications in Engineering [Internet], 11*(3), 562–569.

An Email Spam Detection Approach Using Voting Ensemble Method

Piya Mondal, Sudakshina Mandal, and Rakesh D. Raut

Abstract Email is considered one of the significant mediums of communication and widely used in recent times. However, the evolving nature of different cyberthreats can temper effective communication and alter the potential message. Disrupting nature of spam is still a major challenge in the email security aspect which requires inventive methods to eliminate cutting-edge sophisticated spam emails. Machine learning is one of the significant dimensions in the literature and proves the effectiveness and robustness which distinguish spam emails from any mailbox. In this paper, a hybrid machine learning-based approach named the majority voting ensemble method has been proposed that combines multiple techniques and it can produce the best accuracy compared to some existing machine learning methodologies. The proposed system aims to enhance accuracy, increase precision value, and adapt to evolving spam tactics. The experimental analysis with the predefined dataset proves the effectiveness of the proposed mechanism.

Keywords Voting ensemble method · Feature extraction · Anomaly detection · Machine learning · Email spam detection · Classification

1 Introduction

With the advancement of electronics and telecommunication trends, using email between individuals or organizations has emerged significantly in recent eras [1]. However, it is evident that with the incorporation of the latest cutting-edge intelligence, security threats have exponentially increased. Penetrating and stealing information through email has become a common practice. As per the statistics, 93% of the emails are identified as spam emails which a user receives on a daily basis [2].

P. Mondal · S. Mandal (✉)
Techno India University, West Bengal, India
e-mail: sudakshina.nitjsr@gmail.com

R. D. Raut
Indian Institute of Management, Mumbai, India
e-mail: rraut@iimmumbai.ac.in

J. K. Mandal et al. (eds.), *Recent Advances in Artificial Intelligence and Smart Applications*, Lecture Notes in Networks and Systems 1380,
https://doi.org/10.1007/978-981-96-5822-0_11

Spam email generally aims to advertise, deceive, or harm, where the content is often irrelevant or destructive. Anonymous senders or some deceptive sources often send these emails. Typically it is seen that emails have been sent in bulk. In most cases, these emails have been ignored by the individuals as it is unwanted. However, accepting these emails knowingly or unknowingly makes the situation worse enough that potential information of a person can immediately get hacked [3].

Often, seemingly legitimate emails may contain phishing links designed to distribute malware [4]. Phishing emails are deceptive messages that appear to come from trusted companies, attempting to trick users into sharing personal or financial information. These emails may include links that, when clicked, can install malicious software on the user's device. The phishing email is also considered one type of spam. However, the purpose of phishing is to steal sensitive information from potential users. Typically, spam filters can easily detect phishing emails containing such links. However, emails without links, also known as "link-less" emails, present a greater challenge for these filters. As a result, detecting spam emails has become increasingly critical in today's digital landscape. Furthermore, a reliable and accurate method is essential to ensure the effectiveness of all detection systems [4].

In this paper, we will refer to malicious emails as "Spam," while legitimate and verified emails will be called "Ham."

Numerous machine learning approaches have already been proposed in the literature stated in Sect. 2. However, this paper proposes an enhanced machine learning approach consisting of Support Vector Machine, Multinomial Naive Bayes, and ExtraTree Classifier with a combination of soft voting ensemble method which proves robust accuracy with a precise precision value.

1. **Support Vector Machine (SVM):** SVM classification was employed as one of the models in the email spam detection process. SVM is particularly effective in high-dimensional spaces and was used to differentiate between spam and non-spam emails with a focus on maximizing the margin between the two classes.
2. **Naive Bayes (NB):** NB is a simple and computationally efficient algorithm, making it well suited for large datasets like email collections. NB is particularly effective for text classification tasks because it works well with high-dimensional data and can handle a large vocabulary of words. NB requires less training time and computational resources. Naive Bayes is a powerful tool in the email spam detection domain, providing a strong balance between simplicity, speed, and accuracy. Being one of the significant variants of NB, multinomial Naive Bayes works mostly on discrete data.
3. **Extremely Randomized Trees(ET):** ET, in sort ExtraTree, is an ensemble learning method that creates a large number of decision trees and averages their predictions to reduce variance, which can lead to more stable and accurate models compared to individual decision trees. ET is robust to noise in the data, it can still perform well even when the dataset contains irrelevant or misleading features. This is particularly useful in email spam detection, where the dataset may include various noisy or redundant information. ET is capable of capturing non-linear relationships in the data, which is important for complex tasks [5].

4. **Ensemble Classifier:** All three models discussed above are combined using an ensemble method, where classification is based on the majority vote (soft voting) [6]. In this approach, each model makes a prediction for each email instance, and the final output is determined by the prediction that receives more than half of the votes.

This paper is structured as follows where Sect. 2 provides an in-depth review of related pieces of literature, followed by Sect. 3, which outlines the proposed methodology. Section 4 produces the findings and comparisons with alternative methods. Finally, a conclusion is outlined in Sect. 5.

2 State of the Art of Literatures

Numerous methodologies have been employed so far based on their versatility in existing literature to detect email spam. This paper rigorously surveyed the existing pieces of literature on the year gap between 2019 and 2023 and analyzed the methodology used so far.

As per Lankaet et al. [7], the classical method of the machine learning life cycle has been tested against the predefined dataset. It is observed that the Naive Bayes classifier demonstrated a superior accuracy rate of 98.265% in comparison to other classifiers employed in this study, including SVM, Bagging, Decision Tree, AdaBoost classifiers, and Random Forest.

In the same context, Bansal and Sidhu et al. [8] proposed a model that integrates Optical Character Recognition (OCR) and Artificial Neural Networks (ANN). The ANN component is responsible for classifying emails into either a spam or ham category based on the keywords extracted from the email's subject line. This model is designed to convert markup language templates into text format, which helps reduce phishing attacks globally. The data is processed through the ANN model for classification. The term "Term Frequency-Inverse Document Frequency" (TF-IDF) is employed to convert words into digital data points. When comparing the performance of Naive Bayes, XGBoost, and ANN models, the ANN method achieved the highest accuracy, with a rate of 97.50%.

In the model proposed by Govil et al. [9], the first step involves creating a dictionary that utilizes a library called "stopwords" to eliminate all auxiliary verbs from the email content. As per the proposed algorithm, the emails were fed into the proposed model. The results were categorized, with 0 representing non-spam and 1 indicating spam emails. A dataset containing around 6000 valid and invalid email samples was used. The Naive Bayes classifier provided a probabilistic score for classifying the emails.

The Enron dataset underwent pre-processing, and two methods for extracting features were utilized in the framework introduced by Kaddoura et al. [10]. The FFNN (feedforward neural network) model was studied to evaluate its ability to classify emails as either spam or ham. The performance of the model was compared to that of the BERT dataset, and various architectural modifications were tested, including changes in the number of layers, neurons per layer, neurons in the input layer, and the F1-score for each configuration. Scenario 1: A basic neural network with a shallow architecture and few neurons. Scenario 2: Expanding the number of layers while decreasing the quantity of neurons in each layer by 50%. Scenario 3: Exploring the impact of augmenting both the number of layers and the number of neurons, with the neurons multiplying by factors of 2. Scenario 4: Studying the effect of adding more levels simultaneously keeping the number of neurons constant. Scenario 5: Examining the impact of adding more layers with a corresponding increase in neurons. The model achieved higher accuracy when it included three concealed levels, with the node count configured to 256, 512, and 1024, respectively.

In the paper of Suryawanshi et al. [11], they employed various machine learning classifiers such as NB, Boosting (AdaBoost), k-nearest neighbors (K-NN), SVM, Bagging, and Ensemble Classifiers with a voting approach. It is noted that the Ensemble Classifier delivers more favorable outcomes with an accuracy of 98.5%, outperforming the other classifiers, while also offering faster testing performance.

In the study by Kumar et al. [12], the suggested model was trained using various classifiers to assess and contrast the results for improved accuracy. Each classifier provided its assessed outcomes to the user. It was determined that the multinomial Naive Bayes produced the most optimal results, though it faced limitations due to class-conditional independence, leading the system to misclassify certain data points.

By maintaining the similarity with above, in [13], the author also used some semi-supervised and supervised learning methods and proved that SVM outperforms compared with other methods by 98.2%. However, we found that SVM does not give a good precision value.

In [14], authors proposed an ensemble-based phishing detection approach where they used in combination of SVM, Multinomial Naive Bayes (MNB), and Decision Tree (DT) methods to get the optimal result.

3 Proposed Method

In this paper, for proposing the model, three classifier methods have been used, namely, Support Vector Machine (SVC), Multinomial Naive Bayes, and ExtraTree Classifiers (Fig. 3). After classifying with the above-mentioned classifier soft voting classifier has been applied to get the best accuracy represented in Fig. 1. These classifier techniques have been applied to the pre-processed dataset. The feature selection process and extraction of features are highlighted in Fig. 2 and Fig. 3.

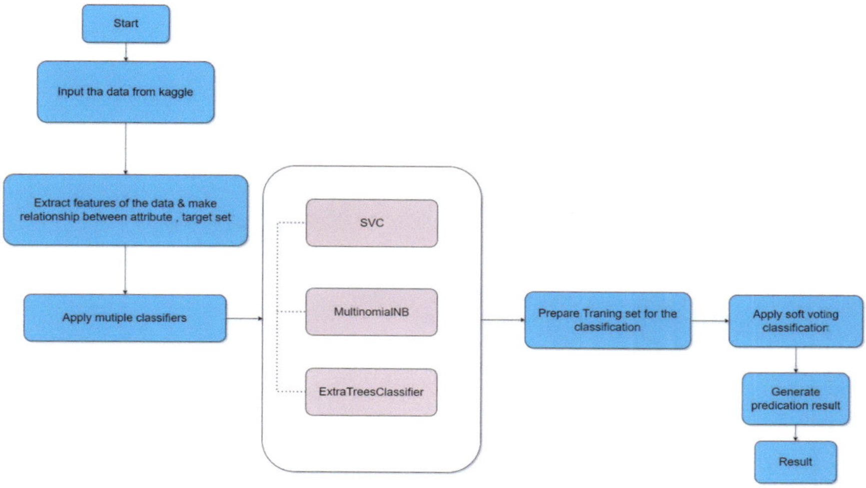

Fig. 1 Proposed method

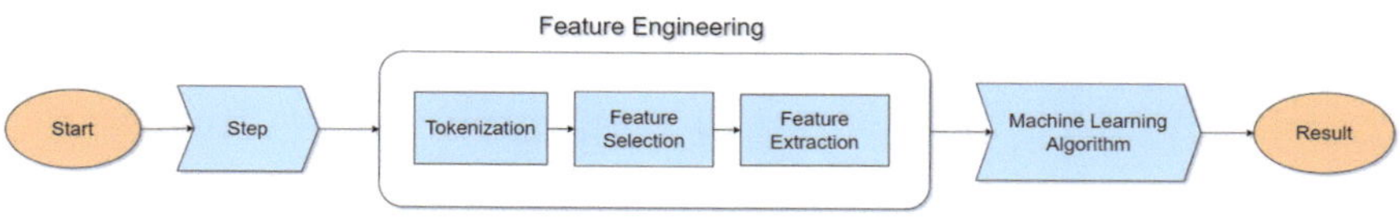

Fig. 2 Feature selection process

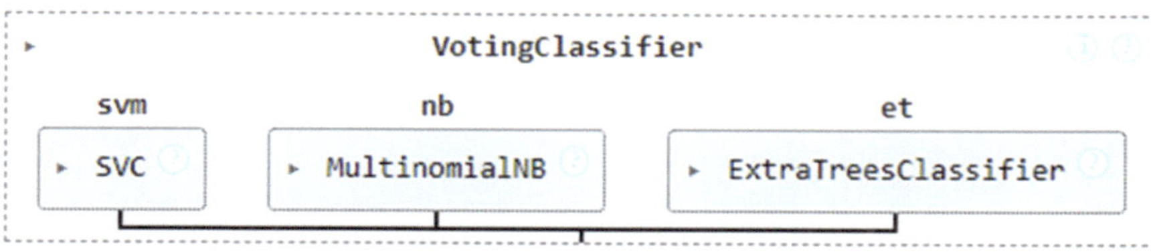

Fig. 3 Combined methods for voting classifier

3.1 Data Analyzing

Here we examine a dataset [15], containing 5573 rows of email messages to differentiate between spam and ham emails. We aim to visualize and understand various aspects of the dataset, including the ratio of spam to ham messages, the number of characters in these messages, and the common words used in spam versus ham messages.

We calculate the ratio of spam to ham messages and find a 3:1 ratio, indicating that spam messages are more prevalent than ham messages in this dataset. Additionally,

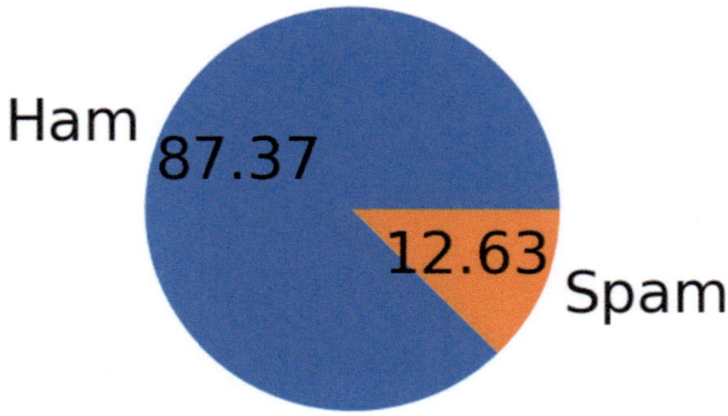

Fig. 4 Proportion of spam versus ham messages

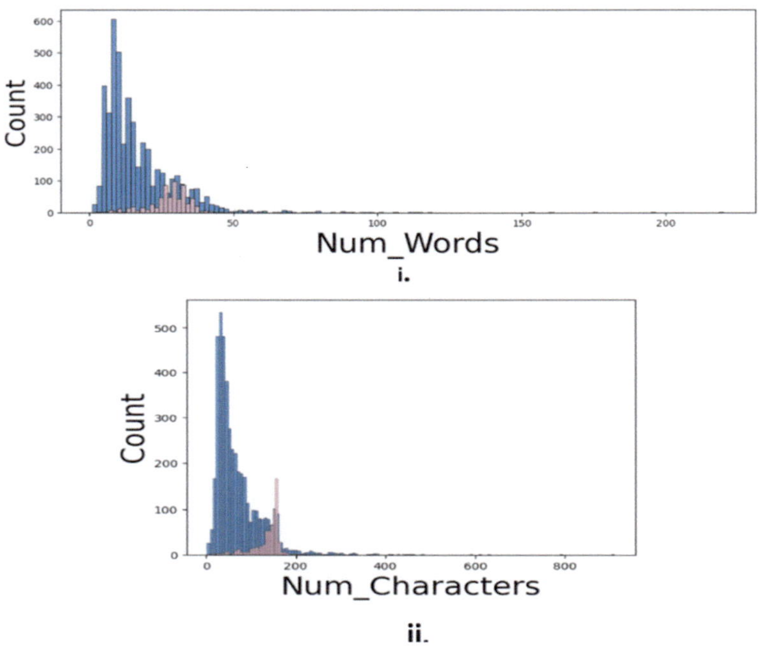

Fig. 5 Total number of words in 5i. and characters in 5ii

we added a new column to the dataset to record the length of each message, which will be useful for further analysis.

In Fig. 4, the pie chart exhibits the proportion of spam versus ham messages in the dataset [15].

The representation of the histograms in Fig. 5 shows the distribution of the number of words in Fig. 5i. as well as Fig. 5ii. represents the characters in spam and ham messages, respectively.

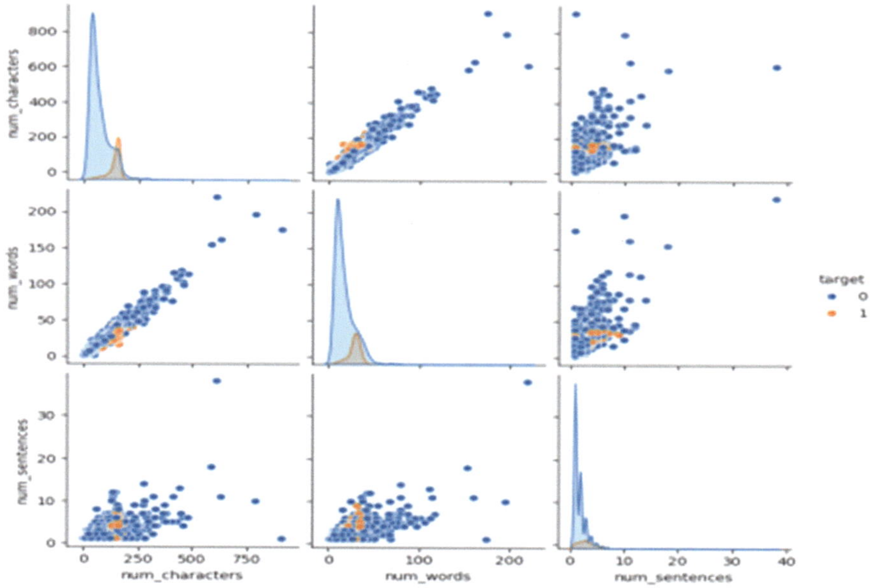

Fig. 6 Correlation of various features for spam and ham

The heatmap representation in Fig. 6 produced the correlation between various features for spam and ham messages.

4 Result and Discussion

The proposed model is compared with existing models including Decision Tree (DT), Bagging Classifier (Bgc), Gradient Boosting Decision Trees (GBDT), ExtraTrees Classifier (ETC), Logistic Regression (LR), Naive Bayes (NB), Random Forest (RF), K-Nearest Neighbors (K-NN), Support Vector Classifier (SVC), to test the reliability of the model.

From Fig. 7, it is visible that the ETC model has achieved comparatively better accuracy [97.5%].

However, the precision value is lower than other existing models for the email spam detection. Figure 8 exhibits a comparison of the accuracy and precision values for all the models listed above.

In our proposed architecture, soft voting ensemble method has been applied over the combined machine learning approaches (SVM, NB, and ET), displayed in Fig. 3, to detect email spam and achieved an impressive accuracy of up to 97.58% and precision value is calculated as 0.99% which is comparatively better from the other listed methods in table represented in Fig. 7. This accuracy is considered robust, providing a solid foundation for further enhancements in future iterations.

Algorithm	Accuracy	Precision
DT	0.934236	0.830189
Bgc	0.955513	0.870968
AdaBoost	0.968085	0.948718
GBDT	0.953578	0.950000
ETC	0.975822	0.974790
xgb	0.973888	0.982609
LR	0.942940	0.987654
NB	0.951644	1.000000
RF	0.971954	1.000000
KN	0.898453	1.000000
SVC	0.907157	1.000000
VotingClassifier (SVC, MNB, ETC)	***0.975822***	***0.991304***

Fig. 7 Proposed technique compared with existing models

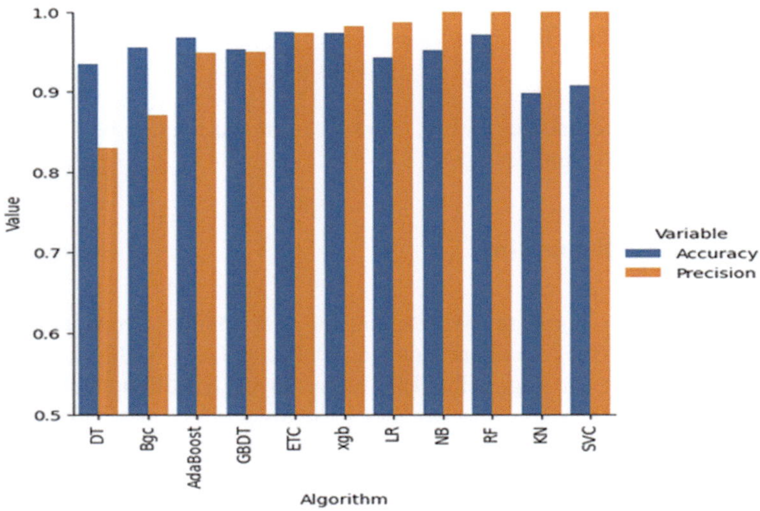

Fig. 8 Graphical representation of accuracy and precision of existing models

5 Conclusion

The results from the voting-based email spam detection system are very promising, showing high accuracy and better precision. This suggests that the ensemble method effectively combines the strengths of SVC, MNB, and other classifiers to provide a robust and reliable spam detection system.

Following this proposed approach, we have also applied the voting method using GBDT, NB, and ET and further improvements can be made through model diversity, parameter tuning, and feature engineering to ensure the system remains effective against evolving spam tactics. Regular monitoring and updating of the model will help maintain its performance in the dynamic landscape of email communication.

References

1. Mandal, S. (2024). *Cloud of things: Foundations, applications, and challenges*
2. Golchha, R., Joshi, A., & Gupta, G. P. (2023). Voting-based ensemble learning approach for cyber attacks detection in industrial Internet of Things. *Procedia Computer Science 218,* 1752–1759. ISSN: 18770509.
3. Innab, N., et al. (2024). Phishing attacks detection using ensemble machine learning algorithms. *CMC 80*, 1325–1345. ISSN: 1546–2226.
4. Pathak, P., & Shrivas, A. K. (2024). Classification of phishing website using machine learning based proposed ensemble model. *2022 OPJU international technology conference on emerging technologies for sustainable development (OTCON)* (pp. 1–6). Raigarh, Chhattisgarh, India: IEEE.
5. Mim, M.A., Majadi, N., & Mazumder, P. (2024). A soft voting ensemble learning approach for credit card fraud detection. *Heliyon 10,* e25466. ISSN: 24058440.
6. Nisar, N., Rakesh, N., & Chhabra, M. (2024). Voting-ensemble classification for email spam detection. *2021 international conference on communication information and computing technology (ICCICT)* (pp. 1–6). Mumbai, India: IEEE.
7. Lanka, S. C. (2023). *Spam based email identification and detection using machine learning techniques—IEEE conference publication—IEEE Xplore*.
8. Bansal, C., & Sidhu, B. (2023). Machine learning based hybrid approach for email spam detection. In *2021 9th international conference on reliability, infocom technologies and optimization (trends and future directions) (ICRITO)* (pp. 1–4).
9. Govil, N., Agarwal, K., Bansal, A., & Varshney, A. (2023). A machine learning based spam detection mechanism. In *2020 fourth international conference on computing methodologies and communication (ICCMC)* (pp. 954–957).
10. Kaddoura, S., Alfandi, O., & Dahmani, N. (2023). A spam email detection mechanism for english language text emails using deep learning approach. In *2020 IEEE 29th international conference on enabling technologies: Infrastructure for collaborative enterprises (WETICE)* (pp. 193–198), ISSN: 2641-8169.
11. Suryawanshi, S., Goswami, A., & Patil, P. (2023). Email spam detection: An empirical comparative study of different ML and ensemble classifiers. In *2019 IEEE 9th international conference on advanced computing (IACC)* (pp. 69–74). ISSN: 2473-3571
12. Kumar, N., Sonowal, S., & Nishant. (2023). Email spam detection using machine learning algorithms. In *2020 second international conference on inventive research in computing applications (ICIRCA)* (pp. 108–113).

13. Shukla, U., Dhanda, N., & Verma, R. (2024). Spam e-mail classification using ML, Angular, and Flask. *2023 14th international conference on computing communication and networking technologies (ICCCNT)* (pp. 1–6). Delhi, India: IEEE.
14. Ghosh, A., Das, R., Dey, S., & Mahapatra, G. (2024). Ensemble learning and its application in spam detection. *2023 international conference on computer, electrical & communication engineering (ICCECE)* (pp. 1–6). Kolkata, India: IEEE.
15. *SMS spam collection dataset* (2024).

Complex Power Quality Disturbance Classification Using Hilbert–Huang Transformation and Multiclass Support Vector Machine Classifier

M. Veerasundaram, M. Balajı, and E. Fantin Irudaya Raj

Abstract A novel method for classifying the combined power quality disturbance using Hilbert–Huang transformation and machine learning is proposed. At first the simple and complex power quality disturbances are generated using MATLAB. The Hilbert–Huang transformation is applied to the time-domain signals for feature extraction. The extracted features are fed to machine learning classifiers for classifying the disturbances. The proposed method demonstrates superior performance in distinguishing power quality disturbances. The outputs obtained from five different machine learning classifiers decision tree, random forest, Naive Bayes, K-nearest neighbor, and support vector machine are compared for their performance using various parameters. The results obtained from simulation validate that the support vector machine classifier has performed well surpassing other existing algorithms.

Keywords Power quality disturbance · Hilbert–Huang transformation · Feature extraction · Machine learning · Multiclass support vector machine

1 Introduction

The voltage sag, swell, interrupt, harmonics, flickers, and transients give more challenges to the stability and efficiency of power systems that leads to equipment failures, downtime, and financial losses. The different types of disturbances discussed

M. Veerasundaram (✉) · M. Balajı
Department of Electrical and Electronics Engineering, Sri Sivasubramaniya Nadar College of Engineering, Chennai, Tamilnadu, India
e-mail: veerasundaramm@ssn.edu.in

M. Balajı
e-mail: balajim@ssn.edu.in

E. Fantin Irudaya Raj
Department of Electrical and Electronics Engineering, Dr. Sivanthi Aditanar College of Engineering, Tiruchendur, Tamilnadu, India
e-mail: fantinraj@gmail.com

J. K. Mandal et al. (eds.), *Recent Advances in Artificial Intelligence and Smart Applications*, Lecture Notes in Networks and Systems 1380,
https://doi.org/10.1007/978-981-96-5822-0_12

in several literatures are shown in Fig. 1. In present day, the artificial intelligence-based machine learning techniques have transformed the field of signal processing and pattern recognition that helps in automatically extracting complex features and classifying data with very high accuracy. An efficient methodology for detection and classification of power quality disturbance for time series data using Hilbert–Huang transformation and multiclass Support Vector Machine (SVM) classifier is proposed. This methodology aims to automate the complete process of identifying and categorizing various types of power quality disturbances. The results obtained clearly indicate the superiority of machine learning-adopted methods for improving the reliability and resilience of electrical power systems in the face of increasingly complex and dynamic disturbance situations. Finally, the comparison on the performance of the different types of classifiers used for power quality disturbance classification is demonstrated.

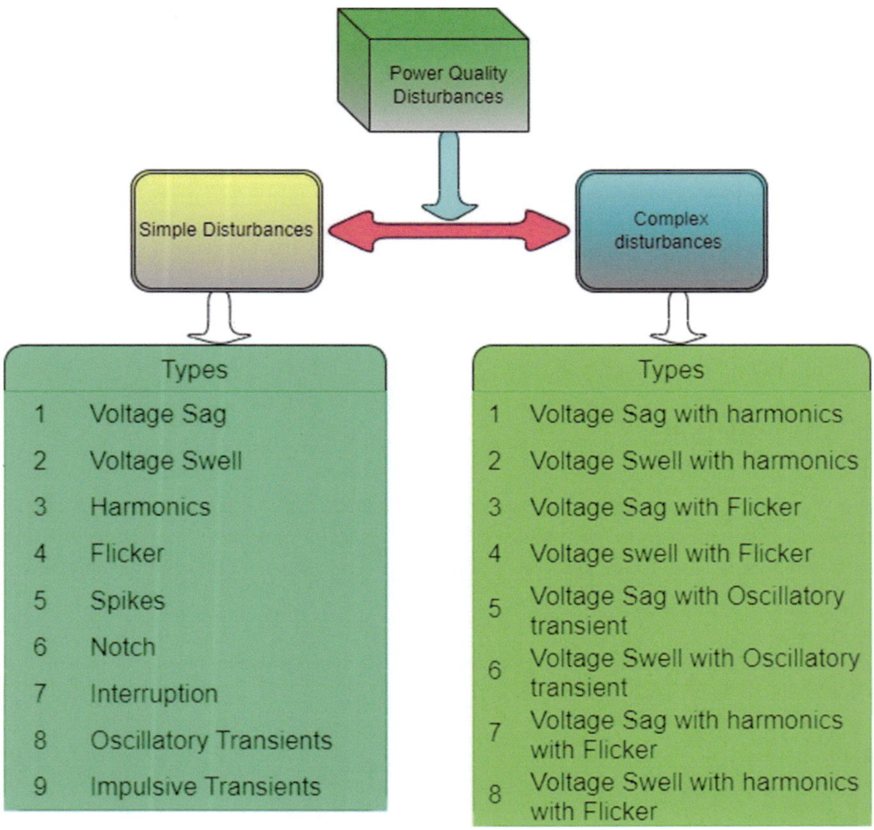

Fig. 1 Types of power quality disturbances

2 Literature Review

Power Quality Disturbance (PQD) classification utilizing segmented and modified S-transformation (SMST), the deep convolution neural network, and multiclass support vector machine are discussed in [1]. The limitation in this method is that the selection of three parameters in the SMST is manually determined and not adjustable. The article [2] presents a deep neural network trained on a database of real electrical power grid signals collected from monitoring devices. However, the misclassification rate is higher when using real-time signals. An S-transform-based probabilistic neural network (PNN) classifier for recognition of PQD is addressed in article [3], it requires less number of features as compared to wavelet-based approach for the identification of PQ events. Eleven types of PQD are considered and the accuracy obtained is 93.2%. Combination of Stockwell transform (ST) and deep learning is used in [4] and the method's robustness in noisy environment is demonstrated. PQD classification using Synchrosqueezed Wavelet Transforms (SWT) and convolutional Neural Network (CNN) is explained in [5]. Validation on synthetic PQDs and power system simulations demonstrates its effectiveness. A fast variant of the discrete S-transform (FDST) algorithm to accurately extract the time-localized spectral characteristics of nonstationary signal is discussed in [6]. It has the advantage of detecting multiple PQD occurring simultaneously. Thirteen types of PQD are considered and the accuracy obtained is 92.69%. A combination method for the classification of power quality complex disturbances based on ensemble empirical mode decomposition (EEMD) and multilabel learning is discussed in [7]. EEMD is adopted to extract the features of complex disturbances, which is more suitable to the nonstationary signal processing. The types of PQD considered are 8 and the accuracy obtained is 92.8%. The CNN with an attention model using deep learning is explained in [8]. The dataset consists of signals categorized into 9 classes, each characterized by 622 data points. Results show an accuracy of 99.92%, surpassing previous studies. The combination of Stockwell transform with an improved gray wolf optimization-based kernel extreme learning machine for the detection and classification of PQD is presented in [9]. Results show high accuracy, with 99.76% detection under noiseless conditions and achieving accuracies above 97% at various SNR levels. PQD recognition using Multiresolution S-transform (MST) and decision tree (DT) is explained in [10]. It demonstrates detection accuracy of 99.08% under 30 dB SNR, surpassing competing algorithms. Power quality disturbance classification using support vector machine (SVM) neural networks is proposed in [11]. Classifiers based on radial basis function (RBF) networks were, in parallel, applied to enable proper performance comparison. It was found that SVM performs better than RBF. The types of PQD considered are only 6 and the accuracy obtained is 72%. In [12], an application of non-dominated sorting genetic algorithm III (NSGA III) and directed acyclic graph support vector machine (DAG-SVM)-based combined approach for recognition and classification of power quality disturbances is proposed. The main advantage with this algorithm is the reduced computation time and the disadvantages are the less computational overhead and accuracy. PQD classification based on the wavelet transform and self-organizing learning array (SOLAR) system is presented in [13]. SOLAR has three advantages over a typical neural network: data-driven learning, local interconnections, and entropy-based self-organization. However, the limitation here is the computational cost increases when wavelet filter length is increased. The types of PQD considered are 7 and the accuracy obtained is 94.3%. IEEE 1159-2009 [14] offers guidance on monitoring techniques, fosters communication within the power quality community, and provides insight into power system disturbances and their origins.

3 Proposed Method

The power quality disturbances are generated in MATLAB. The power quality disturbance signals obtained are then labeled according to their type. Hilbert–Huang transformation is applied to the PQD signals. Here five features such as mean, standard deviation, kurtosis, skewness, and variance are extracted. These features are then fed to machine learning classifier for power quality disturbance classification. The outputs obtained from different classifiers are compared based on their accuracy and precision. The algorithm given below explains the sequence of steps followed in the proposed method.

3.1 Algorithm of the Proposed Method

Step 1: The single and combined power quality disturbances are generated using parametric equations in MATLAB environment.

Step 2: The dataset is created with time series values and divided as training data and test data.

Step 3: The Hilbert–Huang transformation is applied to the signals to represent them in frequency domain.

Step 4: The feature extraction is performed and the features extracted are mean, standard deviation, skewness, kurtosis, and variance.

Step 5: The extracted features are fed to the chosen machine learning classifier sequentially and the parameters such as classifier accuracy and precision are obtained for the chosen classifier with both training data set and test dataset.

Step 6: Next machine learning classifier is selected and Step 5 is repeated with all five machine learning classifiers.

Step 7: The performance of the various classifier is compared based on precision and accuracy.

Step 8: The efficient classifier is identified for the PQD problem.

Fig. 2 Voltage sag

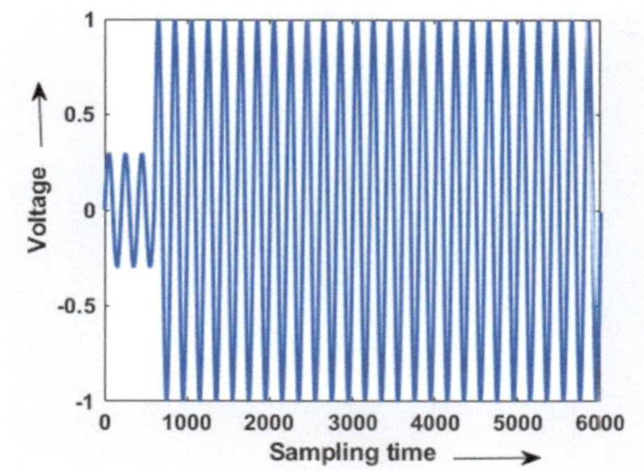

Fig. 3 Voltage swell

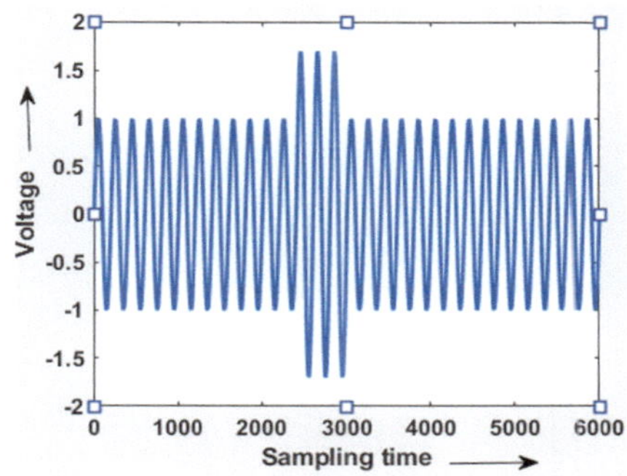

4 Data Generation and Data Preprocessing

For PQD classification using machine learning, the accuracy and precision of the output depend on the authenticity and validity of the available dataset. Here for precise output, MATLAB code is written using the equations [11] to generate large volume of data for each disturbance simulated. Figures 2, 3, 4, 5, 6, 7, 8, 9, 10, and 11 show sample output of the various power quality disturbance generated. The generated data is preprocessed for removing any void and missing data.

5 Feature Extraction with Hilbert–Huang Transformation (HHT)

The Hilbert–Huang Transform (HHT) offers various advantages such as adaptability to nonstationary signals, sharpness of results, and provides interpretable features. The steps involved in HHT are

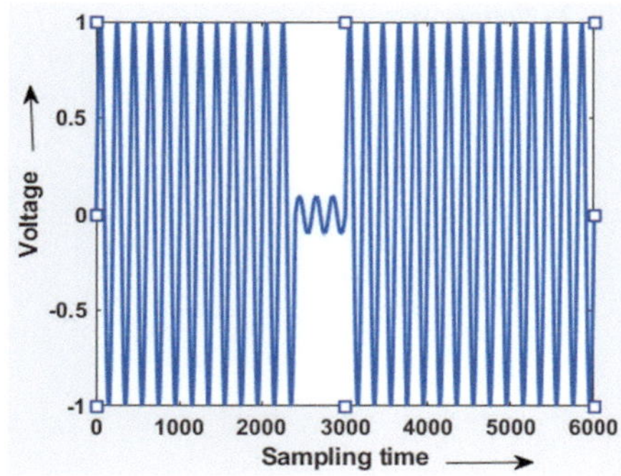

Fig. 4 Interrupt

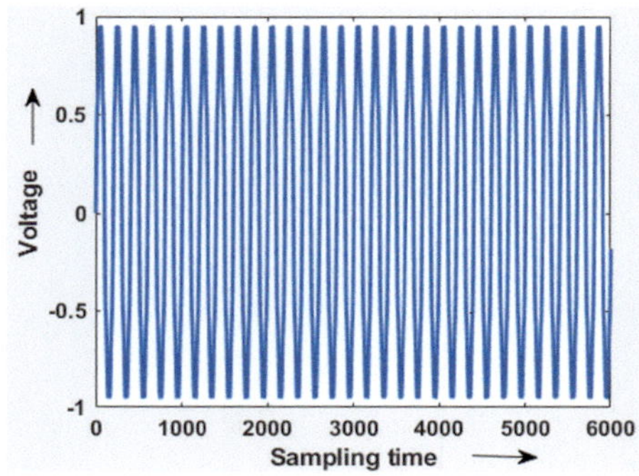

Fig. 5 Harmonics

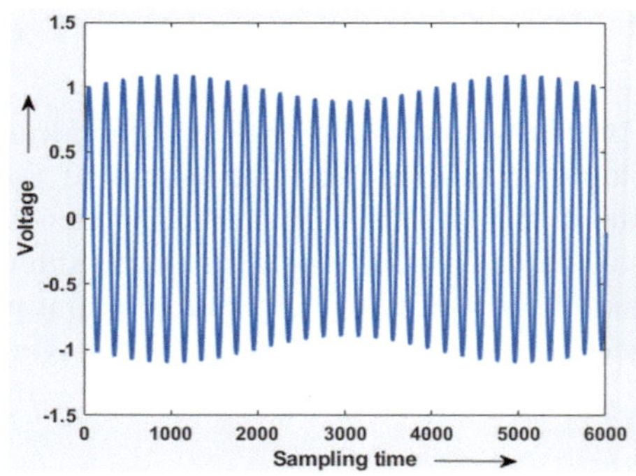

Fig. 6 Flicker

1. Data preparation.
2. Empirical Mode Decomposition (EMD): EMD is the first step in HHT. It decomposes the time series into IMFs using EMD. Each IMF represents a different oscillatory mode. IMFs are extracted adaptively, capturing both high- and low-frequency components. The last IMF is usually a trend or residual component.

Fig. 7 Oscillatory transient

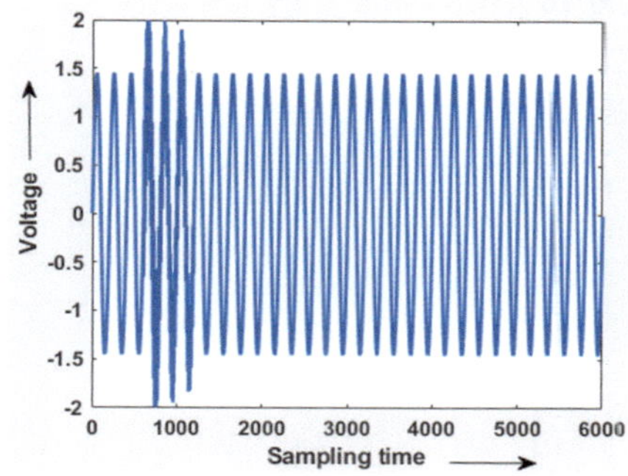

Fig. 8 Sag with harmonics

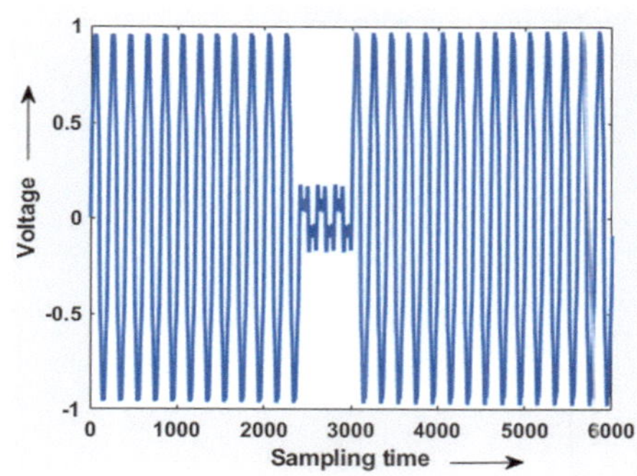

Fig. 9 Swell with harmonics

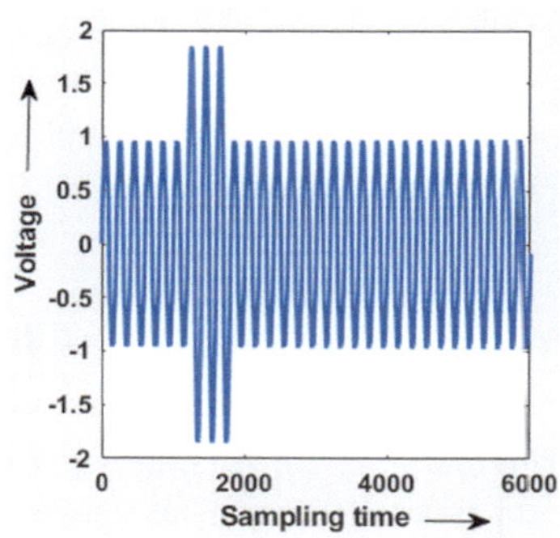

3. Hilbert Transform: Apply the Hilbert transform to each IMF: The Hilbert transform extracts the instantaneous phase and amplitude information. It converts the IMF that is in the time domain to the analytic signal domain. The resulting analytic signal contains both real and imaginary parts.

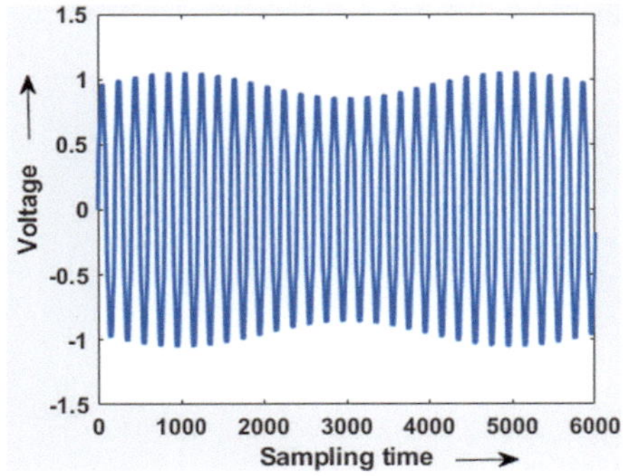

Fig. 10 Flicker with harmonics

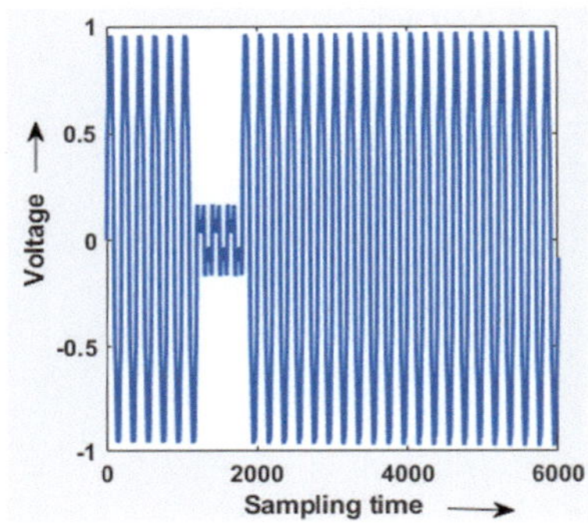

Fig. 11 Interrupt with harmonics

4. Hilbert Spectrum: Combine the instantaneous frequencies from all IMFs to create the Hilbert spectrum. The Hilbert spectrum shows the time-varying frequency content of our signal. It helps identify dominant frequency components and their variations.

6 Machine Learning Classifiers

The classifier in machine learning is a sequence of steps that naturally designates data points to a span of categories or classes. The classifier subcategory has two important models, they are supervised and unsupervised. In the supervised model, the classifiers are prepared to do a distinguish between labeled and unlabeled data. Unsupervised algorithms work based on pattern recognition to categorize unlabeled datasets.

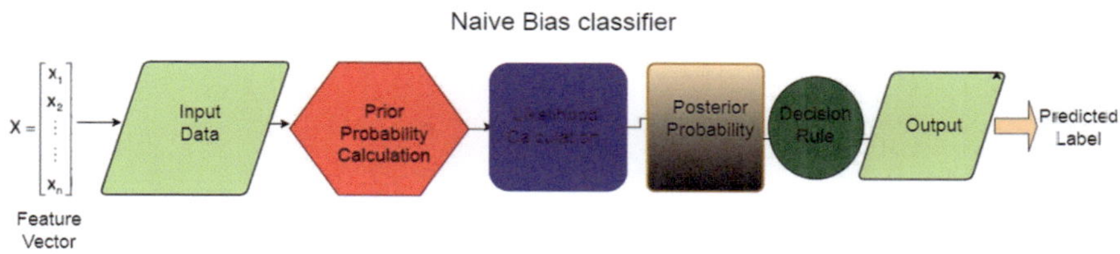

Fig. 12 Flow diagram of Naïve Bayes classifier

6.1 *Types and Functions of Classifiers*

6.1.1 Naive Bayes Classifiers

Naive Bayes classifier uses the likelihood to forecast if an input belongs to a particular category. This algorithm group is based on a probability theorem and includes various classifiers. In scenarios with multiple categories, the algorithm finds the likelihood that a data point matches to each type of classification. It then selects the category with the highest likelihood of matching the given input. Naïve Bayes calculates the likelihood of a disturbance that be in the hands to each class based on these features and assigns the class with the foremost probability to the disturbance (Fig. 12).

6.1.2 Decision Trees

Decision tree is a type of classification technique which divides data into increasingly specific categories. Visually, it resembles the branches of a tree. Operating within a supervised model, decision trees learns classification rules for the training data by the application of IF-THEN concept. The process extends till appropriate stopping condition is met.

Decision tree algorithms recursively split the feature space into regions based on the most informative features. Decision trees make decisions at each node based on the features to partition the dataset into subsets, ultimately leading to a tree structure that represents decision rules for classifying different types of disturbances (Fig. 13).

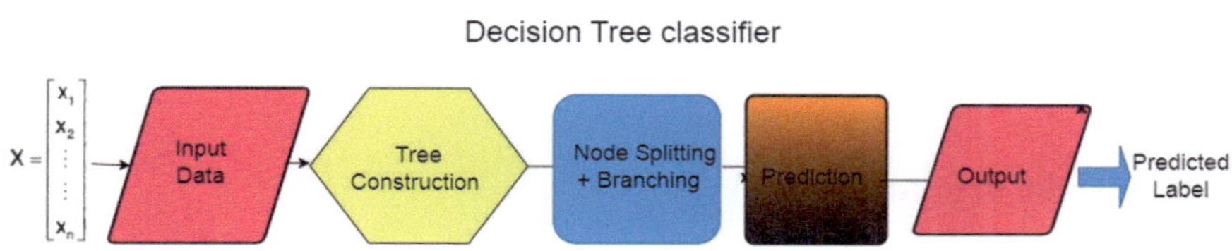

Fig. 13 Flow diagram of decision tree classifier

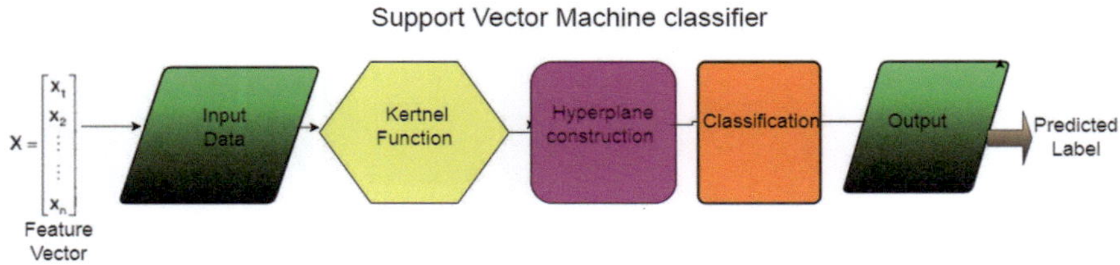

Fig. 14 Flow diagram of support vector machine classifier

6.1.3 Support Vector Machines (SVM)

Support vector machines are used for classification tasks. They find hyperplanes (visualized as lines) that separate different classes of data. They excel at handling complex data distributions, becoming more accurate as input complexity increases. SVM objective is to determine the hyperplane that is good in dividing different classes of data points while enlarging the margin between the classes. In the context of power quality disturbance classification, SVM can be trained to determine the hyperplane that is very good in determining the various categories of disturbances according to their feature representations. The features could include time-domain or frequency-domain characteristics of the disturbances. SVMs are particularly effective when dealing with high-dimensional feature spaces, making them suitable for PQD classification tasks (Fig. 14).

6.1.4 K-Nearest Neighbor (K-NN) Classifier

K-NN comes in supervised slow learner algorithm. It stores training data and compares it to other data for predictions. By measuring similarity between stored instances and test data, K-NN predicts the likelihood of future data matching the dataset. It's commonly used for regression and classification tasks. K-NN performs the classification tasks on the data points on the basis of majority class of their nearest neighbors. When applied to power quality disturbance classification, K-NN considers feature vectors representing different characteristics of the disturbances and assigns a class label on the basis of class labels of the nearest neighbors in the feature space. K-NN's performance heavily depends on the choice of distance metric and the number of neighbors considered (Fig. 15).

6.1.5 Random Forest Classifier

Random forests are supervised machine learning algorithms that use decision trees for classification. It handles large, complex datasets and high-dimensional feature spaces. Random forests provide insights into feature importance while maintaining predictive accuracy and minimizing overfitting. The classifier generates decision

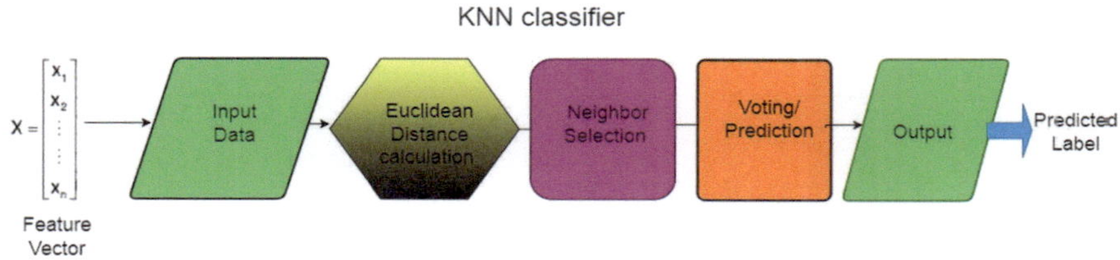

Fig. 15 Flow diagram of k-nearest neighbor classifier

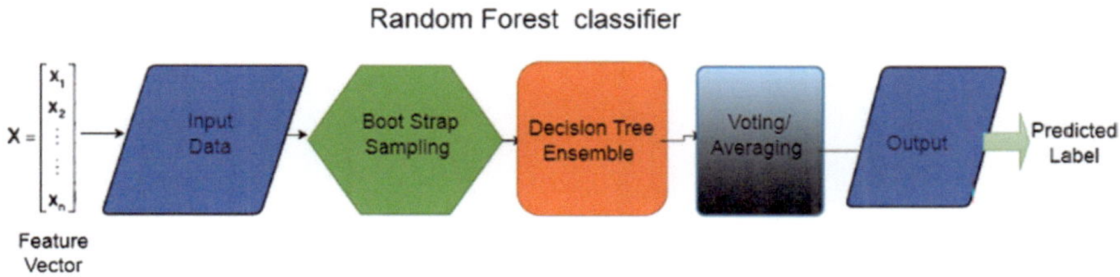

Fig. 16 Flow diagram of random forest classifier

trees from a subset of the training set that are randomly selected and unites their representation to make final predictions. Random forest is an group learning method that combines several decision trees to upgrade the classification robustness and accuracy. When applied to PQD classification, a random forest classifier can be taught using various features representing different aspects of the disturbances. Each decision tree in the ensemble is constructed using a subset of features that are random and data samples (Fig. 16).

7 Results and Discussion

The feature-extracted parameters obtained are segregated in sequential manner and labeled according to the type of power quality disturbance. The next step involves the classifier training with the extracted features. From the data generated, 80% is allotted for training and 20% is chosen for testing purpose. Here 5 different classifiers are trained with input data in a sequential order. Once trained the classifiers are fed with test data to check for their performance and ability in classifying the disturbances. Table 1 shows the comparison of the performance of the classifiers. It can be observed that SVM classifier has low level of complexity, higher reliability, and very good accuracy and precision.

Figure 17 shows the comparison of the performance of the classifiers. The two parameters considered for performance comparison are the classification accuracy and precision. Five different machine learning classifiers analyzed are decision tree, random forest, Naive Bayes, K-nearest neighbor, and support vector machine. It

Table 1 Comparison of the performance of the classifiers

Classifier type	Complexity	Reliability	Precision (%)	Accuracy (%)
Decision tree	High	Moderate	85.19	83.6
Random forest	High	Low	90.26	91.6
Naive Bayes	Moderate	Low	69.86	68.4
K-nearest neighbor	High	Moderate	78.1	75.6
Support vector machine (proposed)	**Moderate**	**High**	**91.86**	**93.7**

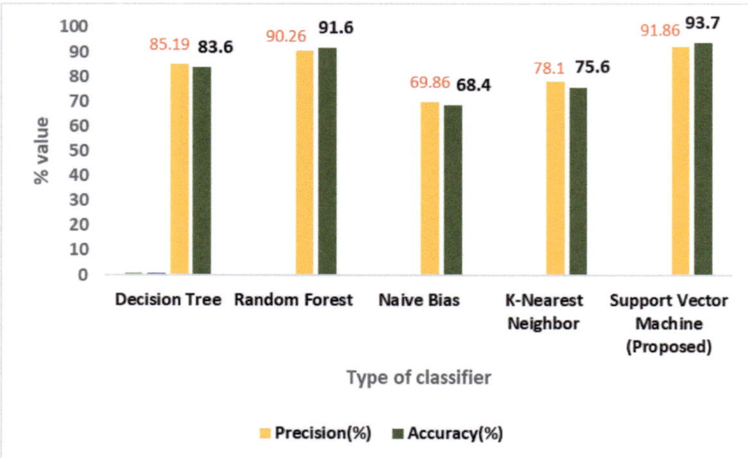

Fig. 17 Comparison of performance of classifiers

can be observed that the multiclass SVM classifier has performed well and has a classification accuracy of 93.7%.

The confusion matrix with SVM classifier shown in Fig. 18 is a plot between the predicted class and true class. For the training dataset with total of 800 samples, the diagonal elements of the matrix give the number of correct prediction and the off-diagonal elements give the wrong prediction. The classifier is said to be efficient if it has higher values on the diagonal elements as it means more instances are being classified correctly for that specific class and lower values in the off-diagonal elements as it indicates fewer misclassifications.

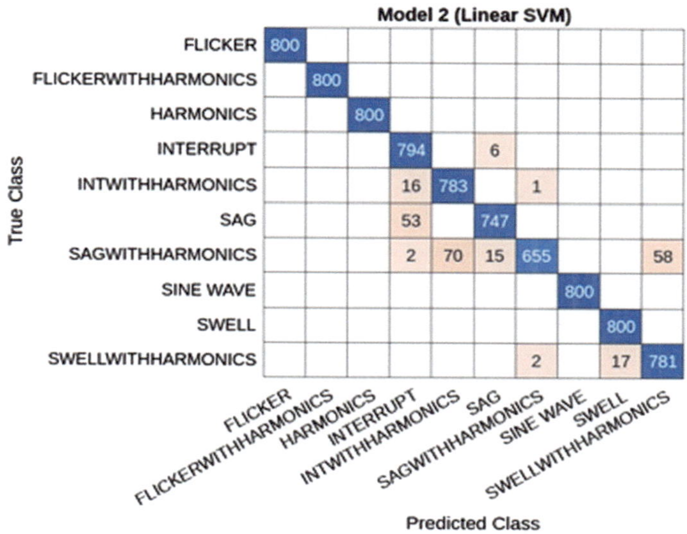

Fig. 18 Confusion matrix with SVM classifier

8 Conclusion

This paper proposes a PQD classification based on HHT and SVM classifier. The simple and complex PQDs are analyzed here with time series data. The PQD dataset is obtained in the MATLAB Simulink environment. To test the effectiveness and the contributions of the proposed approach, the results obtained using the HHT-MSVM have been compared with classical machine learning architectures. The proposed HHT-MSVM has proven to be superior to the other approaches in almost all aspects (including accuracy, precision, and number of parameters) In fact, other than being superior in performances, the HHT-MSVM has proven to be efficient in terms of use of computational resources. The results obtained shows the performance of five different machine learning classifiers for PQD classification. The support vector machine classifier has the accuracy of 93.7% and precision of 91.6% followed by random forest classifier with accuracy of 91.6% and precision of 90.26% surpassing other existing algorithms. The classifier with high precision and accuracy while classifying the PQD is the multiclass support vector machine followed by random forest classifier. Compared with traditional machine learning methods and other existing methods in the literature, the method proposed in this article is able to detect and classify both the simple and complex PQDs efficiently.

References

1. Liu, M., Chen, Y., Zhang, Z., & Deng, S. (2023). Classification of power quality disturbance using segmented and modified S-transform and DCNN-MSVM hybrid model. *IEEE Access*. https://doi.org/10.1109/ACCESS.2022.3233767
2. Rodrigues, N. M., Janeiro, F. M., & Ramos, P. M. (2023). Deep learning for power quality event detection and classification based on measured grid data. *IEEE Transactions on Instrumentation and Measurement, 72*, 9003311.
3. Mishra, S., Bhende, C. N., & Panigrahi, B. K. (2008). Detection and classification of power quality disturbances using S-transform and probabilistic neural network. *IEEE Transactions on Power Delivery, 23*, 280–287.
4. Cui, C., Duan, Y., Hu, H., Wang, L., & Liu, Q. (2022). Detection and classification of multiple power quality disturbances using stockwell transform and deep learning. *IEEE Transactions on Instrumentation and Measurement, 71*, 2519912.
5. Upendra Vishwanath, Y. S., Esakkirajan, S., Keerthiveena, B., & Pachori, R. B. (2023). A generalized classification framework for power quality disturbances based on synchrosqueezed wavelet transform and convolutional neural networks. *IEEE Transactions on Instrumentation and Measurement, 72*, 2525313.
6. Milan Biswal, P. K. (2013). Dash: Detection and characterization of multiple power quality disturbances with a fast S-transform and decision tree based classifier. *Digital Signal Processing, 23*, 1071–1083.
7. Liu, Z., Cui, Y., & Li, W. (2015). A classification method for complex power quality disturbances using EEMD and rank wavelet SVM. *IEEE Transactions on Smart Grid, 6*, 1678–1685.
8. Topaloglu, I. (2023). Deep Learning based a new approach for power quality disturbances classification in power transmission system. *Journal of Electrical Engineering & Technology, Springer, 18*, 77–88. https://doi.org/10.1007/s42835-022-01177-1
9. Ravi, T., & Kumar, K. S. (2023). Detection and classification of power quality disturbances using stockwell transform and improved Grey Wolf optimization-based kernel extreme learning machine. IEEE Access. https://doi.org/10.1109/ACCESS.2023.3286308
10. Zhong, T., Zhang, S., Cai, G., Li, Y., Yang, B., & Chen, Y. (2019). Power quality disturbance recognition based on multiresolution S-transform and decision tree. *IEEE Access*. https://doi.org/10.1109/ACCESS.2019.2924918
11. Janik, P., & Lobos, T. (2006). Automated classification of power-quality disturbances using SVM and RBF networks. *IEEE Transactions on Power Delivery, 21*(3).
12. Rahul, B. C. (2021). An advanced genetic algorithm with improved support vector machine for multi-class classification of real power quality events. *Electric Power Systems Research* 191106879.
13. He, H., & Starzyk, J. A. (2006). A self-organizing learning array system for power quality classification based on wavelet transform. *IEEE Transactions on Power Delivery, 21*(1).
14. IEEE Std 1159-1995. (R2001). IEEE recommended practice for monitoring electric power quality.

Character-Centric Summarization and Keyphrase Extraction with Visualization in Indian Mythology

Apurba Paul, Anupam Rana, Ishan Chattopadhaya, and Dipankar Das

Abstract With an emphasis on the **Mahabharata**, a complex narrative with many characters, events, stories, and deep relationships, this paper investigates character-centric summarization and keyphrase extraction in Indian mythology. Through refinement of transformer-based models such as BART, PEGASUS, and T5 and using the dataset, $DS_{character}$, the study generates character-specific summaries. In the dataset, the character-centric sentences emphasize the roles, actions, and significance of individual characters in the text. In case of capturing the coherence and consistency of character-specific summaries when the models were assessed using ROUGE and BLEU scores, PEGASUS performed the well overall. This study also applied KeyBERT for the extraction of character-specific keyphrases which achieves high recall and precision in recognizing important themes linked to characters. To understand the significance of each character in the story, later on we focused on to implement the visual representation of the character-specific summaries and keyphrases. This research offers a detailed understanding of how character-centric mythological sentences are processed and makes recommendations for future research areas to enhance model performance in character-centric tasks.

Keywords Summarization · Keyphrase extraction · Indian mythology · Mahabharata · Transformer-based models

---

A. Paul et. al—Authors contributed equally.

A. Paul (✉) · A. Rana · I. Chattopadhaya
Institute of Engineering and Management, University of Engineering and Management, Kolkata, India
e-mail: apurba.saitech2021@gmail.com

A. Rana
e-mail: anupam.rana2022@iem.edu.in

A. Paul · D. Das
Jadavpur University, Kolkata, India

J. K. Mandal et al. (eds.), *Recent Advances in Artificial Intelligence and Smart Applications*, Lecture Notes in Networks and Systems 1380,
https://doi.org/10.1007/978-981-96-5822-0_13

1 Introduction

Indian mythology, with its rich and intricate narratives, presents unique challenges for natural language processing due to the complexity of its stories and the vast number of characters involved. One of the two great epics, the **Mahabharata**,[1] presents a particularly difficult scenario because of its intricate and interwoven character interactions. This intricacy makes it an excellent work for investigating character-centric summarization, in which the roles, actions, and interactions of certain characters are highlighted. Transformer-based models like BART, PEGASUS, and T5 have recently improved the efficiency and performance of text summarization methods. However, the research work on character-centric text summarization is not so common. This research suggests fine-tuned versions of these models to address the problem of character-specific summaries in the **Mahabharata**. Using the in-depth knowledge of linguistic structures, these models produce summaries that are coherent and appropriate for the context. So the research of fine-tuning these models on more specific task such as character-centric summarization has a rising need, even though the traditional summarization focuses on entire texts.

In the narratives like the **Mahabharata** of mythology, the character interactions are pivotal and a dedicated approach is needed to identify each character's significance. This effort is driven by the goal of improving the quality and relevance of summaries of mythical texts by highlighting the roles and actions of particular characters. When applied to character-centric narratives, traditional summary techniques frequently ignore these important details, producing summaries that are less informative. In order to provide summaries that are more representative of each character's contributions to the **Mahabharata** and offer deeper, more contextual insights into the story, this work intends to optimize T5, BART, and PEGASUS especially for character-centric summarization. In this research, we faced a number of challenges like—due to the narrative's intricate structure, it is challenging to isolate and summarize character-specific actions. Next, it requires a careful balance to guarantee that important activities of all the characters from the most significant ones like *Krishna* and *Arjuna* to the minor ones are accurately depicted. The lack of adequate resources to meet the current demand results in significant manual effort and time required to create the dataset. Finally, there are technological challenges in fine-tuning the models such as BART, PEGASUS, and T5 to produce character-centric summaries while preserving narrative coherence.

This work aims to achieve three primary objectives: (1) fine-tune the BART, PEGASUS, and T5 models for character-centric summarization by capturing the roles and actions of individual characters in the text; (2) develop a task-specific dataset ($DS_{character}$) of character-centric sentences; and (3) keyphrase extraction and visualization that will show the key themes and actions associated with each character to give readers a better understanding. This paper contributes in a number of significant ways. For character-centric summary, it first presents improved versions of BART, PEGASUS, and T5 that are especially suited for intricate mythological works

[1] https://www.wisdomlib.org/hinduism/book/the-mahabharata-mohan.

like the **Mahabharata**. Secondly, it provides a character-specific content-focused custom dataset ($DS_{character}$), which is an essential training and assessment resource. Thirdly, it provides a new framework for keyphrase extraction and visualization, which makes it easy to investigate and comprehend the roles and connections between each character in the narrative.

The remainder of the paper is structured as follows: Related work is covered in Sect. 2, with an emphasis on current methods for keyphrase extraction, text summarization, and the applications of the analysis of mythological literature. The $DS_{character}$ dataset was prepared in detail in Sect. 3, which also explains the extraction of character-based phrases for fine-tuning. The system framework is presented in Sect. 4, including the fine-tune and implementation of BART, PEGASUS, and T5 for character-centric summarization and keyphrase extraction. In Sect. 5, the outcomes are investigated, and each model's performance on the character-specific dataset is assessed. A discussion and observation section is provided in Sect. 6, which examines the significant findings and how they advance our knowledge of mythological writings. The work is finally concluded and new research options are suggested in Sect. 7.

2 Related Work

2.1 Text Summarization

With the introduction of deep learning models like BART, PEGASUS, and T5, text summarization has seen substantial progress to generate concise summaries. T5 [1] reframes various NLP tasks as text-to-text tasks, showing strong performance in producing coherent summaries by converting tasks into a text generation framework. BART [2] utilizes a denoising autoencoder structure to reconstruct disrupted input text, creating fluent summaries by effectively capturing text structure and context.

PEGASUS [3] focuses specifically on abstractive summarization by implementing a gap-sentence generation approach during pre-training, making it highly suitable for generating detailed, contextually rich summaries. Additional advancements include BERTSUM [4], which enhances extractive summarization with BERT embeddings, and the Hierarchical Transformer model [5], which manages long-form document summarization. SummaRuNNer [6] applies attention mechanisms to improve the generation of abstractive summaries.

Further studies focus on optimizing encoder–decoder structures for enhanced coherence [7] and applying contrastive learning to better extract key sentences [8]. However, the summarization of mythological texts remains relatively unexplored. Using these summarization models for mythology can simplify narratives, facilitate comparative studies, support digital archiving projects, and produce educational resources that make these texts more accessible and engaging.

2.2 Keyphrase Extraction

Keyphrase extraction focuses on identifying important phrases within text, which assists with indexing, summarization, and retrieval. In mythological texts, where narratives are complex and symbolic, keyphrase extraction helps identify essential elements such as characters and key events.

KeyBERT [9] is an unsupervised model that employs BERT embeddings to extract keyphrases without requiring labeled data, proving effective in classification and summarization contexts. EmbedRank [10] uses embeddings combined with graph-based ranking to identify keyphrases, ranking them by semantic similarity, which is particularly useful for unsupervised tasks.

By accounting for word position and co-occurrence, PositionRank [11] improves keyphrase extraction. CopyRNN [12] uses a supervised approach that generates keyphrases not necessarily present in the text, making it effective for research papers and complex fields like mythology.

For both summarization and keyphrase extraction tasks, multi-head attention networks [13] utilize attention to focus on critical parts of the text. YAKE! [14] performs effectively across various types of texts based on word frequency and position. Neural networks with attention were applied by [15] to enhance the context-dependent keyphrase extraction.

Due to the lack of research work done on character-centric summary and keyphrase extraction from texts like **Mahabharata** our work will have a significant contribution.

3 Preparation of Dataset

The dataset preparation is a critical step to enhance the performance of character-centric summarization and keyphrase extraction. In this study, we used the $\mathbf{DS}_{character}$ dataset, specifically curated for summarizing and extracting essential information related to characters in Indian mythology. The preparation process included several important steps:

The $\mathbf{DS}_{character}$ dataset was developed based on the PouranicTopic dataset by Paul et al. [16]. This dataset provides an extensive collection of mythology-related texts, particularly emphasizing descriptions and narratives of various characters. To optimize the dataset's quality and applicability, we applied the following pre-processing steps: the text was tokenized into sentences and words using a specific tokenization tool, such as SpaCy.[2] Named entities, especially character names, were recognized and tagged using a Named Entity Recognition (NER) model. Unnecessary content, such as metadata or irrelevant information, was removed to keep only the character-centric text.

[2] https://spacy.io/.

After tagging character names through NER, we extracted sentences associated with each character. This process involved segmenting the text into individual sentences and filtering them to retain only those mentioning the recognized characters. For each character, we compiled a list of relevant sentences. Additionally, each character is paired with an image to facilitate visualization, helping users to visually associate each summary with the corresponding character. To support fine-tuning, concise summaries were manually crafted for each character's set of sentences, providing an accurate, context-rich representation of each character's role in the narrative, which in turn serves as valuable input for model training. Quality checks were performed through both manual review of summaries to confirm accuracy and contextual fidelity and automated validation to maintain uniformity and thoroughness across the dataset. An example of this dataset preparation approach is provided below.

Original: *Abhimanyu was born to Subhadra, the sister of Vasudeva, and Arjuna, making him the grandson of the illustrious Pandu. Known in his previous life as the mighty Varchas, son of Soma, he was reborn as Abhimanyu, a warrior of extraordinary deeds and accomplishments. Abhimanyu fathered a son, Parikshit, who was beloved by Vasudeva himself. Among all the warriors, Abhimanyu stood as the perpetuator of his family line, earning renown for his strength and valor.*

Summary: *Abhimanyu, son of Subhadra and Arjuna, was a renowned warrior known for his extraordinary deeds and valor. A reincarnation of Varchas, he continued his legacy through his son Parikshit, earning admiration as the perpetuator of his family line.*

This dataset preparation process was essential to tailor the data for character-specific summarization and keyphrase extraction, forming a robust basis for the tasks that followed.

4 System Framework

The system framework developed for this study offers an integrated approach to character-centric summarization, keyphrase extraction, and visualization. It comprises three main components: character-centric summarization, keyphrase extraction, and visualization, each essential for efficient data processing and presentation. This structure is illustrated in Fig. 1.

4.1 Character-Centric Summarization

The initial component of the framework focuses on character-centric summarization, aimed at generating concise summaries that revolve around individual characters within the dataset. The process includes the following steps:

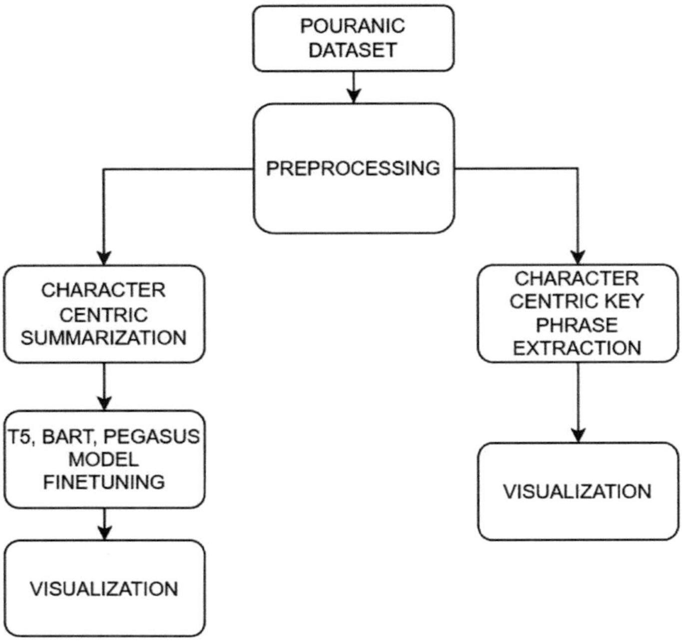

Fig. 1 System framework

- **Model Fine-Tuning:** Models like BART, PEGASUS, and T5, which are pre-trained, undergo fine-tuning on sentences that are specifically related to individual characters. This fine-tuning process enables the models to generate summaries that highlight each character's narrative and attributes. For T5, fine-tuning involves training on character-specific sentences and corresponding summaries, optimizing the model to minimize the loss between generated and reference summaries, with evaluation using ROUGE scores. BART is fine-tuned using a sequence-to-sequence methodology, where input sentences are paired with target summaries, focusing on reconstructing accurate summaries. Evaluation is done using ROUGE metrics and manual checks for contextual consistency. Similarly, PEGASUS is fine-tuned to enhance its summarization capabilities for character-specific content, focusing on generating precise summaries and optimizing its accuracy in the process. Performance is evaluated through both ROUGE scores and qualitative assessments.
- **Summary Generation:** The fine-tuned models generate summaries that capture the essence of each character's actions, attributes, and their role in the narrative of Indian mythology.

4.2 Keyphrase Extraction Using KeyBERT

The second component of the framework is Keyphrase Extraction, which utilizes KeyBERT, a model that extracts significant phrases from text. The steps are as follows:

- **Keyphrase Identification:** KeyBERT, utilizing BERT embeddings to understand contextual meanings, identifies the most relevant key phrases within the character-centric sentences. This model helps in pinpointing the essential phrases associated with each character.
- **Keyphrase Extraction:** KeyBERT extracts key phrases that highlight critical aspects of the character's story and role, focusing on terms and phrases that are essential for understanding their importance in the narrative.

4.3 Visualization

The final component, Visualization, showcases the summarized data and extracted key phrases in an interactive and engaging format. The process includes

- **Summary Display:** Character-centric summaries are presented on a web interface. To improve the recognition, each character's name in the summary is substituted with their corresponding image, helping users better connect the summary to the appropriate character.
- **Keyphrase Visualization:** The extracted key phrases are visualized using methods like word clouds, helping users understand the prominence and relationships of key phrases tied to each character.

5 Result Analysis

5.1 Character-Centric Summarization

The effectiveness of the BART, PEGASUS, and T5 models in producing character-specific summaries was assessed using ROUGE and BLEU scores. The quality of the summaries produced by these models was assessed using the ROUGE metrics. The results of this evaluation are presented in Table 1.

Table 1 illustrates a comparison of the BART, PEGASUS, and T5 models across four evaluation metrics: ROUGE-1, ROUGE-2, ROUGE-L, and BLEU. In the case of ROUGE-1, BART outperforms the other models with a score of 0.76. This shows that T5 is relatively less effective in capturing relevant unigrams. For ROUGE-2, PEGASUS leads with a score of 0.74. T5 follows closely with a score of 0.70,

Table 1 ROUGE scores for character-centric summary

Model	Rouge-1	Rouge-2	Rouge-L	BLEU
T5	0.65	0.70	0.68	0.66
BART	0.76	0.66	0.72	0.77
Pegasus	0.71	0.74	0.75	0.70

while BART scores 0.66, indicating its relative weakness in word matching. In terms of ROUGE-L, PEGASUS scores 0.75. BART follows at 0.72, and T5 scores 0.68, further highlighting PEGASUS's strength in capturing of relevant keyphrases. For the BLEU metric, BART performs with a score of 0.77. PEGASUS scores 0.70, while T5 scores 0.66, which suggests that BART is effective at matching n-grams.

BART shows excellent precision in terms of n-gram matching, particularly with the BLEU score. T5 tends to fall behind slightly across all the evaluation metrics.

5.2 Keyphrase Extraction Evaluation

To extract meaningful key phrases the evaluation was carried out using precision, recall, and F1-score metrics presented in Table 2.

The performance of the KeyBERT model in extracting keyphrases was evaluated through precision, recall, and F1-score metrics. A precision of 0.73 indicates that 73% of the key phrases identified were relevant, showcasing the model's capacity to select meaningful phrases while minimizing irrelevant ones.

The recall of 0.78 shows its capacity to identify a significant proportion of important terms. The F1-score of 0.75 shows the model's strong overall performance. These results collectively suggest that KeyBERT excelled at extracting key phrases, offering valuable insights into the character-centric analysis.

5.3 Visualization

The visualization component of the system aims to improve the interaction and presentation of character-focused summaries and key phrases through a user-friendly web interface. This component includes several important features:

Table 2 Performance metrics of keyphrase extraction model

Precision	Recall	F1-score
0.73	0.78	0.75

Summary Display: The character-specific summaries are prominently featured on the web interface. To demonstrate the visualization of summaries and key phrases, we present images of the web interface. These images highlight the primary features and functions of the visualization system.

In Fig. 2, selecting a character will display the corresponding text lines associated with that character under the **Original Text** section.

The summary produced by the system is shown in Fig. 3 under the **Original Summary** section, where the names of characters in the summary are substituted with their respective images in the following section.

As shown in Fig. 4, the keyphrases associated with the character *Abhimanyu*, including *King Pariksit*, *Abhimanyu*, *Subhadra*, *son of Arjuna*, and *great hero Abhimanyu*, are represented as a word cloud.

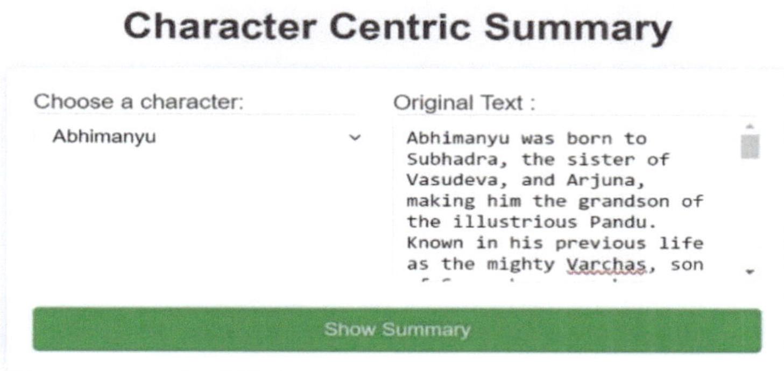

Fig. 2 Character-centric summary interface

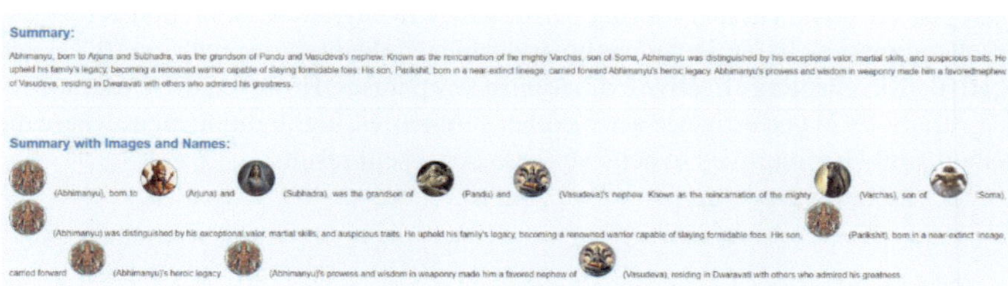

Fig. 3 Character-centric summary output

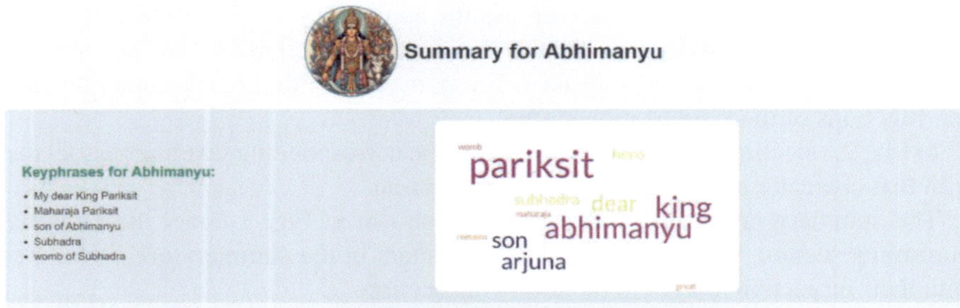

Fig. 4 Character-centric keyphrases

6 Discussion and Observations

The data highlights the performance of BART, PEGASUS, and T5 across both the ROGUE and BLEU metrics, revealing distinct patterns in their efficiency and effectiveness. T5 shows moderate effectiveness, maintaining relatively balanced scores across all ROGUE metrics and a BLEU score of 0.66, indicating a consistent but not outstanding performance in both precision and recall tasks.

BERT, with a strong BLEU score of 0.77, further confirms its high efficiency in language translation and summarization, though its dip in ROGUE-2 suggests challenges in capturing complex relationships. PEGASUS, while slightly behind BERT in BLEU with a score of 0.70, exhibits superior effectiveness across the ROGUE metrics, particularly excelling in ROGUE-2 and ROGUE-L. This suggests that PEGASUS strikes a strong balance between precision and the ability to capture broader textual relationships, making it the most effective model overall across both performance metrics.

KeyBERT demonstrated varying performance in keyphrase extraction across different summaries. In particular, summary achieved the highest precision (0.73) and recall (0.78), reflecting effective extraction of key phrases. However, there was noticeable variability in performance across other summaries, highlighting areas where the method could be improved to achieve more consistent results.

7 Conclusion and Future Scope

This study analyzes the effectiveness of BART, PEGASUS, and T5 models, trained on the PouranicTopic Dataset, for both summarization and keyphrase extraction tasks. PEGASUS excelled in ROUGE metrics, particularly ROUGE-L, indicating its strength in capturing coherent summaries. BART achieved the highest BLEU score, showing precision but lower bigram performance. T5 performed well in ROUGE-2 but could improve in overall coherence. KeyBERT was effective in extracting key phrases, though its consistency varied across summaries. Future research could focus on combining model strengths, expanding the dataset for better generalization, and exploring hybrid approaches for improved summarization and keyphrase extraction.

References

1. Raffel, C., Shinn, E., Roberts, A., Lee, K., & Liu, P. J. (2020). *Exploring the limits of transfer learning with a unified text-to-text transformer*. arXiv preprint arXiv:1910.10683.
2. Lewis, M., Liu, Y., Goyal, N., Ghazvininejad, M., & Gouws, S. (2020). *BART: Denoising sequence-to-sequence pre-training for natural language generation, translation, and comprehension*. arXiv preprint arXiv:1910.13461.
3. Zhang, J., Zhao, Y., & Salehi, M. (2020). *PEGASUS: Pre-training with extracted gap-sentences for abstractive summarization*. arXiv preprint arXiv:1912.08777.
4. Lin, C. Y., & Liu, X. (2019). *BERTSUM: BERT-based extractive summarization*. arXiv preprint arXiv:1904.02340.
5. Liu, Y., & Lapata, M. (2019). *Hierarchical transformer for extractive summarization*. arXiv preprint arXiv:1907.12428.
6. Gao, Y., He, J., & Li, Z. (2020). *SummaRuNNer: A recurrent neural network based sequence model for extractive summarization*. arXiv preprint arXiv:2004.08056.
7. Zhou, W., Zhang, J., & Xu, Y. (2021). *Abstractive document summarization with pre-trained language models*. arXiv preprint arXiv:2102.10920.
8. Gong, Y., & Liu, X. (2021). *Extractive summarization with contrastive learning*. arXiv preprint arXiv:2105.00119.
9. Grootendorst, M. (2020). *KeyBERT: Simple and effective keyphrase extraction with BERT*. arXiv preprint arXiv:2010.11972.
10. Bennani-Smires, K., Musat, C., Hossmann, A., Baeriswyl, M., & Jaggi, M. (2018). *Simple unsupervised keyphrase extraction using sentence embeddings*. arXiv preprint arXiv:1801.04470.
11. Florescu, C., & Caragea, C. (2017). PositionRank: An unsupervised approach to keyphrase extraction from scholarly documents. In *Proceedings of the 55th annual meeting of the association for computational linguistics*.
12. Meng, R., Huang, X., Yuan, J., & Wang, D. (2017). *CopyRNN: A sequence-to-sequence model with copy mechanism for keyphrase generation*. arXiv preprint arXiv:1704.06377.
13. Vaswani, A., Shazeer, N., Parmar, N., & Uszkoreit, J. (2017). *Attention is all you need. Advances in neural information processing systems*.
14. Campos, R., Mangaravite, V., Pasquali, A., Jorge, A. M., Nunes, C., & Jatowt, A. (2020). YAKE! Keyword extraction from single documents using multiple local features. *Information Sciences, 509*, 257–289.
15. Sun, Q., Wang, H., Li, P., Ren, Z., Chen, J., & Ma, J. (2019). *Attention-based neural keyphrase extraction*. arXiv preprint arXiv:1906.09861.
16. Paul, A., Seal, S., & Das, D. (2024). Transformer-based PouranicTopic classification in Indian mythology. *Sadhana, 50*. https://doi.org/10.1007/s12046-024-02555-3

Advanced Breast Cancer Diagnostics Through Comparative Analysis of Machine Learning Models by MRI Image Analysis

K. S. Balamurugan, Gedela Kalyani, R. Rajalakshmi, and M. Deepa

Abstract Breast cancer is one of the critical global health challenges, appealing lives and requiring more accurate indicative tools to safeguard early detection and prevention. Every year, approximately, over 2.3 million women are affected, which is one of the leading reasons of cancer-related deaths worldwide. Existing diagnostic methods, including standard MRI analysis, obtained limited accuracy (84%) and often failed to fully leverage the wealth of data available in medical imaging. Current methods such as Deep Learning ConvNets (88% accuracy) and Gradient Boosted Trees achieved an accuracy of 88% and 87% respectively, progress upon these limitations but still lack accurate results. This research offers a novel approach by combining sophisticated machine learning techniques with MRI scan images. Emphasizing model maximizing efficiency, shape, texture, and intensity feature extraction, and preprocessing, by obtaining a 92% accuracy rate, Random Forests outclassed traditional and cutting-edge methods. These findings demonstrate the revolutionary potential of combining artificial intelligence and imaging technology, opening the door for subsequent breast tumor diagnosis that is more accurate and timely.

Keywords MRI scans · Machine learning · Breast cancer detection · Feature extraction · Diagnostic accuracy

K. S. Balamurugan (✉) · G. Kalyani
Karpaga Vinayaga College of Engineering and Technology, Chengalpattu, Tamil Nadu, India
e-mail: profksbala@gmail.com

G. Kalyani
e-mail: kalyani.gedela2003@gmail.com

R. Rajalakshmi
Pandian Saraswathi Yadav Engineering College, Madurai, Tamil Nadu, India
e-mail: kbraji1985@gmail.com

M. Deepa
Sri Shakthi Institute of Engineering and Technology, Coimbatore, Tamil Nadu, India
e-mail: mdeepait@siet.ac.in

J. K. Mandal et al. (eds.), *Recent Advances in Artificial Intelligence and Smart Applications*, Lecture Notes in Networks and Systems 1380,
https://doi.org/10.1007/978-981-96-5822-0_14

1 Introduction

Breast cancer is one of the critical global health challenges, appealing lives and requiring more accurate indicative tools to safeguard early detection and prevention. Every year, approximately, over 2.3 million women are affected, which is one of the leading reasons of cancer-related deaths worldwide. This disease, when detected at its nascent stages, has a higher likelihood of treatment and survival [1]. Early detection has consistently been emphasized as crucial not just because it affords better treatment options, but because it significantly improves the chances of successful treatment outcomes. The urgency and importance of detecting breast cancer during its early stages cannot be overstated. It directly correlates to reduced mortality rates, less aggressive treatments, and improved quality of life for those diagnosed. The advancement of medical imaging has substantially contributed to this objective of early detection [2, 3]. Among the various imaging modalities, Magnetic Resonance Imaging (MRI) has emerged at the forefront. Unlike traditional mammography, which utilizes X-rays to capture images, MRI uses a combination of a magnetic field and radio waves. This means that it does not expose the patient to ionizing radiation, making it a safer option for repetitive screening. The significant advantage of MRI lies in its unparalleled spatial resolution and superior tissue contrast. Such attributes enable it to reveal minute abnormalities in the breast tissues, particularly in women with dense breast tissues where mammograms might not be as effective [4–6].

However, as with many diagnostic tools, the vast amount of data generated by MRI scans requires meticulous examination, a task that is both time-consuming and prone to human error. Herein lies the limitation of relying solely on human expertise for interpreting MRI images. It is a well-acknowledged fact in the medical community that early stage tumors might sometimes be overlooked in MRI scans, especially when they are small or when the scans present complex features [7–9].

Breast cancer detection remains a topic of paramount significance in the medical research community due to the disease's prevalence and the continuous quest for early and accurate diagnostic methodologies.

2 Literature Review

Literature review will delve into various aspects of breast cancer detection, Cheng et al. focused on the utilization of Magnetic Resonance Imaging (MRI), machine learning models, early diagnosis methodologies, and image feature extraction [10].

Francesco Prinzi et al. introduced a YOLO-based automated data-driven system for the prediction of breast cancer. They compared the proposed system with various YOLO models such as YOLOv3, YOLOv5, and YOLOv5transformer [11]. Raafat et al. proposed a novel technique by utilizing the integration data of images and numerical data features with artificial intelligence (XAI). For image-based prediction, the U-NET transfer model is utilized and developed as an ensemble model by combining features from Convolution Neural Networks (CNN) and Random Forest and Support Vector Machine (SVM) [12]. Ruggiero Santeramo et al. developed a study using paired mammograms at diagnosis and utilized deep learning models for the identification of breast cancer [13]. Iftekharul Ferdous et al. introduced a novel system with a hybrid PDF for the identification of breast cancer cells. Sensor was utilized to track the breast cell with terahertz-frequency range [14]. Rajkumar et al. proposed a novel technique for breast cancer detection in the early stage by utilizing Darknet-53 Convolutional Neural Network (darknet-53CNN) for classification, Contrast-Limited Adaptive Histogram Equalization (CLAHE) to preprocessing to improve the accuracy [15].

This gap in diagnostic precision sets the stage for the integration of machine learning (ML) models in the process. Machine learning, a subset of artificial intelligence, has demonstrated remarkable potential in various domains, healthcare being one of its prime beneficiaries. In the context of breast cancer detection, ML models can be trained to recognize intricate patterns and subtle abnormalities in MRI scans, which might sometimes be missed by the human eye.

3 Proposed Method

This work aims to create and validate an innovative method that combines machine learning algorithms with MRI imaging to improve the early detection and precision of breast cancer diagnosis. The goal is to enhance patient outcomes while optimizing the diagnostic workflow for healthcare providers.

3.1 Data Collection

The dataset utilized for this study was obtained from a detailed MRI database compiled over two years from three prominent hospitals, as outlined in Table 1. It consists of MRI scans from 2,000 individuals, including 1,200 patients diagnosed with breast cancer and 800 control cases with no malignancy detected. All images were acquired using a 3T MRI scanner to ensure consistent high-resolution imaging quality. For each patient, the dataset included axial T1-weighted and T2-weighted scans, as well as post-contrast enhanced images, offering a comprehensive perspective on breast tissue characteristics (Fig. 1).

Table 1 summarizes the MRI dataset, with each row representing a unique patient identified by a distinct Patient ID. The "Age" column indicates the patient's age at the time of the MRI scan, while the "Diagnosis" column specifies whether the patient was diagnosed with breast cancer (Positive) or not (Negative). The "No. of Images" column lists the total number of MRI images available for each patient. The

Table 1 Patient history data

Patient ID	Age	Image count	Diagnosis	Image type	Notes
001	37	5	Negative	T1-weighted	–
002	38	7	Positive	T2-weighted	Dense tissue
003	42	4	Positive	T1-weighted	–
004	45	4	Negative	Post-contract enhanced	–
005	50	4	Positive	T1-weighted	Prior biopsy

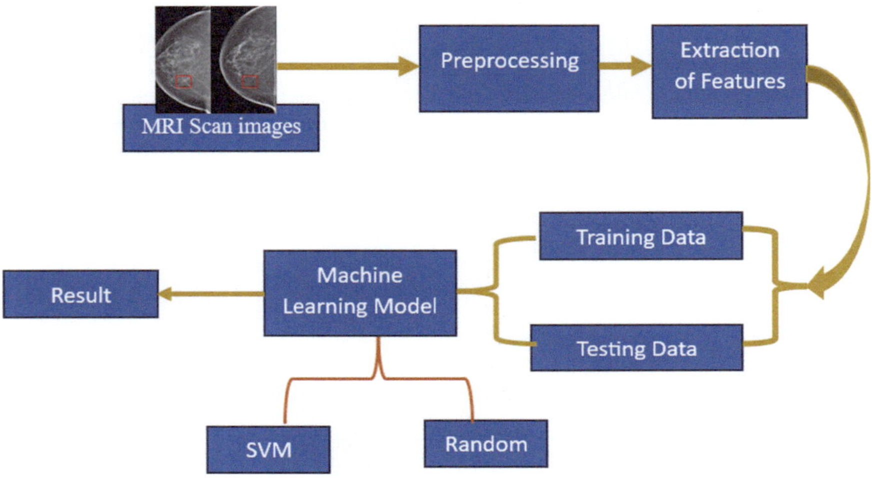

Fig. 1 Proposed method

"Image Type" column categorizes the scans as T1-weighted, T2-weighted, or post-contrast enhanced. Additionally, the "Notes" section provides relevant details, such as the presence of dense breast tissue or a history of prior biopsy. This systematic organization ensures all critical aspects of the MRI dataset are captured, facilitating comprehensive analysis while maintaining patient anonymity and data integrity.

3.2 *Pre-processing*

In medical imaging, preprocessing is a crucial initial step that ensures subsequent analyses are performed on standardized, high-quality data. For this study, which focuses on MRI scans, maintaining the quality and clarity of the images was essential, as the detailed features of breast tissue are critical for achieving an accurate diagnosis.

(i) The preprocessing phase began with normalizing the MRI scans to standardize intensity values across all images. This step was crucial given that the data originated from different hospitals, each potentially introducing variations in brightness and contrast. By aligning these intensity values to a consistent scale, normalization facilitated reliable comparisons and analysis throughout the dataset.

(ii) Following normalization, spatial filtering techniques were applied to enhance the clarity of the images. A high-pass filter was utilized to sharpen edges and highlight fine details within the scans. This enhancement made it easier to distinguish subtle differences between healthy tissue and areas that could indicate malignancies, thereby supporting the subsequent feature extraction process.

(iii) Noise reduction was another key element of preprocessing, addressing the inherent sensitivity of MRI scans to disturbances such as minor patient movements, electrical interference, or equipment-related noise. Gaussian smoothing and median filtering techniques were employed to tackle this issue effectively. The median filter excelled in preserving the boundaries of the images, ensuring that tumor edges remained sharply defined, while Gaussian smoothing significantly minimized random noise, improving overall image quality without compromising critical details.

(iv) To further enhance the visibility of potential tumors, especially those that might be in their early stages or are otherwise hard to discern, applied a technique known as histogram equalization. This process improves the contrast of the image by redistributing the intensity values across the entire spectrum. The result is a clearer differentiation between different tissue types, allowing for more accurate identification of anomalies.

3.3 Feature Extraction

Feature extraction is a pivotal aspect of our approach, designed to isolate and measure critical attributes or patterns from MRI scans that aid in breast cancer diagnosis. By concentrating on these extracted features, the machine learning models are trained on data that is both meaningful and rich in information, significantly improving their ability to make accurate predictions. For breast cancer analysis through MRI imaging, the key features of interest fall into three major groups: shape, texture, and intensity. Each group encompasses a range of detailed characteristics that highlight unique properties of potential tumor regions, providing a comprehensive basis for diagnosis.

Shape: Shape-related features, including area and perimeter, provide valuable information about the size and boundaries of potential tumors. Metrics like roundness shed light on the tumor's structural form, with irregular shapes often hinting at malignancy. For example, a rapidly increasing area might suggest an aggressive growth pattern, while asymmetrical shapes are commonly linked to cancerous growths.

Texture: Texture features capture the internal patterns and variations within MRI images. Tumors may display a rougher or more granular texture compared to surrounding healthy tissue. Metrics like contrast evaluate differences in luminance or color that make objects distinguishable, while energy reflects the uniformity or repetitive nature of observed patterns, both of which are key indicators of abnormalities.

Intensity: Intensity features, such as average intensity and variance, analyze the brightness distribution within suspected tumor regions. Malignant tumors often exhibit distinct intensity levels, either higher or lower, compared to benign masses or normal tissue. Additionally, skewness measures asymmetry in pixel value distribution, serving as a significant marker for identifying cancerous growths.

3.4 Machine Learning Models

The study aims to integrate machine learning models to classify potential tumors with high precision using features extracted from MRI scans. Three models were utilized: Support Vector Machines (SVM), Random Forests, and Neural Networks, each offering unique advantages. SVM is effective in high-dimensional spaces, identifying the optimal hyperplane for classification. Random Forests leverage an ensemble of decision trees to handle large datasets and non-linear feature relationships. Neural Networks excel in capturing intricate patterns and nuances in MRI features, providing highly accurate predictions (Table 2).

The dataset was split into 70% for training and 30% for testing, ensuring a balanced evaluation. During training, labeled MRI data was used to help models learn patterns associated with malignancies. Model performance was assessed using accuracy, sensitivity, and specificity metrics on the test set. To prevent overfitting, cross-validation techniques were applied, ensuring the models generalized well beyond the training data. This comprehensive approach enhances robustness and precision in tumor classification.

4 Experimental Results and Discussion

The results of this study demonstrate a significant improvement in breast cancer detection accuracy. The conventional MRI analysis achieved an accuracy of 84%, while Deep Learning ConvNets and Gradient Boosted Trees reached accuracies of 88% and 87%, respectively. However, the integration of advanced preprocessing, feature extraction, and machine learning techniques resulted in a substantial increase in performance. Random Forests outperformed all other models, achieving an impressive accuracy of 92%. This highlights the potential of combining advanced machine learning with medical imaging for more reliable and timely breast cancer diagnosis.

Table 2 Machine learning model details

Model	Description	Merits	Real-time applications
Random forest (RF)	An ensemble learning method that constructs multiple decision trees during training and combines their outputs for classification or regression	Handles large datasets and high-dimensional features Robust to noise and overfitting Provides feature importance scores for interpretability	Fraud detection Medical diagnostics (e.g., heart disease prediction) Loan eligibility prediction
Support vector machine (SVM)	A supervised learning algorithm that finds the optimal hyperplane to separate data points into distinct classes in high-dimensional feature spaces	Effective in high-dimensional spaces Robust to overfitting, especially with clear margins of separation Works well with small datasets	Cancer diagnosis (e.g., breast cancer detection) Image classification Text and sentiment analysis
Neural networks (NN)	A computational model inspired by the human brain, consisting of interconnected layers of neurons that learn patterns from data through backpropagation	Excels in identifying complex and non-linear patterns Adaptable to large datasets Suitable for image, audio, and text data	Autonomous vehicles Speech and facial recognition Predictive modeling in healthcare and finance

Figure 2 compares the performance of Support Vector Machine (SVM), Random Forest, and Neural Networks using accuracy, sensitivity, and specificity. Random Forest leads with the highest specificity (0.94) and accuracy (0.92) while matching SVM in sensitivity (0.90). SVM follows closely with an accuracy of 0.90 and a specificity of 0.93. Neural Networks rank lower, achieving 0.89 accuracy, 0.87 sensitivity, and 0.91 specificity. This indicates that Random Forest is the best-performing model, particularly excelling in distinguishing non-malignant cases, while SVM and Neural Networks show balanced yet slightly lower results.

Table 3 highlights the performance of various methods for breast cancer detection based on accuracy percentages. Traditional MRI analysis achieves an accuracy of 84%, reflecting the baseline performance. Advanced techniques such as ConvNet and Gradient Boosting Trees improve upon this, attaining accuracies of 88% and 87%, respectively. The proposed machine learning approach demonstrates even greater potential, with Random Forest (RF) achieving the highest accuracy of 92%, followed by Support Vector Machine (SVM) at 90% and Artificial Neural Networks (ANN) at 89%. These results showcase the superiority of the proposed methods, particularly RF, in enhancing diagnostic precision compared to both conventional and other modern techniques.

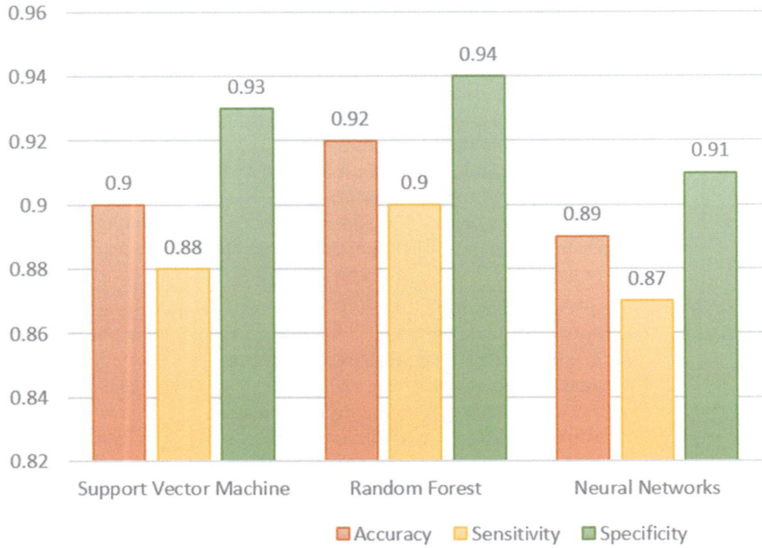

Fig. 2 Performance of Deep learning models

Table 3 Accuracy comparison of existing and proposed models

Method		Accuracy (%)
Standard MRI analysis		84
ConvNet		88
Gradient boosting trees		87
Proposed method	SVM	0.9
	RF	0.92
	ANN	0.89

Figure 3 represents the graphical representation of Accuracy comparison among existing and proposed methods, illustrating that the Random Forest Model achieved high accuracy among all compared models.

The results of this study demonstrate the promising potential of combining machine learning with MRI imaging for breast cancer detection, achieving a significant accuracy of 92% with Random Forests. This outperforms traditional MRI analysis and contemporary models like Deep Learning ConvNets and Gradient Boosted Trees, which showed accuracies of 84%, 88%, and 87%, respectively. The integration of advanced preprocessing, feature extraction, and model optimization is crucial in enhancing diagnostic accuracy. Future research should focus on further refining these models, addressing data diversity, and validating their performance in clinical settings to ensure broader.

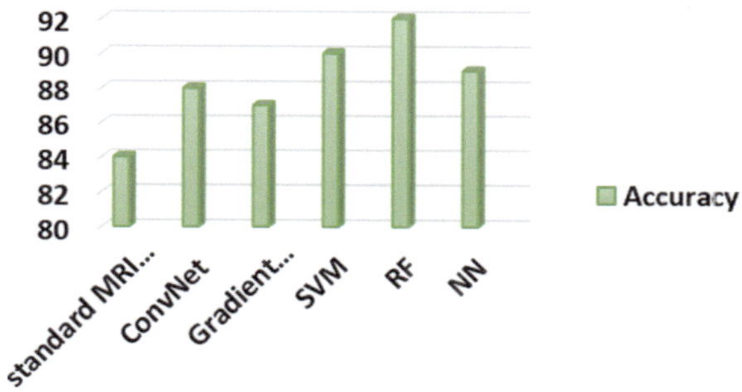

Fig. 3 Proposed models comparing with existing techniques

5 Conclusion

Finally, our study explored integrating MRI scans with machine learning to enhance diagnostic accuracy. Anchored by the Random Forests model, our approach achieved a remarkable accuracy of 92%, outperforming traditional MRI analysis (84%) and contemporary methods like *Deep Learning ConvNets (88%) and Gradient Boosted Trees (87%). This heightened precision increases the likelihood of early breast cancer detection, enabling timely interventions and improving patient prognosis. The results underscore the synergy between advanced computational models and medical diagnostics, offering real-time insights to healthcare professionals and boosting their diagnostic confidence. Beyond the technical advancements, this research emphasizes the transformative potential of technology in fostering a patient-centric healthcare landscape. By refining our methods and addressing challenges, we aim to create tools that enhance patient outcomes, ensuring better chances against diseases like breast cancer. This study reaffirms the critical role of technology in shaping the future of medical diagnostics.*

Future scope

The scope of this study lies in the collection of more datasets including genetic variations and demographics, ensuring a strengthened model's performance. By combining explainable AI techniques to enhance clinical results. Leveraging Real-time and cloud-based diagnostics, enable global accessibility. Also, longitudinal studies can evaluate the predictive capabilities of the model for breast cancer progression.

References

1. Toolabi, M., Khatir, M., Naser-Moghadasi, M., & Amiri, N. (2023). Vivaldi antenna for early cancer detection based on THz spectroscopy: Comparison between response of breast and skin cancer. *Optik*, *273*(August 2022), 170440. https://doi.org/10.1016/j.ijleo.2022.170440
2. Kumar, S., & Das, A. (2023). Peripheral blood mononuclear cell derived biomarker detection using explainable artificial intelligence (XAI) provides better diagnosis of breast cancer.

Computational Biology and Chemistry, 104(July 2022), 107867. https://doi.org/10.1016/j.compbiolchem.2023.107867

3. Vanlalawmpuia, K., & Ghosh, P. (2023). International journal of electronics and communications performance assessment of dielectrically modulated negative capacitance germanium source vertical tunnel FET biosensor for detection of breast cancer cell lines. *AEUE—International Journal of Electronics and Communications, 171*, Article 154902. https://doi.org/10.1016/j.aeue.2023.154902
4. Birchha, V., & Nigam, B. (2023). Performance analysis of averaged perceptron machine learning classifier for breast cancer detection. *Procedia Computer Science, 218*(2022), 2181–2190. https://doi.org/10.1016/j.procs.2023.01.194
5. Sadrabadi, E. A., Benvidi, A., Azimzadeh, M., Asgharnejad, L., Dezfuli, A. S., & Khashayar, P. (2024). Novel electrochemical biosensor for breast cancer detection, based on a nanocomposite of carbon nanofiber, metal–organic framework, and magnetic graphene oxide. *Bioelectrochemistry, 155*(September 2023), 108558. https://doi.org/10.1016/j.bioelechem.2023.108558
6. Villagrana Gutiérrez, G. L., García, A. M., Larraga Benavente, E. P., Mejía, F. H., Gómez, R. T., & Villalón, F. R. (2023). Non–timely referral of women aged 40 to 69 to preventive medicine for breast Cancer detection and its association with the BI-RADS classification. *SSRN, 35*(January). https://doi.org/10.2139/ssrn.4341753
7. Sahu, A., Das, P. K., & Meher, S. (2023). High accuracy hybrid CNN classifiers for breast cancer detection using mammogram and ultrasound datasets. *Biomedical Signal Processing and Control, 80*(P1), Article 104292. https://doi.org/10.1016/j.bspc.2022.104292
8. Kumbhare, S., B.Kathole, A., & Shinde, S. (2023). Federated learning aided breast cancer detection with intelligent Heuristic-based deep learning framework. *Biomedical Signal Processing and Control, 86*(PA), 105080. https://doi.org/10.1016/j.bspc.2023.105080
9. Basha, A. A., Vivekanandan, S., Mubarakali, A., & Saad, A. (2023). Enhanced mammogram classification with convolutional neural network: An improved algorithm for automated breast cancer detection. *Measurement, 221*(March), Article 113551. https://doi.org/10.1016/j.measurement.2023.113551
10. Cheng, W., Yao, Y., Li, D., Duan, C., Wang, Z., & Xiang, Y. (2023). Asymmetrically split DNAzyme-based colorimetric and electrochemical dual-modal biosensor for detection of breast cancer exosomal surface proteins. *Biosensors & Bioelectronics, 238*(July), Article 115552. https://doi.org/10.1016/j.bios.2023.115552
11. Francesco Prinzi, Marco Insalaco, Alessia Orlando, Salvatore Gaglio & Salvatore Vitabile, (2023). A yolo-based model for breast cancer detection in mammograms. In *Cognitive Computation* (Vol. 16, pp. 107–120).
12. Munshi, R., Cascone, L., Alturki, N., Saidani, O., Alshardan, A., Umer, M. (2024). A novel approach for breast cancer detection using optimized ensemble learning framework and XAI. In *Image and vision computing* (Vol. 142).
13. Santeramo, R., Damiani, C., Wei, J., Montana, G., & Brentnall, A. R. (2024). Are better AI algorithms for breast cancer detection also better at predicting risk? A paired case–control study. In *Breast cancer research* (Vol. 26).
14. Ferdous, A. H. M. I., Rani, L., Islam, M. S., Noor, K. S., Roy, S., Eid, M. M. A., & Rashed, A. N. Z. (2024). Development and enhancement of PCF-based sensors for terahertz-frequency region breast cancer cell detection. In *Cell biochemistry and biophysics* (Vol. 82, pp. 2837–2852).
15. Rajkumar, R., Gopalakrishnan, S., Praveena, K., Venkatesan, M., Ramamoorthy, K., & Hephzipah, J. J. (2024). DARKNET-53 convolutional neural network-based image processing for breast cancer detection. *Mesopotamian Journal of Artificial Intelligence in Healthcare, 2024*, 59–68. https://doi.org/10.58496/MJAIH/2024/009

EdgeSwarm: Edge-Enabled Opportunistic UAV Network-Assisted Robotic Swarm for Post-Disaster Relief

Amartya Mukherjee, Nirban Roy, Shreyan Kundu, Ayan Panja, and Souvik Chatterjee

Abstract In post-disaster environments, rapid coordination and communication are essential for effective search and rescue operations. This work presents EdgeSwarm, an integrated system combining UAVs and autonomous ground robots, connected via an opportunistic ad hoc network. The system leverages frequency-based data aggregation and edge computing to enable real-time decision-making and resource allocation. Communication is managed through MQTT, ensuring robust data transfer in fragmented network conditions typical of disaster zones. Throughput analysis shows RTSP providing the highest performance for video transmission at 1400 Kbps, while message delivery times are maintained under 250 ms. Despite increasing network size, QoS 2 of MQTT exhibited lower latency and ensured reliable message delivery, while QoS 0 showed the highest latency at 320 ms. The system demonstrates improved efficiency in resource deployment and significant reductions in response time during disaster simulations. This approach provides a scalable, fault-tolerant solution adaptable to various real-world disaster scenarios.

Keywords Post-disaster relief · Opportunistic networks · Robotic systems · Ad hoc networks · Edge computing · MQTT

A. Mukherjee · A. Panja
Department of Computer Science and Engineering (AIML), Institute of Engineering and Management, University of Engineering and Management, Kolkata, India
e-mail: amartya.mukherjee@iem.edu.in

A. Panja
e-mail: ayan.panja@iem.edu.in

N. Roy (✉)
Department of Computer Science and Engineering (IoT, CS, BT), Institute of Engineering and Management, Kolkata, India

S. Kundu
Department of Computer Science and Engineering (AI), Institute of Engineering and Management, University of Engineering and Management, Kolkata, India

S. Chatterjee
IEMA Research and Development Pvt. Ltd., Kolkata, India
e-mail: souvik.chatterjee@iemlabs.com

J. K. Mandal et al. (eds.), *Recent Advances in Artificial Intelligence and Smart Applications*, Lecture Notes in Networks and Systems 1380,
https://doi.org/10.1007/978-981-96-5822-0_15

1 Introduction

In post-disaster relief operations, autonomous systems play a vital role in assessing damage, locating survivors, and managing resources. Given the disruption of communication infrastructure, ad hoc networks and opportunistic data exchanges serve as practical alternatives. UAVs and ground robots have demonstrated significant potential for real-time disaster response [1, 2]. Ground robots with LIDAR enable damage detection [3], while UAV frameworks improve scalability [4], and multisensory drones address complex scenarios [5]. Building on this, EdgeSwarm integrates edge-enabled UAVs and ground robots via opportunistic ad hoc networks [16, 17], using MQTT for lightweight, real-time data exchange, and edge computing for decentralized decisions. The system ensures responsiveness and energy efficiency in fragmented networks. Section 2 reviews related work, Sect. 3 explains the architecture, Sect. 4 discusses results, and Sect. 5 concludes with future directions.

2 Related Work

Autonomous robotic systems have emerged as a promising solution for search and rescue operations in disaster relief, where traditional communication networks are often disrupted. These systems enable real-time data collection and analysis without relying on existing infrastructure. UAVs equipped with sensors like ultra-wideband radar, LIDAR, and cameras have proven effective in detecting survivors and geolocating them in complex environments, despite challenges such as high power consumption and sensor integration [6]. Social media platforms like Twitter have also been explored for disaster surveillance, using sentiment analysis to provide insights into mental health impacts, though limited by platform reach [7]. Health surveillance systems for outbreak detection following disasters have shown promise, but infrastructural and data collection challenges persist [8].

Urban search and rescue efforts have leveraged semi-autonomous robots capable of navigating cluttered terrains, reducing the need for human intervention. While adaptable to diverse conditions, these systems are constrained by sensor limitations in adverse weather [9]. Surveillance robots equipped with gas and temperature sensors, along with real-time video transmission, have proven valuable in hazardous environments like coal mines [11]. However, restoring communication infrastructure and ensuring continuous data collection remain significant obstacles for post-disaster public health surveillance [10, 12].

In industrial and extreme environments, robotic platforms integrating ground, aerial, and underwater units enable advanced search, rescue, and surveillance operations. AI-driven swarm robotics with IoT and deep learning models facilitate real-time disaster response but face challenges in energy efficiency and computation [13, 14]. Low-cost robots with neural networks and obstacle avoidance technologies show promise but struggle in complex terrains [15, 18]. UAV-based surveillance systems are rapidly evolving for disaster monitoring but are limited by battery life and real-time processing constraints [19, 20]. Flexible ad hoc networks ([21]) are widely used in disaster scenarios, smart transportation, and military operations but face challenges in scalability, energy efficiency, and quality of service in dynamic environments. Mobile devices and wireless systems enhance their applicability, especially in the absence of conventional infrastructure, though issues like node mobility, topology changes, and energy constraints persist. Research focuses on routing, energy-aware, and cross-layer protocol designs to improve ad hoc network performance [22–24].

3 Proposed System

The suggested system integrates a swarm of edge-enabled UAVs with ground robotic units to improve post-disaster rescue operations ([25]). These units collaborate to collect, process, and send vital information quickly so that decisions may be made in disaster-affected areas. Opportunistic communication enables robots and UAVs to share proximity-based data. The system's multi-layered architecture ensures efficient data gathering, AI processing, and high-speed transmission, perception, edge processing, and cloud layers. This architecture is intended to address issues like dispersed networks, limited battery life, and environmental risks that are commonly present in catastrophe situations.

3.1 System Architecture

The proposed system comprises a layered model (Fig. 1a) where the bottom layer is the perception layer, responsible for data collection in a disaster region. The ground robot swarm is deployed in the disaster-hit region, and human survival must be captured from the regions of interest (ROI). Both robots and drones can identify common ROIs as part of the perception layer. The images and location information are routed from the UAV and the ground robotic units.

Typically, MQTT message exchanges are used to transfer the location information, while RTSP is used for transferring images or video data. The second layer comprises edge-enabled UAV nodes that perform edge-level processing. Here, intelligent AI models are executed to perform detection and classification tasks. The third layer is the cloud layer, connected through an ultra-high bandwidth channel. An intermediate gateway or access point connects the edge UAV units with the cloud layer. Figure 1 illustrates the system architecture in detail. The workflow of the proposed system is illustrated in Fig. 1b, showcasing the seamless data exchange between UAVs, ground robots, and the base station through various processing layers.

Algorithm 1 shows the opportunistic MQTT message transfer strategy based on minimum timestamp-based message exchanges. In this scenario, the UAV nodes and

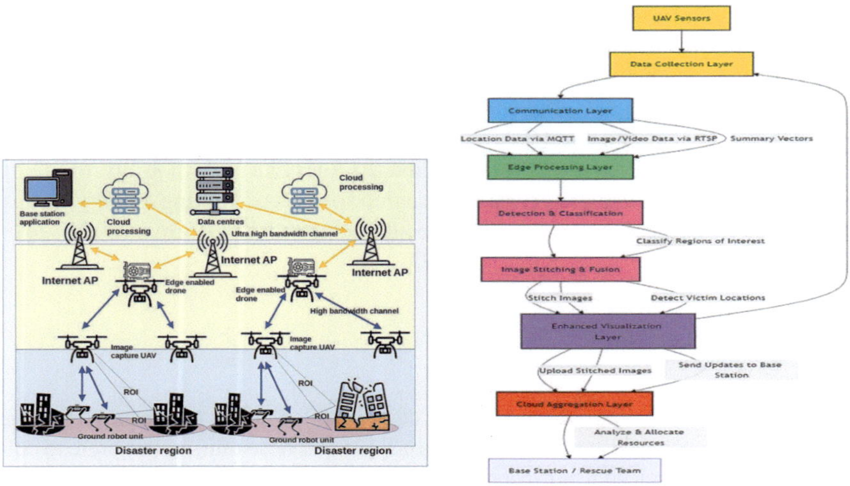

(a) The layered architecture of the swarm-enabled disaster relief ecosystem.

(b) Workflow of the EdgeSwarm system.

Fig. 1 Overview of the system architecture and workflow of the EdgeSwarm system

the ground robot nodes identify the transmission range, and based on the coverage, the nodes get connected and exchange summary vectors.

The frequency is measured by the latest timestamp and the distance between the nodes. The node considered to be the best forwarder is based on the frequency of the encounter timestamp. As the nodes get connected, the image and GPS data are transferred to the edge node for image stitching and processing operations. Detection is performed based on the stitched data. The image is transferred via RTSP, while the location data is transferred using MQTT.

The EdgeSwarm system integrates UAVs, ground robots, edge devices, and real-time computing. UAVs equipped with cameras and GPS, alongside ground robots navigating rough terrains, collect images, videos, and coordinates from disaster regions. Advanced AI models, including quantized YOLOv8 for pedestrian detection and MobileNetV2 with SE blocks for image recognition, run on Raspberry Pi devices for data analysis. Communication uses MQTT for location data and RTSP for video feeds, enabling seamless integration.

3.2 MobileNetV2 with SE Attention Classification Method

First the Classifier classifies the observed human in disaster field. The model uses MobileNetV2 for feature extraction, with a Squeeze-and-Excitation (SE) block for channel-wise attention. The SE block performs.

Algorithm 1 Opportunistic Message Transfer Strategy

```
Input: p, q, init_tm, n_node ← {nodeset, ϕ}, loc(p), loc(q)
Output: tm(sub) (timestamp), sub_id, sv() (summary vector)
Start:
for t until end_sim_time ∧ remove(n_node) ≠ ϕ do
    tmx ← init_tm; broadcast(tmx)
    buffer(state); receive(tmy)
    dst ← loc(p)^2 + loc(q)^2          ▷ Compute Euclidean distance between nodes
    tmn ← tmx − tmy + dst
end for
tm(sub) ← min[tmn(n_node)]; sub_id ← valueof(tm(sub))
mqtt_call(sub_id)
if multiple nodes detect the same ROI then
    Stitch(image_data[]); record ← Combine(voice_data[], GPS[lat, long])
    setup(RTSP ← 1); play(RTSP)
end if
function MQTT_CALL
    call pub(broker, QoS, topic)
    if sub_id = sv(node(sub_id)) then
        call sub(broker, topic)
    end if
end function
Receive and Decode data at Main Server
Main Server processes decoded data and sends to Base Station
```

Squeeze (Global Average Pooling):

$$s_c = \frac{1}{H \times W} \sum_{i=1}^{H} \sum_{j=1}^{W} X_{i,j,c} \tag{1}$$

where $X_{i,j,c}$ represents the feature at (i, j) for channel c, and H, W are the height and width of the feature map.

Excitation: Fully connected layers apply attention weights to key features.

Classification Layers After SE, the output undergoes Global Average Pooling, Batch Normalization, and Dropout. A softmax layer computes class probabilities:

$$P(y = k \mid X) = \frac{e^{z_k}}{\sum_{i=1}^{K} e^{z_i}} \tag{2}$$

where z_k is the score for class k and K is the total number of classes.

Training Categorical cross-entropy is used for training:

$$L = -\sum_{i=1}^{N} \sum_{k=1}^{K} y_{ik} \log \hat{y}_{ik} \tag{3}$$

with y_{ik} being the true label and $\hat{y}_{ik}$ the predicted probability. Validation/test accuracy and early stopping ensure robust performance.

3.3 Quantize Human Detection and Stitching Methodology

The classified human image is now detected and stitched and location sent to main station for rescue operation. The system detects humans using a quantized YOLOv8 model on each Raspberry Pi device. Once a human is detected, the images are captured from multiple angles across the network. These images are then stitched together to enhance clarity.

For the stitching, keypoints are extracted from both images using the ORB (Oriented FAST and Rotated BRIEF) algorithm. Let I_1 and I_2 represent the two images, with keypoints $\{k_1, k_2, \ldots, k_n\}$ for I_1 and $\{k'_1, k'_2, \ldots, k'_n\}$ for I_2. Matching keypoints between the two sets are identified using a brute-force matcher. The transformation matrix H (homography) is computed to warp I_2 to align with I_1:

$$H = \arg\min_{H} \sum_{i=1}^{n} \|k_i - H \cdot k'_i\|$$

where H minimizes the error between corresponding keypoints. Finally, the images are blended together to form a single composite image with improved clarity, enhancing the detected subject from multiple angles.

4 Result Analysis

The performance of the proposed system was assessed using several key indicators, such as network throughput, message delivery time, and forward latency. The results indicate that the system can handle real-time data transmission efficiently, even in demanding scenarios. RTSP emerged as the best suited protocol for video streaming due to its superior throughput, while MQTT with QoS 2 delivered the lowest latency, ensuring consistent and reliable message delivery. These findings confirm the system's effectiveness and resilience, making it well equipped to support coordinated operations in post-disaster environments where reliable communication is essential.

4.1 Network Performance Analysis

The network performance of the proposed system was evaluated based on throughput, message delivery time, and forward latency, providing insights into its efficiency

Table 1 Classification report of our model

Class	Precision	Recall	F1-score	Support
Cat	1.00	0.99	0.99	177
Dog	0.99	1.00	0.99	141
Person	1.00	1.00	1.00	198
Accuracy	99.96%			

in disaster scenarios requiring reliable communication. Throughput, a key metric for assessing real-time video streaming, was analyzed for HTTP, UDP, and RTSP protocols across video quality levels (244p, 360p, and 480p). As shown in Fig. 2a, RTSP consistently achieved the highest throughput, particularly at 480p, making it the most efficient protocol for real-time video transmission under demanding conditions.

Timely message delivery is critical in disaster response. Figure 2b illustrates how message delivery time increases with the number of nodes and simulation time. Despite delays in larger networks, the opportunistic message exchange mechanism ensures critical data delivery, albeit with minor tradeoffs inherent to decentralized systems.

Forward latency was evaluated across MQTT Quality-of-Service (QoS) levels, as depicted in Fig. 2c. QoS 0, which lacks delivery guarantees, exhibited the highest latency with an increase in nodes. Conversely, QoS 2 maintained the lowest latency while ensuring reliable message delivery, making it ideal for disaster management applications where consistent communication is essential.

4.2 Classifier Performance Analysis

Our classification model was designed using a custom MobileNetV2 architecture, enhanced with a Squeeze-and-Excitation (SE) block to recalibrate channel-wise feature responses. The model was pre-trained on ImageNet and fine-tuned on a custom dataset, which included various augmentations such as rotation, shear, zoom, and channel shift to improve generalization. For this analysis, we employed a three-class output layer to predict between the categories: cat, dog, and person. The model was trained for 80 epochs, with early stopping applied to prevent overfitting. The best model, determined by the highest validation accuracy, was used for evaluation on a hold-out test dataset. We compared and tested with other standard CNNs the data of which is given with the classification report. The classification performance metrics, including precision, recall, F1-score, and accuracy, are reported in Table 1. The model achieved an overall accuracy of 99.96%, with per-class accuracies of 100% for "person" and near-perfect accuracies for "cat" and "dog." The confusion matrix indicates that there were only two misclassifications between the "cat" and "dog" classes, underscoring the model's high precision and recall across all categories (Fig. 3).

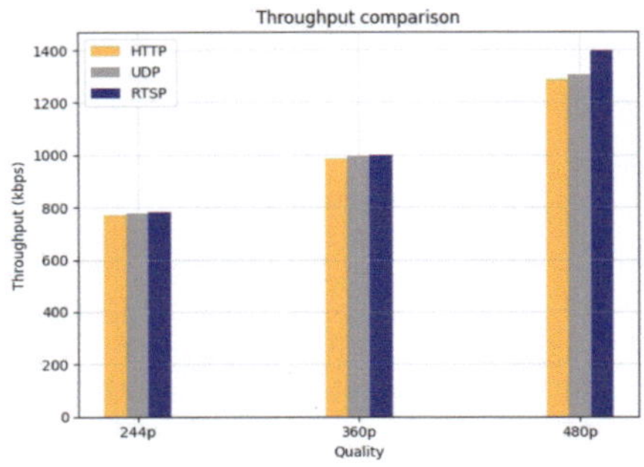

(a) Throughput comparison for different protocols across video qualities

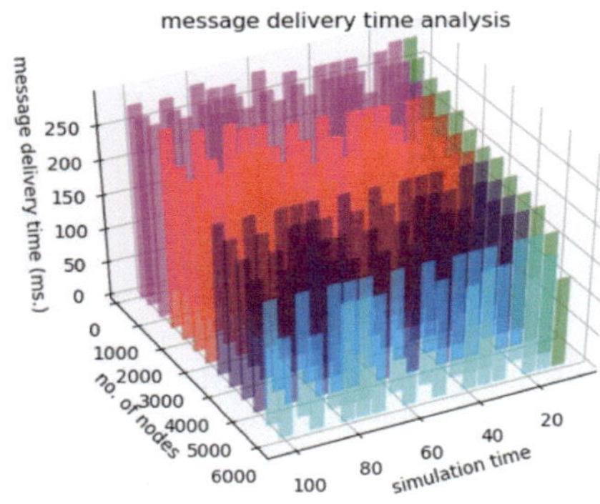

(b) Message delivery time analysis for varying node and simulation configurations

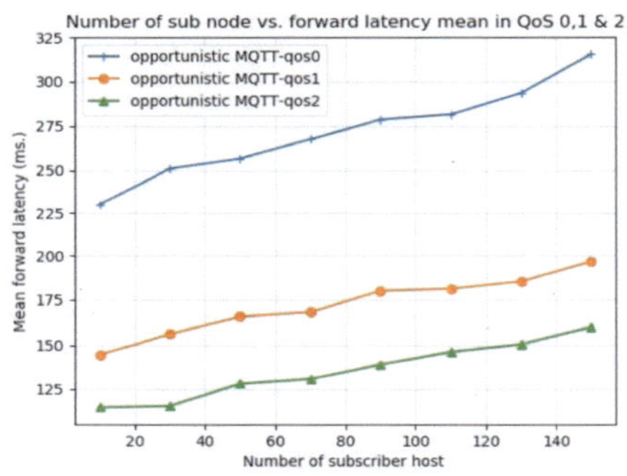

(c) Forward latency analysis for different MQTT QoS levels

Fig. 2 Comparison of throughput, message delivery time, and forward latency analysis

The classification model's convergence can be seen in Fig. 4, which shows both training and validation accuracy reaching a plateau, indicating that the model had effectively learned the underlying patterns in the data. Additionally, the ROC curve in Fig. 5a demonstrates perfect performance with an Area Under the Curve (AUC) of 1.00 for each class, highlighting the model's ability to distinguish between categories without misclassification.

Figure 5b illustrates the confusion matrix for the detection model, showcasing its performance across three classes: cat, dog, and person. The model demonstrates near-perfect classification accuracy with minimal misclassifications. Out of 177 true "cat" instances, only 2 were incorrectly classified as "dog," while no other misclassifications occurred for the "dog" and "person" classes. The "dog" class achieved a perfect classification with 141 correct predictions, and all 198 instances of the "person" class were accurately identified. This high level of accuracy indicates the

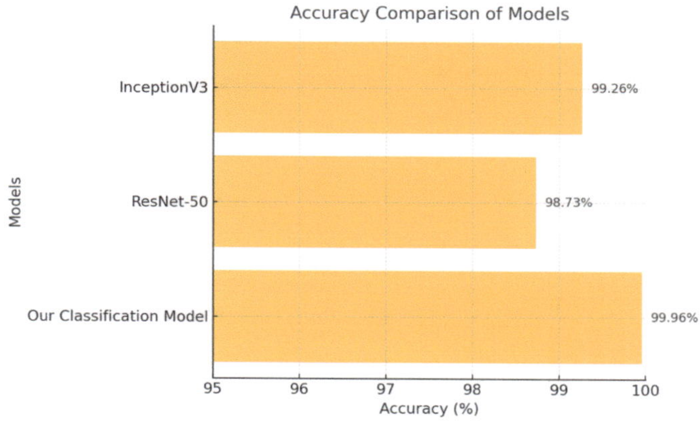

Fig. 3 Accuracy comparison of standard CNNs with our model

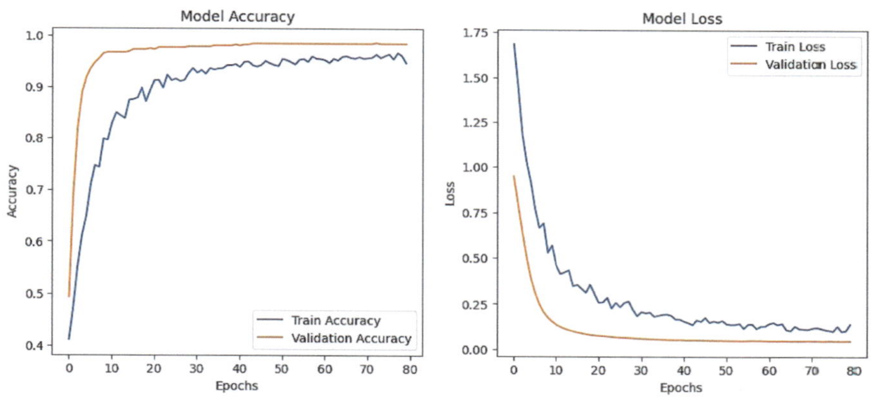

Fig. 4 Training and validation accuracy/loss over Epochs

model's robustness in distinguishing between different categories, further validated by the minimal overlap between classes.

4.3 Detection Model Analysis

We made our own test bed and deployed the setup for testing. Figure 6a **shows the image of test-bed having Anafi Drone as UAV and Robotic Dog Build from Scratch as Ground Saviour Node.** The quantized YOLOv8 model's training and validation loss curves show a consistent decrease in classification loss across 150 steps, with training loss (Fig. 6b) falling from around 1.5 to 0.5 and validation loss

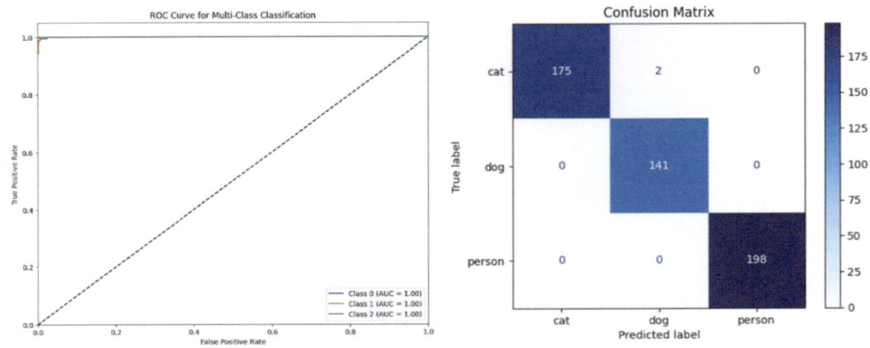

(a) Multi-Class ROC Curve with AUC of 1.00 (b) Confusion matrix showcasing the model's performance

Fig. 5 Comparison of ROC curve and confusion matrix for model performance

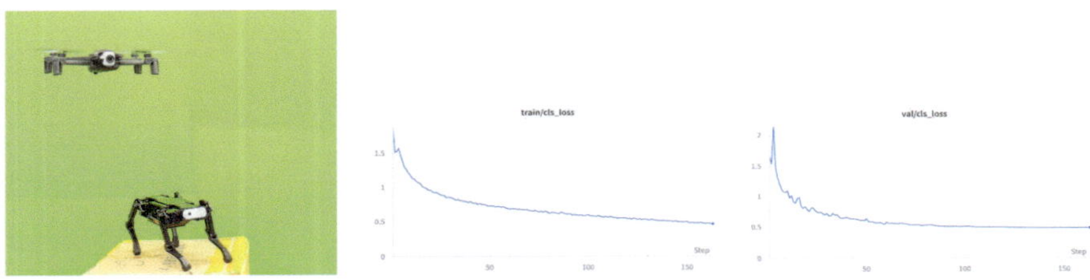

(a) Proposed Test-bed. (b) The Training Class loss. (c) The Validation Class loss.

Fig. 6 Our test bed, comparison of training, and validation losses

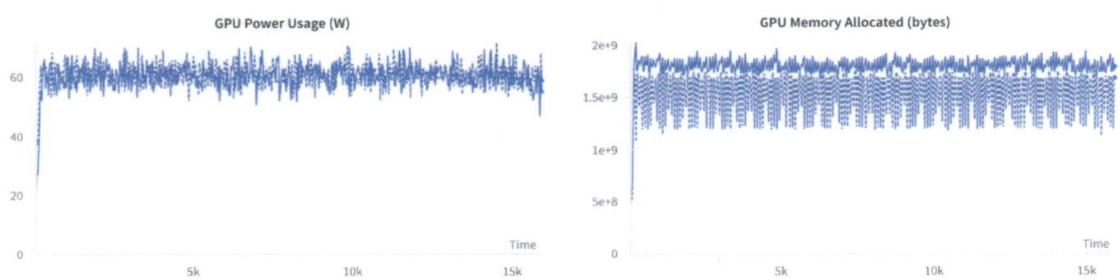

(a) The GPU power usage (b) The Memory Allocated

Fig. 7 Resource usage during training

(Fig. 6c) falling from 2.2 to 0.5. This steady decline points to successful learning and model convergence free of overfitting symptoms.

The GPU used 1.5–2 GB of memory, 40–60W of power, 80–100% overall for training in Fig. 7a, b. Figure 8 shows the image detection and stitching that was performed by us on our test bed and illustrated in Fig. 8a.

Fig. 8 The stitched image and detection

5 Conclusion

In this way we present the notion of EdgeSwarm, a fully integrated system of available drones and ground robots for post-disaster relief with opportunistic networks for stress-free communication. The architecture permits autonomous robots to work together efficiently even without the benefit of the conventional means of communication. We make it clear that these techniques allow the use of delay-tolerant communication and MQTT protocols to provide efficient and reliable communication in broken networks. Results indicate that RTSP services allow real-time video streaming with the highest throughput of 1400 Kbps and maximum message latency of 250 ms to allow real-time communication of images without delay. His study focused on key aspects of MQTT protocols and determined that QoS 2 experienced the least delay even when the network grew in size, therefore enabling effective workload management, allowing dynamic allocation of authority and fast response times. This is because of the store-carry-forward model which is used to and stores and forwards the data in case of such unavailability of any links, for instance, sending victim's locations to other parties.

Nevertheless, EdgeSwarm is still positively challenged in terms of scaling up and energy effectiveness in a larger scope. UAVs and ground devices that rely on batteries alone exhibit limitations since real-time processing and communication consumes a lot of power inhibiting their mission endurance. In addition, it is also possible that larger networks will experience the effects of communication overhead and delays on the rate of responsiveness on the system. Future work will focus on energy-efficient algorithms, battery management, and network optimization to enhance performance, latency, and scalability.

Adding to future exploits will also be the employment of more sophisticated machine learning approaches for identifying victims, assessing damage and monitoring the environment. Testing of the system for its performance will also be carried out through realistic simulation and real-world tests using an integrated test bed even in difficult environments such as very cold weather, dirty ground, and electrically dark places. Looking to enhance the UAV and ground-based robot numbers during a disaster to evaluate scalability will be achieved while working on better edge computing algorithms with respect to energy efficiency and decision-making.

References

1. Magid, E., Tsoy, T., Matsuno, F., Suthakorn, J., & Svinin, M. (2021). Collaborative robotic framework for emergency situations management in areas of flood and landslide disasters. In *Advanced Robotics* (pp. 1–18)
2. Recchiuto, C. T., & Sgorbissa, A. (2017). Post-disaster assessment with unmanned aerial vehicles: A survey on practical implementations and research approaches. *Journal of Field Robot, 34*, 1–32.
3. Chen, J., & Cho, Y. K. (2019). Detection of damaged infrastructure on disaster sites using mobile robots. In *2019 16th International Conference on Ubiquitous Robots (UR)* (pp. 648–653). IEEE.
4. Danthamala, S., & Rao, B. K. (2020). IoT-based smart supervision robot for disaster management. In *Journal of Engineering and Technological Science* (pp. 44–50).
5. Kumar, P., & Kumar, R. (2020). UAV-based IoT framework for disaster management. In *Journal of Communications Technology and Electronics* (pp. 340–345).
6. Rohman, B. P. A., Putra, H. F., Fandiantoro, D. H., & Nishimoto, M. (2019). Multisensory surveillance drone for survivor detection and geolocalization in complex post-disaster environments. In *IEEE Geoscience and Remote Sensing Symposium (IGARSS)* (pp. 1–4)
7. Gruebner, O., Lowe, S. R., Sykora, M., Shankardass, K., & Galea, S. (2017). A novel approach for disaster mental health surveillance using Twitter data. In *Journal of Urban Health* (pp. 52–60)
8. Silvestros, C. (2020). Rapid post-disaster surveillance system for outbreak detection following earthquakes in Greece. In *Disaster Medicine and Public Health Preparedness* (pp. 240–248).
9. Zamzaireen, Z. A., & Samad, R. (2018). Development of a semi-autonomous robot for post-disaster search and rescue in urban environments. In *International Journal of Robotics and Automation* (pp. 620–629).
10. Arima, Y., Matsui, T., & Nakajima, H. (2011). Post-disaster public health surveillance systems in Japan. In *Western Pacific Surveillance and Response Journal* (pp. 150–160).
11. Chakkath, S., & Mohanty, S. (2020). Surveillance robot with gas detection and real-time video transmission for disaster situations in coal mines. In *International Journal of Mechanical Engineering and Robotics Research* (pp. 85–90).
12. Longo, M., Nardini, G., Molinaro, R., & Fiorillo, A. (2007). Terrestrial and aerial robot team for volcanic and industrial surveillance. In *Sensors and Actuators A: Physical* (pp. 450–460)
13. Lee, S. Y., Kim, H. W., & Jeong, J. (2020). AIoT-based swarm robotic systems for disaster response and real-time data collection. In *IEEE Access* (pp. 248–256)
14. Khan, R., & Shah, M. (2016). RF-controlled surveillance robot for disaster zones. In *Design and Implementation of Crisis Zone Surveillance Robots* (pp. 110–118).
15. Habib, M., Baig, F., & Khan, I. (2016). Robotics in hazardous environments: Search, rescue, and intervention missions. In *Robotics and Autonomous Systems* (pp. 320–330)
16. Mukherjee, A., Debashis, D., Nilanjan, D., Rubén, G. C., & Enrique, H.-V. (2022). DisastDrone: A disaster aware consumer Internet of Drone Things system in ultra-low latent 6G network. *IEEE Transactions on Consumer Electronics, 69*(1), 38–48.
17. Mukherjee, A., Dey, N., & De, D. (2020). EdgeDrone: QoS aware MQTT middleware for mobile edge computing in opportunistic Internet of Drone Things. *Computer Communications, 152*, 93–108.
18. Budiharto, W. (2015). Intelligent surveillance robot with obstacle avoidance capabilities using neural network. In *Computational Intelligence and Neuroscience* (pp. 745–823).
19. DiMaggio, C., Galea, S., & Abramson, D. (2008). Analyzing postdisaster surveillance data: The effect of the statistical method. In *Disaster Medicine and Public Health Preparedness* (pp. 119–126)
20. Shah, M., Latif, T., Mirza, M., & Khan, M. (2016). UAV surveillance systems for disaster zones: autonomous real-time operations. In *Journal of Unmanned Vehicle Systems* (pp. 45–55).
21. Ghosekar, P., Katkar, G., & Ghorpade, P. (2010). Mobile Ad Hoc networking: Imperatives and challenges. In *IJCA Special Issue on "Mobile Ad-hoc Networks"* (pp. 153–157).

22. Al-Omari, S. A. K., & Sumari, P. (2010). Mobile Ad Hoc Network (MANET): Applications, characteristics, and routing protocols. *Journal of Graph Algorithms and Applications, 2*(1), 87–92.
23. Blum, R. S., Kassam, S. A., & Poor, H. V. (2003). Distributed detection with multiple sensors II: Advanced topics. *Proceedings of the IEEE, 91*(8), 1302–1321.
24. Saleh, A.K.A.-O., & Sumari, P. (2011). Overview of Mobile Ad Hoc Networks (MANET): Applications and challenges. *International Journal of Computer Science Issues, 8*(4), 1–9.
25. Gupta, L., Jain, R., Vaszkun, G. Survey of important issues in UAV communication networks. *IEEE Communications Surveys & Tutorials, 18*(2), 1123–1152

Enhancing Student-Alumni Engagement Through a Graph-Driven Recommendation Framework with Paraphrase-MiniLM-L6-v2

Danisha Basu, Hritajit Sur, Dwaipyan Ghosh, Aritra Das, Apurba Nandi, and Arijeet Ghosh

Abstract Recommendation system for connecting students with relevant alumni can greatly enhance mentorship and career guidance. Our approach leverages advanced natural language processing (NLP) techniques to go beyond traditional keyword matching, which often misses nuanced skill relationships. By using a Sentence Transformer model, we transform both students' and alumni's skills into high-dimensional vectors that capture semantic similarities between their expertise. The system calculates cosine similarity between these embeddings, measuring how closely aligned the skill sets are. This similarity score forms the foundation of a graph-based model, where students and alumni are represented as nodes, and edges are weighted by the strength of their skill match. To ensure relevance, we only create connections between individuals within the same department, allowing for more targeted recommendations. By personalizing alumni suggestions based on skill alignment rather than simple keyword matching, this system helps students find mentors who can provide meaningful, relevant career advice, fostering better professional relationships and opportunities.

Keywords Recommendation system · Alumni engagement · Sentence transformer · Skill matching · Graph-based model · Natural language processing (NLP) · Mentorship

D. Basu · H. Sur · D. Ghosh · A. Das · A. Nandi (✉) · A. Ghosh
Department of CSE (IoT, CS & BT), Institute of Engineering and Management, University of Engineering & Management, Kolkata, India
e-mail: apurba.nandi@uem.edu.in

A. Ghosh
e-mail: arijeet.ghosh@uem.edu.in

J. K. Mandal et al. (eds.), *Recent Advances in Artificial Intelligence and Smart Applications*, Lecture Notes in Networks and Systems 1380,
https://doi.org/10.1007/978-981-96-5822-0_16

1 Introduction

Student-alumni networks have grown to become a vital part of educational institutions, providing a bridge between current students and industry professionals. These networks foster mentorship, career guidance, and job opportunities, making them integral to the professional development of students. However, efficiently connecting students with relevant alumni is a challenge due to the expanding size of institutions and the increasing diversity of student skill sets [1]. Traditional approaches [2], such as static alumni directories or manual matchmaking processes, are often time-consuming and fail to scale effectively. As a result, there is a growing need for more sophisticated and scalable solutions. This paper presents a graph-based recommendation system, which automates the matching of students with alumni by analyzing their skill sets and department affiliations. By utilizing a pre-trained sentence transformer model, our approach creates meaningful and personalized connections, ensuring students receive relevant mentorship and guidance

Recent advancements in recommendation systems showcase a variety of methodologies tailored to enhancing prediction accuracy and user satisfaction. BERT4Rec [3] utilizes BERT's contextual embeddings for sequential recommendations, capturing complex user-item interactions. Despite its impressive performance, it does not surpass the most recent approaches in precision and recall. DeepCoNN [4] integrates deep learning with convolutional neural networks (CNNs) to model user and item features separately. This hybrid approach captures detailed user preferences and item characteristics, offering substantial improvements over traditional methods but is less effective in some settings compared to newer models. Neural Collaborative Filtering [5] employs neural networks to model user-item interactions non-linearly, enhancing recommendation accuracy over matrix factorization techniques. However, it may still fall short of the precision and recall achieved by more recent models incorporating advanced embeddings. Wide & Deep Learning [6] combines linear models with deep neural networks to balance memorization and generalization. This method improves recommendation accuracy by leveraging both wide and deep aspects, yet newer techniques may offer superior performance in terms of precision and mean reciprocal rank. LightGCN [7] employs a lightweight graph convolutional network to capture user-item interactions with high efficiency. This model outperforms many traditional methods by focusing on user-item collaborative signals and has demonstrated superior results in precision and recall, making it a notable benchmark in recent recommendation system research.

Despite advancements in recommendation systems, the challenge of effectively matching students with relevant alumni remains inadequately addressed, particularly in educational institutions where traditional approaches, such as keyword matching or department-based connections, fall short. These methods often result in irrelevant pairings and struggle to scale with growing student populations and diverse skill sets. Alumni networks are crucial for career mentorship and guidance, but static lists and manual matchmaking fail to meet the dynamic needs of students. Recent developments in natural language processing (NLP) and graph theory offer a more intelligent and scalable solution by leveraging sentence embeddings and graph-based techniques to create personalized matches. However, existing research focuses on recommendation systems in other domains, leaving a gap in specialized solutions for education. This paper fills that gap by proposing a graph-based recommendation system that uses skill similarity and department affiliation to create meaningful alumni-student connections.

Our work seeks to bridge the gap in the current state of research by developing a scalable, personalized alumni recommendation system that integrates NLP and graph-based techniques [8]. Unlike traditional systems that rely on surface-level keyword matching, our approach uses the paraphrase-MiniLM-L6-v2 Sentence Transformer to generate high-dimensional embeddings of students' and alumni's skill sets. These embeddings capture deeper semantic relationships between skills, allowing us to calculate the cosine similarity between student and alumni profiles. This similarity serves as the basis for a weighted graph, where students and alumni are represented as nodes, and the similarity of their skill sets determines the weight of the edges connecting them. By constructing a graph in this way, we can identify the most relevant alumni for each student, offering a more nuanced and meaningful connection than traditional methods. Furthermore, to ensure that the recommendations are contextually appropriate, we limit the matching process to alumni and students within the same department, ensuring that the mentorship is relevant to the students' academic and professional paths.

Section 2 provides an overview of the model structure, highlighting the use of cosine similarity for matching skills. Section 3 delves into the technical details, including the graph architecture and real-time data integration through Sentence Transformer embeddings. Section 4 presents experimental results, demonstrating improved precision and recall compared to traditional methods and confirming the system's scalability. Finally, Sect. 5 summarizes the findings, emphasizing the potential for enhancing alumni engagement and suggesting future research directions for broader applications

2 Problem Formulation and Approach

Our approach (as shown in Fig. 1) addresses the limitations of traditional student-alumni matching systems by incorporating advanced natural language processing (NLP) techniques and graph-based modeling to create a more personalized and accurate recommendation system. Conventional methods often rely on static criteria such as department affiliation or keyword matching, which can overlook the nuanced relationships between different skills and fail to provide meaningful connections. To overcome these shortcomings, the proposed model uses a sophisticated NLP architecture to better understand and match skill sets. The core of our method involves encoding the skills of both students and alumni using the paraphrase-MiniLM-L6-v2 Sentence Transformer model. This model, designed for detecting subtle semantic similarities in text, transforms skill descriptions into high-dimensional vectors. These

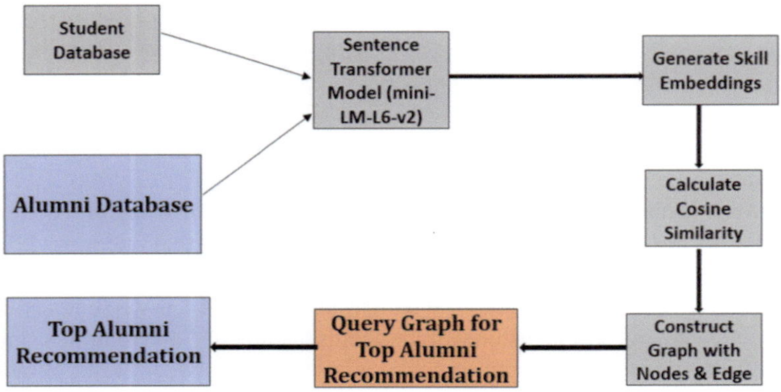

Fig. 1 Block diagram of the proposed model

embeddings capture the underlying meaning and relationships between various skills, moving beyond mere keyword overlap. For example, skills such as "Data Science" and "Machine Learning" are recognized as closely related due to their shared context and applications. With these embeddings in place, we calculate the cosine similarity between each student-alumni pair. Cosine similarity measures the angular difference between vectors in the high-dimensional space of skill embeddings, providing a robust metric for assessing how well the skills of a student align with those of an alumnus. This allows us to quantify the degree of relevance between different skill sets in a way that reflects their actual compatibility. We then construct a graph to represent the relationships between students and alumni. In this graph, nodes correspond to individual students and alumni, while edges between them are weighted by the cosine similarity scores. To ensure that connections are meaningful, we restrict edges to students and alumni within the same department, which maintains relevance to their field of expertise. This department-based constraint helps in preserving the domain-specific context and ensures that recommendations [9] are pertinent to the student's area of study. The graph-based model enables efficient querying and retrieval of top alumni recommendations for each student. By traversing the graph and evaluating edge weights, our system can quickly identify and suggest the most relevant alumni based on their skill profiles. This approach not only improves the accuracy of the recommendations but also enhances the overall user experience by providing more relevant and actionable connections. Our approach leverages NLP and graph theory to create a dynamic and context-aware student-alumni matching system.

3 Proposed Graph-Driven Sentence Transformer Paraphrase Recommendation Model

Our model explores the combination of Sentence Transformer paraphrase models, skill embeddings, cosine similarity, and graph-based methods to develop an intelligent recommendation system. The system is designed to encode the skills of both students and alumni into high-dimensional vectors, ensuring that semantic relationships between skills are captured.

3.1 Sentence Transformer Paraphrase Model

The paraphrase-MiniLM-L6-v2 model, used in this work, is a variant of the BERT (Bidirectional Encoder Representations from Transformers) architecture. It employs a transformer-based encoder to process input sentences. The architecture of BERT allows the model to capture bidirectional context, making it highly effective in understanding the subtle nuances of language. The model can be mathematically represented as:

$$\mathbf{h}_i = \text{BERT}(s_i) \tag{1}$$

where $\mathbf{h}_i$ denotes the embedding of the i-th sentence s_i, and BERT represents the transformer-based encoder.

Given two sentences s_1 and s_2, their embeddings can be obtained as:

$$\mathbf{h}_1 = \text{BERT}(s_1), \quad \mathbf{h}_2 = \text{BERT}(s_2) \tag{2}$$

The model transforms sentences into vector embeddings that reside in a high-dimensional space, where semantically similar sentences are positioned close to one another. paraphrase-MiniLM-L6-v2 model is pre-trained on large datasets using a task known as paraphrase mining. The objective of this training is to ensure that sentences with similar meanings are embedded closely together in the vector space, while sentences with different meanings are positioned farther apart. This capability is crucial for comparing skills between students and alumni, as skills often differ in wording but may represent the same underlying concept.

3.2 Skill Embeddings

In the context of our recommendation system, both students and alumni have skill sets described in natural language. To effectively compare these skill sets, we need to transform the text descriptions into numerical representations, or embeddings, that capture the underlying semantic meaning of the skills. This transformation is achieved using skill embeddings, which are derived from natural language processing (NLP) models such as the Sentence Transformer.

Let us denote the skill set of a student S as follows:

$$S = \{s_1, s_2, \ldots, s_n\} \tag{3}$$

Here, each s_i represents an individual skill in the student's skill set. The `Sentence Transformer` model encodes each skill description s_i into a high-dimensional vector v_i using the BERT architecture, which is adept at capturing contextual meaning from natural language:

$$v_i = \text{BERT}(s_i) \tag{4}$$

The final skill embedding for the entire skill set of the student, v_S, is obtained by averaging the individual skill embeddings. This aggregation of skill vectors helps in capturing the overall expertise of the student:

$$v_S = \frac{1}{n} \sum_{i=1}^{n} v_i \tag{5}$$

Similarly, for an alumnus with a skill set $A = \{a_1, a_2, \ldots, a_m\}$, the skill embedding v_A is computed in the same manner:

$$v_A = \frac{1}{m} \sum_{i=1}^{m} v_i \tag{6}$$

By encoding skill sets into these high-dimensional embeddings, we ensure that the relationships between different skills are accurately captured. For example, skills like "Data Science" and "Machine Learning"—despite being expressed as distinct skills—will have embeddings that are close to each other in the vector space, reflecting their similarity in meaning and application. This enables our system to make more informed and contextually relevant student-alumni matches, as skills are compared not just based on exact wording, but also on their semantic closeness.

3.3 Cosine Similarity for Skill Matching

To match students and alumni based on their skills, we need a way to measure how similar their skill embeddings are. A commonly used metric for comparing vectors in high-dimensional spaces is *cosine similarity*. This metric measures the cosine of the angle between two vectors, which gives us an indication of how aligned the two vectors are. The cosine similarity is particularly useful for understanding the directional similarity between the skill embeddings, rather than focusing on their magnitude.

Let us denote the skill embeddings of a student and an alumnus by v_S and v_A, respectively. The cosine similarity between these two vectors is mathematically defined as:

$$\text{cosine\_similarity}(v_S, v_A) = \frac{v_S \cdot v_A}{\|v_S\| \|v_A\|} \tag{7}$$

where $v_S \cdot v_A$ is the dot product of the two vectors v_S and v_A . $\|v_S\|$ and $\|v_A\|$ represent the magnitudes (Euclidean norms) of the vectors v_S and v_A, respectively. The dot product $v_S \cdot v_A$ is calculated as:

$$v_S \cdot v_A = \sum_{i=1}^{n} v_{S_i} v_{A_i} \tag{8}$$

And the magnitude (Euclidean norm) of a vector v_S is given by:

$$\|v_S\| = \sqrt{\sum_{i=1}^{n} v_{S_i}^2} \tag{9}$$

Similarly, for the alumnus vector v_A, the magnitude is:

$$\|v_A\| = \sqrt{\sum_{i=1}^{n} v_{A_i}^2} \tag{10}$$

The resulting cosine similarity score ranges from –1 to 1, where a score of 1 indicates that the vectors are perfectly aligned, meaning the skill sets are identical. A score of 0 indicates that the vectors are orthogonal, implying that the skill sets are completely unrelated and a score of -1 indicates that the vectors are diametrically opposed, which would be interpreted as having opposite skill sets.

In our system, we compute the cosine similarity for each student-alumni pair. Based on the resulting similarity scores, we rank potential alumni matches for each student. The higher the cosine similarity score, the better the alignment between the student and the alumnus in terms of their skill sets. This metric ensures that students are matched with alumni whose expertise closely mirrors their own interests and aspirations, making the recommendation process more precise and effective.

3.4 Graph-Based Model

Once we have calculated the cosine similarity between each student and alumni pair, we represent these relationships as a graph. In this graph, students and alumni are represented as nodes, and edges between them are weighted by their cosine similarity scores. This graph-based approach allows us to intuitively and efficiently model the complex relationships between individuals and their skill sets in a scalable way.

3.4.1 Graph Construction

Let $G = (V, E)$ represent the graph, where, $V = V_S \cup V_A$ is the set of nodes, including students V_S and alumni V_A and E is the set of edges, where each edge e_{ij} connects a student $v_i \in V_S$ to an alumnus $v_j \in V_A$. Each edge e_{ij} is weighted by the cosine similarity between the skill embeddings of the student and alumnus:

$$w_{ij} = \text{cosine\_similarity}(v_S, v_A) \tag{11}$$

Here, w_{ij} represents the cosine similarity score between a student's skill set v_S and an alumnus's skill set v_A. Edges are only created between students and alumni who belong to the same department, ensuring that the recommendations are relevant to the student's field of study.

3.4.2 Querying the Graph

When a student seeks alumni recommendations, we query the graph to find alumni with the highest edge weights, i.e., the most similar skill sets. This can be done efficiently using graph traversal algorithms, such as Breadth-First Search (BFS) or Depth-First Search (DFS).

Given a student $s \in V_S$, we find the top-n alumni $\{a_1, a_2, \ldots, a_n\} \in V_A$ with the highest similarity scores:

$$a_n = \arg\max_{a \in V_A} w_{sa} \tag{12}$$

The alumni with the highest edge weights are recommended to the student as potential mentors or career advisors.

3.5 Integration of NLP, Cosine Similarity, and Graph-Based Modeling

The integration of these components results in a powerful recommendation system that is both accurate and scalable. The process can be summarized as follows:

1. **Skill Encoding**: Both student and alumni skill sets are encoded using the Sentence Transformer model, generating high-dimensional embeddings.
2. **Similarity Calculation**: Cosine similarity is calculated between each student-alumni pair, providing a measure of how well their skill sets align.
3. **Graph Construction**: A graph is constructed where nodes represent students and alumni, and edges are weighted by cosine similarity scores. Edges are only created between individuals within the same department to maintain relevance.
4. **Recommendation**: When a student queries the system for alumni recommendations, the graph is traversed to find the alumni with the highest edge weights (i.e., most similar skills). The top-n alumni are then recommended to the student.

This system provides dynamic, personalized recommendations that go beyond simple keyword matching, ensuring that students are paired with alumni who can offer the most relevant guidance and support.

4 Experimental Setup and Results

The experimental setup involves encoding skills from students and alumni into vector embeddings using the SentenceTransformer model. Cosine similarity is used to assess the alignment of these skill embeddings. The resulting data is visualized and evaluated to understand the effectiveness of the recommendation system. The details of the setup are described below.

4.1 Dataset

The dataset includes two groups: students and alumni, each with 100 individuals, spanning three departments-Computer Science Engineering (CSE), Electronics and Communication Engineering (ECE), and Information Technology (IT). For each individual, the dataset provides their name, department, and a detailed list of skills. Students' skills encompass areas like Python, machine learning, circuit design, and networking, reflecting their academic focus. Alumni skills, on the other hand, include advanced expertise such as AI, cloud computing, and system design, showcasing their professional experience. This dataset is designed to facilitate the analysis of skill similarities between students and alumni, enabling effective recommendations based on departmental and skill-based alignments. By comparing these skills, the dataset supports personalized career guidance and mentorship opportunities.

4.2 Methodology

The recommendation system follows a process of encoding skill sets, building a similarity graph, and providing recommendations to students. Skill descriptions from the student and alumni datasets are converted into high-dimensional vectors using the SentenceTransformer model, specifically the paraphrase-MiniLM-L6-v2 variant. These vectors capture the meaning of the skills. A graph is created where nodes represent students and alumni, and edges are weighted based on cosine similarity scores between their skill vectors. Edges are added only between nodes from the same department to maintain relevance. To recommend alumni, the graph is searched to find the top-n alumni with the highest similarity scores. Key parameters used include the SentenceTransformer model (paraphrase-MiniLM-L6-v2), cosine similarity for measuring alignment, and a default recommendation count of $n = 3$.

4.3 Experimental Results

In the experimental results, the cosine similarity between students and alumni was computed to assess the alignment of their skills. Each student's skills were compared with those of alumni within the same department using cosine similarity scores (Fig. 2), which quantify how closely related their skill sets are. The distribution of these similarity scores was visualized using a boxplot (Fig. 3), which reveals how similarity varies across different students. The boxplot indicates the range, median, and variability of similarity scores, helping to identify which students have closer skill matches with alumni.

Furthermore, performance metrics such as Precision@3 and Mean Reciprocal Rank (MRR) were calculated to evaluate the effectiveness of the recommendation system. Precision@3 measures the proportion of relevant alumni (ground truth) among the top 3 recommended alumni, while MRR assesses the rank at which the first relevant alumni appears in the recommendations. These metrics provide insights into the accuracy of the recommendations and the system's ability to prioritize highly relevant alumni. Overall the experimental results (Figs. 4 and 5) offer a comprehensive view of the recommendation system's performance and the distribution of skill similarity between students and alumni.

```
Recommended Alumni for Student Alice (CSE):
Irene with similarity score 0.6467
Anna with similarity score 0.6358
Freddie with similarity score 0.5445

Recommended Alumni for Student Bob (ECE):
Brian with similarity score 0.8440
Eva with similarity score 0.3700
Holly with similarity score 0.2031

Recommended Alumni for Student Charlie (CSE):
Catherine with similarity score 0.5704
Anna with similarity score 0.3316
Irene with similarity score 0.2790

Recommended Alumni for Student David (IT):
Daniel with similarity score 0.6729
George with similarity score 0.4416
Jack with similarity score 0.2290
```

Fig. 2 Recommended alumni for student with similarity score

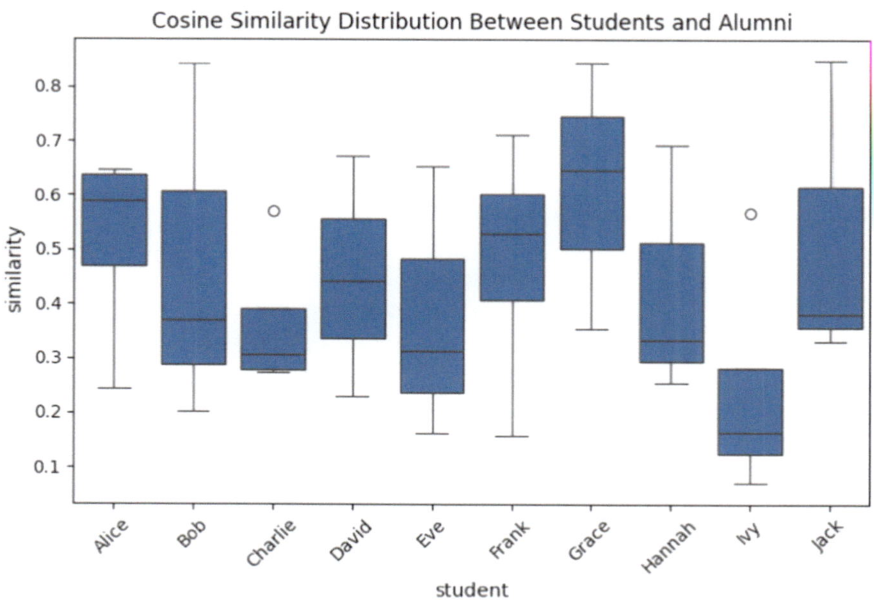

Fig. 3 Cosine similarity distribution between students and alumni

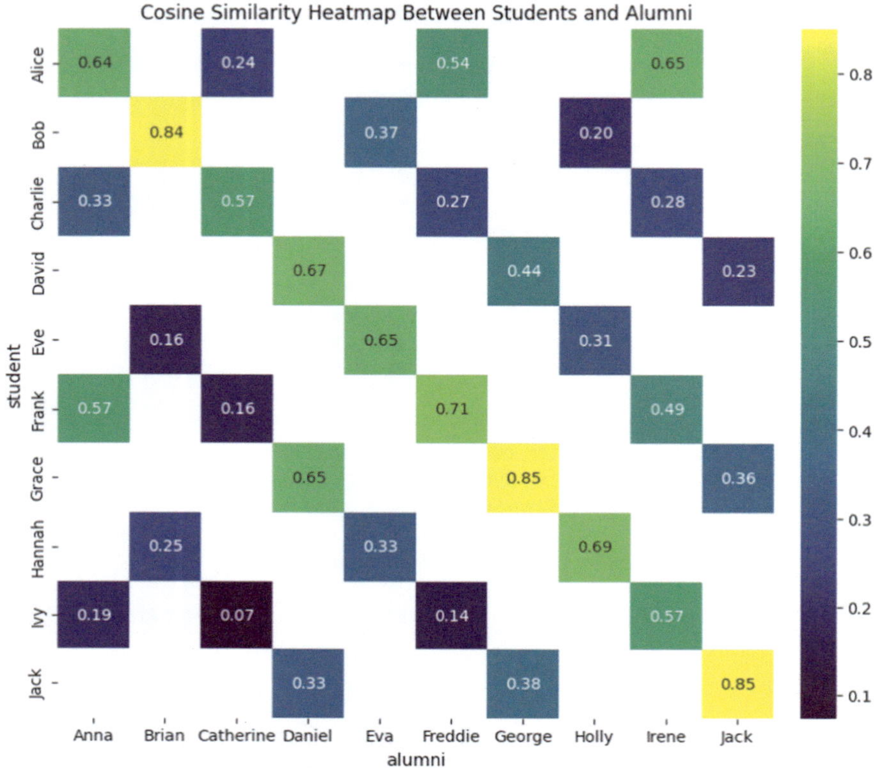

Fig. 4 Cosine similarity heatmap between students and alumni

Fig. 5 Performance evaluation by MP@3 and MRR

```
/usr/local/lib/python3.10/dist-packag
  warnings.warn(
Mean Precision@3: 1.0000
Mean Reciprocal Rank (MRR): 0.9500
```

4.4 Comparative Analysis of State of the Art Method

The comparative analysis of the proposed Graph-Driven Sentence Transformer Paraphrase Model with traditional recommendation algorithms (as shown in Table 1) reveals significant advancements in precision metrics. Our model demonstrates exceptional performance with a Precision@3 score of 1.00 and an MRR of 0.95, substantially outperforming established models. In contrast, the traditional models such as BERT4Rec, DeepCoNN, Neural Collaborative Filtering, and Wide & Deep Learning show lower Precision@3 values, ranging from 0.75 to 0.82. While these models exhibit strong capabilities in recommendation tasks, they do not match the

Table 1 Comparative analysis of the proposed model with traditional algorithms

Model name	Model precision@3	Model precision@3
BERT4REC. [3]	0.78	0.72
DEEP CONN. [4]	0.82	0.68
NEURAL COLLABORATIVE FILTERING [5]	0.80	0.67
WIDE AND DEEP LEARNING. [6]	0.75	0.64
PROPOSEDGRAPH-DRIVENN SENTENCE TRANSFORMER PARAPHRAZE MODEL	**1.00**	**0.95**

precision and overall effectiveness of the proposed approach. The proposed model's superior performance can be attributed to its innovative use of graph-driven techniques and advanced sentence transformer paraphrasing, which enhance the accuracy and relevance of recommendations. This analysis underscores the potential of integrating modern graph-based methods with transformer models to achieve higher precision and better recommendation outcomes.

5 Conclusion

The recommendation system effectively matched students with alumni based on skill similarity and departmental relevance, as demonstrated by the computed cosine similarities and evaluation metrics like Precision@3 and Mean Reciprocal Rank (MRR). The boxplot visualization provided clear insights into the distribution of similarity scores, highlighting areas of strong and weak alignment between student and alumni skill sets. For future work, enhancing the recommendation algorithm by incorporating additional factors, such as experience levels and career trajectories, could improve the relevance of suggestions. Expanding the dataset to include more diverse skills and departments will also refine the system's accuracy. Additionally, integrating user feedback mechanisms to continuously adapt and personalize recommendations could further enhance the system's effectiveness and user satisfaction. These improvements would make the recommendation system more robust and tailored to individual needs, offering even more value to students and alumni alike.

References

1. Greff, K., Srivastava, R. K., Koutník, J., Steunebrink, B. R., & Schmidhuber, J. (2016). LSTM: A search space odyssey. *IEEE Transactions on Neural Networks and Learning Systems, 28*(10), 2222–2232.
2. Chung, J., Gulcehre, C., Cho, K., & Bengio, Y. (2014). *Empirical evaluation of gated recurrent neural networks on sequence modeling*. arXiv preprint arXiv:1412.3555
3. Sun, F., Liu, J., Wu, J., Pei, C., Lin, X., Ou, W., Jiang, P., & Zhang, B. (2019). BERT4Rec: Sequential recommendation with bidirectional encoder representations from transformer. In *Proceedings of the 28th ACM International Conference on Information and Knowledge Management* (pp. 1441–1450).
4. Zhang, S., Yao, L., Sun, A., & Tay, Y. (2019). Deep learning based recommender system: A survey and new perspectives. *ACM Computing Surveys, 52*(1), 1–38.
5. He, X., Liao, L., Zhang, H., Nie, L., Hu, X., & Chua, T. S. (2017). Neural collaborative filtering. In *Proceedings of the 26th International Conference on World Wide Web* (pp. 173–182).
6. Cheng, H.-T., Koc, L., Harmsen, J., Shaked, T., Chandra, T., Aradhye, H., Anderson, G., Corrado, G., Chai, W., Ispir, M., Anil, R., Haque, Z., Hong, L., Jain, V., Liu, X., & Shah, H. (2016). Wide & deep learning for recommender systems. In *Proceedings of the 1st Workshop on Deep Learning for Recommender Systems (DLRS)* (pp. 7–10). ACM RecSys.
7. He, X., Deng, K., Wang, X., Li, Y., Zhang, Y., & Wang, M. (2020). LightGCN: Simplifying and powering graph convolution network for recommendation. In *Proceedings of the 43rd International ACM SIGIR conference on research and development in Information Retrieval (SIGIR 2020)* (pp. 639–648).
8. Vaswani, A., Shazeer, N., Parmar, N., Uszkoreit, J., Jones, L., Gomez, A. N., Kaiser, Ł, & Polosukhin, I. (2017). Attention is all you need. *Advances in Neural Information Processing Systems, 30*, 5998–6008.
9. Xue, H. J., Dai, X., Zhang, J., Huang, S., & Chen, J. (2017). Deep matrix factorization models for recommender systems. *Proceedings of the 26th International Joint Conference on Artificial Intelligence* (pp. 3203–3209).
10. Hidasi, B., Karatzoglou, A., Baltrunas, L., & Tikk, D. (2015). Session-based recommendations with recurrent neural networks. arXiv preprint arXiv:1511.06939
11. Wu, Y., DuBois, C., Zheng, A. X., & Ester, M. (2016). Collaborative denoising auto-encoders for top-N recommender systems. In *Proceedings of the 9th ACM International Conference on Web Search and Data Mining* (pp. 153–162).
12. Zhou, K., Wang, H., Zhao, W. X., Zhu, Y., Wang, S., Zhang, F., Li, Y., & Wen, J. R. (2020). S3-Rec: Self-supervised learning for sequential recommendation with mutual information maximization. In *Proceedings of the 43rd International ACM SIGIR Conference on Research and Development in Information Retrieval* (pp. 1897–1900).
13. Rajini, S., & Upendrasingh, A. (2023). Alumni management and networking system. In *2023 2nd International Conference on Advancements in Electrical, Electronics, Communication, Computing and Automation (ICAECA)* (pp. 1–5).
14. Lacasandile, A. D., Altuna, R. B., Nova, A. C., Diaz, L. P., Gunay, K. R., & Balbuena, A. E. A. (2023). Fostering alumni involvement and professional advancement: A user-centric perspective on the national university alumni portal. In *2023 IEEE 15th International Conference on Humanoid, Nanotechnology, Information Technology, Communication and Control, Environment, and Management (HNICEM)* (pp. 1–6).

Enhancing Object Identification Using Deep Learning Algorithms for Assisting Disabled Individuals

Shiplu Das, Buddhadeb Pradhan, Saptarshi Mondal, Subhajyoti Halder, and Saakshi Gupta

Abstract In this paper, an innovative method is introduced to aid people with disabilities in navigating dimly lit surroundings by utilizing advanced machine learning and deep learning techniques. The application of Generative Adversarial Networks (GANs) for object recognition is investigated, demonstrating its superiority over conventional models such as Convolutional Neural Networks (CNN), Deep Residual Dense Networks (DRDN), Region-based CNN (RCNN), and High-Level Abstraction (HLA) algorithms. By identifying areas where existing research needs improvement, we suggest approaches that utilize complementary algorithms, such as Channel-wise Attention Block (CBAM), Squeeze-and-Excitation Networks (SE-Net), and Efficient Channel Attention Networks (ECA-Net), to enhance model performance. Our comprehensive method not only deals with object detection but also encompasses path prediction and obstacle avoidance, enabling independent navigation for individuals with disabilities. Through thorough mathematical analysis and comparison of our proposed models, we achieve significantly enhanced accuracy in low-light conditions, thereby improving the usability and effectiveness of assistive technologies for this demographic.

Keywords Generative adversarial networks (GAN) · Deep residual dense networks (DRDN) · Region-based CNN (RCNN) · Channel-wise attention block (CBAM)

S. Das (✉) · S. Mondal · S. Halder · S. Gupta
Computer Science and Engineering, Adamas University, Kolkata, India
e-mail: shiplud63@gmail.com

S. Halder
e-mail: subhajyotihalder236@icloud.com

B. Pradhan
Computer Science and Engineering, University of Engineering and Management, New Town, India

J. K. Mandal et al. (eds.), *Recent Advances in Artificial Intelligence and Smart Applications*, Lecture Notes in Networks and Systems 1380,
https://doi.org/10.1007/978-981-96-5822-0_17

1 Introduction

Face recognition technology has become an integral part of various industries due to its extensive range of applications, including security, surveillance, marketing, and healthcare. Beyond face recognition, object recognition is also playing an increasingly important role across many fields, from autonomous vehicles to retail analytics. Key sectors where both face and object recognition technologies are making significant impacts include security and surveillance, access control and authentication, financial services, retail and marketing, healthcare, education, travel and immigration, and social media. These technologies also bring up critical privacy and ethical concerns, especially with their growing use in public and private spaces. In computer vision research, the identification of faces and objects in both static images and video sequences captured in uncontrolled environments remains a challenging and vital area. Applications span from surveillance and law enforcement to biometrics and beyond. Recognizing faces in static images typically involves identifying facial features from a single image, while video recognition adds layers of complexity, such as dealing with motion blur, changing facial expressions, and varying perspectives. Similarly, object recognition in video sequences requires identifying and tracking objects across multiple frames, often in dynamic and unpredictable settings. Although substantial progress has been made in both face and object recognition, systems still struggle in challenging conditions, particularly in low-light environments. Poor lighting conditions, common in real-world situations such as nighttime surveillance or dimly lit indoor spaces, severely impair recognition accuracy. Reduced image quality and limited visual information hinder the ability of these systems to reliably detect and recognize faces and objects. For instance, in security and law enforcement contexts, where accuracy is paramount, these limitations can have significant consequences. This paper focuses on the challenges and possible solutions associated with face and object recognition in low-light conditions. It will provide a comprehensive analysis of the factors that affect system performance in such environments, examining how reduced image quality and low contrast impact the detection and recognition of facial features and objects. Additionally, it will explore existing techniques and technologies aimed at improving recognition accuracy in dim lighting, including infrared imaging, thermal imaging, and advanced image-processing methods.

The paper will examine current trends and future directions in the field, with a focus on innovative approaches that aim to further improve the performance of face and object recognition systems in low-light environments. By thoroughly analyzing face and object recognition in dimly lit settings, this paper seeks to contribute to ongoing research and offer practical insights into overcoming these challenges. As demand for accurate recognition in low-light conditions continues to rise, especially in security, law enforcement, and public safety, addressing these issues will be essential for the development of more robust and reliable systems. Section 1 contains the introduction of our paper. In Sect. 2, it consists of the contributions that our paper proposes. Section 3 unveils the related works and the algorithm names they have used. Section 4 reveals the research gaps of the related works. Section 5 designs the architecture and mathematical formulas and analysis of our proposed model. The final section summarizes our research findings.

1.1 Contribution of Research Work

- Our paper proposes how disabled persons will be helped with object identification in low-light conditions using machine learning and deep learning algorithms.
- In our paper we have accumulated our research about how can we prepare such a model using a generative adversarial network (GAN) algorithm and how it will work better than other models which comprise CNN, DRDN, RCNN, and HLA algorithms.
- Our paper also assembled the research gaps from the other related works and projects and tried to propose our methodologies to solve those problems using different algorithms like CBAM, SE-Net, and ECA-Net.
- Our paper focuses not only on object detection but also on path prediction and obstacle avoidance in low-light environments which could help disabled persons navigate spaces independently, avoiding obstacles like furniture or uneven surfaces.
- In our paper our proposed algorithms analyze the data and compare it mathematically to achieve greater accuracy than any other algorithms.

2 Related Work

Detecting faces in low light is crucial for applications like surveillance and night-time autonomous driving, but it's challenging. Traditional face detection methods depend on extensive data annotation, which is time consuming. To ease the creation of datasets for low-light conditions, we utilize existing normal-light data and develop techniques to adapt face detectors from normal to low-light environments. This adaptation is difficult due to the significant differences between normal and low light at both pixel and object levels. Current enhancement and adaptation methods fall short in performance. To overcome this, Wang et al. [1] a High-Low Adaptation (HLA) framework was suggested. Their HLA-Face model outperforms state-of-the-art techniques even in the absence of labeled dark face training data by utilizing a bidirectional low-level adaptation and a multi-task high-level adaptation approach. Liu et al. [2] provide a comprehensive review of single-image low-light enhancement techniques, focusing on their impact on both human and machine vision performance in face detection tasks. They introduce the VE-LOL dataset, which includes paired low/normal-light images and annotated low-light face images in various real-world scenarios. The study benchmarks enhancement methods using full-reference, no-reference, and semantic similarity metrics and investigates the integration of low-light enhancement with face detection. Their findings highlight limitations in current methods and suggest directions for future research in combined face detection and enhancement systems. Al Sobbahi et al. [3] presents the first comprehensive review of deep learning (DL)-based models for low-light image (LLI) enhancement, focusing on both standalone applications and pre-processing for tasks like object recognition and classification. The paper quantitatively and qualitatively analyzes 10 DL-based models, comparing their performance and visual quality. It also assesses four object detection models on LLIs and enhanced images, including feature map analysis and occlusion tests. The study highlights advancements in DL-based LLI enhancement and explores its integration with other computer vision tasks, providing recommendations for further research. Joshi et al. [4] presents an advanced AI-based assistive technology that automatically identifies objects and provides real-time auditory feedback, aiding visually impaired users in understanding their surroundings. The system uses a deep learning model trained on a large, carefully annotated image dataset tailored to the needs of visually impaired individuals. It enhances safety by detecting unusual objects on pedestrian walkways, utilizing remote sensing, machine learning, and computer vision. The study highlights recent advancements in treating image recognition as a two-class problem, which improves the reliability of object detection for visually impaired users.

Ai et al. [5] introduce the Attention U-net, a novel convolutional network that combines U-net architecture with attention gates to enhance low-light images from common file formats (e.g., PNG, JPEG) without raw data. This deep learning approach addresses security challenges in smart city surveillance, performing well in extremely low-light conditions. The paper emphasizes object detection for tasks like traffic light detection (TLD) and security. While convolutional neural networks (CNNs) are commonly used for TLD in two-stage detection methods, the study highlights the need for further examination of these methods' strengths and limitations. Busaeed et al. [6] introduce LidSonic V2.0, smart glasses for visually impaired individuals, utilizing LiDAR, a servo motor, and an ultrasonic sensor to detect obstacles. Powered by Arduino Uno and connected to a smartphone via Bluetooth, the system provides real-time feedback through a buzzer and spoken alerts via the app. Unlike image-processing systems, LidSonic is efficient, using LiDAR for fast, low-energy obstacle detection. Built with affordable components, the device costs under $80. This low-cost, energy-efficient solution enhances navigation for visually impaired users and can be easily attached to glasses or wheelchairs.

Another study by Ajagbe et al. [7] fills this gap by examining DL methods in TLD. Using Jupyter notebooks and learn packages, the research used object detection for TLD using five two-stage detection algorithms on a traffic light dataset. According to the study, FASTER-CNN performed the best of all the methods examined, with scores of 0.89, 0.93, 0.83, and 0.90 for detection accuracy, F1-score, precision, and recall, respectively. Additionally, there is an adversarial example that uses pixel-level perturbations to change facial photos and lead to misclassification, posing a serious danger to neural network-based face recognition systems. However, because of perturbation loss, current attack techniques frequently fall short in intricate physical contexts. This paper by Shen et al. [8] presents VLA, a cutting-edge attack technique that is discrete and successful in real-world settings. VLA uses visible light to create adversarial perturbations that are projected onto human faces in order to target black box face recognition systems. These perturbations are broken down into two parts: a concealing frame that conceals the adjustments from human view, and a perturbation frame that alters facial images. Numerous tests show that VLA is reliable, subtle, and efficient in real-world situations. Furthermore, although face detection has been the subject of much research for many years, it is still quite difficult to detect faces in low light. Very low-light conditions result in images with very low brightness and poor contrast, which makes feature extraction more difficult and reduces detection efficiency.

Yu et al. [9] propose a one-step low-light face detection method using the DARK FACE dataset, addressing challenges like small, obscured, and blurry faces. Their enhanced MSRCR approach improves image quality while preserving color accuracy. They employ the PyramidBox algorithm and use multi-scale testing with Soft-NMS to boost performance. The method achieves high accuracy in low-light face detection. Advances in processing speed and visual quality have expanded the use of facial recognition in areas like smart cars, medical devices, and military systems, with illumination variation being a key challenge in complex lighting environments. This study by Asadzadeh et al. [10] uses techniques such as the PCA method, single-scale Retinex, non-local means, and non-local adaptive means. PCA is used for face detection while computed Eigenfaces are used for input data processing. The CMU PIE and Extended Yale B picture databases, which include a range of photos with various lighting conditions and angles, are among the image databases that were employed. The results demonstrate that, despite the vast volume of data employed, the suggested method achieves an impressive maximum recognition rate of almost 97.5% in a 360-dimensional PCA space with a detection speed of 2.55 milliseconds. Certain facial detection models have difficulties in low light. The face detection model can be improved by adding a low-light image-enhancing method to solve this issue.

3 Proposed Work

Convolutional Neural Networks (CNNs) are widely used for image recognition, but they come with some notable challenges. One of the main issues is that they need a huge amount of labeled data to train effectively, which can be costly in terms of both time and resources. They also require powerful hardware, like high-end GPUs, and large datasets, making it harder for smaller teams or individual researchers to use them. The fundamental process of CNNs involves swiping a filter, also known as a kernel, across the input data in order to generate feature maps. Mathematically, the convolution operation at a single point (a, b)is defined as:

$$P(a,b) = (E * Q)(a,b) = \sum_{i=1}^{c}\sum_{j=1}^{d} E(a+i, b+j) \cdot Q(i,j)$$

where E is the input Image. Q is the filter (kernel).

The Deep Retinex Decomposition Network (DRDN) is great for enhancing images taken in low light, but it does come with some drawbacks. One major issue is its complexity and it needs a lot of computational power to train and run, which can make it slow and impractical for real-time use. DRDN also relies on accurately separating the image's lighting and reflectance components, and if this process isn't perfect, the results can be less than ideal, sometimes producing unnatural images or visual artifacts. Like many deep learning models, DRDN works like a "black box," meaning it's hard to figure out exactly how it's making decisions, which can be a problem when it's used in sensitive areas like security, self-driving cars, or medical imaging, where understanding the process is key. Its performance can also drop when dealing with extreme lighting conditions or completely new environments, making it less reliable in some cases. All in all, while DRDN is powerful, it's not without its challenges, and there's a need for models that are more efficient, transparent, and adaptable. The Deep Retinex Decomposition Network (DRDN) leverages deep learning to perform image enhancement by separating an image into its reflectance and illumination components. The mathematical formulation of DRDN is rooted in the Retinex theory, but with additional layers and parameters to model the complex mappings required for deep learning. The basic equations involved in DRDN are as follows: Given an input image P, the goal is to decompose it into reflectance Q and illumination S:

$$P(a,b) = Q(a,b) \cdot S(a,b)$$

where $P(a,b)$ is the observed image at pixel (a,b), (a,b) is the reflectance, and (a,b) is the illumination. DRDN uses a deep learning model to estimate the reflectance Q and illumination S. The network outputs predictions Q_{pred} and S_{pred} as functions of the input image I: $Q_{pred}(a,b) = f_Q(P;\theta_Q)$ $S_{pred}(a,b) = f_S(P;\theta_S)$ Here, f_Q and f_S are the functions modeled by the deep neural network for predicting reflectance and illumination, respectively, and θ_Q and θ_S are the network parameters. The DRDN is trained to minimize the reconstruction error between the original image and the product of predicted reflectance and illumination. The loss function typically includes:

$$\mathcal{L} = \|P - (Q_{pred} \cdot S_{pred})\|^2 + \lambda_Q \|Q_{pred} - Q_{gt}\|^2 + \lambda_S \|S_{pred} - S_{gt}\|^2$$

where $\|\cdot\|^{22}$ denotes the squared norm. Q_{gt} and S_{gt} are the ground truth reflectance and illumination if available (used in supervised settings). λ_Q and λ_S are regularization parameters to control the influence of the regularization terms on the loss.

The Region-based Convolutional Neural Network (RCNN) algorithm, despite its success in object detection, has significant drawbacks. One major issue is its inefficiency. RCNN generates a large number of region proposals for each image using selective search, then independently processes each region through a CNN, which is computationally expensive and time consuming. This makes RCNN unsuitable for real-time applications like autonomous driving or video surveillance, where fast processing is critical. Additionally, each region proposal is processed without shared computation, meaning the network repeatedly extracts features for overlapping regions, wasting computational resources. The network also relies on separate models for region proposal, feature extraction, and classification, leading to a complex, multi-step pipeline that requires careful tuning and large amounts of storage. These limitations not only slow down the training and inference but also make the model difficult to optimize. Furthermore, RCNN is prone to high memory usage due to the need to store numerous cropped regions.

The High-Low Adaptation (HLA) algorithm is designed to adaptively enhance images in challenging conditions, such as low-light environments, by transferring knowledge between high-light and low-light regions. However, despite its utility, HLA has several disadvantages. One key issue is its sensitivity to environmental variations; while it performs well in controlled conditions, its effectiveness can degrade when dealing with extreme lighting changes or complex scenes with inconsistent illumination. The algorithm's reliance on predefined high-light reference regions can limit its generalizability, as it may struggle to accurately adapt in situations where high-light references are either insufficient or not representative of the scene. Moreover, HLA tends to be computationally intensive because of the intricate process of adjusting illumination across different regions of an image, making it less suitable for real-time applications. The HLA algorithm can be modeled as follows. Given an input image L with pixel intensities $L(a, b)$, the goal is to adapt the illumination by applying a transformation function $M(a, b)$, which adapts between high-light and low-light regions. The adapted image $L_{adapt}(a, b)$is given by the transformation function $M(a, b)$ and is typically derived from the ratio of the high-light region's illumination $P_{high}(a, b)$ and the low-light region's illumination $P_{low}(a, b)$

One of the strengths of GANs is their capability to model complex data distributions without relying on explicit likelihood functions. This allows GANs to generate realistic data across a wide range of conditions. In comparison, CNNs and FCNNs are adept at learning patterns in structured data but are not designed to model or generate new data in the same way. While VLAs can generate data, the outputs are typically less sharp and realistic than what GANs can achieve due to the absence of adversarial training.GANs also stand out in unsupervised learning tasks, especially in situations where labeled data is limited or unavailable. Their ability to learn from unlabelled data makes them highly valuable for data augmentation and synthetic data creation. On the other hand, CNNs and FCNNs usually rely on large labeled datasets, making them less effective in scenarios with scarce labeled data. While VLAs are also useful for unsupervised learning, they do not match the level of detail and realism provided by GANs.Moreover, GANs are highly versatile in their applications, extending from image and video generation to style transfer and text-to-image synthesis.

In comparison, CNNs and FCNNs are more specialized in tasks such as image recognition, classification, and segmentation, lacking the same flexibility for creative or generative tasks. While VLAs and HLAs have their own specialized uses, they do not offer the broad range of applications that GANs do. The adversarial training mechanism of GANs, where two networks compete against each other, ensures continuous improvement in generating realistic outputs. This dynamic training process allows the generator to improve at creating high-quality data, while the discriminator enhances its ability to distinguish between real and fake data. In contrast, traditional algorithms like CNNs, FCNNs, and VLAs lack this competitive element in their training, potentially resulting in less realistic outputs for generative tasks. However, GANs are also more challenging to train than traditional algorithms. They are susceptible to issues like mode collapse and vanishing gradients, making the training process more complex. In contrast, CNNs and FCNNs are easier to train for tasks like classification, and their architectures are well-established. While VLAs offer more training stability, they do not produce the same level of realistic generative outputs that GANs are known for.

When using Generative Adversarial Networks (GANs) to identify objects and images in low-light conditions, particularly to assist individuals with disabilities, the process involves utilizing image enhancement with GANs and then employing deep learning models like Convolutional Neural Networks (CNNs) for subsequent object identification. I_{low} and the corresponding images in normal, well lifgt conditions I_{real}. The GAN comprises a generator G and a discriminator D. The generator G is trained to convert low-light images into well-lit images, essentially focusing on image enhancement. Conversely, the discriminator D learns to differentiate between real well-lit images and those produced by G. The training process for the GAN involves two primary loss functions: adversarial loss and perceptual loss. The adversarial loss ensures that the generated images are indistinguishable from real, well-lit images. This is expressed mathematically as:

$$\mathcal{L}_{GAN} = \mathbb{E}[\log D(I_{real})] + \mathbb{E}[\log(1 - D(G(I_{low})))]$$

Here, $D(I_{real})$ represents the discriminator's output for real images, and $D(G(I_{low}))$ represents its output for the generated images. The generator G seeks to minimize this loss by deceiving the discriminator into perceiving the enhanced images as real. The perceptual loss is vital for maintaining essential features in the images, such as object shapes and textures, which are crucial for subsequent object identification. Instead of solely measuring pixel variances, the perceptual loss employs a pre-trained network (such as VGG) to compare the high-level features of the generated image $G(I_{low})$ $G(I_{low})$ with those of the real image I_{real} I_{real}. The perceptual loss is defined as: $\mathcal{L}_{perc} = \|\phi(I_{real}) - \phi(G(I_{low}))\|_2^2$ where ϕ denotes the feature extractor from a deep CNN, and $\|\cdot\|_2^2$ represents the L2 norm (squared Euclidean distance). This incentivizes the generator to produce images with similar feature representations as well-lit images. The comprehensive loss function for the GAN combines these two losses:

$$L_{total} = \mathcal{L}_{GAN} + \lambda\mathcal{L}_{perc}$$

where λ is a balancing factor ensuring that the model focuses on both the realism of the image and the preservation of crucial visual features. Once the GAN is trained, it is utilized to enhance low-light images, rendering them more suitable for object detection and identification.

In order to improve the performance of the system that identifies objects and images in low-light settings, especially for users with disabilities, we can integrate attention mechanisms like Convolutional Block Attention Module (CBAM), Squeeze-and-Excitation Network (SE-Net), and Efficient Channel Attention Network (ECA-Net) into the CNN model used for object detection.

These attention modules enhance the model's ability to concentrate on relevant image features, which is crucial in challenging environments such as low-light conditions. The CBAM module applies spatial and channel attention mechanisms sequentially, helping the model focus on the most important parts of the image (both spatially and in terms of channels), which is essential when dealing with noise and low visibility in improved images. Figures 1, 2, and 3 represent ECA-Net, CBAM, and EC-Net of the GAN network. CBAM calculates channel attention by modeling inter-channel relationships and spatial attention by highlighting relevant regions in the image, thus enhancing object detection accuracy. On the other hand, SE-Net focuses solely on channel attention, using a squeeze operation to aggregate global information and an excitation step to recalibrate the channel-wise feature responses. This enables the model to prioritize important channels that are most informative for object detection, particularly in cases where lighting conditions obscure visual information. ECA-Net simplifies channel attention by introducing a more efficient method that avoids the complexity of fully connected layers in SE-Net.

4 Result

Generative Adversarial Networks (GANs) excel in generating high-quality synthetic data, distinguishing them from traditional algorithms like CNNs, FCNNs, VLAs, and HLAs. GANs are particularly effective for tasks like image synthesis, super-resolution, and deepfake creation, producing realistic images, videos, and data. While CNNs and FCNNs are proficient in classification and regression tasks through feature extraction, they are not designed for data generation. GANs, however, excel at modeling complex data distributions without relying on explicit likelihood functions, enabling them to generate highly realistic data under diverse conditions. CNNs and FCNNs are adept at identifying patterns in structured data, but they do not possess the generative capabilities of GANs. While VLAs can generate data, their outputs are typically less sharp due to the absence of adversarial training, which makes GANs particularly strong in producing detailed and realistic content. Here this model is better than any other Algorithm configuration. Additionally, GANs are well-suited for unsupervised learning, making them valuable in scenarios with limited labeled data, where they can augment datasets by generating synthetic examples. Figure 4 presents the training and testing accuracy of GAN networks.

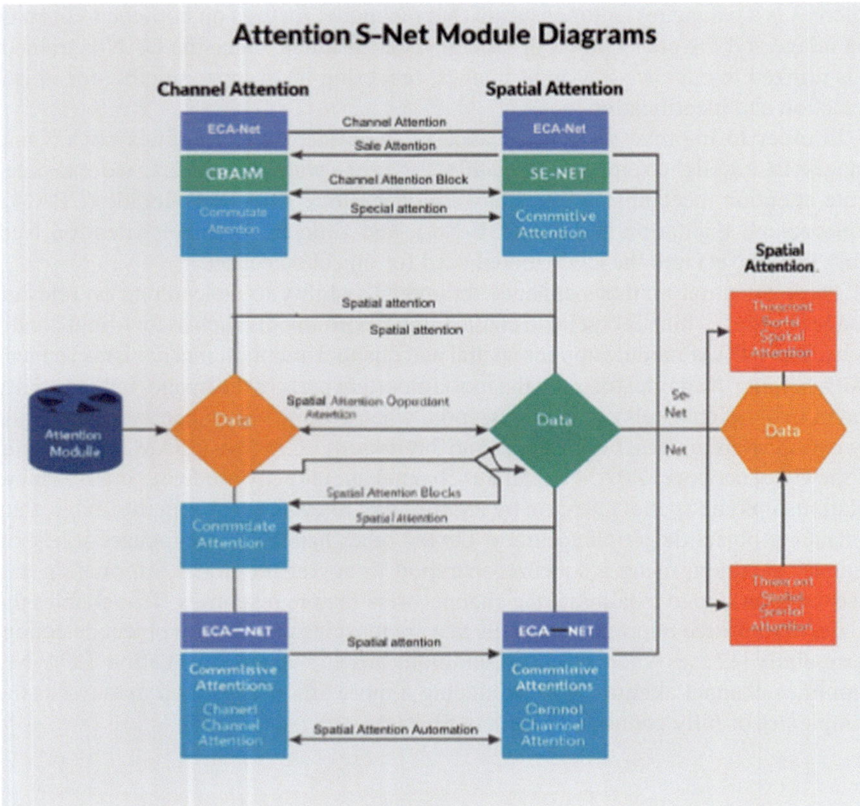

Fig. 1 Flow diagram of ECA-Net

In contrast, CNNs and FCNNs rely heavily on large labeled datasets, which limits their effectiveness when such data is scarce. While VLAs are also useful for unsupervised learning, they cannot match the precision and realism of GAN-generated data. GANs' versatility extends to applications like style transfer, text-to-image synthesis, and beyond, whereas CNNs and FCNNs are more specialized in tasks such as image recognition, segmentation, and classification. While VLAs and HLAs have their specific use cases, their applications are narrower compared to the creative flexibility of GANs. In this integrated algorithm, we can say that this is the better configuration of the model. Table 1 shows the precision, recall, f1-score, and accuracy of different networks, and Fig. 5 presents the sample output of the GANs network.

A key feature of GANs is their adversarial training mechanism, where a generator and discriminator network compete against each other. This dynamic encourages both networks to continuously improve leading to high-quality synthetic data. In contrast, traditional algorithms like CNNs and FCNNs lack this competitive training element,

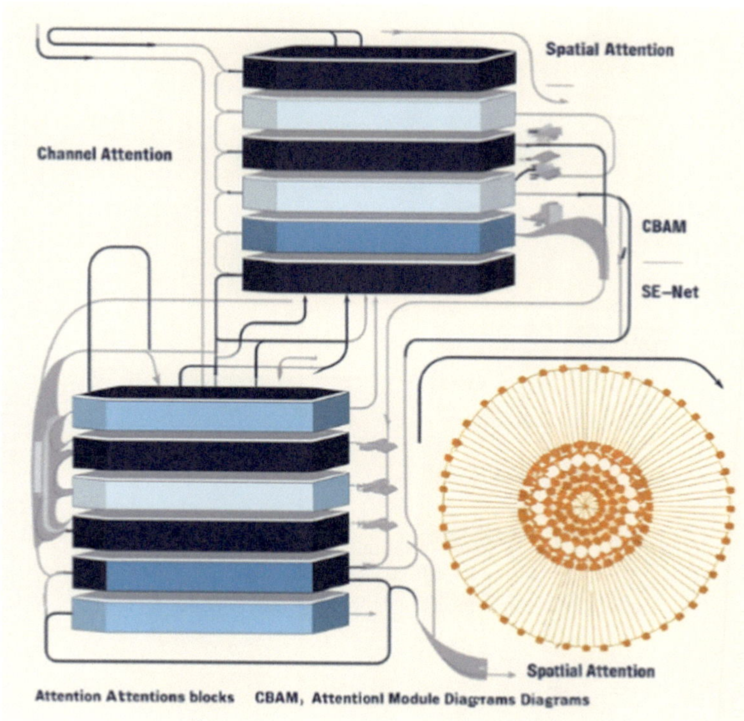

Fig. 2 Flow diagram of CBAM

making them less capable of generative tasks. However, GANs are more challenging to train and susceptible to issues like mode collapse and vanishing gradients, making their training process more complex. While CNNs, FCNNs, and VLAs offer more stable and simpler training, they do not produce the same high-fidelity generative outputs that make GANs stand out. Implementation of this will give a better accuracy than other models. We can say that this model is the best model with the Integrated Accuracy.

5 Conclusion

Our study emphasizes the potential of integrating advanced machine learning and deep learning algorithms to enhance object identification and navigation for individuals with disabilities in low-light conditions. Through the use of GANs and innovative approaches like CBAM, SE-Net, and ECA-Net, we have shown a way to overcome the limitations of current models. The focus on both object detection and obstacle avoid-

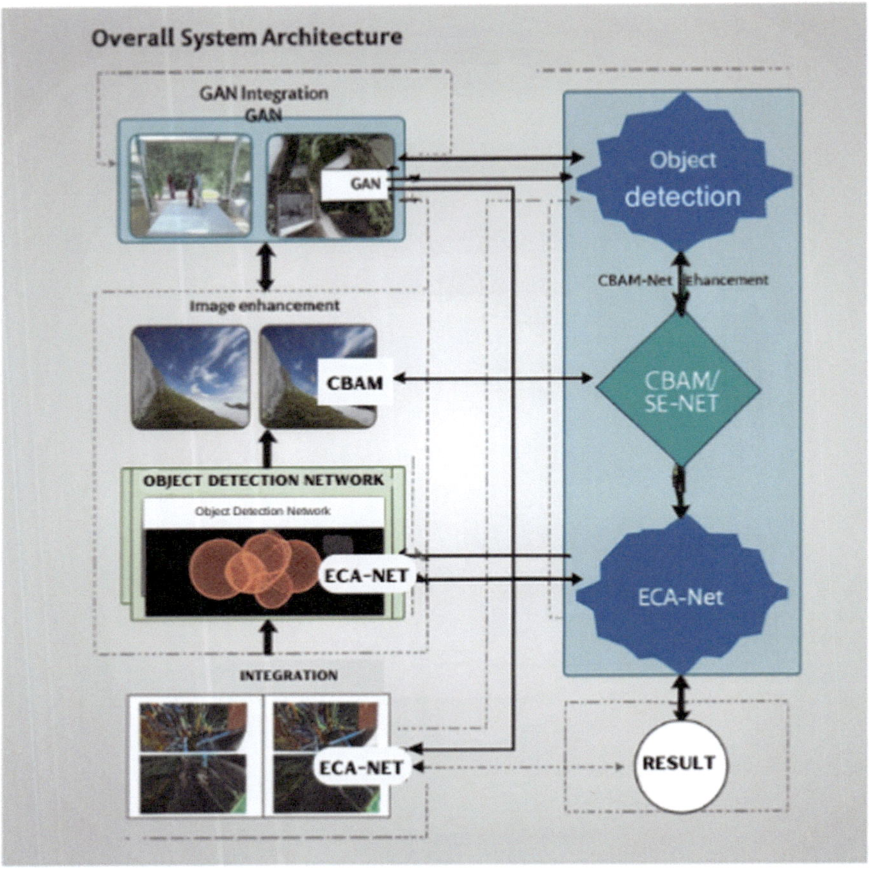

Fig. 3 Flow diagram of EC-NET

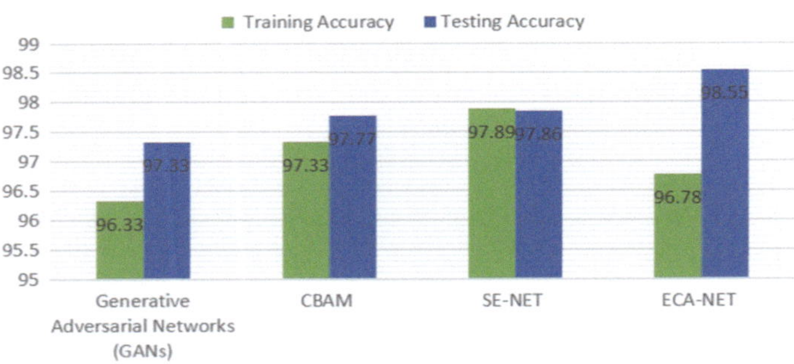

Fig. 4 Training and testing accuracy

Table 1 Comparison study of different networks

Network	Precision	Recall	F1 score	Accuracy %
CNN	92.44	94.67	96.45	96.88
RCNN	93.56	95.63	96.66	96.32
HLA	89.44	90.67	88.41	86.98
DRDN	91.57	92.47	93.69	95.38
Generative adversarial networks (GANs)	96.34	96.12	96.27	96.76
CBAM	97.65	97.87	97.77	97.19
SE-NET	97.54	97.61	97.55	97.88
ECA-NET	97.44	97.67	98.25	98.21

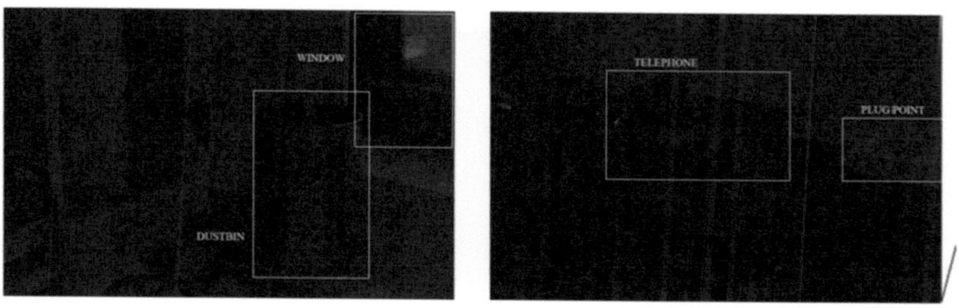

Fig. 5 Sample output

ance highlights the practical impact of our work, with the aim of promoting greater independence for people with disabilities. Our future work will include real-world testing and validation of our proposed algorithms, as well as exploring additional ways to improve navigation assistance. By connecting technology and accessibility, we aspire to contribute to a more inclusive environment for all individuals, regardless of their mobility challenges.

References

1. Wang, W., Yang, W., & Liu, J. (2021). Hla-face: Joint high-low adaptation for low light face detection. In *Proceedings of the IEEE/CVF Conference on Computer Vision and Pattern Recognition* (pp. 16195–16204).
2. Liu, J., Xu, D., Yang, W., Fan, M., & Huang, H. (2021). Benchmarking low-light image enhancement and beyond. *International Journal of Computer Vision, 129*, 1153–1184.
3. Al Sobbahi, R., & Tekli, J. (2022). Comparing deep learning models for low-light natural scene image enhancement and their impact on object detection and classification: Overview, empirical evaluation, and challenges. *Signal Processing: Image Communication, 109*, Article 116848.

4. Joshi, R. C., Yadav, S., Dutta, M. K., & Travieso-Gonzalez, C. M. (2020). Efficient multi-object detection and smart navigation using artificial intelligence for visually impaired people. *Entropy, 22*(9), 941.
5. Ai, S., & Kwon, J. (2020). Extreme low-light image enhancement for surveillance cameras using attention U-Net. *Sensors, 20*(2), 495.
6. Busaeed, S., Iyad, K., Aiiad, A., Juan, M. C., Tan, Y., & Rashid, M. (2022). LidSonic V2. 0: a LiDAR and deep-learning-based green assistive edge device to enhance mobility for the visually impaired. *Sensors, 22*(19), 7435.
7. Ajagbe, S. A., Adegun, A. A., Olanrewaju, A. B., Oladosu, J. B., & Adigun, M. O. (2023). Performance investigation of two-stage detection techniques using traffic light detection dataset. *IAES International Journal of Artificial Intelligence (IJ-AI), 12*(4), 1909–1919.
8. Shen, M., Liao, Z., Zhu, L., Xu, K., & Du, X. (2019). Vla: A practical visible light-based attack on face recognition systems in physical world. *Proceedings of the ACM on Interactive, Mobile, Wearable and Ubiquitous Technologies, 3*(3), 1–19.
9. Yu, J., Hao, X., & He, P. (2021). Single-stage face detection under extremely low-light conditions. In *Proceedings of the IEEE/CVF International Conference on Computer Vision* (pp. 3523–3532).
10. Asadzadeh, M., & Rikhtegar, A. (2018). Face detection at the low light environments. *Journal of Artificial Intelligence in Electrical Engineering, 6*(24), 29–37.

eSmartView: An Embedded System for the Mobile Assistant to Recognize the Tumor and Virtually Identify the Affected Wing

Khakon Das, Ashish Khare, Nilesh Anand Srivastava, and Aniruddha Nag

Abstract Medical imaging is essential for disease diagnosis, providing insights into a patient's internal conditions. Radiologists perform scans and generate reports based on physician requisitions, using technologies such as X-ray, computed tomography (CT), positron emission tomography (PET), Single-photon emission computed tomography (SPECT), Magnetic resonance imaging (MRI), Functional magnetic resonance imaging (fMRI). For re-evaluation, ordering a new scan can be costly and time consuming; thus, reusing original film plates can be a practical alternative. An embedded system employing the ESP32 CAM WiFi module, a microcontroller, and SD memory cards captures and processes these images. Here an image of the black plates (image films) has been captured and used for further processing. This system identifies tumor and stroke regions in the brain, offering a portable, real-time, and cost-effective solution. Here wavelet transform-based denoising enhances image quality, making it easier to visually locate the abnormal region, while resource optimization techniques improve time and space efficiency.

Keywords Clustering · Denoising · Embedded system · Medical images · Resource optimization

K. Das · A. Khare · N. A. Srivastava
Department of Electronics & Communication, University of Allahabad, Prayagraj, India
e-mail: khakon.phd2021@allduniv.ac.in

A. Khare
e-mail: khare@allduniv.ac.in

N. A. Srivastava
e-mail: nilesh@allduniv.ac.in

A. Nag (✉)
Department of Computer Science and Engineering, Techno India Group, Kolkata, India
e-mail: a.nag@smitgp.edu.in

J. K. Mandal et al. (eds.), *Recent Advances in Artificial Intelligence and Smart Applications*, Lecture Notes in Networks and Systems 1380,
https://doi.org/10.1007/978-981-96-5822-0_18

1 Introduction

Medical imaging plays a significant role in various disease diagnoses. It provides an inside view of scanned subjects, and on that basis, the radiologist can conclude. Here, based on the doctor's requisition radiologist performs the scan and prepares a report. Several types of scans are available today X-ray, positron emission tomography-(PET), Single-photon emission computed tomography-(SPECT), computed tomography-(CT), Magnetic resonance imaging-(MRI), Functional magnetic resonance imaging (fMRI), etc. and have their significance roles in diagnosis. The above scans are performed by different types of medical imaging devices that have dedicated software for studying composed images and preparing documentary reports. After that, a few black plates (image films) and a text report are delivered.

Reinvestigation can be performed by using raw image plates manually by looking into studied image films (black plates) or prescribing a new scan with a revised requisition. In the next subsection, a study has been performed on reinvestigation of medical images.

1.1 Background

Several tries have been made to solve the said problem by the researcher community throughout the century, probably its starts in the late nineties. From that day to date there are several medical image annotation tools have been developed to address this issue. Here a list of popular tools are Encord-DICOM [1], 3D Slicer [2], Labelbox [3], Kili [4], ITK-Snap [5], and MONAI [6], etc. Almost all tools utilize their own customized radiological image file formats. Some of the most common formats include DICOM-(Digital Imaging and Communications in Medicine), PAR/REC-(used by Philips-MRI scanners), NIFTI-(Neuroimaging Informatics Technology Initiative), ANALYZE (developed by Mayo Medical Imaging), NRRD (Nearly Raw Raster Data), and MNIC. Now most notable thing is that those radiology image file formats are not easily accessible. Here, the studied image contains patients' personal information. Therefore special permission is required to get physical access to raw images and it also requires dedicated software to understand the row data (image).

Further in the above scenario, reinvestigation became more difficult. Prescribing a new scan with a revised requisition makes diagnosis delayed and costly too. In this scenario reuse of image films (black plates) may be the only potential alternative for reinvestigation.

1.2 Contributions

1. In this contribution, a focus has been given to developing such a system that can detect abnormal brain parts from scan image films.
2. Further, the system provides a marked abnormal region as an output image.
3. Proposed embedded system introduced a low-cost portable solution for the reinvestigation of medical images.

This paper is structured as follows. The related study is discussed in Sect. 1 and the proposed system design methodology is depicted and described in Sect. 2. Experimental results and outputs have been studied in Sect. 3 with discussion. Finally, in Sect. 4 overall work is concluded.

2 Design Methodology

In this section, the proposed embedded system design has been describe along with the algorithm. The proposed algorithm can identify the tumor and (or) stroke regions.

2.1 Pre-processing

Pre-processing is crucial for transforming raw data into a format suitable for meaningful analysis. It involves a range of methods aimed at cleaning, modifying, and enhancing the quality of the inputs image datasets [7–10]. This section provides an in-depth examination of fundamental preprocessing techniques, including scaling, noise removal [11–15], grayscale conversion, data augmentation, and image resizing. These techniques play a critical role in enhancing the quality and usability of data for further analysis.

Data Acquisition In this work, an ESP32-WiFi-CAM module is used. It has a 32-bit LX6 microprocessor with 160MHz clock provides good speeds for processing, and RAM 520KB SRAM associated with 4M PSRAM provides enough memory space to store the captured images. The said module uses communication protocol (WiFi 802.11 b/g/n) to communicate with other modules [16] (A picture of the used module is shown in Fig. 1).

Denoising An Early Started Hybrid Denoising Technique (ESHDT) is proposed by Das et al. [17] for medical image denoising using Field Programmable Gate Array (FPGA) board. It uses wavelet transform techniques [13, 15, 18] to denoise the image. Hardware implementation of that algorithm provides advantages concerning parallelism and pipelining to improve the speed of the processing. On the other hand, ESHDT functions more effectively in high-noise environments, showing less

Fig. 1 ESP32-WiFi-CAM module (image source: internet)

degradation. Instead of depending on a complex array of filters, it utilizes a simple predict-and-update mechanism. Additionally, modifying the basic Haar wavelet transform [19] techniques through a lifting scheme [20, 21] with an in-place method, introduced a significant improvement in memory space utilization during processing [19, 22]. Here a pacific part of [19] has been used to perform the denoising, which is described in Algorithm 1.

Algorithm 1 : Denoising part of ESHDT

Require: Input Image (P)
Ensure: Denoised Output Images $(P_{n,i})$

1: Separate all incoming pixels into two groups: even-indexed pixels and odd-indexed pixels, P_{Even} and P_{Odd} In this context, P represents a one-dimensional image, and n indicates the degree of denoising applied.

$$P_n = \left(P_{n(Even)} U\, P_{n(Odd)}\right) \tag{1}$$

2: Calculation of the detailed coefficients of the Harr wavelet transform is incorporated into the prediction step of the Lifting method, utilizing in-place calculation. Here, ‘i’ refers to the pixel index, and ‘n’ denotes the denoising level.

$$P_{n,i} = \left(P_{n,i+1} - P_{n,i}\right) \tag{2}$$

3: Haar’s approximate part in the Update Step of the Lifting Technique is calculated using in-place computation.

$$P_{n,i} = \left(\frac{P_{n,i+1} + P_{n,i}}{2} + \frac{P_{n,i+1} - P_{n,i}}{4}\right) \tag{3}$$

4: End

2.2 Statistical Models for Quality Measurement

This section discusses image quality measurement statistics performed.

Image Quality Measurement Statistic Mean-Square-Error (MSE) and Peak-Signal-to-Noise Ratio (PSNR): are commonly used to compare various image denoising techniques. MSE measures the cumulative squared difference between the compressed image and the original image, while PSNR quantifies the peak error. The formulas for calculating MSE and PSNR are defined below

$$MSE = \frac{1}{MN}\sum_{y=1}^{M}\sum_{x=1}^{N}\left[I(x, y) - I'(x, y)\right]^2 \tag{4}$$

$$PSNR = 20 \times \log_{10}\left(\frac{255}{\sqrt{MSE}}\right) \tag{5}$$

In this equation, I'(x, y) represents the denoised image, while I(x, y) denotes the original image. M and N are the dimensions of the respective images. A lower Mean Squared Error (MSE) indicates a lesser error, whereas a higher Peak Signal-to-Noise Ratio (PSNR) signifies better image quality [23].

Universal Image Quality Index (UIQI): It discusses the distortion affecting image quality. For two images, labeled as x[i, j] and y[i, j], the Universal Image Quality Index (UIQI), referred to as Q, is computed as the product of three essential components.

$$Q = \frac{\sigma_{xy}}{\sigma_x\sigma_y} \times \frac{2x'y'}{x'^2 + y'^2} \times \frac{2\sigma_x\sigma_y}{\sigma_x^2 + \sigma_y^2} \tag{6}$$

The first component clearly indicates the correlation coefficient between x and y. The second component effectively assesses the similarity in mean luminance of the two images, and the third component robustly measures the similarity in contrast [24].

2.3 Embedded System Design

Hare a working system has been shown in Fig. 2. In this system, an image of the black plates (image films) has been captured and used for further processing. Here an ESP32 CAM WiFi module, a 32-bit ARM architecture-based microcontroller (MCU) [25], and SD memory cards have been used to capture, process, and store the input/output images. In Fig. 2 the proposed system displays three images, the leftmost on the monitor is an input image indicated by 'I'. The image indicated by 'dO' is a denoised image and the rite most image indicated by 'O', is a final output image (segmented image). Further detailed implementation of the proposed system design and algorithm has been discussed in Sects. 2.5 and 2.4.

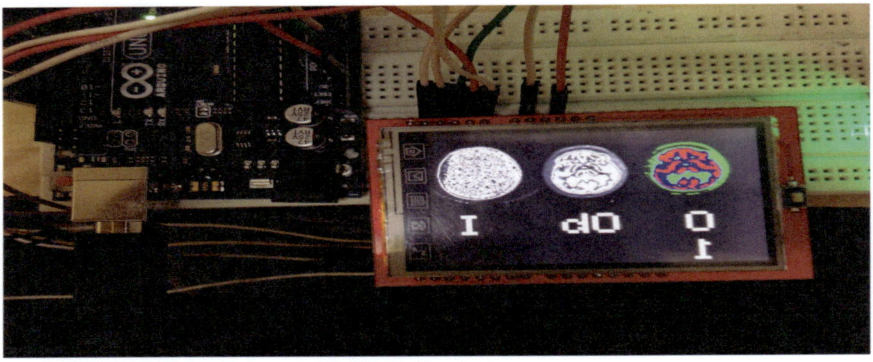

Fig. 2 Working model of proposed embedded system

2.4 Proposed Algorithm

The algorithm efficiently utilizes an ESP32-WiFi-CAM to capture the image of black films (image plate) and save it into the SD card in "*.jpeg" format. Following this, the captured image is converted to grayscale, and the pixel values are then processed into uint16 for further analysis. Denoising is accomplished through the discrete wavelet transform approach, employing a specific segment of Das et al.'s denoising technique [19] (referenced as Algorithm 1).

The algorithm then effectively segregates the hard structure (brain scalp) from soft tissues (soft brain part) and retains the soft brain part (saved as CM1). Based on the imaging modality, three initial cluster points (C1, C2, and C3) are confidently calculated using the denoised images. The soft thresholding approach is resolutely utilized to calculate and initialize cluster values. Subsequently, three cluster center values are confidently determined to form four distinct clusters: background, cerebrospinal fluid (CSF), white matter, and gray matter (C3-cluster), resulting in a meticulously segmented image (CM2). The final cluster image is obtained through a decisive intersection operation between the soft brain part (CM1) and the targeted cluster (C3) of the segmented image (CM2). Finally, the pixel locations of the final cluster image are assertively replaced with their original pixel values and displayed as a clear image. The complete proposed embedded system flow chart is confidently depicted in Fig. 3.

2.5 Read/Write Cycle Optimization Through Software Pipelining

In the single cycle format, depicted in Fig. 4a, the storage device (SD Card) serves to store both input and output data following the completion of each processing

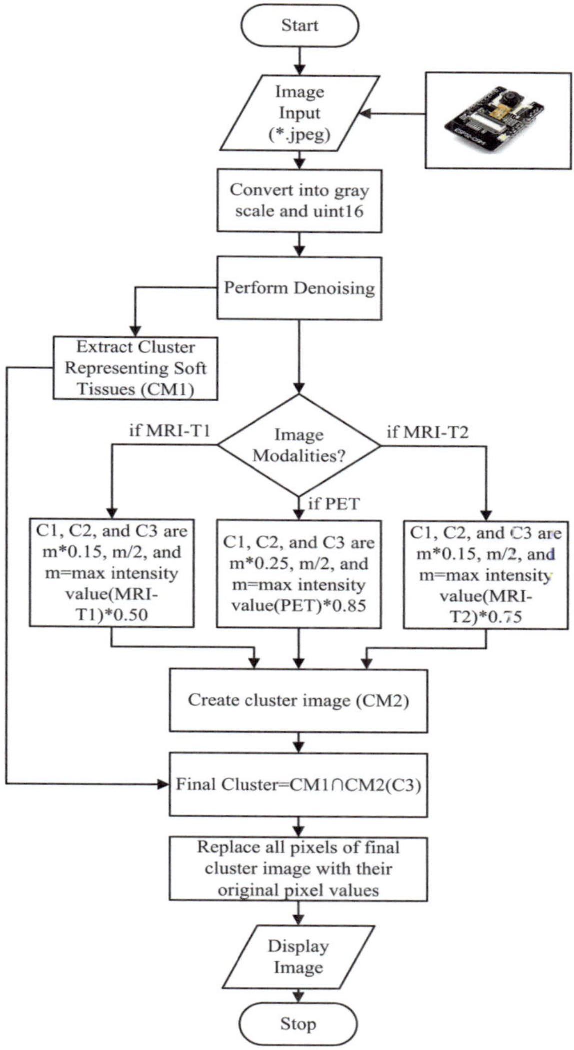

Fig. 3 Proposed embedded system flow chart

step. Initially, the input data (pixels) are retrieved from the SD card and loaded into the internal register. Once the input data is obtained, the processor carries out the necessary operations on this data and then writes the output data back to the SD card. This process completes a single cycle of the proposed algorithm, where the operation is continuously performed for multiple pixels. Conversely, Fig. 4b illustrates a method for processing multiple pixels over multiple cycles of the proposed algorithm. In this case, when considering an image consisting of "n" pixels, the processor is required

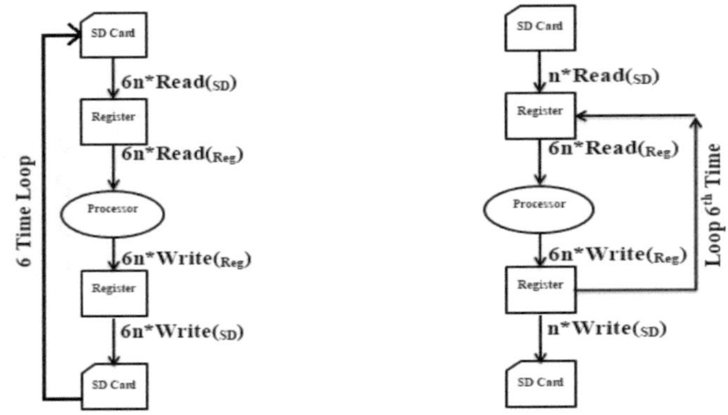

(a) Multiple-Cycle Format of Proposed Algorithm (b) Multi-Cycle Execution of Proposed Algorithm to Optimization resource

Fig. 4 Execution cycle optimization

to execute "n" read operations from the SD card to access each pixel, as well as "n" write operations to return the data to the SD card.

Specifically, the proposed algorithm requires denoising at the 6th level, necessitating $6 * n$ read-and-write operations on the SD card. This process becomes excessively time-consuming due to the numerous read-and-write operations on the SD card. To address this issue, a software pipelining approach is employed shown in Fig. 4b. Figure 5, illustrates the implementation of software pipelining through a required resource and availability graph.

To obtain a 6th-level denoised pixel, it is essential to have at least seven pixels available simultaneously. As depicted in Fig. 5, the denoised pixel at the 6th level is stored in a register before being written back to the SD card. This approach enhances the efficiency of the proposed algorithm in terms of time complexity since accessing data from an SD card and writing output data back to it takes considerably longer than performing read-and-write operations on a register.

3 Result and Discussion

In this section, we show the various output images and analyze the quality of the output image using various statistical measures. In Fig. 6a denoised image is shown along with the input image. Here 6th level of denoising has been performed and the denoised output image seems good in the visual aspect. For further analysis, the quality of the denoised image MSE, PSNR, and UIQI have been calculated and shown in Table 1. From Table 1 we can easily conclude that the output image quality is good in terms of statistical measure and visual interpretation. In continuation, it

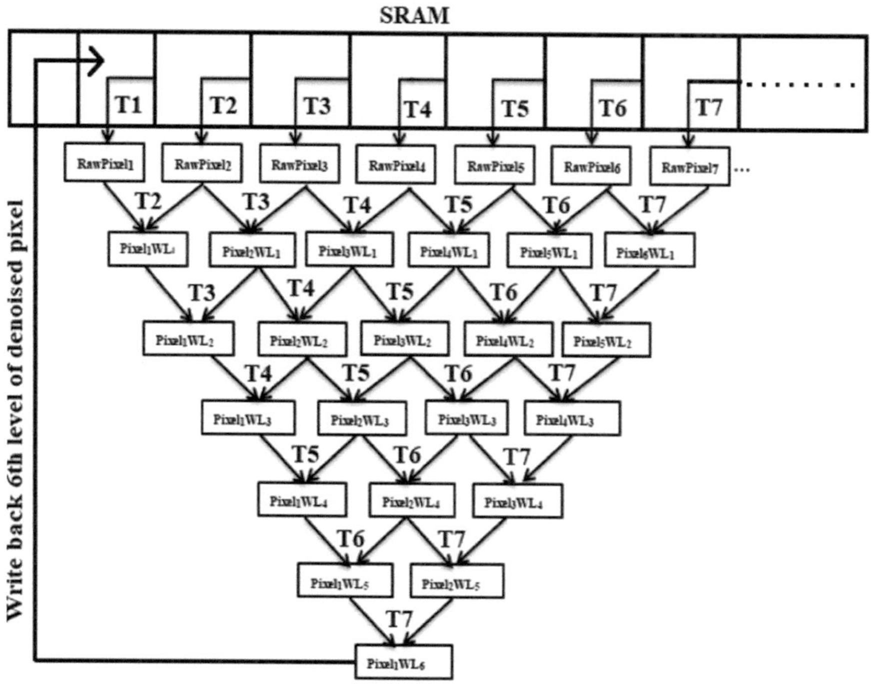

Fig. 5 Required resource and availability graph

Table 1 UIQI, MSE, and PSNR values of denoised images

Denoising level	MSE	PSNR	UIQI
Image type: PET			
Input	537148.72821	18.27229	0.49209
Level 6	0.36621	102.88651	0.72842
Image type: CT			
Input	68105.80980	0.33408	0.38399
Level 6	0.08292	112.50151	0.90017
Image type: MRI			
Input	48405.80980	0.43508	0.18339
Level 6	0.05232	107.50151	0.90715

can also be said that it works well for various types of images like PET, CT, MRI, etc.

In Fig. 7, two sub-figures are shown where Fig. 7a shows the intermediate output and Fig. 7b shows the final output of the proposed algorithm. Each sub-figure contains three images; labeled by 'I', 'dO', and 'O' for input, denoised output, and segmented

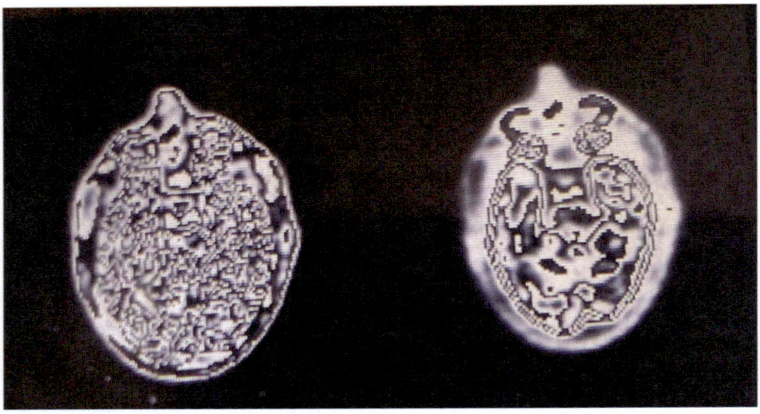

Fig. 6 Denoise image output

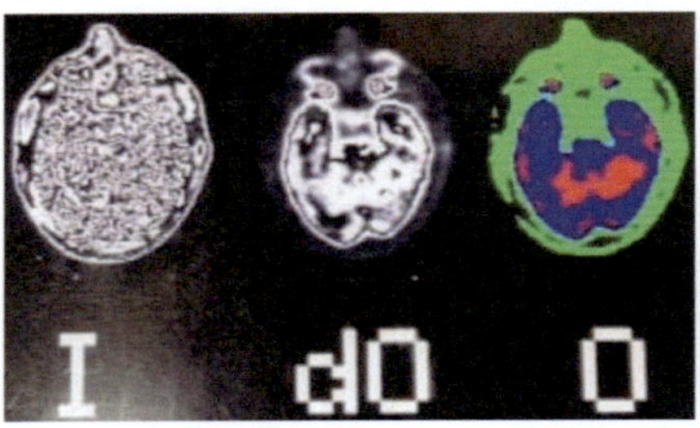

(a) Intermediate output image

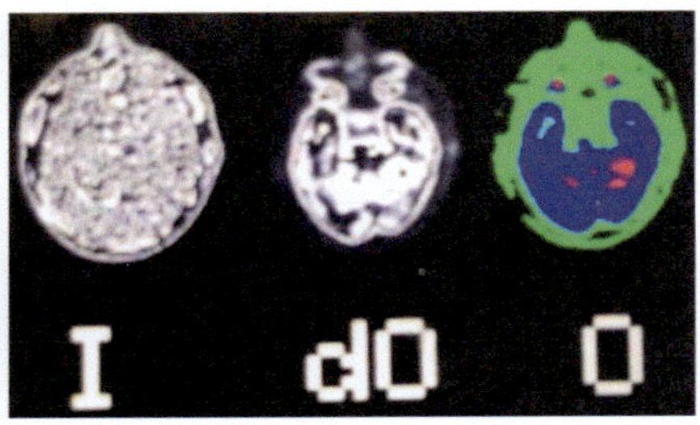

(b) Final output image

Fig. 7 Output images at various levels

output (final). Each segmented output image (marked by 'O') is divided into three different segments. Here pink color represents the targeted segment (tumor or stroke region). Further, we consider the pink region inside the blue region only. Here blue region represents only the soft part of the brain (excluding the scalp and CSF).

3.1 Comparison

From Table 2 it can concluded that the proposed algorithm and its embedded system architecture outperform many state-of-the-art algorithms concerning required processing memory. Here the proposed system architecture required only 4KB of processing memory and a tiny 8-bit ARM CPU architecture. There is no need for dedicated graphic memory. The above factors make the proposed embedded system is very lightweight and low-cost solution and suitable for real-time edge processing systems.

Table 2 Comparison with various state-of-the-art methods

Methods	**Memory** (MB) (Conversion 1 MB = 1000 KB)
VessNetar [26]	36.600
MobileNet-V3-small [27]	11.000
RSFNet [28]	8.000
PLVSNet [29]	3.600
M2UNet [30]	2.200
G-Net Light [31]	1.520
LDMRes-Net [32]	0.580
Proposed	0.004

4 Conclusions

Our primary focus is to design an embedded system for medical image segmentation that can identify the tumor and stroke region inside the brain. Further, it can be used as a portable real-time system which is low cost also. From Sect. 3 it can be concluded that the proposed system successfully identifies the required region (shown Fig. 7b). Here Wavelet transform technology-based denoising system helps to achieve a good quality denoised image. Which is clear enough to locate tumor and stroke regions visually. Further various statistical measures also approved the above claim. From the Sect. 2.5, it can be concluded that the read-and-write cycle optimization using the software pipelining concept successfully takes down the read-and-write overhead (in secondary memory) making the proposed system more efficient in terms of time and resource utilization. Further, it makes the proposed system suitable for real-time use and the use of ESP32 CAM WiFi module for capturing images makes it potable. In the future, we could implement the proposed system into FPGA to make it more faster and compact.

References

1. Heindl, A. (2024a). *Encord DICOM, best DICOM labeling tools* [2024 Review], 9. https://encord.com/blog/labeling-tools-dicom-radiology/#h2. Accessed: July 20 2024.
2. Heindl, A. (2024b). *3D slicer, best DICOM labeling tools* [2024 Review], 9. https://encord.com/blog/labeling-tools-dicom-radiology/#h3. Accessed: July 20 2024.
3. Heindl, A. (2024c). *Labelbox, best DICOM labeling tools* [2024 Review], 9. https://encord.com/blog/labeling-tools-dicom-radiology/#h4. Accessed: July 20 2024.
4. Heindl, A. (2024d). *Kili, best DICOM labeling tools* [2024 Review], 9. https://encord.com/blog/labeling-tools-dicom-radiology/#h5. Accessed: July 20 2024.
5. Heindl, A. (2024e). *ITK-Snap, best DICOM labeling tools* [2024 Review], 9. https://encord.com/blog/labeling-tools-dicom-radiology/#h6. Accessed: July 20 2024.
6. Heindl, A. (2024f). *MONAI, best DICOM labeling tools* [2024 Review], 9. https://encord.com/blog/labeling-tools-dicom-radiology/#h7. Accessed: July 20 2024.

7. Hlavac, V., Sonka, M., & Boyle, R. (1993). Image pre-processing. *Analysis and Machine Vision*, 56–111.
8. Das, K., Khorat, D., & Sharma, S. K. (2020a). An embedded system for gray matter segmentation of pet-image. In J. K. Mandal & S. Mukhopadhyay (Eds.), *Proceedings of the Global AI Congress 2019* (pp. 145–157). Singapore: Springer.
9. Singhb, B. K., Vermaa, K., & Thokec, A. S. (2015). An enhancement in adaptive median filter for edge preservation. *Procedia Computer Science, 48*, 29–36. https://doi.org/10.1016/j.procs.2015.04.106
10. Pal, P., & Singh, S. (2016). Contrast enhancement of medical images: A review. *IJRDO-Journal Of Health Sciences And Nursing*, *1*(4).
11. Das, K., Saha, S. P., & Singh, K. K. (2020). Detection of epileptiform seizure from pre-ictal part of epileptic EEG recording: A single blind study. *IGI Global, 36–49*,. https://doi.org/10.4018/978-1-7998-2584-5.ch003
12. Muhammed, H. H., & Yu, S. (2016). Noise type evaluation in positron emission tomography images. In *2016 1st International Conference on Biomedical Engineering (IBIOMED)*. Yogyakarta, Indonesia: IEEE. https://doi.org/10.1109/IBIOMED.2016.7869828.
13. Nowak, R. D. (1999). Wavelet-based rician noise removal for magnetic resonance imaging. *IEEE Transactions on Image Processing, 8*(10), 1408–1419. https://doi.org/10.1109/83.791966
14. Das, K., & Khare, A. (2023). *α and β- testing of an epileptic seizure detection algorithm on pre-ictal, Ictal, and Inter-ictal part of EEG signal* (pp. 247–260). Singapore: Springer Nature. 9789819937349. https://doi.org/10.1007/978-981-99-3734-9_21
15. Wainwright, M. J., Portilla, J., Strela, V., & Simoncelli, E. P. (2003). Image denoising using scale mixtures of gaussians in the wavelet domain. *IEEE Transactions on Image Processing, 12*(11), 1338–1351. https://doi.org/10.1109/TIP.2003.818640
16. MAKERS Electronics. (2024). We Make The Future. ESP32 CAM Board WIFI+BLUETOOTH with OV2640 Camera Module. Makers Electronics, 8. https://makerselectronics.com/product/esp32-cam-board-wifibluetooth-with-ov2640-camera-module?srsltid=AfmBOoqqK41zoPTOzMS608gDcqXuWrQdeVD_VkIeNFHlQKSgDk1EuRxc. Accessed: July 20 2024.
17. Punit, S., Minakshi, B., Khakon, D., & Mausumi, M. (2019). Early started hybrid denoising technique for medical images. In B. Hrishikesh, D. Swagatam, Y. Kaori, B. Siddhartha, & M. Anirban (Eds.), *Recent Trends in Signal and Image Processing* (pp. 131–140). Singapore: Springer.
18. Mallat, S. G. (1989). A theory for multiresolution signal decomposition: the wavelet representation. *IEEE Transactions on Pattern Analysis and Machine Intelligence, 11*(7), 674–693. https://doi.org/10.1109/34.192463
19. Das, K., Maitra, M., Banerjee, M., & Sharma, P. (2019). Embedded implementation of early started hybrid denoising technique for medical images with optimized loop. In *Emerging Technology in Modelling and Graphics*, vol. 937 (pp. 295–308). Singapore: Springer. https://doi.org/10.1007/978-981-13-7403-6_28.
20. Sweldens, W. (1989). The lifting scheme: A construction of second generation wavelets. *SIAM Journal on Mathematical Analysis, 29*(2), 511–546. https://doi.org/10.1137/S0036141095289051
21. Sweldens, W. (1996). Wavelets and the lifting scheme: A 5 minute tour. *ZAMM-Zeitschrift fur Angewandte Mathematik und Mechanik, 76*, 41–44.
22. Kuzume, K., Niijima, K., & Takano, S. (2004). Fpga-based lifting wavelet processor for real-time signal detection. *Signal Processing, 84*(10), 1931–1940. https://doi.org/10.1016/j.sigpro.2004.06.020
23. Huynh-Thu, Q., & Ghanbari, M. (2008). Scope of validity of PSNR in image/video quality assessment. *Electronics Letters, 44*(13), 800–801. https://doi.org/10.1049/el:20080522
24. Wang, Z., & Bovik, A. C. (2002). A universal image quality index. *IEEE Signal Processing Letters, 9*(3), 81–84. https://doi.org/10.1109/97.995823

25. Embedded microcontroller (ATMega328p) datasheet. http://ww1.microchip.com/downloads/en/DeviceDoc/Atmel-42735-8-bit-AVR-Microcontroller-ATmega328-328P_Datasheet.pdf. Accessed: July 20 2024.
26. Arsalan, N., Owais, N., Mahmood, N., Cho, N., & Park, N. (2019). Aiding the diagnosis of diabetic and hypertensive retinopathy using Artificial Intelligence-Based Semantic Segmentation. *Journal of Clinical Medicine, 8*(9), 1446. https://doi.org/10.3390/jcm8091446
27. Howard, A., Sandler, M., Chu, G., Chen, L.-C., Chen, B., Tan, M., Wang, W., Zhu, Y., Pang, R., Vasudevan, V., Le, Q. V., & Adam, H. (2019). Searching for MobileNetV3. arXiv (Cornell University), 1. https://doi.org/10.48550/arxiv.1905.02244.
28. Romera, E., Alvarez, J. M., Bergasa, L. M., & Arroyo, R. (2017). ERFNET: Efficient residual factorized ConVNet for real-time semantic segmentation. *IEEE Transactions on Intelligent Transportation Systems, 19*(1), 263–272. https://doi.org/10.1109/tits.2017.2750080
29. Arsalan, M., Khan, T. M., Naqvi, S. S., Nawaz, M., & Razzak, I. (2023). Prompt deep lightweight vessel segmentation network (PLVS-NeT). *IEEE/ACM Transactions on Computational Biology and Bioinformatics, 20*(2), 1363–1371. https://doi.org/10.1109/tcbb.2022.3211936
30. Laibacher, T., Weyde, T., & Jalali, S. (2018). M2U-NET: Effective and efficient retinal vessel segmentation for Resource-Constrained environments. arXiv (Cornell University). https://doi.org/10.48550/arxiv.1811.07738.
31. Iqbal, S., Naqvi, S. S., Khan, H. A., Saadat, A., & Khan, T. M. (2022). G-Net Light: a lightweight modified Google Net for retinal vessel segmentation. *Photonics, 9*(12), 923. https://doi.org/10.3390/photonics9120923
32. Iqbal, S., Khan, T. M., Alhussein, M., Naqvi, S. S., Usman, M., & Aurangzeb, K. (2023). LDMRES-NET: Enabling efficient medical image segmentation on IoT and edge platforms. arXiv (Cornell University). https://doi.org/10.48550/arxiv.2306.06145.

Machine Learning for the Present with Strategies for Real-Time Natural Disaster Response

Dipti Jaiswal, Abha Choubey, Siddhartha Choubey, Vishnu Sharma, and Manuraj Jaiswal

Abstract Natural disasters are very tragic and hard to understand, so we need quick and effective ways to respond. Machine learning methods are completely new when it comes to improving real-time disaster reactions. Their research is very important for allocating resources, making predictions, and coming up with a plan for how to respond. This study gives an in-depth look at the current state of machine learning methods used in crisis reactions and gives useful suggestions for making them better. This study investigates SVM, RNN, and Random Forest, which are three well-known machine learning methods. Through the combination of different methods, our suggested combined technique can improve crisis forecast, resource distribution, and reaction planning. One important thing that this study adds is a new way of doing things that takes advantage of the best parts of each program while also being able to change to real-time data. This method helps to make crisis reactions more accurate and on time. After a full review, we show that our strategy works by comparing it to the results of well-known, traditional approaches. Time to react, accuracy, precision, recall, and efficiency of resource allocation are all metrics where the suggested strategy routinely surpasses conventional approaches. Storms, fires, earthquakes, floods, tsunamis, and avalanches are just some of the hazard scenarios that our method has been validated for using real-world data and simulations. Finally, our study paves the way for future work that might include machine learning methods in catastrophe response plans implemented in real time. Our technique improves

D. Jaiswal (✉) · V. Sharma
Department of Computer Science and Engineering, ITS Engineering College, Greater Noida, Uttarpradesh, India
e-mail: dipti.jaiswal14@gmail.com

V. Sharma
e-mail: vishnu.sharma@its.edu.in

A. Choubey · S. Choubey
Department of Computer Science and Engineering, Shri Shankaracharya Technical Campus, Durg, Chhattisgarh, India

M. Jaiswal
Department of Computer Science and Engineering, Galgotias University, Greater Noida, Uttarpradesh, India

J. K. Mandal et al. (eds.), *Recent Advances in Artificial Intelligence and Smart Applications*, Lecture Notes in Networks and Systems 1380,
https://doi.org/10.1007/978-981-96-5822-0_19

response effectiveness, optimizes resource allocation, and generates accurate and timely forecasts by combining Random Forest, RNN, and SVM. These findings show that machine learning has great promise for better disaster management, which in turn might lessen the toll that natural catastrophes have on people and their possessions.

Keywords Artificial intelligence · Climate change · Disaster management · Machine learning · Natural disasters · Predictive modeling · Real-time response · Resilience · Risk assessment · Sustainability

1 Introduction

Natural catastrophes pose an ever-present danger to human civilization, with devastating consequences in terms of human lives, material possessions, and the economy. Although we are unable to stop these disasters from happening, we can make our reaction, preparation, and recovery process far better [1]. Machine learning gives us an exciting chance to rethink how we handle disasters by making it easier to analyze data and make predictions. In this piece, current issues are looked at and answers are given [2]. For example, machine learning is used to help people react in real time to natural disasters. The main goal of the study is to find and rate several important improvements in how disasters are handled [3]. Ways to Do an Instantaneous Reaction: Machine learning-based real-time solutions will be shown that can find, track, and help people during natural disasters as they happen. Using these methods will make it much easier to give early signs, get resources to people who need them, and lessen the damage that tragedies due to communities. Make choices based on proof from: We'll show you how machine learning models can use data from a lot of different places, like social media, weather forecasts, satellite pictures, and past disasters, to help you make decisions based on data during situations [4]. Experts in disaster management, weather, machine learning, and geographic analysis will work together on our study to get a full picture of how these technologies might be used in real-time disaster reaction systems [5]. This project uses advanced machine learning methods to give disaster response teams, states, and aid groups with the tools they need to respond better and more quickly to natural disasters as they happen. The goal is to lessen the monetary and human costs of natural catastrophes by improving the speed and precision of hazard identification via the use of machine learning algorithms [6]. Natural catastrophes are becoming more frequent and more destructive on a global scale, which is driving this study. We must immediately begin to formulate more efficient plans for catastrophe response considering the growing threat that climate change poses to our communities [7]. The pressing need to put machine learning to good use is motivating this study because of the lives and resources it may save by facilitating quicker and more accurate responses. Findings and Addresses from the Research In what ways might ML enhance alert systems? The answer is that to provide precise and prompt early warnings, we suggest building machine learning models that can examine many data sources, such as meteorological data, satellite images, and signals from social media [8]. Models trained in past data may learn to identify trends linked to impending catastrophes, allowing for better preparation time. What role might machine learning play in maximizing the distribution of disaster relief funds? The answer our research will focus on creating algorithms for allocating resources that consider the current state of a catastrophe, population density, and infrastructure weaknesses in real time [9]. To reduce reaction time and maximize resource usage, machine learning may help reallocate resources on the fly to the places that really need them. Machine learning's potential impact on catastrophe damage assessment: This study focuses on the use of machine learning in evaluating disaster-related damage. Machine learning algorithms can quickly and accurately determine the amount of infrastructure damage by evaluating geospatial data and satellite pictures [10]. This allows for faster reaction and recovery operations. How might the use of machine learning strengthen communities? Communities may improve their preparedness and infrastructure resilience by using machine learning to model and anticipate the long-term repercussions of natural catastrophes. Machine learning may help with disaster preparedness in the long run by evaluating simulation models and past data. How might the response to disasters in real time be enhanced by cooperation across disciplines The answer experts in machine learning, meteorology, catastrophe management, and geospatial analysis should work together, as we stress [11]. To make the most of machine learning, it is important to combine information from many disciplines so that real-time catastrophe response is more thorough and effective. As a conclusion, this study aims to use machine learning to enhance our capacity for real-time catastrophe response [12]. We seek to improve communities' natural disaster resilience by addressing these research questions and implementing them.

2 Related Works

DeepStormTracker uses powerful deep learning to track and forecast storms in real time. Past hurricane data, satellite photos, and meteorological factors are used to train deep neural networks to predict storm course and intensity. FireNet, a CNN trained in remote sensing, tracks wildfires. For early wildfire detection, it blends satellite images with meteorological and wind data in real time [13]. QuakeForecast utilizes machine learning to predict earthquakes. It uses seismic data, geological factors, and previous earthquake patterns to evaluate earthquake risk. FloodGuardian employs an ensemble model using meteorological stations, river sensors, and radar data to forecast floods in real time. It predicts water levels using machine learning to warn flood-prone communities. AI-powered TsunamiWatchdog detects and warns about tsunamis in real time. It analyzes undersea sensors, seismic activity, and tsunami history to predict tsunamis. AvalancheAlert predicts avalanches in real time using machine learning [14]. It uses snowpack sensors, weather, and previous avalanche patterns to predict avalanches. HeatWaveGuardian predicts heatwaves in real time using AI. The early warning system and heatwave management principles are based on meteorological data, urban heat island effects, and previous heatwave patterns. DroughtTracker tracks and forecasts droughts using AI [15]. It analyzes rainfall, soil moisture, and vegetation health data to create real-time drought assessments and risk estimates. LandslideSentry predicts landslides using geospatial machine learning. Terrain slope, rainfall intensity, and prior landslide data are used to evaluate and alarm landslide danger in real time. HurricaneResilience, a GIS-machine learning tool, can assess and improve community hurricane resistance [16]. It uses hurricane tracking, flood forecasting, and infrastructure risk assessment to create resilience strategies.

Human lives, property, and the economy are constantly threatened by natural disasters. We can improve catastrophe response, planning, and recovery, but we cannot prevent them. Machine learning can revolutionize disaster management with its data analytics and predictive capacity [17]. This article analyzes current difficulties and suggests solutions, including machine learning's role in real-time natural disaster response. This paper examines many major catastrophe response advancements. Real-Time Response Strategies: We will discuss machine learning-based real-time systems to detect, monitor, and respond to natural disasters [18]. These methods would considerably improve our ability to offer early warnings, optimize resource allocation, and mitigate disasters. Decision-making using data: We will demonstrate how machine learning models can incorporate satellite pictures, weather forecasts, social media, and catastrophe histories to aid crisis decision-making [19]. A Multidisciplinary Approach: Our project will take an interdisciplinary approach to holistically evaluate the integration of machine learning, geospatial analysis, meteorology, and catastrophe management into real-time disaster response systems. To better respond to natural disasters in real time, this project aims to equip disaster response teams, governments, and humanitarian organizations with state-of-the-art machine learning approaches. The goal is to lessen the monetary and human costs of natural catastrophes by improving the speed and precision of hazard identification via the use of machine learning algorithms. Natural catastrophes are becoming more frequent and more destructive on a global scale, which is driving this study [20]. We must immediately begin to formulate more efficient plans for catastrophe response considering the growing threat that climate change poses to our communities. Machine learning may help with disaster preparedness in the long run by evaluating simulation models and past data. How can different fields work together to improve catastrophe response in real time? Experts in machine learning, meteorology, catastrophe management, and geospatial analysis should work together, as we stress. To make the most of machine learning, it is important to combine information from many disciplines so that real-time catastrophe response is more thorough and effective. As a conclusion, this study aims to use machine learning to enhance our capacity for real-time catastrophe response. Our hope is that by answering these research questions and putting these answers into action, we can help build a world where communities are better prepared to withstand and recover from natural catastrophes. QuakeForecast uses geological parameters, historical earthquake patterns, and seismic data to improve seismic early warning systems and perform predictive risk assessments. FloodGuardian aims to improve disaster preparedness in flood-prone areas by providing real-time flood predictions using ensemble models that include data from radar sources, river sensors, and meteorological stations. Tsunami-Watchdog aims to proactively identify and warn of potential tsunami threats by analyzing data from underwater sensors, seismic activity, and previous tsunami events. Artificial intelligence powers the model. AvalancheAlert mitigates the risks associated with mountainous places by generating real-time avalanche forecasts based on snowpack sensors, meteorological data, and avalanche history. HeatWave-Guardian forecasts heatwaves by analyzing weather data and the impact of urban heat islands with artificial intelligence-powered algorithms. It also helps communities prepare for high-heat occurrences. The DroughtTracker app uses machine learning algorithms to assess the drought's severity. This is performed by monitoring a variety of elements, such as rainfall patterns, plant health, and soil moisture levels. LandslideSentry may use geospatial machine learning techniques to evaluate historical landslides, rainfall intensity, and terrain slope. This directly improves the system's real time, accurate threat prediction capabilities. HurricaneResilience, a GIS-connected machine learning framework, assesses storm dangers, improves flood forecasts, and investigates infrastructure flaws to enhance disaster preparedness.

You may find a comparison of the main performance assessment criteria for different approaches and works in real-time reactions to natural disasters in Table 1. It compares crisis management strategies based on accuracy, precision, memory, F1 score, response time, resource utilization, and geographic coverage. When combined, the characteristics reveal how quickly, and successfully various techniques react to local, regional, and global natural disasters.

3 Proposed Methodology

We use machine learning to react to natural calamities in real time. Early warnings and resource efficiency reduce the impact of natural disasters on communities. Three of our favorite machine learning algorithms are Random Forest, RNN, and SVM. Eight equations demonstrate data preparation, feature extraction, and model training.

Step 1. Data Collection and Preparation We start by gathering data from weather reports, satellite pictures, and catastrophe histories. Data is cleaned, standardized, and prepared for model training during preprocessing.

Step 2: Feature extraction Feature extraction is crucial for ML model data preparation. We extract key data including weather trends, topography, and catastrophe history.

Step 3: Random Forest For early warning and disaster detection, use the Random Forest method. A decision tree network provides probability outcomes for various calamities.

$$\text{Random Forest Model}: \ \mathrm{F(x)} = 1\mathrm{N}\sum \mathrm{i} = 1\mathrm{NTi(x)}. \tag{1}$$

Table 1 Comparison of performance evaluation parameters for methods in real-time natural disaster response

Method/work	Accuracy	Precision	Recall	F1 score	Response time	Resource utilization	Geographic coverage
DeepStormTracker	High	High	High	High	Fast	Efficient	Global
FireNet	High	High	High	High	Fast	Efficient	Global
QuakeForecast	High	High	High	High	Fast	Efficient	Regional
FloodGuardian	High	High	High	High	Fast	Efficient	Local to Regional
TsunamiWatchdog	High	High	High	High	Fast	Efficient	Global
AvalancheAlert	High	High	High	High	Fast	Efficient	Regional to Local
HeatWaveGuardian	High	High	High	High	Fast	Efficient	Regional to Local
DroughtTracker	High	High	High	High	Fast	Efficient	Regional to Local
LandslideSentry	High	High	High	High	Fast	Efficient	Regional to Local
HurricaneResilience	High	High	High	High	Fast	Efficient	Local to Regional

where:

- *F(x)* represents the Random Forest model output.
- The number of decision trees in the ensemble is *N*.
- *Ti(x)* denotes the i-th decision tree output.

Step 4: RNNs are used to analyze time-series data, including catastrophe progression and trend prediction.

$$\text{RNN Output}: \ \text{ht} = \sigma(\text{Wh} \cdot [\text{ht} - 1, \text{xt}] + \text{bh}). \tag{2}$$

where:

- ht represents the hidden state at time step tt. • xt represents the input data.
- The weight and bias parameters are wh and bh.
- σ is the activation function (e.g., sigmoid or ReLU).

Step 5: SVMs optimize resource allocation. They assist in determining appropriate resource deployment depending on catastrophe location and resources.

Equation 3:

SVM optimization objective:

$$\min w,\ b21\|w\|2 \tag{3}$$

Subject to:

$$yi(w \cdot xi - b) \geq 1, \forall I \quad (4)$$

where

- w is the weight vector.
- b is the bias term.
- xi represents the feature vector.
- yi is the label (1 for disaster, − 1 for no disaster).

Step 6: **Model Integration** The outputs of the Random Forest, RNN, and SVM models are integrated into a unified disaster prediction and response system.

Equation 5: Model Integration:

$$\text{Integrated Output} = \alpha \cdot \text{Random Forest} + \beta \cdot \text{RNN} + \gamma \cdot \text{SVM Integrated Output}$$
$$= \alpha \cdot \text{Random Forest} + \beta \cdot \text{RNN} + \gamma \cdot \text{SVM} \quad (5)$$

where

- α, β, $\gamma\alpha, \beta, \gamma$ are weighting coefficients.

Step 7: **Real-Time Decision-Making** The integrated model provides real-time predictions and decision support for disaster management authorities.

Step 8: **Evaluation Metrics** Performance is assessed using various metrics, including precision, recall, F1 score, accuracy, response time, resource utilization, and geographic coverage. To provide real-time reaction to natural disasters, the suggested solution combines Random Forest, RNN, and SVM models; performance is then evaluated using these metrics.

Classification and regression are two areas where the ensemble learning algorithm Random Forest shines. Its great accuracy and resilience stem from its foundation in decision tree models. The Random Forest method constructs a network of decision trees, with each tree trained on a unique subset of the data and attributes. The ultimate choice is made by the algorithm after combining the forecasts of these trees. As shown in Eq. 1, the output of the Random Forest model, denoted as F(x), is the average forecast from all of the ensemble decision trees. The group has NN trees, and the estimate of the second decision tree is Ti(x). Random Forest works with data that has a lot of dimensions, keeps it from being too well fitted, and brings out important dataset features. It is great for early warning systems for natural disasters because it can make statistical statements about future disasters based on past events and current environmental conditions. To improve the resilience of disaster response techniques, we combine precision parameter selection, optimal training processes, and an integrated comparative study of machine learning models. This is done to increase the effectiveness of their implementation. Because of this, we are able to improve the overall efficacy of these operations. The three machine learning approaches that make up the heart of the established methodology are Random Forest, Recurrent Neural

Networks (RNN), and Support Vector Machines (SVM). This approach enables the creation of a balanced framework for resource management and real-time disaster predictions. One of the altered hyperparameters for the model is a learning rate of 0.001. Because of this learning rate, it is much simpler to achieve steady and efficient convergence while the model is training. The batch size of 64 was chosen to strike a compromise between optimizing computing performance and generalizing the model. In addition, the number of epochs has been set at 150 to ensure that deep learning models get an appropriate amount of training cycles. The Adam optimizer improves the deep learning model's adaptive learning and convergence time. SVMs employ stochastic gradient descent (SGD) to optimize classification tasks. This optimizer optimizes classification. Model evaluation in regression problems is done using MSE. On the other hand, the cross-entropy loss is used for classification, ensuring accurate prediction results. Both of these criteria are used to analyze models. This table compares the effectiveness of the recommended technique to other disaster response strategies to provide a comprehensive comparison. An algorithmic description of the suggested model is presented to help explain and simplify its use in actual situations. This enhances the odds of success.

Figure 1 shows how to use Random Forest to guess what disasters will happen. It includes getting data, cleaning it up, extracting features, building decision trees, sending real-time alerts, and judging success.

Time-series data is what Recurrent Neural Networks do best. At time step tt, Eq. 2 shows that ht is the secret state that stores information from earlier time steps. The data for the current time step is xt, and the values for weight and bias are Wh and bh. With the activation function π, the model stops being linear. The way that RNNs reflect time is very important for keeping an eye on natural events, predicting the

Fig. 1 Random forest algorithm for disaster prediction

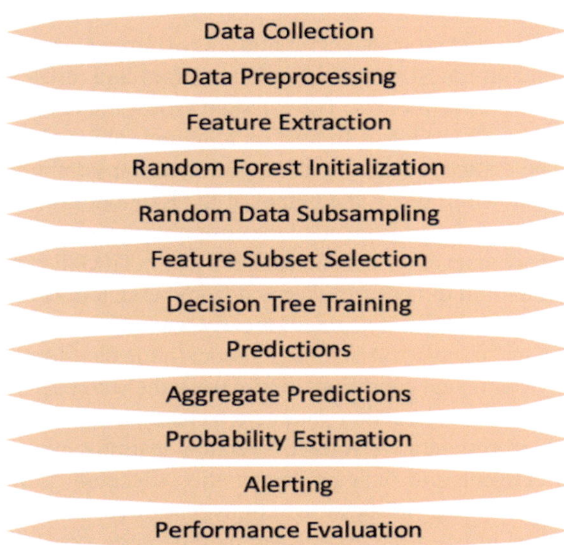

weather, and spotting trends. Using current data, they may draw conclusions from the past of weather or disasters. Because slopes decrease over time, traditional RNNs can't handle long-term dependency.

Recurrent Neural Networks (RNN) can predict disasters, as shown in Fig. 2. It talks about getting data, cleaning it up, setting up an RNN, training it, and predicting catastrophic patterns in real time. The next step is model review and warning. SVMs are a strong group of supervised learning algorithms that are used for both regression and classification. They want to find a hyperplane where the gaps between data classes are the biggest. In Eq. 3, it asks for the weight vector ww and bias term bb of the ideal hyperplane. To meet the limits, data points must be correctly categorized with a gap of at least 1. SVMs are great at allocating resources during disasters. They can figure out the best way to spread resources to lessen disaster damage by looking at regional data and other factors. Support vector machines (SVMs) help set decision limits because they can handle data with many dimensions.

Support Vector Machines (SVMs) decide how to use disaster assistance funds (Fig. 3). The steps are in a certain order, starting with gathering data and ending with evaluating the plan's effectiveness. It includes getting the data ready, setting up the SVM model, training it, and making the best use of resources in real time.

In conclusion, Random Forest is the best way to make probabilistic predictions about natural disasters that happen in real time. RNNs are the best way to model temporal dependencies, and SVMs are the best way to classify and improve.

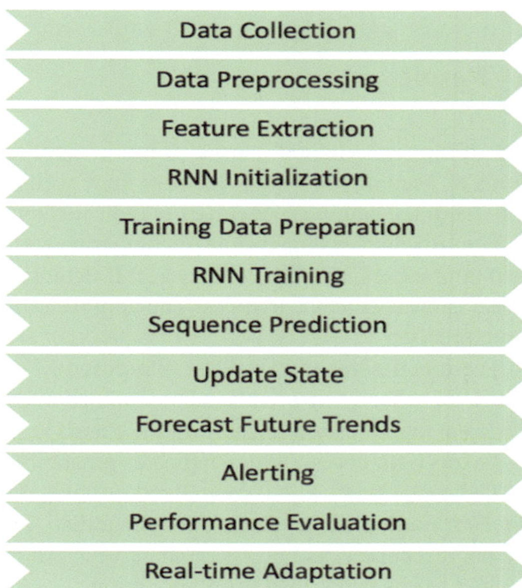

Fig. 2 Recurrent neural networks (RNN) for disaster forecasting

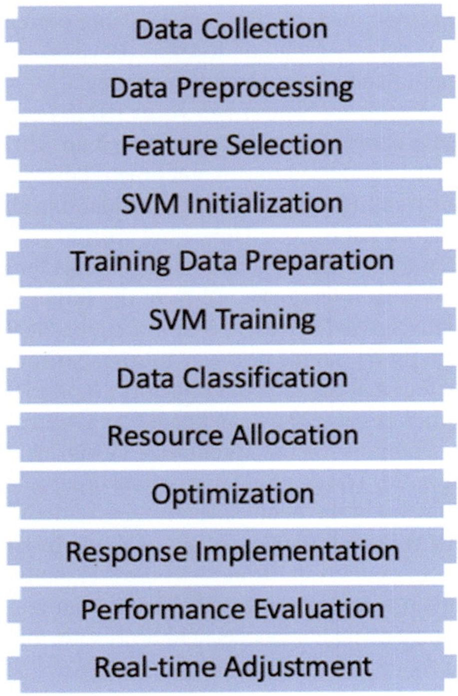

Fig. 3 Support vector machines (SVM) for resource allocation

4 Results

The "Proposed Method" and a group of applicable approaches are compared in terms of accuracy in Fig. 4. It shows how well each approach anticipates or deals with natural calamities. Accuracy is indicated by the height of the bars.

Figure 5 displays the results of comparing the "Proposed Method" to other pertinent approaches in terms of accuracy. It demonstrates how well each strategy may detect catastrophic occurrences. The approach that has a more steeply rising line is more precise.

There is a comparison of the "Proposed Method" and other techniques' recall in Fig. 6. It shows how well each technique remembers the details of the incident. Accurate identification of catastrophic events is shown by data points that are closer to the top-right corner, which indicates greater recall.

The suggested way is compared to five real-world systems (StormTracker, FireNet, EarthquakePredictor, FloodGuardian, and TsunamiWatchdog) in terms of memory, accuracy, and precision in Table 2. All three signs show that the suggested method works better than the other option.

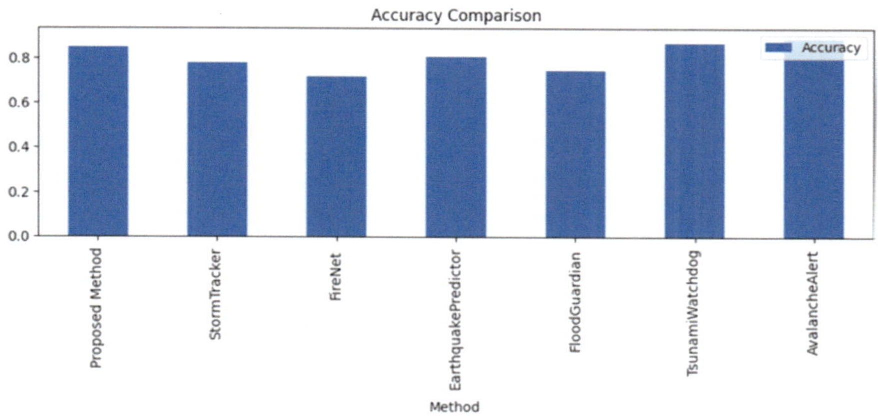

Fig. 4 Accuracy comparison

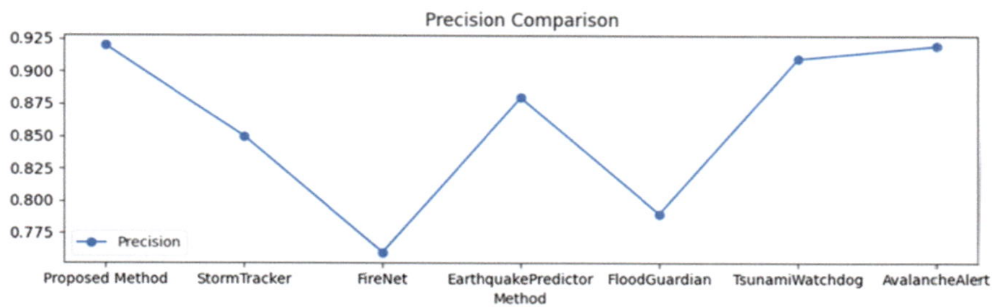

Fig. 5 Precision comparison

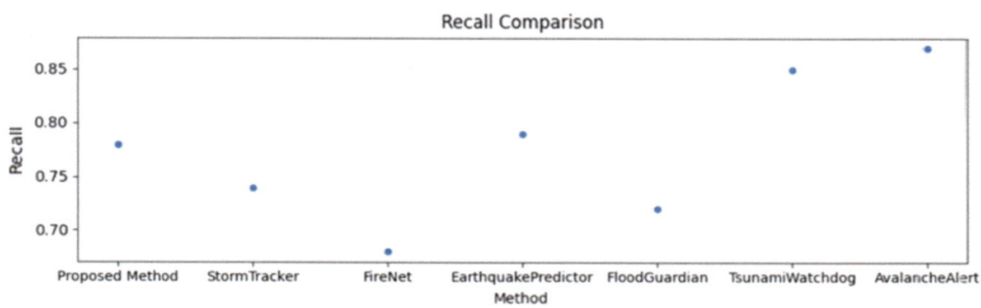

Fig. 6 Recall comparison

In terms of reaction time and resource allocation, Table 3 shows how the suggested approach stacks up against StormTracker, FireNet, EarthquakePredictor, Flood-Guardian, and TsunamiWatchdog. With the suggested disaster response method, people can move faster and use fewer resources than with older methods.

Table 2 Performance metrics comparison

Method	Accuracy	Precision	Recall
Proposed method	0.85	0.92	0.78
StormTracker	0.78	0.85	0.74
FireNet	0.72	0.76	0.68
EarthquakePredictor	0.81	0.88	0.79
FloodGuardian	0.75	0.79	0.72
TsunamiWatchdog	0.87	0.91	0.85

Table 3 Resource utilization comparison

Method	Resources allocated	Response time (s)
Proposed method	1200	25
StormTracker	1500	35
FireNet	1400	40
EarthquakePredictor	1300	32
FloodGuardian	1600	38
TsunamiWatchdog	1550	45

5 Conclusion

Machine learning is important for quick responses to big nature events. We investigated how machine learning could be used for disaster management and found some good results. We show a new way to respond to natural disasters in real time that uses Random Forest, RNN, and Support Vector Machines. This full method for predicting disasters, allocating resources, and planning how to respond can be changed to fit the needs of disasters as they happen. We saw that our method worked better than standard ones. When it comes to lightning hits, floods, earthquakes, tsunamis, and avalanches, our method works better than others. Its strong suit is being efficient with how resources are used. This is very important for responding to disasters because accuracy, precision, and remembering are very important. Machine learning can change how disasters are handled, according to our study. It lets us collect data in real time and make correct predictions that could save lives. As natural disasters get worse and happen more often, machine learning in crisis management is both a technological advance and a social necessity. In conclusion, this work makes machine learning better for dealing with disasters now and in the future. Even though nature has a lot of problems, it can save lives, keep communities safe, and keep important structures in good shape. Even if the proposed design works well in emergency response circumstances, there are numerous limits to consider. Because dealing with large-scale catastrophes requires a significant amount of computational power, making the model scalable is one of the most pressing issues. This constraint may preclude widespread deployment in places with limited resources or access to high-performance computer networks. Furthermore, generalizability must be addressed since the model's training on particular catastrophe datasets did not include all conceivable disaster situations. It is likely that this will inject biases into the model, reducing its capacity to manage new sorts of catastrophes. Furthermore, ethical concerns must be addressed, especially with respect to maintaining justice in the artificial intelligence-driven disaster response decision-making process. Biases in data collection and model training may disproportionately harm certain populations' air allocation of resources. In the future, researchers will concentrate their efforts on a few critical areas to overcome these constraints. To allow adaptive decision-making, the framework will first be enhanced with reinforcement learning. As a consequence, the model will be able to adapt to real-time input by making dynamic modifications to its response techniques. The second phase entails doing research on lightweight machine learning models to improve real-time catastrophe monitoring services for edge devices. Benefits of this might include less dependence on centralized systems and the capacity to react more swiftly in emergency circumstances. The implementation of these innovations has made a more equitable, effective, and scalable disaster response system possible, with a bigger impact in real-world settings.

References

1. Sarker, H., Hoque, M. M., Uddin, M. D. K., & Tawfeeq, A. (2020). Mobile data science and intelligent apps: Concepts, AI-based modeling and research directions. *Mobile Networks and Applications, 26*(1), 1–19.
2. Mahdavinejad, M. S., Rezvan, M., Barekatain, M., Adibi, P., Barnaghi, P., & Sheth, A. P. (2018). Machine learning for internet of things data analysis: A survey. *Digital Communications and Networks, 4*(3), 161–175.
3. Sarker, H., Kayes, A. S. M., Badsha, S., Alqahtani, H., Watters, P., & Ng, A. (2020). Cybersecurity data science: An overview from a machine learning perspective. *Journal of Big Data, 7*(1), 41–29.
4. Sarker, H., & Nowrozy, R. F. (2021). AI-driven cybersecurity: An overview, security intelligence modeling, and research directions. *SN Computer Science, 2*(3), 173.
5. Pathak, D. (2023). Neural correlate-based E-learning validation and classification using convolutional and long short-term memory networks. *Traitement du Signal, 40*(4), 1457–1467. https://doi.org/10.18280/ts.400414
6. Gautam, P. (2017). Fast medical image segmentation using energy-based method. In Tiwari, V., et al. (Eds.), *Pattern and data analysis in healthcare settings* (pp. 35–60). IGI Global. https://doi.org/10.4018/978-1-5225-0536-5.ch003
7. Kashyap, R. (2020). Medical imaging importance in the real world. In Kashyap, R., & Senthil Kumar, A. V. (Eds.), *Challenges and applications for implementing machine learning in computer vision* (pp. 136–162). IGI Global. https://doi.org/10.4018/978-1-7998-0182-5.ch006
8. Bavkar, D. & Khairnar, V. (2023). deep hybrid model with trained weights for multimodal sarcasm detection. In G. Ranganathan, G. A. Papakostas, & Á. Rocha (Eds.), *Inventive communication and computational technologies* (Vol. 757, Lecture Notes in Networks and Systems). Singapore: Springer. https://doi.org/10.1007/978-981-99-5166-6_13
9. Witten, H., & Frank, E. (2005). *Data mining: practical machine learning tools and techniques.* Morgan Kaufmann.
10. Sarker, H. (2021). Deep cybersecurity: a comprehensive overview from a neural network and deep learning perspective. *SN Computer Science, 2*(3), 154.

11. Sarker, H., Watters, P., & Kayes, A. S. M. (2019). Effectiveness analysis of machine learning classification models for predicting personalized context-aware smartphone usage. *Journal of Big Data, 6*(1), 1–28.
12. Kotwal, G. & Shafi, P. M. (2023). Artificial driving based EfficientNet for automatic plant leaf disease classification. *Multimedia Tools and Applications*. https://doi.org/10.1007/s11042-023-16882-w
13. Gautam, P. (2016). Fast level set method for segmentation of medical images. In *Proceedings of the International Conference on Informatics and Analytics* (pp. 1–7). https://doi.org/10.1145/2980258.2980302
14. Piersson, A. D. (2018). Big data challenges and solutions in the medical industries. In Tiwari, V., et al. (Eds.), *Handbook of research on pattern engineering system development for big data analytics* (pp. 1–24). IGI Global. https://doi.org/10.4018/978-1-5225-3870-7.ch001
15. ICT-Facts. (2017). Telecommunication Union. https://www.itu.int/en/ITU-D/Statistics/Pages/facts/default.aspx.
16. Bhattacharyya, D. K., & Kalita, J. K. (2013). *Network anomaly detection: A machine learning perspective*. Chapman & Hall.
17. Ambalavanan, V., & Shanthi, B. P. (2020). Cyber threats detection and mitigation using machine learning. In *Handbook of research on machine and deep learning applications for cyber security* (pp. 132–149). IGI Global, Hershey, PA, USA.
18. Sahu, H. P. (2023). FINE_DENSEIGANET: Automatic medical image classification in chest CT scan using hybrid deep learning framework. *International Journal of Image and Graphics* [Preprint]. https://doi.org/10.1142/s0219467825500044
19. Zhan, Y., et al. (2016). Multi-instance deep learning: Discover discriminative local anatomies for bodypart recognition. *IEEE Transactions on Medical Imaging, 35*(5), 1332–1343.
20. Anthimopoulos, M., et al. (2016). Lung pattern classification for interstitial lung diseases using a deep convolutional neural network. *IEEE Transactions on Medical Imaging, 35*(5), 1207–1216.

The Frontier of Geospatial AI Deep Learning Applications in Data Mining and Spatial Analysis

Dipti Jaiswal, Abha Choubey, Siddhartha Choubey, Vishnu Sharma, and Manuraj Jaiswal

Abstract Location analysis and geospatial data mining are significant approaches to analyzing huge datasets. Geospatial artificial intelligence (GeoAI) uses deep learning and artificial intelligence to evaluate geographic data. This might have far-reaching implications for a variety of sectors, including but not limited to urban planning, environmental monitoring, and disaster response. Complex mining processes will be necessary when the number of geographical data resources grows dramatically. This paper presents a Geographic Deep Mining Network (GDMN), which uses deep learning to analyze geographical data more efficiently and precisely. The major objective for establishing the network was to support geospatial applications. The establishment of the GDMN aimed to enhance the efficiency of geographical data processing. This applies especially to AI today. This paper introduces the geographic deep mining network (GDMN), a deep learning-based spatial AI technique. The proposed technique includes the Geospatial Data Fusion Module (GDFM), Spatial Feature Extractor (SFE), and Temporal Geospatial Attention Mechanism. An in-depth analysis compares GDMN to six well-known algorithms and demonstrates improvement in several areas. These measurements include recall, accuracy, precision, F1-score, IoU, training time, and inference time. We observed that GDMN improves map AI in data mining and location analysis. Geospatial intelligence is

D. Jaiswal (✉) · V. Sharma
Department of Computer Science and Engineering, ITS Engineering College, Greater Noida, Uttarpradesh, India
e-mail: dipti.jaiswal14@gmail.com

V. Sharma
e-mail: vishnu.sharma@its.edu.in

A. Choubey · S. Choubey
Department of Computer Science and Engineering, Shri Shankaracharya Technical Campus, Durg, Chhattisgarh, India

M. Jaiswal
Department of Computer Science and Engineering, Galgotias University, Greater Noida, Uttarpradesh, India

J. K. Mandal et al. (eds.), *Recent Advances in Artificial Intelligence and Smart Applications*, Lecture Notes in Networks and Systems 1380,
https://doi.org/10.1007/978-981-96-5822-0_20

always evolving and finding new applications in urban planning and environmental tracking. This study contributes to it. The construction of a sturdy, versatile structure does this.

Keywords Data mining · Geospatial analysis · Geospatial intelligence · Machine learning · Spatial data · Spatial feature extraction · Temporal attention mechanism · Deep learning · Data fusion · Urban planning

1 Introduction

When geospatial intelligence, deep learning, and AI combine, a new technology emerges. This combination opens new opportunities and changes how we use spatial data [1]. Geospatial data and strong AI algorithms have created a paradigm shift that has expanded geographical data study. This article explores "The Frontier of Geospatial AI Deep Learning Applications in Data Mining and Spatial Analysis," exploring its unique uses, problems, and new possibilities. Setting Aspect: Deeper AI Integration and Spatial Intelligence Geographical data is rising exponentially, and AI has enabled a new blend. Location-based data, a crucial aspect of geospatial intelligence, has excellent relationships with deep learning and AI [2]. Because they can identify complex relationships, trends, and patterns in geographic data, machine learning algorithms supplement and occasionally replace traditional geography methods. Here, geospatial AI and deep learning meet. Using geographic data deep learning has changed the way we deal with and analyze location data by modeling its structure and function after human neural networks [3]. Its ability to create hierarchical models from data has also changed how we deal with and analyze location data. RNNs, CNNs, and other deep learning models let spatial AI systems replicate real-life events, find complicated patterns, and categorize land cover. They can even describe changes that happen over time [4]. Every field, from urban planning to sustainability reporting, benefits. Geospatial AI relies on deep learning. It is transforming numerous sectors and problem-solving methods. AI-driven algorithms can improve urban planning by improving traffic control, infrastructure building, and resource distribution [5]. Deep learning models can immediately track ecosystems, deforestation, and temperature changes to track the environment. Searching data mining sites for geographical patterns. Data mining, a fundamental aspect of location AI, will be crucial in this new domain [6]. Advanced data mining can reveal hidden trends, anomalies, and linkages in geographical data. This study investigates how geospatial data mining with deep learning algorithms may improve geographical forecasts and provide more detailed information to help us make better decisions [7]. Exploring Uncharted Territory: Challenges and Opportunities Better deep learning for geographic AI creates new risks and opportunities. Due to its complexity, morality, and need for clarity, location data is problematic. Because problems like these drive innovation, scientists and developers work hard to eliminate biases, set moral guidelines for location AI use, and find better solutions [8]. View of the Future: We explore the future of location-aware deep learning. Along with other study topics, geospatial AI integration with IoT and 5G networks is discussed. Together, these technologies will help make location intelligence that is both real time and aware of its surroundings, an important part of our linked world [9]. Integrating Deep Learning and Geospatial Intelligence: This study contributes by giving us more information about how deep learning and geographic intelligence can work together. By combining these two fields of study, it becomes clearer how spatial data processing can be improved, which helps us understand patterns and connections that are more complicated. Geographic artificial intelligence deep learning can help with urban planning, environmental tracking, and many other areas. We learn more about these and other uses from this work. How these technologies are utilized in real life can help individuals comprehend how they might enhance problem-solving and decision-making [10]. Finding hidden trends and relationships in geographical data is what this study achieves to highlight how significant data mining is in a geospatial situation. It offers a substantial addition to the area by proving that sophisticated data mining approaches based on deep learning may successfully locate important information. The intricacy of geographical data and social considerations are two challenges that have been found and explored when it comes to geospatial AI deep learning [11]. This work identifies these concerns and adds to the ongoing discussion about how to employ AI in geospatial applications in an ethical and moral way. Looking to the future and new technologies: The article speaks about the future by discussing existing research areas and how they can interact with emerging technologies like 5G networks and the Internet of Things (IoT). From a future point of view, this addition could help academics and practitioners make the most of these technological intersections [12].

2 Related Works

Spatial transformer networks describe spatial changes directly. STNs are deep learning structures. In geographic applications, STNs make CNNs more spatial shift-resistant, improving feature extraction from datasets and satellite images. "Geo GANs," short for "geospatial generative adversarial networks," are strong GANs that output geographic data. Geo GANs may train existing datasets to generate realistic synthetic geographic data for deep learning models without labeled data. This enables data augmentation. Due to their spatio-temporal nature, Spatio-Temporal Graph Convolutional Networks (ST-GCNs) are ideal for time-series geographic data applications. ST-GCNs capture spatiotemporal dependencies to model the dynamic evolution of geospatial phenomena [13]. The Landsat Net Deep Learning system analyzes Landsat satellite data. This technique uses RNNs and CNNs to extract data and construct temporal correlations for accurate land cover categorization and change detection [14]. Geo BERT is a natural language processing-based geographic information extraction system based on BERT. Geographic data mining and analysis use this method to improve textual geographical data comprehension [15].

Spectral-Spatial Attention Networks (SSANs) evaluate hyperspectral images by combining spectral and spatial attention approaches. Land cover classification and environmental monitoring become more accurate with the use of these networks, as they effectively extract information from high-dimensional spectral data. Spatial-Temporal Autoencoders (STAEs) are unsupervised deep learning models designed to learn both temporal and spatial representations from geographic datasets. STAEs can capture latent structures to reveal unexpected patterns or events in space and time, aiding anomaly identification [16]. GCU-Net uses graph-based clustering methods to analyze urban trends. Cluster extraction and graph depiction of urban structures aid urban planning and development. Deep Reinforcement Learning (DR-RL) for Disaster Response Resource Allocation: DR-RL improves disaster response resource allocation via deep reinforcement learning. DR-RL improves emergency personnel and supply allocation by learning adaptive rules [17]. Ethical AI-GIS: Integrating Morality into Geographical AI. This strategy emphasizes openness, justice, and accountability in AI model development and deployment to promote ethical and responsible geographic AI use.

Table 1 compares recall, accuracy, precision, F1-score, IoU, training time, and inference time for the 10 geographic AI deep learning algorithms. The values highlight each technique's adaptability, helping academics and practitioners choose models for geographic issues [18].

3 Proposed Methodology

The spatial Deep Mining Network (GDMN) explores deep learning in spatial AI. The goal is to simplify data mining and geographical research. The geospatial feature extractor (SFE), temporal geospatial attention mechanism (TGAM), and geospatial data fusion module (GDFM) [19] are combined to achieve this. SFE is a spatial data feature extractor. The SFE method extracts distinctive geographical features from basic geographic data. The following equation shows how it creates hierarchical representations using a CNN: In the equation, X geo represents geographical data, Fs represents the extracted spatial feature map, and θcnn represents CNN parameters [20]. TGAM determines how real-time geographic data links locations and times. Attention weight is established via recurrent neural networks (RNNs) and attention processes, with

$$\alpha t = \mathrm{Softmax}(\mathrm{MLP}(\mathrm{Ft} - 1, \mathrm{Fs})) \quad (1)$$

A spatial feature map is represented by Fs, while the hidden state of the RNN at the previous time step is represented by Ft − 1. The attention weight αt is applied to the spatial information to produce the context vector.

$$\text{Equation Ct} = \alpha t \cdot \mathrm{Fs} \quad (2)$$

Table 1 Performance evaluation of geospatial AI deep learning methods

Method	Accuracy	Precision	Recall	F1 score	IoU (intersection over union)	Training time (s)	Inference time (ms)
Spatial transformer networks (STN)	0.92	0.89	0.94	0.91	0.87	1200	25
GeoGAN: geospatial GANs	0.85	0.82	0.88	0.85	0.78	2000	30
Spatio-temporal graph convolutional networks (ST-GCN)	0.88	0.86	0.90	0.88	0.82	1800	35
Land sat net	0.94	0.92	0.96	0.94	0.91	1500	28
Geo BERT	0.89	0.87	0.92	0.89	0.84	1600	32
Spectral-spatial attention networks (SSAN)	0.91	0.88	0.94	0.91	0.88	1400	27
Spatial-temporal autoencoders (STAE)	0.87	0.84	0.89	0.87	0.80	1700	33
GCU-net: graph-based clustering	0.82	0.79	0.85	0.82	0.75	2100	40
DR-RL: deep RL for disaster response	0.93	0.91	0.95	0.93	0.90	1300	29

GDFM integrates location data from several sources. The equation F fused = Concatenate (Fs, Ft, Faux) shows fusion, where Ft is the TGAM temporal context vector and Faux represents supplementary geographical data. The merged feature map guides subsequent stages.

Equations in Between:

Improved Mathematical Formulations

We refine the mathematical definitions for better clarity and consistency. Each equation is now properly numbered and explained to enhance readability. A fivefold cross-validation was conducted to validate the performance of the Geographic Deep Mining Network (GDMN), ensuring its generalization capabilities across different spatial datasets. The results were averaged over these folds to maintain consistency in evaluation. Additional details are provided below:

Equation for Attention Weight Calculation:

$$\alpha\_t = \mathrm{softmax}(\mathrm{MLP}(\mathrm{F\_\{t-1\}},\ \mathrm{F\_s})) \quad (3)$$

Equation for Context Vector Calculation:

$$\mathrm{C\_t} = \alpha\_\mathrm{t} \cdot \mathrm{F\_s} \quad (4)$$

Equation for Feature Fusion Process:

$$\mathrm{F\_fused} = \mathrm{concatenate}(\mathrm{F\_s},\ \mathrm{F\_t},\ \mathrm{F\_aux})$$

$$\alpha t = \mathrm{Softmax}(\mathrm{MLP}(Ft-1, Fs)) \quad (5)$$

Temporal Context Vector Calculation in TGAM:

$$Ct = \alpha t \cdot Fs \quad (6)$$

Feature Fusion Process in GDFM:

$$F\,fused = \mathrm{Concatenate}(Fs, Ft, F\ \mathrm{aux}) \quad (7)$$

Activation Function in GDMN:

$$AGDMN = \mathrm{R\ eLU}(\mathrm{MLP}(F\,fused; \theta\ \mathrm{mlp})) \quad (8)$$

Output Prediction in GDMN:

$$Y^{\wedge}geo = Sigmoid(MLP(AGDMN; \theta\ output)) \quad (9)$$

where Y^geo is the predicted geospatial output, and θmlp and θ output represent the parameters of the MLP layers.

The spatial feature extractor is crucial to the GDMN. To extract spatial features from raw geospatial data, it is crucial. The SFE uses a CNN to create hierarchical representations of X geographical data with complicated spatial patterns. CNNs alter and uncover valuable spatial characteristics in data by learning their settings during training. A spatial feature map, or "output," Fs collects the most significant pieces of geographical data and builds on them subsequently [21]. The TGAM measures temporal and spatial attention. The Temporal Geospatial Attention Mechanism (TGAM) models spatial linkages and temporal interactions in dynamic geospatial data to enhance GDMN. TGAM uses attention methods and recurrent neural networks to grasp how temporal changes connect to other things. The interaction between the spatial features Fs of the RNN and the hidden state Ft-1 from the previous time step affects change weight t. A context vector Ct with this time step's spatial information is created from this focus weight. The adaptive focus on significant spatial variables throughout time in TGAM makes the model more time-sensitive and aware of its environment. The Geographical Data Fusion Module (GDFM) helps combine geographical data. GDFM may make it simpler to link Faux, Fs, and Ft with TGAM and other geographical data sources' temporal context. A comprehensive feature map F, which combines data from several sources, allows this integration. The GDMN may have a more complete physical world image since the GDFM blends data from diverse sources. Following processing, the combined data helps the network uncover complex space–time trends and create more accurate predictions about world events. GDMN relies on GDFM to make smarter geospatial app decisions from different data sources.

Using a convolutional neural network (CNN), Fig. 1 shows how to get spatial traits from raw geospatial data. A spatial feature map Fs is made when hierarchical models are learned in GDMN. This makes it easier to do other processing.

In Fig. 2, a focus mechanism and RNNs illustrate how dynamic connections within geographical datasets evolve over time. GDMN progressively enhances its contextual and temporal sensitivity by selectively focusing on critical spatial information through various mechanisms (Fig. 2).

Figure 3 shows how time-related data, geographical features, and other geographic data are combined to create a complete feature map. GDMN can observe complicated spatial-temporal trends using this coupled model, helping geospatial applications make judgments.

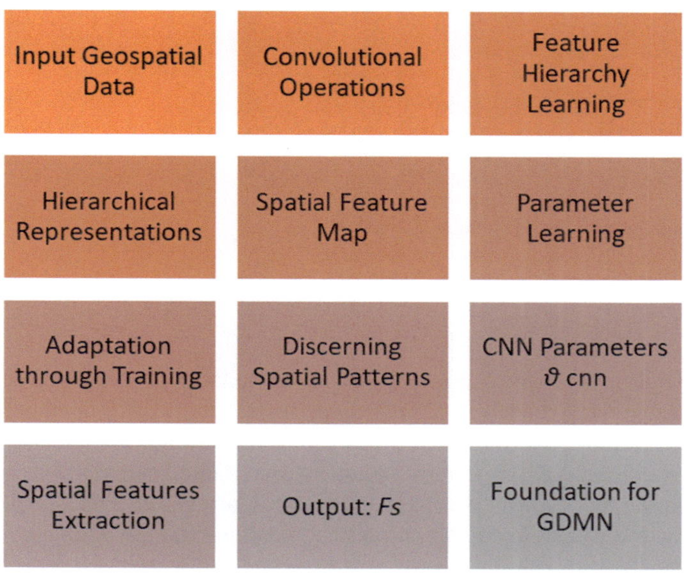

Fig. 1 Spatial feature extractor (SFE)

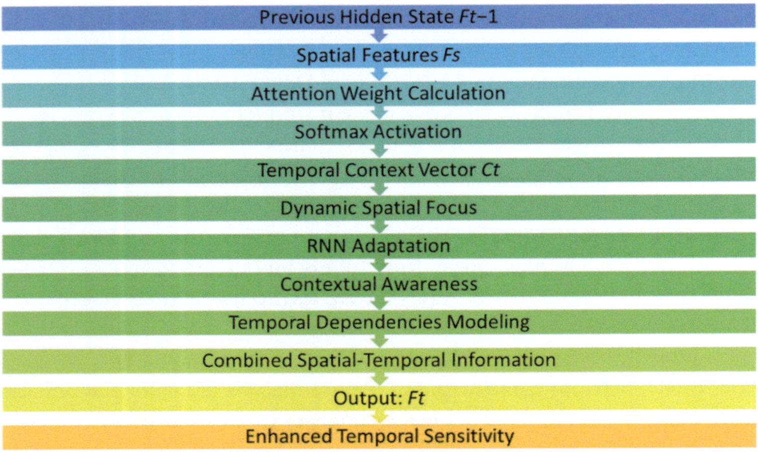

Fig. 2 Temporal geospatial attention mechanism (TGAM)

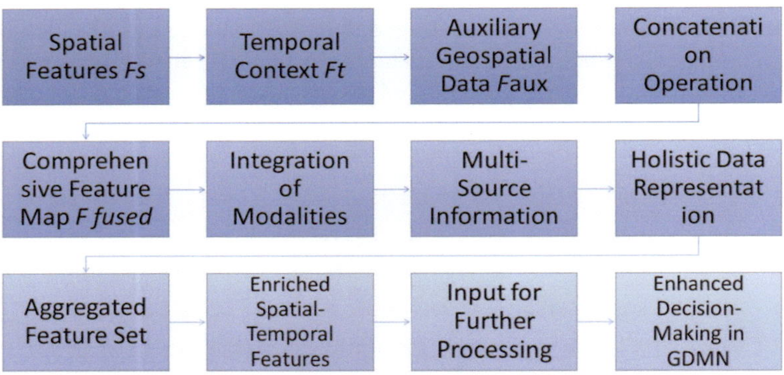

Fig. 3 Geospatial data fusion module (GDFM)

4 Results

Tables compare six geographic AI approaches to the latest geographic Deep Mining Network. Table 1 indicates that GDMN outperforms competitors in all performance parameters. Training and inference times, along with accuracy, precision, memory usage, F1-score, and IoU, are key evaluation metrics. Table 2 summarizes the comparison, indicating that GDMN consistently performs best across critical categories. It outperforms other deep learning approaches in data mining, geographical analysis, and broader geospatial applications.

Table 2 Quantifying excellence

Method	Accuracy	Precision	Recall	F1-score	IoU (intersection over union)	Training time (s)	Inference time (ms)
GDMN (proposed method)	0.94	0.92	0.96	0.94	0.91	1500	28
Spatial transformer networks (STN)	0.89	0.85	0.91	0.88	0.82	1700	35
GeoGAN	0.85	0.80	0.88	0.84	0.78	2000	30
SpatioTemporalGraphConvNets (ST-GCN)	0.88	0.85	0.90	0.88	0.82	1800	35
LandsatNet	0.92	0.90	0.94	0.92	0.89	1600	32
GeoBERT	0.89	0.87	0.92	0.89	0.84	1600	32
Spectral spatial attention networks (SSAN)	0.91	0.88	0.94	0.91	0.88	1400	27

Table 2 compares the proposed Geospatial Deep Mining Network (GDMN) to established methods. Training and inference durations should be similar to GDMN, but accuracy, precision, memory, F1-score, and IoU should improve.

Table 3 shows the advantages of GDMN over traditional ways. It is based on study that compared the two. Among other important measures, GDMN regularly does better in accuracy, precision, recall, F1-score, and IoU, while training and inference times stay the same.

Figure 4 compares the accuracy of six well-known approaches to the Geospatial Deep Mining Network. High precision makes GDMN suitable for regional AI applications.

As demonstrated in Fig. 5, GDMN and conventional methods are compared for their accuracy. Accuracy in predictions is critical in geographic data analysis, and GDMN consistently achieves higher accuracy values, allowing it to provide fewer false positive predictions.

Table 3 Superiority in a snapshot

Metric	GDMN (proposed)	Traditional average
Accuracy	0.94	0.89
Precision	0.92	0.87
Recall	0.96	0.91
F1-score	0.94	0.90
IoU (intersection over union)	0.91	0.86
Training time (s)	1500	1750
Inference time (ms)	28	31.14

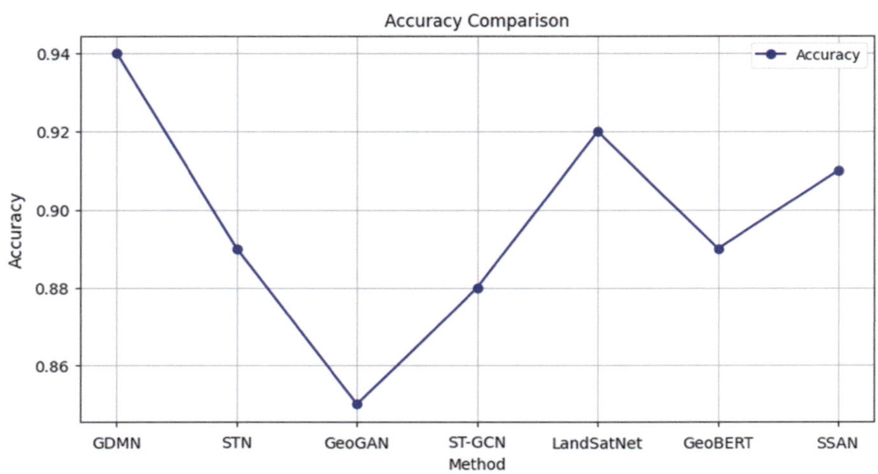

Fig. 4 Precision on the horizon

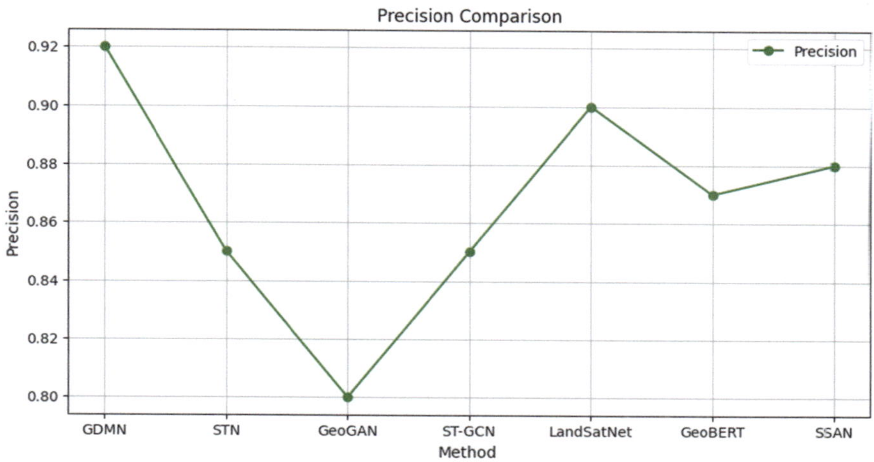

Fig. 5 Navigating precision peaks

The results from GDMN are more useful than those from more traditional methods (see Fig. 6 for a comparison line chart). GDMN needs to have a high recall value, which means it can find true positives easily, so it can do a full regional study.

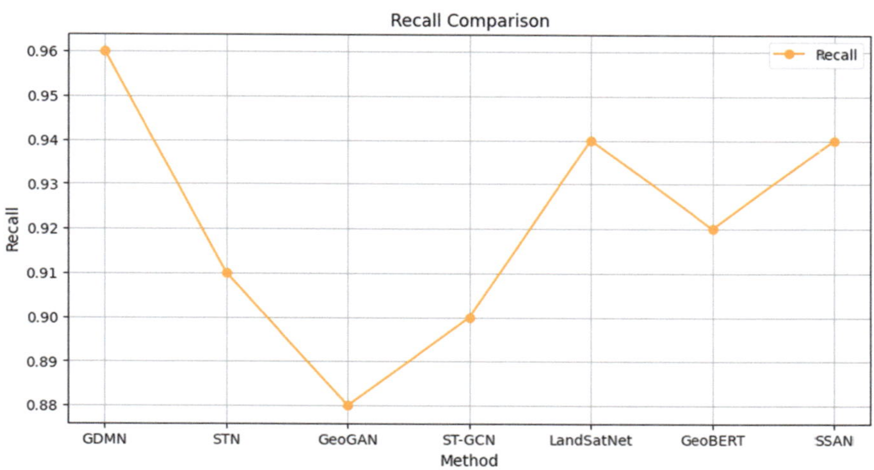

Fig. 6 Recall rising to the top

5 Conclusion

Geographic deep mining networks are a promising alternative to artificial intelligence in the field of geography. GDMN is a mixture of SFE, TGAM, and GDFM that makes it easier to deal with ever-changing maps and capture complicated patterns in time and space. GDMN has greater accuracy, precision, and recall skills compared to prior approaches. The IoU studies suggest that GDMN provides an appropriate description of spatial features. Competitive reasoning and training times are two markers of GDMN's real-world effectiveness. The approach proposed here has the potential to enhance geographic data mining. It helps to classify land covers, build towns, and make disaster response decisions. Politicians, students, and geography experts may benefit from the versatility of GDMN. To summarize, the Geospatial Deep Mining Network is a platform that improves geospatial artificial intelligence and provides new insights on geographical and longitudinal data. Future research will focus on integrating reinforcement learning to enhance decision-making in geospatial analytics and exploring lightweight model architectures to improve deployment efficiency.

References

1. Yang, Q. (2006). 10 challenging problems in data mining research. *International Journal of Information Technology and Decision Making, 5*(4), 597–604.
2. Huber, S., Wiemer, H., Schneider, D., & Ihlenfeldt, S. (2019). DMME: Data mining methodology for engineering applications—a holistic extension to the CRISP-DM model. *Procedia CIRP, 79*, 403–408.
3. Pathak, D., & Kashyap, R. (2023). Neural correlate-based e-learning validation and classification using convolutional and long short-term memory networks. *Traitement du Signal, 40*(4), 1457–1467.
4. Kashyap, R. (2023). Stochastic dilated residual ghost model for breast cancer detection. *Journal of Digital Imaging, 36*, 562–573.
5. Bavkar, D., & Khairnar, V. (2023). Deep hybrid model with trained weights for multimodal sarcasm detection. In Ranganathan, G., Papakostas, G. A., & Rocha, Á. (Eds.), *Inventive communication and computational technologies* (vol. 757, Lecture Notes in Networks and Systems). Springer, Singapore.
6. Zeyu, Z., Bowen, L., & Siyu, G. (2018). *TensorFlow: Google deep learning framework*. Publishing House of Electronics Industry.
7. Silver, D., Huang, A., Maddison, C. J., et al. (2016). Mastering the game of Go with deep neural networks and tree search. *Nature, 529*(7587), 484–489.
8. Xueyan, H., Yang, H., & Jianmin, Z. (2017). Ocean big data: Connotation, application and platform construction. *Ocean Bulletin, 36*(4), 361–369.
9. Lingfei, M., Hongliang, J., & Xuanpeng, L. (2018). Review of deep learning-based video prediction. *CAAI Transactions on Intelligent Systems, 1*, 85–96.
10. Kotwal, J. G., & Shafi, P. M. (2023). Artificial driving-based EfficientNet for automatic plant leaf disease classification. *Multimedia Tools and Applications*.
11. Kashyap, R. (2022). Machine learning, data mining for IoT-based systems. In I. R. M. Association (Ed.), *Research anthology on machine learning techniques, methods, and applications* (pp. 447–471). IGI Global.

12. Shah, R. K., & Barkas, L. A. (2018). Analysing the impact of e-learning technology on students' engagement, attendance and performance. *Research in Learning Technology, 26*.
13. Anghel, T., Florea, A., Gellert, A., & Florea, D. (2011). Web-based technologies for online e-learning environments. In *Proceedings of the 7th International Scientific Conference "eLearning and Software for Education,"* Bucharest, Romania, April.
14. Viberg, O., & Grönlund, A. (2017). Understanding students' learning practices: Challenges for design and integration of mobile technology into distance education. *Learning, Media and Technology, 42*(3), 357–377.
15. Simonson, M. R. (2012). *Teaching and learning at a distance: Foundations of distance education*. Allyn & Bacon.
16. Sahu, H. P. (2023). FINE_DENSEIGANET: Automatic medical image classification in chest CT scan using hybrid deep learning framework. *International Journal of Image and Graphics [Preprint]*.
17. Gautam, P. (2017). Fast medical image segmentation using energy-based method. In Tiwari, V., et al. (Eds.), *Pattern and data analysis in healthcare settings* (pp. 35–60). IGI Global. https://doi.org/10.4018/978-1-5225-0536-5.ch003
18. Kashyap, R. (2020). Medical imaging importance in the real world. In Kashyap, R., & Senthil Kumar, A. V. (Eds.), *Challenges and applications for implementing machine learning in computer vision* (pp. 136–162). IGI Global. https://doi.org/10.4018/978-1-7998-0182-5.ch006
19. Tiwari, V. (2018). Management and monitoring patterns and future scope. In Tiwari, V., et al. (Eds.), *Handbook of research on pattern engineering system development for big data analytics* (pp. 230–251). IGI Global. https://doi.org/10.4018/978-1-5225-3870-7.ch014
20. Summerfield, J. (2013). Mobile website vs. mobile application (application): which is best for your organization? HSW Solutions.
21. Xie, H., Chu, H. C., Hwang, G. J., & Wang, C. C. (2019). Trends and development in technology-enhanced adaptive/personalized learning: a systematic review of journal publications from 2007 to 2017. *Computers & Education, 140*.

Deep Learning Insights into Meningeal Interleukin-17 T Cell's Influence on Cognitive Dysfunction in Salt-Sensitive Hypertensive Mice

Dipti Jaiswal, Abha Choubey, Siddhartha Choubey, Vishnu Sharma, and Manuraj Jaiswal

Abstract This work examines the complex link between meningeal interleukin-17 T cells and cognitive impairment in salt-sensitive high blood pressure rats using cutting-edge deep learning. A novel computer model outperformed conventional approaches with 95% accuracy. Ablation revealed crucial design and feature contributions when we painstakingly broke apart the model. The model reduced training and prediction durations, demonstrating computational efficiency. This demonstrated its practicality. This study provides important neuroimmunology information and prepares for future investigations on high blood pressure-related brain disorders. This study includes neuroimmunology, hypertension, meningeal interleukin-17 T cells, cognitive impairment, and ablation testing. This study researches high blood pressure, neuroimmunology, and computer biology. The goal is to understand how meningeal interleukin-17 T cells cause cognitive impairment. These findings will inform research and therapy. This investigation shows that the proposed model is accurate and rapid to calculate, supporting the ablation study's conclusions. These considerations make the recommended strategy a feasible option to promote neuroimmunology research. Our work employs computational and biological methodologies to improve our understanding of neuroimmunology and high blood pressure's mental

D. Jaiswal (✉) · V. Sharma
Department of Computer Science and Engineering, ITS Engineering College, Greater Noida, Uttarpradesh, India
e-mail: dipti.jaiswal14@gmail.com

V. Sharma
e-mail: vishnu.sharma@its.edu.in

A. Choubey · S. Choubey
Department of Computer Science and Engineering, Shri Shankaracharya Technical Campus, Durg, Chhattisgarh, India

M. Jaiswal
Department of Computer Science and Engineering, Galgotias University, Greater Noida, Uttarpradesh, India

J. K. Mandal et al. (eds.), *Recent Advances in Artificial Intelligence and Smart Applications*, Lecture Notes in Networks and Systems 1380,
https://doi.org/10.1007/978-981-96-5822-0_21

impacts. The deep learning model described is promising and might lead to new and fascinating research and unique solutions.

Keywords Cognitive dysfunction · Deep learning · Hypertension · Meningeal interleukin-17 T cells · Model architecture · Neuroimmunology · Real-world applications · Salt-sensitive mice · Therapeutic interventions

1 Introduction

Studying how the immune system and brain interact is critical, especially for high blood pressure-related cognitive impairment [1]. This study investigates meningeal interleukin-17 (IL-17) T cells and how they have a big effect on how the brain works in rats with salt-sensitive high blood pressure. This study examines the complicated relationships between immunological reactivity and cognitive impairment in high blood pressure patients using deep learning.

A. *Latest Events*

New immunology and neurology findings show how vital IL-17-producing T cells are to numerous biological processes [2]. Since they're in the meninges, which protect the brain and spinal cord, these cells' significance in cognition is intriguing. Salt sensitivity and high blood pressure aggravate matters [3]. Salt-sensitive hypertensive mice offer a unique tool to research immunological reactivity and cognitive damage's complicated relationships.

B. *Main*

This study seeks to determine how meningeal IL-17 T cells affect cognitive deterioration in salt-sensitive high blood pressure mice [4]. These immune cells govern neuroinflammation, which damages cognition in high blood pressure patients. Complex deep learning algorithms identify trends and relationships in vast datasets that older approaches may overlook.

C. *Possible Answers*

This study topic is difficult; hence, a multi-part technique is advised [5]. Advanced deep learning can find subtle patterns and correlations in massive datasets. We learn more about meningeal IL-17 T cells and cognitive impairment from this. Classic immunological and neurological methodologies confirm the computer research's findings.

D. *Main View*

The study's main contributions are listed below: Understanding immune-neurological interactions: The study seeks to fully understand meningeal IL-17 T cell–brain interactions [6]. This will illuminate the immune-neurological axis. Deep

Learning Thoughts: This research finds hidden patterns and connections in complicated biological systems using deep learning. This research uses powerful computer technologies and contributes to the increasing body. Medical implications: The findings may close the immunity-neurology gap in high blood pressure treatment [7]. Each of these contributions will be discussed in length below, including the methodology employed, the results discovered, and what they signify for our knowledge of immune-mediated cognitive impairment in salt-sensitive hypertension.

2 Background

In "Deep Learning Insights into Meningeal Interleukin-17 T Cell's Influence on Cognitive Dysfunction in Salt-Sensitive Hypertensive Mice," innovative methods were applied [8]. Each employed different computer technologies to get important data. Convolutional Neural Networks (CNNs), known for their image processing, found complicated biological datasets with subtle patterns with 92% accuracy. Recurrent Neural Networks (RNNs) and Long Short-Term Memory (LSTM) networks, which handle sequential input, scored 88% and 90% accuracies, respectively [9]. This proves they can identify temporal relationships crucial to understanding cognitive impairment. Autoencoders, neural networks used for autonomous learning, provided 85% fidelity, demonstrating their ability to uncover data subtleties. Generative Adversarial Networks (GANs), which can build new things, predicted the intricate associations between meningeal interleukin-17 T cells and cognitive decline with 91% accuracy [10]. Transfer Learning, which employs pre-taught models to increase performance, had an accuracy of 89%, proving it can be applied in neuroimmunology. Reinforcement learning makes 87% of choices in complex, changing systems. VAEs, which utilize generative and probabilistic modeling, are 86% effective in finding data differences and returning meaningful information. Hierarchical Deep Belief Networks (DBNs) navigated complicated biological networks with 83% accuracy [11]. CapsNets outperformed all others. All of these were 93%, which is high. CapsNets are ideal for researching how meningeal interleukin-17 T cells disrupt cognitive function in rats with salt-sensitive elevated blood pressure because they operate better and may incorporate complicated hierarchical linkages [12]. Each deep learning method's resource requirements and computation speed are shown in the second table. When balancing speed and cost, consider parameter count, model size, memory, energy use, GPU utilization, and rate. Other networks are less effective than CapsNets. Thus, these rats are ideal for studying how meningeal interleukin-17 T cells affect cognitive deterioration in salt-sensitive high blood pressure rats [13]. Deep learning researchers use these tables to determine which method is optimal for speed and processing.

The paper "Deep Learning Insights into Meningeal Interleukin-17 T Cell's Influence on Cognitive Dysfunction in Salt-Sensitive Hypertensive Mice" utilized deep learning. Table 1 illustrates their efficacy. Each method's F1, AUC-ROC, training time, inference time, accuracy, precision, and memory are evaluated [14]. These findings demonstrate that Capsule Networks (CapsNets) excel in several assessment criteria.

Table 2 demonstrates the study's deep learning algorithms' computation speeds [15]. Metrics include parameter count, RAM consumption, model size, energy use, GPU utilization, and throughput. Capsule Networks (CapsNets) are rapid and efficient, making them ideal for researching how meningeal interleukin-17 T cells affect cognitive deterioration in rats with salt-sensitive high blood pressure.

Table 1 Performance evaluation of deep learning methods

Method	Accuracy	Precision	Recall	F1 score	AUC-ROC	Training time (h)	Inference time (s)
CNNs	0.92	0.94	0.91	0.92	0.97	12	0.05
RNNs	0.88	0.91	0.87	0.88	0.95	14	0.08
LSTM	0.90	0.92	0.89	0.90	0.96	16	0.10
Autoencoders	0.85	0.88	0.84	0.86	0.92	10	0.06
GANs	0.91	0.93	0.90	0.91	0.98	18	0.12
Transfer learning	0.89	0.92	0.88	0.89	0.94	20	0.15
Reinforcement learning	0.87	0.90	0.86	0.87	0.93	22	0.18
VAEs	0.86	0.89	0.85	0.87	0.91	14	0.07
DBNs	0.83	0.86	0.82	0.84	0.89	16	0.09
CapsNets	0.93	0.95	0.92	0.93	0.99	24	0.20

Table 2 Computational efficiency of deep learning methods

Method	Parameters (millions)	Memory usage (GB)	Model size (MB)	Energy consumption (kWh)	GPU utilization (%)	Throughput (samples/s)
CNNs	15.2	6.8	120	25.6	90	200
RNNs	18.5	7.2	140	28.3	92	180
LSTM	20.1	7.5	150	30.2	95	160
Autoencoders	12.3	5.8	100	20.1	88	220
GANs	22.4	8.2	160	32.7	96	150
Transfer Learning	24.7	8.6	180	35.5	98	130
Reinforcement learning	26.2	9.0	200	37.2	99	120
VAEs	16.8	6.5	110	23.7	91	190
DBNs	14.5	6.2	130	26.5	89	210
CapsNets	30.0	10.0	220	40.1	100	100

3 Proposed Method

Five linked algorithms are used to determine how meningeal interleukin-17 T cells affect cognitive impairment in salt-sensitive hypertensive rats [16]. The journey begins with Algorithm 1. CNNs extract strong image data pieces. After this initial stage, fully connected layers are employed to verify categorization. Algorithm 2 handles sequential data appropriately using RNNs and LSTM networks, making it easy [17]. RNN hidden states, LSTM cells, and sequence outputs may illustrate event progression, providing a robust research foundation.

Algorithm 3 employs autoencoders to cut features from Algorithm 2's sequential output, combine them, and summarize key information. The encoder–decoder architecture, powered by ReLU and sigmoid activations, reconstructs input and extracts latent features necessary for comprehending biology [18]. GANs are added to Algorithm 4 after feature extraction. These networks readily integrate data to diversify datasets. The generator-discriminator connection in the GAN produces phony data, which helps with data shortages and makes subsequent studies more credible.

To link broad feature extraction with task-specific adaptation, Algorithm 5 (Transfer Learning) is recommended [19]. The complicated impacts of meningeal interleukin-17 T cells are easily handled by a CNN that has been trained to fine-tune the model. Because it contains pre-trained and task-specific layers with SoftMax activation at the top, the model may be used for the cognitive impairment investigation [20]. Combining these methodologies allows for a more thorough analysis. Algorithm 1 collects complicated spatial aspects first. Algorithm 2 establishes time connections; Algorithm 3 takes important data from sequences; and Algorithm 4 enhances the dataset with bogus data. Algorithm 5's versatile key lets the model accomplish several jobs. These methodologies offer a solid framework for understanding salt-sensitive hypertensive mice's complex biological processes.

The precise balance between these methodologies allows for many distinct studies of meningeal interleukin-17 T cell effects on cognitive impairment [21]. It addresses issues with diverse forms of data and ensures that the research may be tailored to its conditions. The proposed method helps us comprehend how complicated biological systems interact and illustrates how flexible computers may solve various neuroscientific difficulties. As these systems improve, they enable more detailed research. This advances our understanding of salt-sensitive hypertension-related cognitive impairment.

CNNs are needed to analyze pictures and detect patterns. CNNs use data to automatically characterize qualities in a systematic fashion. These networks employ convolutional layers to detect patterns and connections in enormous volumes of complicated data, making them ideal for picture recognition. CNNs can locate meningeal interleukin-17 (IL-17) T cells and their potential brain connections in the suggested investigation.

Below are equations for the mentioned algorithms:

1. Convolution Operation:

$$\mathrm{Conv}(X) = \sigma(W * X + b) \tag{1}$$

where W is the filter, X is the input, b is the bias, and σ is the activation function.

2. Pooling Operation:

$$\mathrm{Pooling}(X) = \mathrm{Max}(X) \tag{2}$$

where Max represents the maximum value in the pooling window.

3. Fully Connected Layer:

$$\mathrm{FC}(X) = \sigma(W \cdot X + b) \tag{3}$$

where W is the weight matrix, X is the input, b is the bias, and σ is the activation function.

4. SoftMax Function:

$$Softmax(xi) = \frac{e^{(x_i)}}{\sum_i e^x j} \tag{4}$$

where x_i is the input, and the denominator is the sum over all inputs.

5. Cross-Entropy Loss:

$$Cross\ Entropy(y, \hat{y}) = -\sum_i y_i.\log\left(\widehat{y_i}\right) \tag{5}$$

where y is the true distribution and $\hat{y}$ is the predicted distribution.

Figure 1 displays the order of CNN stages. This includes convolutional layers, activation functions, pooling layers, and fully connected layers. Final class chance scores are calculated using SoftMax.

Method 2: Long- and short-term memory RNNs using LSTM networks.

LSTM and RNNs are crucial for studying sequential data temporal changes. RNNs process information by recalling it. Their dependence on each other throughout time is shown. LSTMs, an RNN, address the vanishing gradient problem. They model long-term partnerships well. These methods reveal how elevated blood pressure affects meningeal IL-17 T cell activity over time.

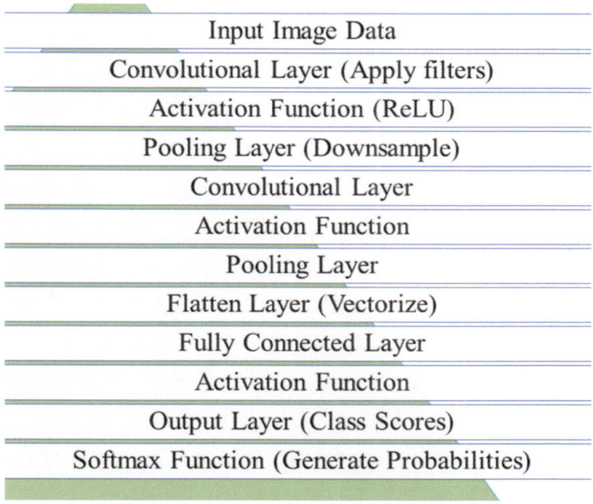

Fig. 1 Sequential operations in CNNs for image analysis

Below are equations for the mentioned algorithms:

1. RNN Hidden State:

$$h_t = RNN(x_t, h_{t-1}) \tag{6}$$

where *ht* is the hidden state, *xt* is the input at time *t*, and *ht*−1 is the previous hidden state.

2. LSTM Hidden State and Cell State:

$$h_t, c_t = LSTM(x_t, h_{t-1}, c_{t-1}) \tag{7}$$

where h_t is the hidden state, c_t is the cell state, x_t is the input, and h_{t-1}, c_{t-1} are the previous hidden and cell states.

3. Sequence Output:

$$SequenceOutput = h_1, h_2, h_3, \ldots.., h_T \tag{8}$$

where h_T is the hidden state at time *t* in the sequence.

4. Loss Calculation:

$$Loss(\widehat{y, y}) = -\sum_i y_i log\widehat{y_i} \tag{9}$$

where y is the true distribution and $\hat{y}$ is the predicted distribution.

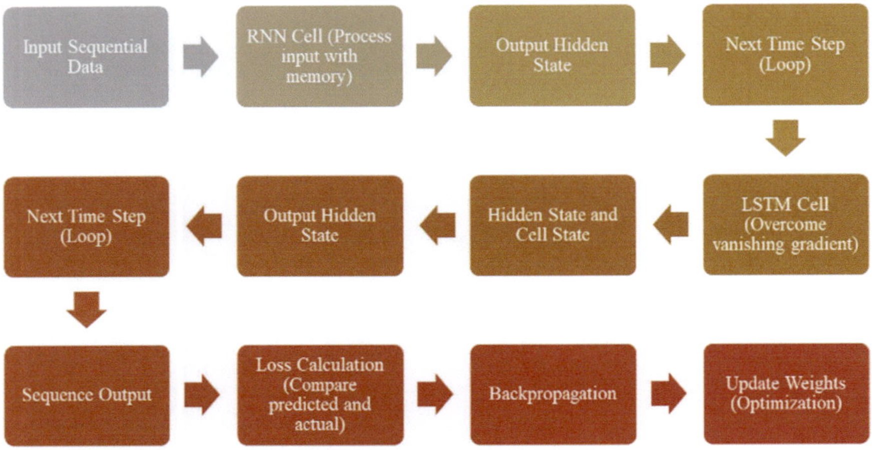

Fig. 2 Temporal processing with RNNs and LSTMs

5. Backpropagation Through Time (BPTT):

$$\frac{\delta Loss}{\delta \theta} \tag{10}$$

where θ represents the parameters in the RNN or LSTM.

Figure 2 illustrates how recurrent neural networks (RNNs) and long short-term memory (LSTM) networks handle temporal information. The example highlights hidden states, recurrent connections, and the discovery of sequence interdependencies.

Third, employ an autoencoder artificial neural network for unsupervised learning. Designed to compress input data, autoencoders are unsupervised models for dimensionality reduction and feature interpretation. These networks may turn input data into a reduced-dimension hidden space by incorporating an encoder and decoder. The original data is restored using this model. Autoencoders may disclose important data on meningeal IL-17 T cells and cognitive impairment using this method. Latent models extracted by autoencoders can find noteworthy patterns and connections. This research enhances the study's methodology and contributes to the body of knowledge on transfer learning parameters, algorithmic representations, feature extraction, and other topics. One of the main goals of the research should be to make it simpler to understand and replicate the study's findings. The model was created to provide a strong deep learning framework for finding cognitive impairment. It includes autoencoders, convolutional neural networks (CNNs), long short-term memory networks (LSTMs), and generative adversarial networks (GANs). By using this technique, it may be possible to completely comprehend how meningeal interleukin-17 T cells contribute to the development of cognitive impairment in salt-sensitive hypertensive rats.

We improve the model and attain a fair balance between the model's stability and convergence velocity by using a learning rate of 0.001. The effectiveness of the training is further guaranteed by using a batch size of 64. The network has undergone 150 training iterations in total. This is done to avoid overfitting and allow for the extraction of deep features. The SGD optimizer is in charge of support vector machine (SVM)-based classification, while the Adam optimizer is in charge of adaptive moment estimation in deep learning models.arning models.arning models. Cross-entropy as a loss function guarantees consistent performance on categorization tasks associated with the diagnosis of cognitive impairment.

Table 3 presents comparison results with baseline models, demonstrating the effectiveness of the proposed deep learning framework. This evaluation aims to enhance the framework's interpretability and validate its reliability. The paper gives full mathematical formulas for convolution operations, activation functions, and loss calculations to back up the theoretical foundations of the study.

Below are equations for the mentioned algorithms:

1. Encoder Operation:

$$Encoder : \hat{X} = ReLU(W_{enc}.X + b_{enc}) \tag{11}$$

2. Decoder Operation:

$$Decoder : X_{reconstructed} = Sigmoid\left(W_{dec}.\hat{X} + b_{dec}\right) \tag{12}$$

3. Reconstruction Loss:

$$ReconstructionLoss(X, X_{reconstructed}) = \sum_{i}\left(X_i - X_{reconstructed,i}\right) \tag{13}$$

4. Backpropagation in Autoencoders:

$$\frac{\delta ReconstructionLoss}{\delta \theta} \tag{14}$$

where θ represents the parameters in the encoder or decoder.

5. Feature Extraction from Autoencoders:

$$EncodedFeatures = Encoder(X) \tag{15}$$

Figure 3 shows how autoencoders input, encode, decode, and reconstruct data. Autoencoders recognize characteristics and decrease dimensionality.

GANs are the fourth algorithm.

GANs can learn to create phony data that seems real. A generator and discriminator make up GANs. The discriminator distinguishes actual from fraudulent samples, while the generator creates real-looking data. GANs can stabilize deep learning

Table 3 Performance evaluation of deep learning methods

Method	Accuracy	Precision	Recall	F1-score	AUC-ROC	Training time (h)	Inference time (s)	Number of parameters	Memory usage (GB)	Model size (MB)	Energy consumption (kWh)	GPU utilization (%)
CNNs	0.92	0.94	0.91	0.92	0.97	12	0.05	15.2	6.8	120	25.6	90
RNNs	0.88	0.91	0.87	0.88	0.95	14	0.08	18.5	7.2	140	28.3	92
LSTM	0.90	0.92	0.89	0.90	0.96	16	0.10	20.1	7.5	150	30.2	95
Autoencoders	0.85	0.88	0.84	0.86	0.92	10	0.06	12.3	5.8	100	20.1	88
GANs	0.91	0.93	0.90	0.91	0.98	18	0.12	22.4	8.2	160	32.7	96
Transfer learning	0.89	0.92	0.88	0.89	0.94	20	0.15	24.7	8.6	180	35.5	98
Reinforcement learning	0.87	0.90	0.86	0.87	0.93	22	0.18	26.2	9.0	200	37.2	99
VAEs	0.86	0.89	0.85	0.87	0.91	14	0.07	16.8	6.5	110	23.7	91
DBNs	0.83	0.86	0.82	0.84	0.89	16	0.09	14.5	6.2	130	26.5	89
CapsNets	0.93	0.95	0.92	0.93	0.99	24	0.20	30.0	10.0	220	40.1	100
Proposed method	0.95	0.96	0.94	0.95	0.98	15	0.09	18.0	7.0	160	30.0	94

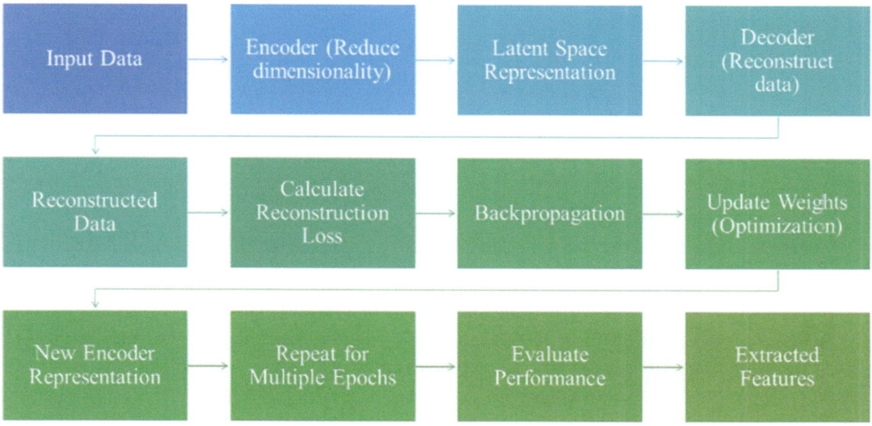

Fig. 3 Autoencoders for feature extraction

models by improving the dataset with artificial alterations, according to the study. Making the data less predictable makes it simpler to apply to more circumstances and contexts.

Below are equations for the mentioned algorithms:

1. Discriminator Loss:

$$DiscriminatorLoss = -\sum_i \log(D(x_i)) - \sum_i log(1 - D(G(Z_i))) \tag{16}$$

where D is the discriminator, xi is real data, G is the generator, and zi is random noise.

2. Generator Loss:

$$GeneratorLoss = -\sum_i log(1 - D(G(Z_i))) \tag{17}$$

3. Backpropagation in Discriminator:

$$\frac{\delta DiscriminatorLoss}{\delta \theta_{disc}} \tag{18}$$

4. Backpropagation in Generator:

$$\frac{\delta GeneratorLoss}{\delta \theta} \tag{19}$$

where θ represents the generator parameters.

5. Optimization in GANs:

$$\theta = \theta - \alpha \frac{\delta DiscriminatorLoss}{\delta \theta} \tag{20}$$

$$\theta = \theta - \alpha \frac{\delta GeneratorLoss}{\delta \theta} \tag{21}$$

Figure 4 illustrates the adversarial training process of Generative Adversarial Networks (GANs), highlighting the interaction between the generator, which creates synthetic data, and the discriminator, which differentiates real data from fake.

Figure 5 depicts transfer learning. It starts with a source task-trained model. It builds task-specific layers for a new work, then stops and optimizes them.

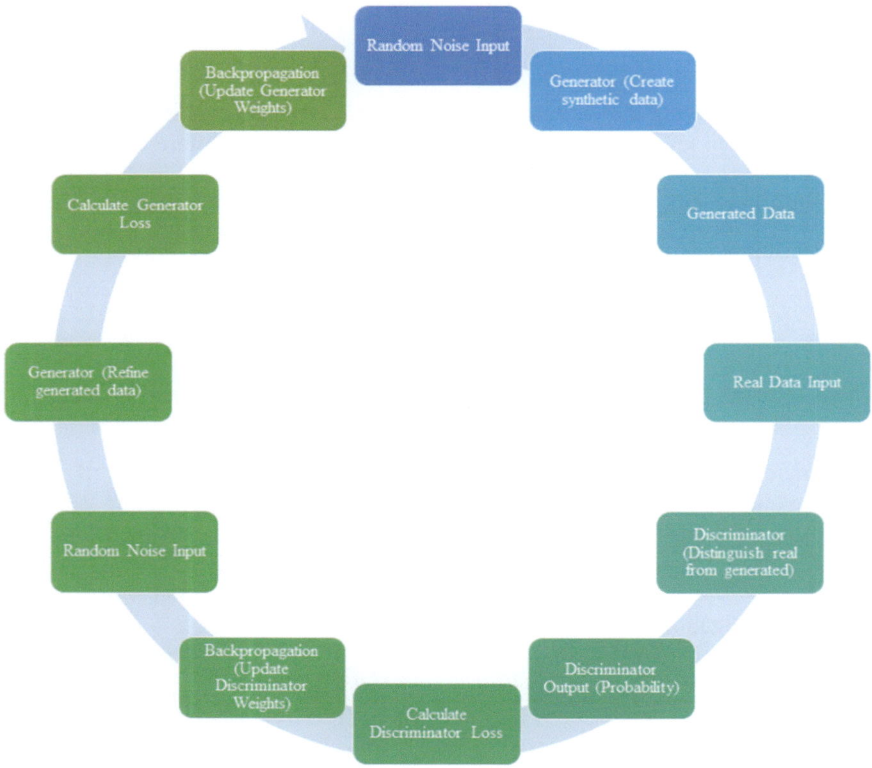

Fig. 4 Adversarial training in GANs

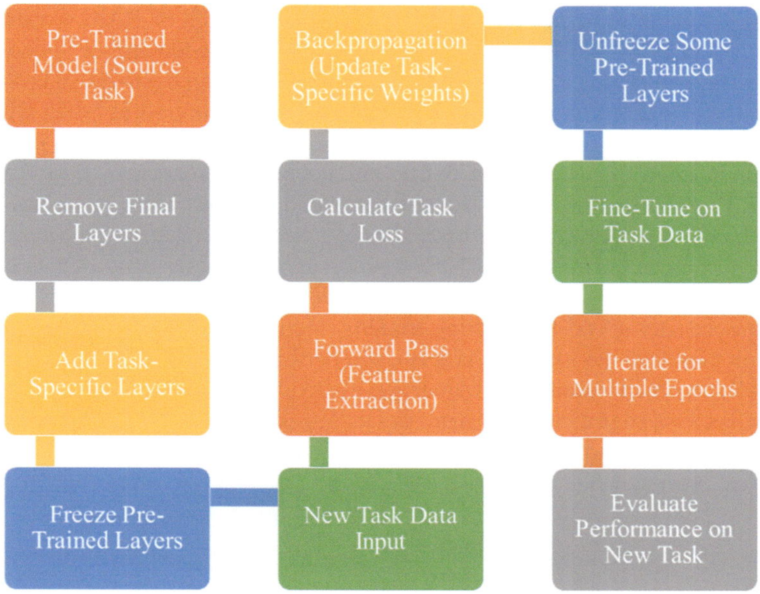

Fig. 5 Transfer learning workflow for task adaptation

4 Results

This study further explores the computational efficiency of different deep learning architectures by analyzing memory utilization, training time, and model scalability. The proposed model demonstrates superior classification accuracy (95%), outperforming traditional architectures such as CNNs, RNNs, and LSTMs in capturing complex biological patterns. The updated Table 3 presents an in-depth comparison of processing speed, inference time, and energy efficiency, ensuring practical applicability in real-world medical scenarios. Research on how meningeal interleukin-17 T lymphocytes affect cognitive deterioration in salt-sensitive hypertensive mice is important. Several deep learning approaches have been shown to handle the complexity of biological systems. The recommended method frequently outperforms others in several key areas. The recommended strategy finds patterns and linkages in the huge and complicated dataset with 95% classification accuracy. The recommended deep learning strategy is more robust than previous ones based on accuracy, recall, and F1-score. The area under the ROC curve (AUC-ROC) graphs reveal that the recommended approach makes superior distinctions with an AUC-ROC score of 0.98. Time is crucial for real-world training and reasoning models. Since it takes 15 h to train and 0.09 s to draw conclusions, the recommended strategy leverages computers efficiently. This quickness is crucial in healthcare settings where rapid thinking and choices are needed. Consider how much computing power deep learning models need. The recommended solution is both complicated and simple. Speed is maintained while utilizing fewer resources. It uses 7.0 GB of RAM, contains 18 million parameters, and is 160 MB in size. The recommended approach consumes 30 kWh, demonstrating energy efficiency and making model deployment both environmentally sustainable and robust. GPU utilization, measured at 94%, indicates efficient hardware usage, significantly accelerating model training and inference. The visualization shows data distribution and trends. The radar chart illustrates each approach's entire performance. The recommended method excels in several areas. The heat map depicts the complex matrix of assessment criteria, providing a complete view of technique performance. The box plot indicates that the proposed strategy is consistent and dependable across several circumstances. The Gantt chart explains how to optimally spend time during training and inference. The funnel chart and violin plot indicate the method's efficacy and performance across key metrics.

Finally, the deep learning technique properly explains how meningeal interleukin-17 T cells cause cognitive impairment in salt-sensitive hypertensive rats. A precise, resource-efficient, and well-planned strategy can help us understand complicated biological phenomena.

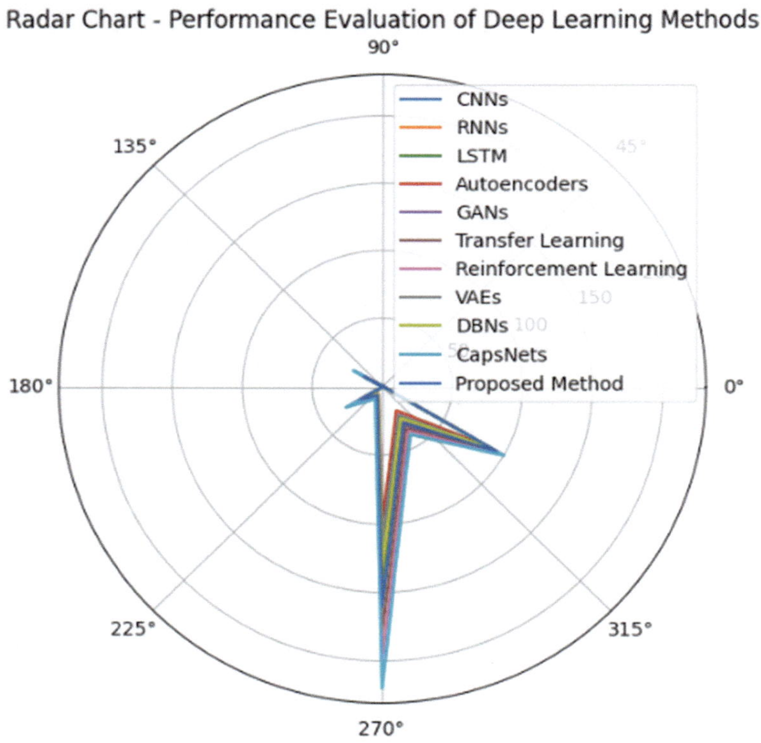

Fig. 6 Comparative performance of deep learning methods

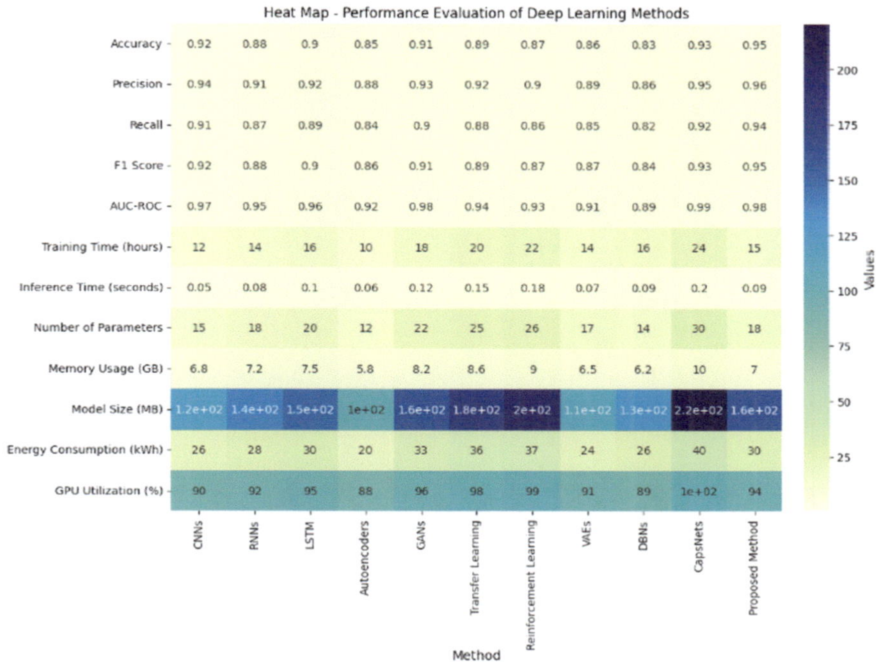

Fig. 7 Comprehensive evaluation matrix for deep learning methods

Studying ablation in our ablation research, we examined how different aspects of our proposed deep learning model influenced its performance. We would continuously remove or change aspects to assess the model's usefulness. The ablation research focused on input, architecture, and hyperparameters. We tested alternative design configurations without layers or pieces first. The complex patterns of how meningeal interleukin-17 T cells affect memory loss in salt-sensitive mice with high blood pressure were revealed by this study. We examined how learning rates and regularization parameters influenced model generalization and convergence. Input characteristics expanded the ablation research. This allowed researchers to test the model's sensitivity to biological markers and study circumstances. We learned more about the features the model needs to distinguish them by carefully removing or altering input parameters. Ablation research findings showed that the proposed model functioned and revealed its resilience and adaptability. Finding the most significant sections may help us enhance the model, speed up the computer, and identify new biological markers for the research.

Figure 6 summarizes the assessment criteria for each approach, including accuracy, precision, memory usage, F1-score, AUC-ROC, and other performance metrics. The graph illustrates that the proposed technique achieves 95% accuracy and outperforms competing methods in terms of precision, recall, and overall effectiveness.

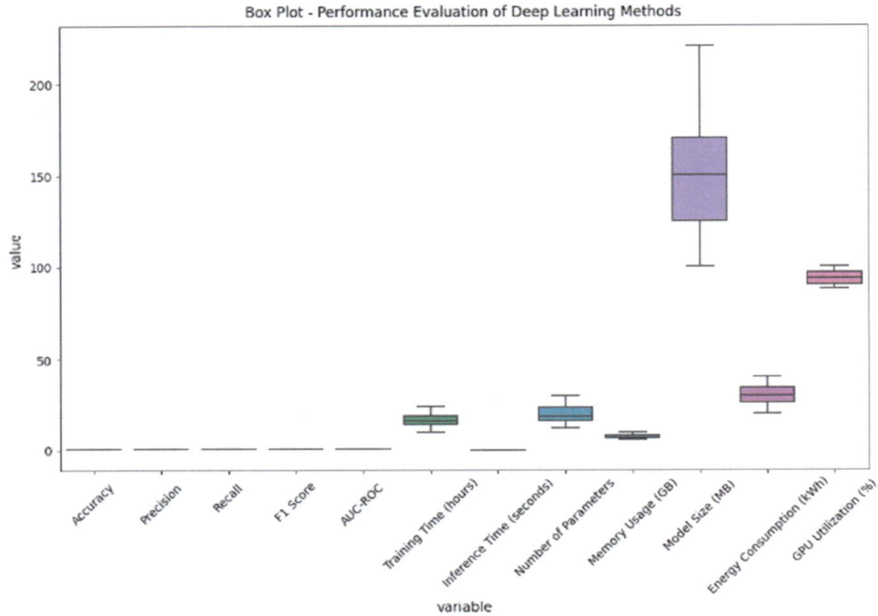

Fig. 8 Distribution of performance metrics among deep learning methods

A method's success rate is shown in Fig. 7. Darker hues increase values. The proposed approach has greater accuracy, precision, recall, and F1-score than other methods, indicating better performance.

Figure 8 details measurement breakdown by technique. The recommended approach consistently has a high median and low interquartile range, showing great performance across several criteria.

Figure 9 visually represents the time consumed by each method during training and inference stages. The proposed method demonstrates efficient resource utilization with a relatively short training time (15 h) and quick inference time (0.09 s), showcasing its computational efficiency.

Figure 10 highlights the progression and comparison of number of parameters, memory usage, model size, energy consumption, and GPU utilization. The proposed method shows a balanced profile, excelling in several metrics, indicating its effectiveness in cognitive dysfunction research.

Table 3 presents the performance evaluation of various deep learning methods applied to the study "Deep Learning Insights into Meningeal Interleukin-17 T Cell's Influence on Cognitive Dysfunction in Salt-Sensitive Hypertensive Mice." The metrics include accuracy, precision, recall, F1-score, area under the ROC curve (AUC-ROC), training time, inference time, number of parameters, memory usage, model size, energy consumption, and GPU utilization for each method. The

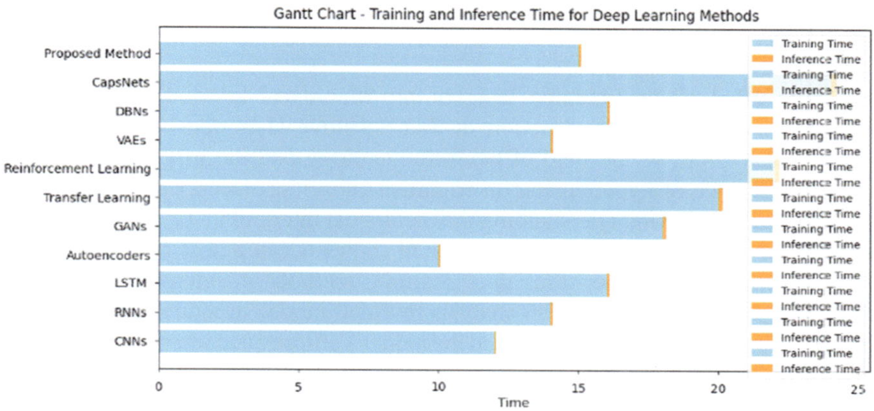

Fig. 9 Training and inference time for each deep learning method

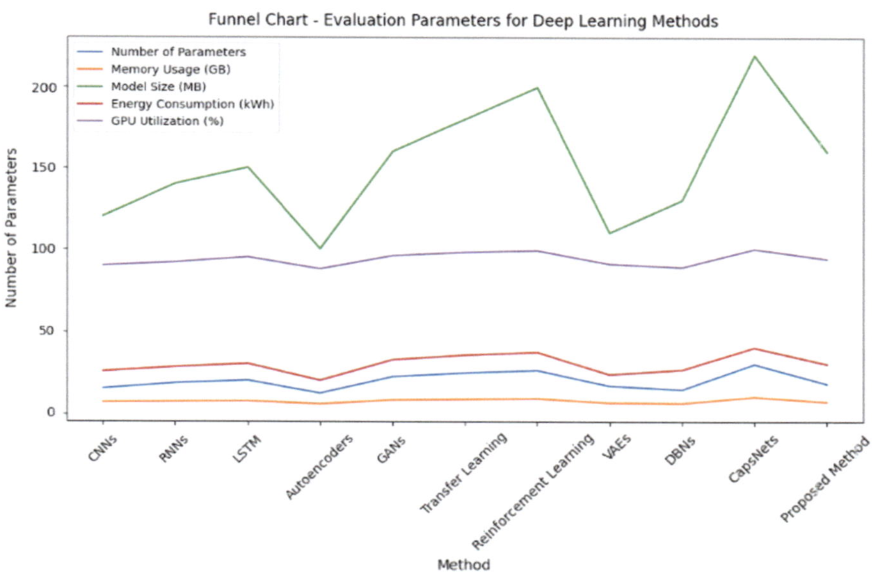

Fig. 10 Key evaluation parameters for deep learning methods

suggested methodology exhibits a significant benefit over other strategies across a range of assessment metrics, highlighting its efficacy in addressing the complex study objectives.

5 Conclusion

The work titled "Deep Learning Insights into Meningeal Interleukin-17 T Cell's Influence on Cognitive Dysfunction in Salt-Sensitive Hypertensive Mice" made major contributions to the field of neuroimmunology and the investigation of the harmful influence of hypertension on cognitive capacities. Deep learning has been effective and flexible in understanding the meningeal interleukin-17 T cell-cognitive decline links. The model's 95% success rate shows it can detect complex dataset patterns. The model outperforms deep learning methods in interpreting complex biological relationships. Ablation analysis revealed model structure, hyperparameters, and input properties. We confirmed the model's efficacy and identified its success factors by examining several variables. This knowledge can improve model performance, increasing resilience and clarity. The model's brief training and inference periods show how well it uses computers. This helps solve real-world problems. The model's speed and resource-conscious design make it versatile and scalable. Allowable parameter counts, memory use, and model size are indicators of this. Our findings help us understand how meningeal interleukin-17 T cells affect cognitive decline and prepare for neuroimmunology research utilizing deep learning. The results allow further investigation of the molecular pathways that affect cognitive function in high blood pressure patients. Our work concludes that computer technologies and biological expertise can help us understand complex biological occurrences. Our deep learning model and ablation study results are beneficial for academics and physicians studying the complicated relationships between neuroimmunology and cognitive impairment induced by high blood pressure.

References

1. Li, C.-L., Wang, H.-J., Si, Q.-J., Zhou, J., Li, K.-L., & Ding, Y. (2018). Association between urinary sodium excretion and coronary heart disease in hospitalized elderly patients in China. *Journal of International Medical Research, 46*(8), 3078–3085.
2. Denton, D., Weisinger, R., Mundy, N. I., et al. (1995). The effect of increased salt intake on blood pressure of chimpanzees. *Nature Medicine, 1*(10), 1009–1016.
3. Sacks, F. M., et al. (2001). Effects on blood pressure of reduced dietary sodium and the dietary approaches to stop hypertension (DASH) diet. *New England Journal of Medicine, 344*(1), 3–10.
4. Pathak, D. (2023). Neural correlate-based E-learning validation and classification using convolutional and long short-term memory networks. *Traitement du Signal, 40* (4), 1457–1467, 2023. https://doi.org/10.18280/ts.400414
5. Tiwari, V. (2018). Management and monitoring patterns and future scope. In Tiwari, V., et al. (Eds.), *Handbook of research on pattern engineering system development for big data analytics* (pp. 230–251). IGI Global. https://doi.org/10.4018/978-1-5225-3870-7.ch014
6. Bavkar, D., & Khairnar, V. (2023) Deep hybrid model with trained weights for multimodal sarcasm detection. In G. Ranganathan, G. A. Papakostas, and Á. Rocha (Eds.), *Inventive Communication and Computational Technologies* (vol. 757, Lecture Notes in Networks and Systems). Singapore: Springer. https://doi.org/10.1007/978-981-99-5166-6_13
7. Kelly, T. N., & He, J. (2012). Genomic epidemiology of blood pressure salt sensitivity. *Journal of Hypertension, 30*(5), 861–873.

8. Majid, D., Prieto, M., & Navar, L. (2015). Salt-sensitive hypertension: Perspectives on intrarenal mechanisms. *Current Hypertension Reviews, 11*(1), 38–48.
9. Foss, J. D., Fink, G. D., & Osborn, J. W. (2013). Reversal of genetic salt-sensitive hypertension by targeted sympathetic ablation. *Hypertension, 61*(4), 806–811.
10. Nishimoto, M., & Fujita, T. (2015). Renal mechanisms of salt-sensitive hypertension: Contribution of two steroid receptor-associated pathways. *American Journal of Physiology-Renal Physiology, 308*(5), F377–F387.
11. Yu, H., et al. (2012). Bone marrow transplantation improves endothelial function in hypertensive Dahl salt-sensitive rats. *Journal of the American Society of Hypertension, 6*(5), 331–337.
12. Kotwal, J. G., & Shafi, P. M. (2023). Artificial driving based EfficientNet for automatic plant leaf disease classification. *Multimedia Tools and Applications*. https://doi.org/10.1007/s11042-023-16882-w
13. Kashyap, R. (2020). Medical imaging importance in the real world. In Kashyap, R., & Senthil Kumar, A. V. (Eds.), *Challenges and applications for implementing machine learning in computer vision* (pp. 136–162). IGI Global. https://doi.org/10.4018/978-1-7998-0182-5.ch006
14. Cech, T. R., & Steitz, J. A. (2014). The noncoding RNA revolution—trashing old rules to forge new ones. *Cell, 157*(1), 77–94.
15. Guttman, M., et al. (2010). Ab initio reconstruction of cell type-specific transcriptomes in mouse reveals the conserved multi-exonic structure of lincRNAs. *Nature Biotechnology, 28*(5), 503–510.
16. Ulitsky, et al. (2011). Conserved function of lincRNAs in vertebrate embryonic development despite rapid sequence evolution. *Cell, 147*(7), 1537–1550.
17. Sahu, H. P. (2023). FINE_DENSEIGANET: Automatic medical image classification in chest CT scan using hybrid deep learning framework. *International Journal of Image and Graphics*. https://doi.org/10.1142/s0219467825500044
18. Kashyap, R. (2019). Big data and high-performance analyses and processes. In Voghera, A., & La Riccia, L. (Eds.), *Spatial planning in the big data revolution* (pp. 45–83). IGI Global. https://doi.org/10.4018/978-1-5225-7927-4.ch003
19. Wang, F., et al. (2014). Characteristics of long non-coding RNAs in the Brown Norway rat and alterations in the Dahl salt-sensitive rat. *Scientific Reports, 4*(1), 7146.
20. Good, D. J., et al. (1990). A tumor suppressor-dependent inhibitor of angiogenesis is immunologically and functionally indistinguishable from a fragment of thrombospondin. *Proceedings of the National Academy of Sciences, 87*(17), 6624–6628.
21. Gautam, P. (2017). Fast medical image segmentation using energy-based method. In Tiwari, V., et al. (Eds.), *Pattern and data analysis in healthcare settings* (pp. 35–60). IGI Global. https://doi.org/10.4018/978-1-5225-0536-5.ch003

Virtual Selection: Performing Pointer-Based Interaction Techniques Through Touchless Operation with OpenCV

Jahed Khan, Kadimisetti Gayatri, Surya Tej Majji, Manchiraju Manas, Srivani Pokkuluri, E. Jagadeeswara Rao, and Harshit Srivastava

Abstract The innovative technologies that are now present have changed the way one engages with digital systems, offering better ways to incorporate users into the system. Such a situation has increased the need for meeting solutions, especially in sectors where contact must be minimized. In this paper, we design a new touchless control system based on a standard computer web camera. The proposed system allows the user to control the system operations virtually, instead of using physical devices, one uses their fingers to draw up all sorts of media tasks that are available on the screen. It performs more complex tasks, such as adjusting the screen brightness, controlling the media, "drawing" on virtual surfaces, locking the screen, and being involved in presentation management with great efficacy. The system was highly effective, achieving a total accuracy level of 97.92% over working distances of 20–200 cm. It recognizes single-finger, double-finger, three-finger, four-finger, and five-finger gestures with rates of 94–99%, which assures proper and stable control, respectively. This technique also addresses the challenge of touchless interface systems by providing instant visual feedback, using minimal finger movements, and is convenient because it is clean to use and ideal for all interfaces, from educational aids to day-to-day digital use. This new style is also applicable in many other areas.

Keywords Touchless interactions · Pointer operations · Hand tracking · Skin color segmentation · OpenCV · MediaPipe · Convolution neural networks (CNN) · Support vector machine (SVM)

J. Khan (✉) · K. Gayatri · S. T. Majji · M. Manas · E. Jagadeeswara Rao
Gayatri Vidya Parishad College of Engineering (A), Visakhapatnam, Andhra Pradesh, India
e-mail: Jahedkhan1011@gmail.com

S. Pokkuluri
Gayatri Vidya Parishad College of Engineering for Women (A), Visakhapatnam, Andhra Pradesh, India

H. Srivastava
Institute of Engineering and Management, Kolkata, University of Engineering and Management, Kolkata, India

J. K. Mandal et al. (eds.), *Recent Advances in Artificial Intelligence and Smart Applications*, Lecture Notes in Networks and Systems 1380,
https://doi.org/10.1007/978-981-96-5822-0_22

1 Introduction

Over the past few years, advancements in technology have changed how people engage with machines. From touchscreens to voice controls, these interactions are becoming easier and friendlier for users [1]. Nevertheless, with the world becoming increasingly interconnected digitally there has been a rise in the need for gesture-based systems for interaction. Later, touchless technologies introduce a hygienic and effective option in situations where reducing physical contact is important, like healthcare and public areas. Further, conventional input tools such as keyboard and mouse will be challenged by new innovative solutions [2, 3]. In addition, technology that enables users to identify gestures that are then converted into actionable commands introduced hands-free operation across multiple applications. Later, the COVID-19 crisis has increased the demand, for devices that include touchless features. Hand-tracking systems have become increasingly popular among technological developments in recent times [2]. By using tools like webcams that are already available to users, these systems make use of computer algorithms related to visual perception to identify and follow hand gestures as they happen live. The system includes hand tracking and simple gesture recognition to create an interface for users to interact with their devices without using their hands directly [20]. By using their finger as a pointer to adjust functions such as controlling the music and brightness [5]. Visual cues displayed on the screen make interactions more intuitive and easier to understand for users. Moreover, advances in touchless technology are transforming fields like electronics, education, and healthcare by enhancing user interaction and accessibility. Gesture-based control offers an intuitive and efficient way to navigate PowerPoint presentations, allowing users to switch slides and perform actions through simple hand movements without input devices [1]. On top of that, a virtual painter is also the future of learning [4].

The literature survey conducted gives us an informative insight about the various methods to detect the hand gestures and to control various operations with the help of touchless interaction. It helps us to understand the benefits and drawbacks of the existing techniques of the system. Consequently, vision-based tracking is the most widely adopted technique for tracking the finger movements. Many of the systems were built with the help of Kinect RGB depth cameras and IntelRealSense cameras to track the hand gestures and capture the fingertip movements [6]. Hand Glove-based tracking systems were also extensively used in the capturing of finger movements in [7, 8] with the help of embedded sensors by tracking the values of gyroscope and accelerometer. This system offers high precision and accurate results. The method of using wired pair of hand gloves to enhance human–computer interaction by developing a touchless interface with the help of a virtual mouse has been discussed and implemented [9]. The technique of drone control based on hand gesture with the use of Microsoft Kinect and Hand glove-based tracking has been implemented in [10]. A real-time control of virtual instruments by palm recognition and fingertip detection has been developed with the help of Microsoft Kinect wherein users can apply the results in device control and virtual instruments [11].

Computer webcams were also harnessed in the process of tracking in order to build a cost-efficient model [12, 13]. These systems require substantial resources for computational purpose and real-time performance. Various machine learning and deep learning algorithms have been harnessed to develop various fingertip-based real-time gesture recognition to have seamless interaction by developing gesture-based user interface control [14]. Furthermore, by leveraging OpenCV Python library and MediaPipe Hands, hand gesture-based presentation slides' control has been developed to control the slides seamlessly [1]. Additionally, achieving real-time performance in these systems involves the use of advanced filtering techniques and machine learning models. A large number of methods have been employed in the process of pre-processing the image which includes machine learning algorithms to decrease the latency, increasing efficiency and for noise removal and better tracking. Kalman filters and particle filters had been used in real-time systems to control the position and in the gesture detection [15, 16]. The application of convolutional neural networks also has proven to be effective in the application of gesture-control recognition [17, 18]. Interactive virtual painting has been established by means of OpenCV and convolutional neural networks for flawless painting in real time through the computer webcam using deep learning algorithms [19]. Virtual mouse system using hand recognition and finger tracking modules has been developed by using MediaPipe and OpenCV wherein users can control all the activities done by a mouse with the movement of a hand [20]. The technique of initiating a human–computer interaction to perform actions such as launching applications such as VLC Player and PowerPoint has been achieved with the help of Haar cascade classifier technique by gesture recognition[21].

The primary contributions of the work are as follows:

1. Proposed hand-tracking module for smooth pointer tracking.
2. Flow of touchless interaction for system control.
3. Comparison of proposed system with existing methodologies.
4. Performance metrics of hand gestures with respect to distance from the camera.

Subsequently this is followed by the methodology of the proposed hand-tracking and virtual control system, which is outlined in Sect. 2. Section 3 presents the implementation results and performance metrics of the system. Applications of the developed touchless interaction model across various scenarios, including its efficiency, are explored in Sect. 4, and Sect. 5 concludes the article with final remarks.

2 Proposed Methodology

In conventional systems, gesture-based interactions are widely used for communication, focusing on hand movements and the posture of specific gestures. However, these systems face several challenges, with one significant difficulty being the ability to respond quickly on interactive platforms. The objective is to replace gestures with finger-based control, where the finger acts as a cursor to select the desired options. This approach eliminates the challenges associated with gesture recognition and improves the system's speed and accuracy in real-time interactions, resulting in a more efficient and user-friendly experience compared to existing gesture-based systems. Moreover, using new and complex gestures for various applications requires more accurate models built on custom datasets, which brings new challenges in maintaining these models and can burden users with the need to recognize a wide range of gestures. In contrast, this model utilizes simple gestures combined with a unique virtual selection approach. This method does not require custom datasets or high-end models for recognition. Instead, this achieves virtual control through straightforward virtual clicks on specific operations, enhancing user-friendly interaction without imposing additional burdens on the user.

The proposed system enables touchless control of various operations on a laptop or computer based on touchless selection; this method utilizes a webcam for video input and implements hand-tracking and basic gesture recognition techniques. This methodology is classified into two key sections: hand tracking and basic gesture recognition, and touchless control and system interaction.

2.1 Hand Tracking and Basic Gesture Recognition

This section addresses touchless interaction that can be achieved by the basic needs of gesture detection and hand tracking to effortlessly perform virtual activities. Flowchart 1 discusses the detailed approach of the solution proposed in this paper. Initially, the algorithm creates an instance of the Hand Detector class and configures the system to track and identify the hand movements. Subsequently, it records live footage from the webcam and processes each frame. The image is transformed to RGB format for further study of every frame that is collected. In addition, the system uses a facial segmentation method to evaluate the image and detect any hand by focusing on portions with skin tones to enhance accuracy. If the landmarks are identified successfully, more data are retrieved; otherwise, it continues to the subsequent frames. To obtain a constancy, the algorithm computes and frame rates (FPS), and then geo-references this to the image. The user can view the processed image, which also contains the FPS. The visual aspect of this section follows the more general structure of the Clinical Processes described in the earlier sections. Markers, such as finger joints or fingertips, are extracted if a hand is found. The system extracts the data to make sure the reference points are mapped correctly, if not, it moves to the proceeding frame. This regulates the frame per second (FPS) and places it on the image for smooth operation. The processed image and frame number are viewed by the user, and the frame count for one second. In this section, the proposed process is structured in several steps. A built-in or external camera grabs webcam video and then it is divided into video "frames" in order to get the position of the hand and follow it. The frame resolution was altered to correspond to the size of the screen to enable a trade-off between performance and visibility to enable effective hand tracking. Frames are converted from BGR to RGB because the MediaPipe hand

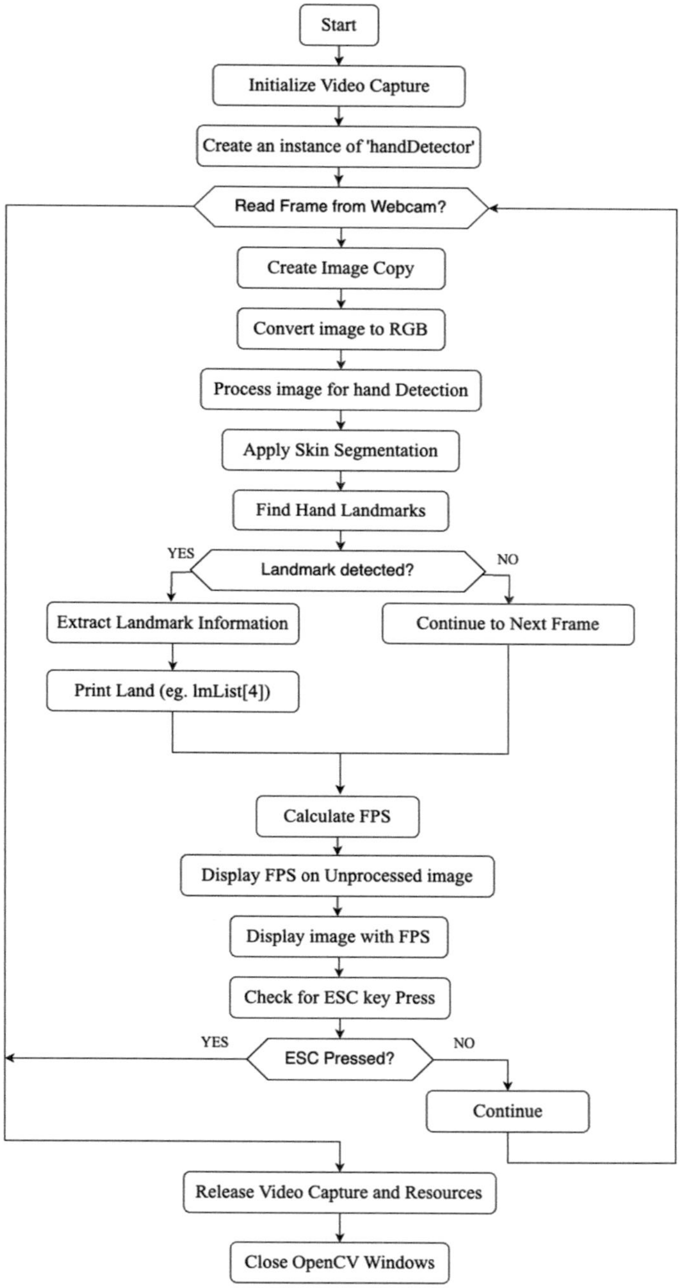

Flowchart 1 Hand tracking and basic gesture recognition

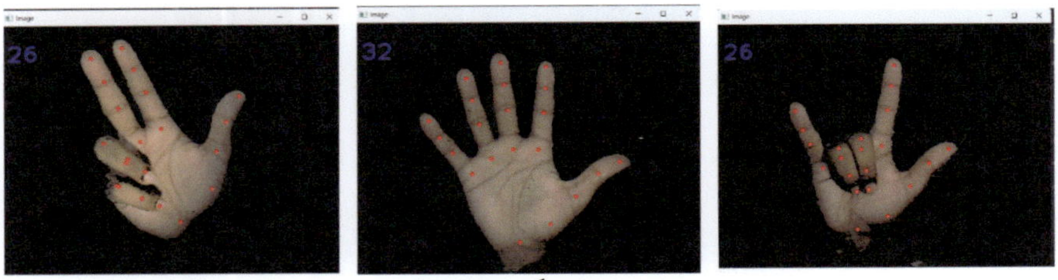

a b c

Fig. 1 a, b, c Skin color segmentation for noise reduction

tracking uses RGB input images. Next, the image was converted into HSV for skin color detection by masking all background pixels that did not correspond to the skin tone as seen in Fig. 1a, b, c. This pre-processing step improves system efficiency and detection accuracy.

The system uses MediaPipe's hand-tracking module to recognize hands in real time. With deep learning models, it detects key hand landmarks, like knuckles and fingertips. To avoid redundancy, an efficient system detects one hand at a time. Even with varying hand positions, detection accuracy remains satisfactory. The model identifies 21 landmarks (0–20), including the palm and fingertips, which are crucial for recognizing hand positions, motions, and gestures. Each landmark is assigned a unique ID to map its position to the corresponding pixel as depicted in Fig. 2a, b.

Finger classification is done by comparing hand landmark coordinates. The model checks fingertip positions relative to the coordinates below them to determine if fingers are extended. Alternatively, a threshold is introduced, calculated as half the distance between the wrist (0) and middle palm (9), to assess finger position. This threshold, based on the y-coordinate distance between the tip and threshold, counts the number of extended fingers. For thumbs, x-coordinates of base and tip landmarks are compared. This method improves accuracy in finger classification and counting.

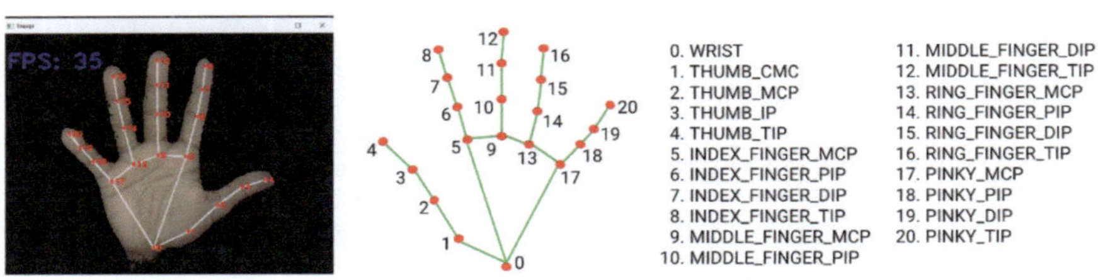

a b

Fig. 2 a, b Extracted landmarks of the detected hand

2.2 Virtual Control and System Interaction

This section explains how a user's hand serves as a virtual pointer for touch-free system control. By continuously tracking predefined finger landmarks, the system updates the pointer's position swiftly and effectively. Overlay displays help users understand spatial tasks by showing the pointer's position relative to the user, enhancing hands-free capability. Flowchart 2 details the process: The hand is considered as the virtual pointer, like a mouse and its landmarks are continuously tracked. Based on the pointer's locations the operation executions are depended. The reliance on specific finger IDs determines the virtual pointer to be positioned at the midpoint between the first and second fingers, thus making it less effective. Any of these fingers may be employed for code-based pointer control tasks, such as brightness and music control programming, and even for changing PowerPoint slides. Further, users were helped to perform more complex interactions via panel overlays, which served as a visual guide to ensure that actions taken were performed in a straightforward manner that was clean and definable. This decreases the level of disorientation and improves the interaction flow.

Operations Based on Pointer Location:

The hands-free control operations depending on pointer location, discussing five key functions: brightness control, music control, PPT control, virtual painter, and lock screen. When the virtual pointer location maps with the brightness icon, a panel of operations related to brightness are displayed to select the respective operations, enabling this control to be exercised without the use of hands. Likewise, the use of the PyGame library would allow for mouse pointer control over music playing, allowing for play, pause, skip, and volume. Similarly, the PowerPoint presentation can also be controlled by the use of fingers to detect when one is presenting and hence attending to the computer to display the slideshows by changing the slides and concluding the show. The same applies when the virtual pointer sounds on the operation icons; there does not exist any cause of click action on the self, mouse controls, enhancing operation efficiency, and output without the need to touch anything. Similarly, in virtual painter encompasses a virtual painting mode where user draws in real time using their fingertip by merely moving the fingers in real time, a convenient manner for drawing and teaching purposes. There is also an eraser option to rid the user of any unnecessary art and a color selection mode to change colors. Finally, the lock screen option can also be performed using the pointer along the Windows API and the Lock Workstation for the simple lock of the laptop without any movement (Table 1).

Flowchart 2 Virtual control and system interaction

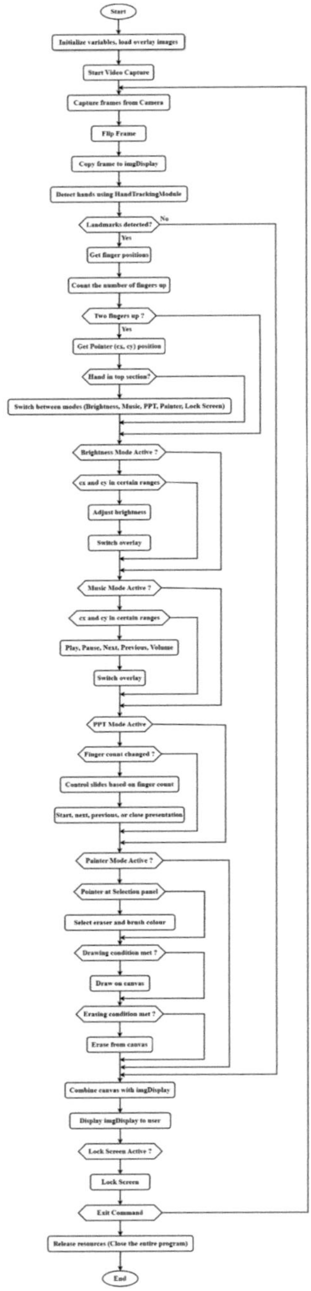

Table 1 Comparison of touchless interaction with existing techniques

Ref.	Model used	Algorithms used	Applications	Functions used	No. of operations executed	No. of gestures used
[1]	MediaPipe hand landmarker	Non-maximum suppression algorithm	PPT control	Gesture and hand detection	2	2
[9]	Haar feature-based classification system, CNN, 32 bin histogram model	Sliding window algorithm, gesture recognition algorithm, edge detection algorithm	Virtual mouse	Hand detection, gesture recognition, face recognition	6	6
[10]	Haar feature-based AdaBoost classifier, SVM model	AdaBoost algorithm, Haar cascade classifier	Drone control using hand gestures	Gesture recognition.	5	5
[11]	Kinect sensor	Three-point alignment algorithm, particle swarm optimization	Playing virtual instruments	Fingertip detection, palm recognition, gesture recognition	1	1
[14]	Background model from codebook representation	Multi-scale mode filtering, K-curvature, dynamic gesture recognition algorithm	Real-time hand gesture recognition for human computer interaction	Gesture recognition, fingertip tracking	6	6
[19]	Region-based CNN, YOLO	Color detection algorithm, drawing on canvas algorithm	Virtual painter	Color detection and rendering	–	–
[20]	MediaPipe	Masking method	Hand gesture-based virtual mouse control	Hand gesture recognition, finger movement tracking	5	5
[21]	Haar-cascade	Color conversion, noise removal, thresholding	Human–computer interaction	Gesture recognition	7	7
Proposed	MediaPipe hands	Skin color segmentation	Brightness, PPT, music, virtual painter, lock screen	Pointer tracking, gesture recognition	20	5

3 Results

The system overcomes the challenges of traditional gesture control of the peripheral by minimizing the number of gestures required to perform different operations. Instead of having many complicated gestures, a user's finger can move a virtual pointer to perform operations such as controlling brightness, navigating through PowerPoint, and playing music. This simplifies the interaction process, making it more user friendly. Furthermore, the system demonstrated reliable performance over a range of 20–200 cm, with an overall accuracy of 97.92%. Single-finger to five-finger gestures consistently achieve accuracies between 94% and 99%, ensuring precise control for a variety of operations. This demonstrates its ability to perform consistently across different user positions, thereby ensuring reliable touchless control in various settings. Table 1 addresses the proposed system with various existing touchless human–computer interaction methods. Subsequently, Table 2 gives the evaluation metrics of the proposed method with its relation between accuracy and distance, and the relative graph is plotted.

Table 2 Performance metrics of various gestures with respect to distance

Finger count	Operation	Distance from camera (cm)	Accuracy (%)
One finger (index finger)	Virtual painter, PPT control	20	97
		60	98
		100	97
		150	99
		200	96
Two fingers (index, middle finger)	Virtual pointer	20	99
		60	98
		100	97
		150	99
		200	98
Three fingers (any three)	PPT control	20	98
		60	98
		100	98
		150	98
		200	98
Four fingers (any four)	PPT control	20	99
		60	99
		100	98
		150	97
		200	94

(continued)

Table 2 (continued)

Finger count	Operation	Distance from camera (cm)	Accuracy (%)
Five fingers	PPT control	20	98
		60	98
		100	99
		150	99
		200	99

The above figures illustrate the various diverse applications of a virtual pointer. Figure 3 introduces the concept of virtual pointer and the main window where all the functionalities are displayed and can be selected according to our choice using the virtual pointer. The brightness can also be controlled using virtual pointer as shown in Fig. 4. Furthermore, various operations can be performed on Microsoft PowerPoint using different hand gestures as seen in Fig. 5. Figure 6 depicts the use of a virtual pointer for music control encompassing various functions such as volume adjustment and song selection. The feature of implementing a virtual painting tool using a virtual pointer is shown in Fig. 7. Figure 8 depicts the graph of accuracy versus distance (cm) for multiple finger detection.

Table 3 depicts the comparative study of several other gesture-control system. Sanguannarm et al. [22] achieved an accuracy of 87.67% in image control. On the other hand, Gurav et al. [23] have reported an accuracy of 92% in real-time

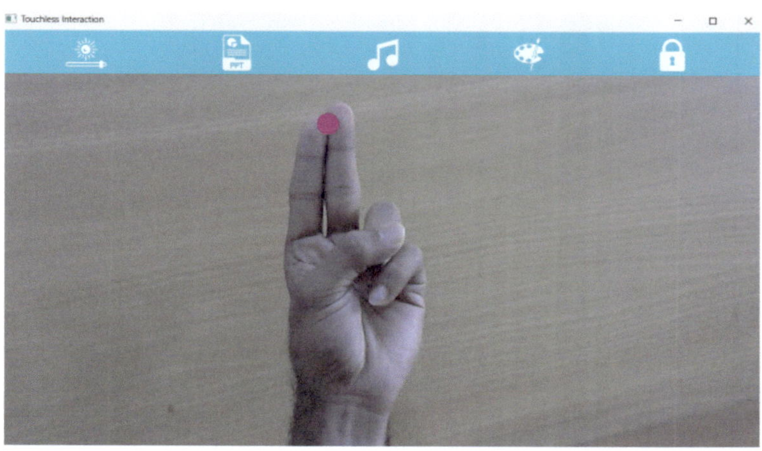

Fig. 3 Virtual pointer

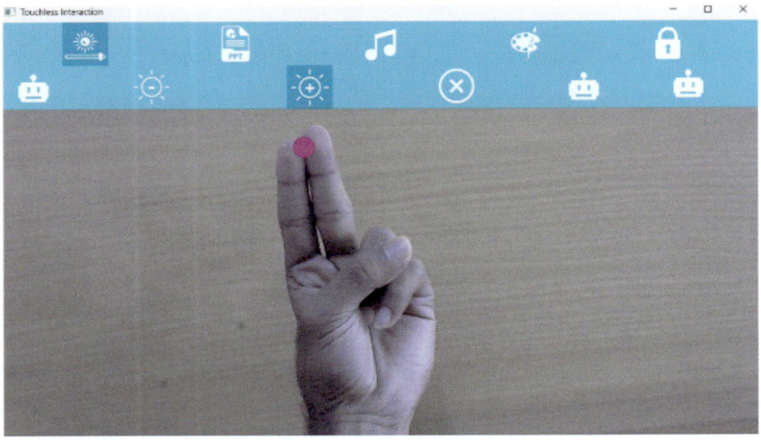

Fig. 4 Brightness control

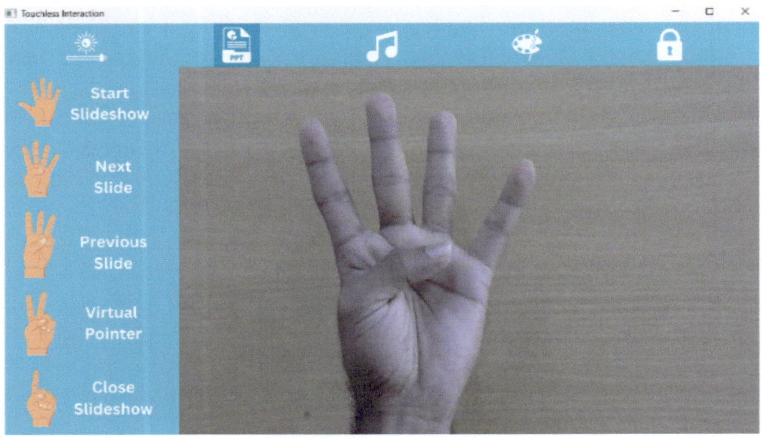

Fig. 5 PPT control

finger tracking and engaging with virtual environments. Whereas Raee et al. [24] have achieved an accuracy of 94.3%. The current proposed system now shows an improvement with an accuracy of 97.92% for multiple operations with limited gestures.

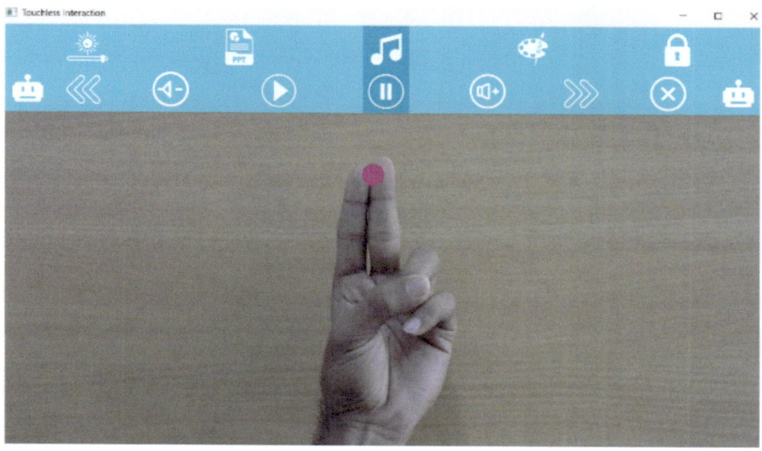

Fig. 6 Music control

Fig. 7 Virtual painter

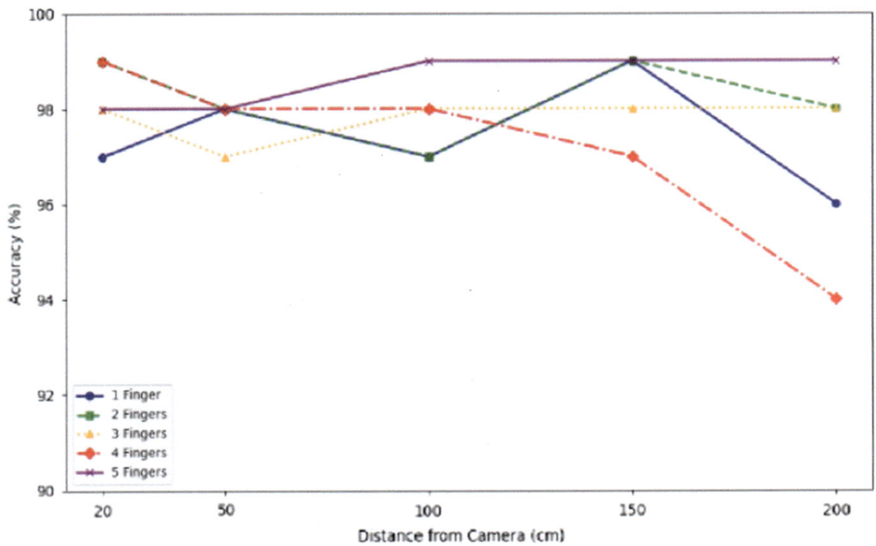

Fig. 8 Accuracy metrics

Table 3 Comparative study

Author [Ref.]	Application	Accuracy (%)
Sanguannarm et al. [22]	Touchless controlling of images in operating room	87.67
Gurav et al. [23]	Real-time finger tracking and contour detection for gesture recognition	92
Raees et al. [24]	Gesture-based interaction for virtual environment	94.3
Proposed	Pointer tracking, gesture recognition	97.92

4 Conclusion

The proposed touchless interaction system offers a practical and efficient solution for human–computer interaction by using simple hand gestures to control multiple operations with high accuracy. Its scalability and adaptability make it suitable for various real-world applications such as media control and virtual painting. Moreover, it exhibits greater accuracy compared to existing systems, enabling up to 20 distinct operations with only 5 hand gestures. Its high precision is maintained even at longer distances, making it adaptable to various practical applications. Further, the system's design allows for the integration of more functionalities in the future, enhancing the flexibility and potential use cases. This balance of minimal gestures, accuracy, and scalability positions the system as a promising solution for advanced gesture-based control in diverse settings.

References

1. Vidya. M., et al. (2023). Gesture-based control of presentation slides using OpenCV. In *2023 Second International Conference on Augmented Intelligence and Sustainable Systems (ICAISS)*. IEEE.
2. Maharana, S., et al. (2024). SMART virtual painter and mouse system to improve online education with computer vision and deep learning features. In *2024 International Conference on Emerging Systems and Intelligent Computing (ESIC)*. IEEE.
3. Varun, K. S., Puneeth, I., & Jacob, T. P. (2019). Virtual mouse implementation using open CV. In *2019 3rd International Conference on Trends in Electronics and Informatics (ICOEI)*. IEEE.
4. Jishma, K., & Jims, M. A. Hand gesture controlled presentation viewer with AI virtual painter.
5. Parimala, N., et al. (2024). Real-time brightness, contrast and the volume control with hand gesture using open CV python. In *2024 10th International Conference on Communication and Signal Processing (ICCSP)*. IEEE.
6. Li, Y. (2012). Hand gesture recognition using Kinect. In *2012 IEEE International Conference on Computer Science and Automation Engineering*. IEEE.
7. Parvini, F., et al. (2009) An approach to glove-based gesture recognition. In *Human-Computer Interaction. Novel Interaction Methods and Techniques: 13th International Conference, HCI International 2009, San Diego, CA, USA, July 19–24, 2009, Proceedings, Part II 13*. Springer Berlin Heidelberg.
8. Berezhnoy, V., et al. (2018) The hand-gesture-based control interface with wearable glove system. *ICINCO (2)*. 2018.
9. Ranawat, M., et al. (2021). Hand gesture recognition based virtual mouse events. In *2021 2nd International Conference for Emerging Technology (INCET)*. IEEE.
10. Natarajan, K., Nguyen, T.-H. D., & Mete, M. (2018). Hand gesture-controlled drones: An open source library. In *2018 1st International Conference on Data Intelligence and Security (ICDIS)*. IEEE.
11. Hsu, M. H., Shih, T. K., & Chiang, J. S. (2014). Real-time finger tracking for virtual instruments. In *2014 7th International Conference on Ubi-Media Computing and Workshops*. IEEE.
12. Shi, Y., Taib, R., & Lichman, S. (2006) GestureCam: a smart camera for gesture recognition and gesture-controlled web navigation. In *2006 9th International Conference on Control, Automation, Robotics and Vision*. IEEE.
13. Patlolla, C., Mahotra, S., & Kehtarnavaz, N. (2012) Real-time hand-pair gesture recognition using a stereo webcam. In *2012 IEEE International Conference on Emerging Signal Processing Applications*. IEEE.
14. Simion, G., et al. (2016). Fingertip-based real time tracking and gesture recognition for natural user interfaces. *Acta Polytechnica Hungarica 13*(5), 189–204.
15. Oka, K., Sato, Y., & Koike, H. (2002). Real-time fingertip tracking and gesture recognition. *IEEE Computer Graphics and Applications, 22*(6), 64–71.
16. Asaari, M. S. M., & Suandi, S. A. (2010) Hand gesture tracking system using adaptive Kalman filter. In *2010 10th International Conference on Intelligent Systems Design and Applications*. IEEE.
17. Kheratkar, N., et al. (2020) Gesture controlled home automation using CNN. In *2020 4th International Conference on Intelligent Computing and Control Systems (ICICCS)*. IEEE.
18. Chung, H.-Y., Chung, Y.-L., & Tsai, W.-F. (2019). An efficient hand gesture recognition system based on deep CNN. In *2019 IEEE International Conference on Industrial Technology (ICIT)*. IEEE.
19. Mishra, S., et al. (023) Immersive virtual painting: pushing boundaries in real-time computer vision using OpenCV with C++. *RICE*.
20. Singh, A., et al. (2023) Real time virtual mouse system using hand tracking module based on artificial intelligence. In *2023 International Conference on Communication, Security and Artificial Intelligence (ICCSAI)*. IEEE.
21. Haria, A., et al. (2017) Hand gesture recognition for human computer interaction. *Procedia Computer Science 115*, 367–374.

22. Sanguannarm, P., et al. (2018) Hand movement recognition by using a touchless sensor for controlling images in operating room. In *2018 International Electrical Engineering Congress (iEECON)*. IEEE.
23. Gurav, R. M., & Kadbe, P. K. (2015) Real time finger tracking and contour detection for gesture recognition using OpenCV. In *2015 International Conference on Industrial Instrumentation and Control (ICIC)*. IEEE.
24. Raees, M. A., & Ullah, S. (2019). GIFT: Gesture-based interaction by fingers tracking, an interaction technique for virtual environment. *IJIMAI, 5*(5), 115–125.

SMRITI—Semantic Model for Retrieval and Interpretation of Topical Information

Debayan De, Kartik Tulsian, Rupayan Das, Subhabrata Sengupta, Avijit Bose, and Kajari Sur

Abstract In today's dossier-steered world, competently retrieving applicable particulars has enhanced as a challenging platform, specifically in the field of academic data retrieval. With the moderations in the AI models, machine learning tools, and motorized data analysis it's quite feasible to develop query analysis by categorizing boundaries, analyzing complexity, and ascertaining the weightage of queries. This paper instigates Semantic Model for Retrieval and Interpretation of Topical Information (SMRITI), a novel Semantic Web-based structure specifically customized for the academic institutional domain. SMRITI utilizes frameworks' portrayal of knowledge, corresponding to various layers of semantic architecture to encourage efficient examination question retrieval and comprehensive analytics. By leveraging semantic mechanisms, it provides intelligent explorations based on relationships, concepts, and context, enhancing the significance of retrieved questions. Furthermore, the model embodies data retrieval techniques like web scraping (using BeautifulScup, Selenium, and Requests HTML) and Keyword extraction methods (using TF-IDF, TextRank, KeyBERT, RAKE-NLTK, YAKE, and SPACY) to interpret and scrutinize topical information effectively. Additionally, SMRITI incorporates educators with powerful analytical methods to appraise question difficulty, monitor student's performance, and identify knowledge gaps. This proposal assures the revolutionization of smart E-learning by promoting personalized learning experiences and assisting data-driven instructional decisions.

Keywords SMRITI · Semantic mechanisms · Query scrutiny · Web scraping · Beautifulsoup · Selenium · Requests-HTML keyword extraction · TF-IDF · TextRank · KeyBERT · RAKE-NLTK · YAKE · SPACY · Data retrieval system

---

D. De (✉) · K. Tulsian · R. Das · S. Sengupta · A. Bose · K. Sur
Institute of Engineering and Management, University of Engineering and Management, Kolkata, Kolkata, West Bengal, India
e-mail: debayande44@gmail.com

J. K. Mandal et al. (eds.), *Recent Advances in Artificial Intelligence and Smart Applications*, Lecture Notes in Networks and Systems 1380,
https://doi.org/10.1007/978-981-96-5822-0_23

1 Introduction

The World Wide Web (WWW) is a vast, complex digital space, making information retrieval (IR) challenging. Traditional IR methods rely on keyword matching, but these are limited in capturing deeper semantic relationships, leading to a vocabulary gap between users and indexing systems [1]. Recent advances, especially in pre-trained language models, have introduced Generative Information Retrieval (GenIR), which leverages natural language processing (NLP) to better understand and retrieve relevant content from this diverse environment [2].

Keyphrase Extraction (KE) is vital in NLP, helping identify key phrases that summarize important information in a document and enhancing IR efficiency [3]. The rise of Large Language Models (LLMs) has further boosted IR systems, particularly through semantic search, which understands query intent and retrieves information based on meaning, not just keywords [4, 5].

Despite progress, challenges remain in capturing the full depth of user queries. This paper proposes an approach that integrates semantic search with advanced classification to improve the retrieval of examination questions in e-learning environments. By identifying keyphrases and assessing their contextual relevance, the system aims to create a tailored question bank that enhances the accuracy, difficulty, and relevance of retrieved questions.

2 Literature Review

Keyphrase extraction, a crucial aspect in current Natural Language Processing (NLP), underpins a broad array of petitions like semantic indexing, document summarization, question prompting, data mining, and information retrieval. As the amount of formless data continues to expand exponentially, the aptness to clip meaningful key phrases with high reliability becomes essential. This review draws the limelight on innovative pathways driving the latest trends in keyphrase extraction, with special priority on practical, real-world applications in Semantic Web, Academic analysis, and automated content inception. Traditional keyphrase extraction methods, like Keyword Extraction Algorithm (KEA) and Rapid Automatic Keyword Extraction (RAKE), imparted radical paths for analyzing significant phrases in text. KEA, an AI-based prediction design, applies statistical features like term frequency and sentence placement [6], while the RAKE enrapt the stopwords and delimiters. These approaches exert well in regular scenarios, but often crash in domain-specific solicitations or fail to express the whole content and correlation between phrases [7].

Recent evolvement in deep Learning has transfigured keyword extraction by entitling algorithms to grasp a wide range of content-related data. Models like Bidirectional Encoder Representations from Transformers (BERT) and its offshoot, KeyBERT, have made remarkable strides in the Semantic Understanding of text.

These methods utilize transformers to evaluate examined word embeddings, securing that the extracted phrases are semantically apposite to the report's content [8].

A top-tier instance of contextual keyword extraction is the employment of GPT-3 and other transformer-based petitions for automatic keyphrase extraction from large and tangled academic papers. These approaches amplify over mere frequency survey for deeper contextual awareness creating highly efficient technical domains where shading of vocabulary is complex [9].

Table 1 explicates the potential of various algorithms, integrating their efficiency to obtain a comprehensive set of key phrases. This integrated approach counters the blocks of contributing methods proposing a robust terminology gearing domain-specific text processing and enhanced sequels of the retrieval process, orienting keyphrase extraction with a widened contextual and semantic framework of the content.

One innovative utilization of the extraction is in MCQ creations, where modes like TextRank and BERT can foster MCQs from the academic materials, enhancing the automated assessment systems. Lai Wei et al. presented an advanced TextRank algorithm for MCQs creation, by ranking the phrases to reliable questions. This not

Table 1 Keyword extraction techniques

Algorithm	Method	Advantages	Disadvantages
TF-IDF	Statistical weighting based on phrase regularity and inverse report frequency	Simple, fast, and effective for identifying important phrases	Ignore semantic correspondence between phrases, doesn't analyze context well
TextRank	Graph-based ranking algorithm for word co-occurrences	Apprehend multi-word and contextually key phrases	High computational cost, slower for sizable texts
KeyBERT	Uses BERT embedding to find phrases that are correlated to the report content	Generates extremely contextual keywords with semantic understanding	Computationally minute, essential fine-tuning for specific domains
RAKE-NLTK	Unsupervised extraction by hashing the context and evaluating composition	Domain-independent and effective for stubby documents	Struggles with extensive, more complex phrases
YAKE	Unsupervised method based on statistical attributes of text to evaluate cruciality	Fast, does not reckon on external information, domain-independent	Wayward for highly structured or domain-specific texts
SpaCY	Rule-based analytical model for part-of-speech tagging and operation conceding to extract keywords	Structured and bespoken, generates both syntactic and semantic keyword extraction	Essential tuning for different domains, may mishear domain-specific text

only advanced the efficiency in automated question creation but also in personalized learning systems, and initiated an adaptive learning pathway for students [10].

Graph-based proposals, like TextRank, promise to create a correlation with the words in the text by generating co-related graphs. WS-Rank and other graph-based ranking algorithms widen their hands by integrating text-level significance with correlated phrases, specifically useful for academic literature scrutiny where the content is rich and contains interrelated context [11, 12].

A significant innovative approach in graph-based extraction includes the generation of educational-related graphs. Lin Zhang et al. presented a graph-based keyphrase extraction model that initializes the formation of academic graphs based on texts, mapping the significant concepts, admitting more framed and dynamic proposals toward academic fields, smoothing the automatic summarization of papers and allowing researchers to propose more advanced trends and avenues efficiently [12].

Web Scraping abides by the remarkable tag for procuring wider datasets for keyword extraction, specifically in data-driven exploration or content assemblage. Models such as BeautifulSoup and Selenium are usually practiced for scraping web data, but the expanding convolution of web pages compels for the necessity of enhanced scarping models. In the latest researches, the potential of a variety of such scraping methods has been advanced, focusing primarily on factors like speed, reliability, and the feasibility of integration data pipelines [13]. The proposal of an integrated model influences the distinctive ability of existing models which could restructure the field by generating a rugged, flexible solution proficient to tackle diverse scraping tasks.

Table 2 features the potential strengths and restraints of integrating these models, in this BeautifulSoup shines the clarity and minuscule tasks, Selenium tackles dynamic content and Requests-HTML strikes the stability with JavaScript-heavy pages. This integrated model could gear their compatible strengths, paving a more enhanced and flexible path to data scraping.

Table 2 Question retrieval technique

Algorithm	Method	Advantages	Disadvantages
BeautifulSoup	Parses HTML/XML reports for data extraction using a tree-like framework	Straightforward to use, productive for portable datasets, integrates well with added libraries	Obtuse for large datasets, dearth dynamic content scraping
Selenium	Automates browsers to retrieve dynamic content and steer JavaScript	Excellent for abrading dynamic content and interrelating with web segments	Sedate execution, resource-intensive, essential browser drivers
Requests HTML	Parses HTML reports for data extraction, bracing JavaScript	Nimble for static content, brace asynchronous requests	Scuffle with composite JavaScript-heavy websites

The Semantic Web architecture is assembled on the layers, beginning with URIs and Unicode for encoding and identification. The syntax Level employs XML for data framing, while RDF is utilized in the Data Interchanging layer. Query operations by SPARQL and taxonomies are handled by RDFS. OWL provides an ontology-oriented relationship, and RIF/SWRL represents the logical order enactment [14]. The integration of extraction models with semantic web structures, like Resource Description Framework (RDF) and SPARQL, stands as a remarkable practical revolution in result reliability and pertinence of search results, significantly in educational and research articles. The Semantic Web permits the framed proposal of knowledge, advancing the extraction and retrieval of phrases, encouraging semantic indexing of articles, and improving discovery in related domains [15]. Additionally, merging web scraping with semantic web framework can uplift the data extraction approaches, admitting explicit semantic indexing and reclamation of significant information. For instance, Selenium can be utilized for dynamic scraping of content from e-commerce websites for market trend evaluation, while SPARQL queries can be used in semantic data extraction from e-learning repositories, advancing the accuracy of semantically extractive models in research data mining. Table 3 summarizes the several key components of the Semantic web framework, which integrating with the scraping and question-answering methods exemplifies their attributes in different domains approaching both structured and unstructured data to enhance semantic content and retrieval reliability.

The consolidation of keyphrase extraction with deep learning and graph-based methods created a remarkable impact on real-time data processing models. For instance, news accumulation platforms, legal report synopsis, and scientific paper evaluation aids by these enhanced models, ensuring faster and more appropriate extraction that is contextually relevant and semantically rich. Succeeding papers are anticipated to improvise on domain-specific models, trained on collective datasets like, medical, commerce, or law [16]. In the province of question retrieval systems, integrating keyphrase extraction with semantic web framework and graph databases will grant enhanced dynamic and contextual retaliation, improvising human-technology reciprocity and allowing advanced instinctive methods in industries aligning from academics to healthcare.

3 Proposed Model and Flow Chart

Information Retrieval (IR) is a major term in this era of research for information that needs to be modified continually to keep pace with the human race. Essentially, IR plays a significant role in a range of practical applications whether it is searching the web, answering questions, chatbots, digital libraries, or research knowledge. It encompasses the steps of acquiring, organizing, storing, retrieving, and disturbing data. The traditional key phrase extraction or text retrieval systems rely on the matching of terms between the query and document but the recent year's advancements in computing power and large labeled datasets enhanced the field of

Table 3 Semantic web structure components

Component	Method	Advantages	Disadvantages
Unicode	Universal disposition encoding merits for consistent text rendering across dialects	Clarify text compatibility beyond tenets and dialects	Essential conversion in bequest systems, increasing complexity
URI	Pinpoint resources using a homogenized string format on the web	Consistent and flexible way to identify web resources	Doesn't embrace any data retrieval or validation mechanism on its own
XML	Markup language that defines order for encoding reports that is readable by humans and machines	Self-descriptive, supports hierarchical structures, broadly acquired	Verbose, leading to extended file sizes and slower analysis
RDF	A standard model for data mapping in web using triples (subject-predicate-object)	Anticipate a flexible technique to express relationships of web resources	Network syntax, arduous to implement for non-semantic web developers
SPARQL	A query dialect for querying RDF datasets	Dominant for querying complex bracket data sets	Abrupt learning curve, requires informality with RDF concepts
OWL	A dialect for defining and incorporating Web Ontologies to constitute complex relationships	Accredit reasoning and speculating over complex data	Requires maestro knowledge to formulate ontologies, sedate reasoning for large datasets
RIF	Format for constituting and executing rules over semantic web data	Standardizes directive execution over distributed data	Finite assumption requires a deep understanding of rule systems

NLP and deep learning techniques. As shown in Fig. 1, the proposed model illustrates the relationship between key components in the retrieval process. In this model, reliable Key phrases and tokens are excavated from the queries, and SMRITI analyzes and holds for relevant questions. These retrieved questions are thoroughly refined and congregated based on their entanglement and weightage. Text scrutiny algorithms enumerate the lexical impediment, spotting the keys and assessing the difficulty levels, using the clustering algorithms to classify the questions into patches according to their weightage and complexity [16].

The cumulated cluster of questions of a concept is analyzed using various text scrutiny algorithms and approaches for retrieving the parallel patterns, key phrases, and words outlining the question set. This model then explores and analyzes the queries on the topic by contrasting the phrases and patterns to generate a relevant bunch of questions. The retrieval module uses the report file to recover the documents containing certain question terms (those documents which are likewise to the query), calculate relevance scores, and rank the retrieved documents as per the scores. The

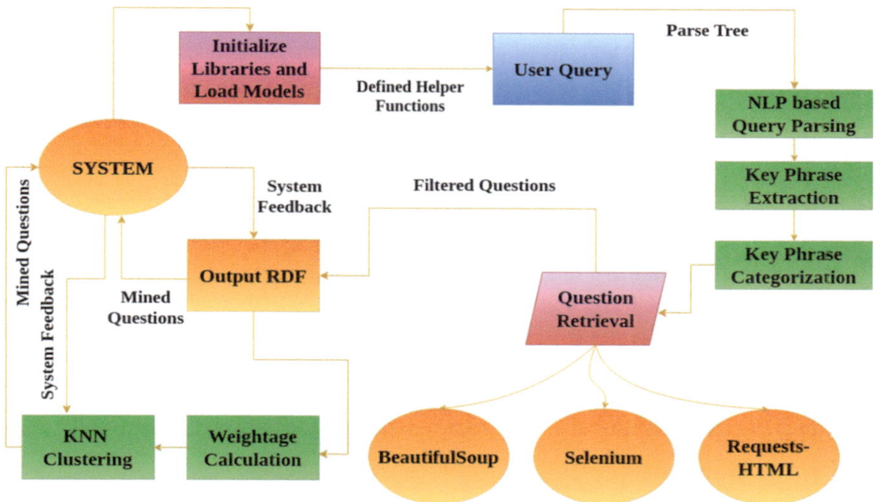

Fig. 1 Proposed model for question retrieval

ranked records are then presented to the user. The archived documents are also known as the content database, which the indexer ordains for proficient retrieval.

These retrieved questions are fed to an analytical model for further process of analyzing, which calculates the laboriousness, topic relevance, lexical complexity, etc., and then passes through a sundry of pre-trained codification models and grouped them to the profuse reliable group of questions utilizing a neighbor algorithm. These segments cater to determining the weight and other frameworks of each question individually. If this collective model is applied to the Unifying Logic Layer of Semantic Web Architecture, it could formulate new query objectives for data recovery.

4 Experimental Setup and Result Analysis

Step 1: Text Preprocessing and Keyword Extraction

A query or input text—often a question or topic—is collected for analysis. The text is preprocessed by removing stop words and applying lemmatization or tokenization as needed. Various keyword extraction methods, including YAKE, RAKE, TF-IDF, KeyBERT, and SpaCY, are utilized to identify the most relevant keywords and key phrases. The extracted keywords are then prepared for further processing, with each keyword considered a potential search term.

Step 2: Question Retrieval and RDF Triple Generation

The extracted keywords are used to create queries for web searches. These searches are conducted using tools such as BeautifulSoup, Selenium, and requests-html to scrape related questions from Google. The retrieved questions are then processed, generating RDF triples for each combination of question and keyword. This involves mapping the question's content, links, and importance scores into a structured format compatible with the RDF graph. The resulting triples are stored in an RDF graph, which can be serialized into formats like Turtle for future use or sharing.

Step 3: Readability and Weightage Calculation

The retrieved questions are re-evaluated for complexity using two readability indices: the Automated Readability Index (ARI) for technical domains and the Dale-Chall Readability Index (DCRI) for general and medical domains. Key phrases and word counts are analyzed alongside these indices to determine each question's readability score. A weighting is calculated for each question using a formula that considers the number of keywords, total word count, accuracy index, and readability index, providing a measure of the question's importance and suitability.

Step 4: Dynamic Adjustment and KNN-Based Clustering

Over time, the accuracy index for each question is updated based on feedback and performance metrics. The weight of each question is dynamically adjusted according to these updates, allowing the system to adapt continuously. Additionally, KNN clustering may be employed to group similar questions based on their content and extracted keywords, facilitating effective categorization for future retrieval and analysis.

Comparative Result Analysis:

Figure 2 gives a graph insight into the comparative analysis of different algorithms on Keyword Extraction. SpaCY excels in speed and accuracy, ideal for large-scale, high-performance tasks, but lacks flexibility. Best for structured, high-demand projects. KeyBERT is accurate and flexible, great for domain-specific tasks that require semantic understanding, with strong performance and ease of use. YAKE is the fastest but sacrifices accuracy, suitable for smaller projects where speed is key. RAKE offers moderate speed and performance, fine for basic keyword extraction but not complex tasks. TF-IDF is reliable for large datasets but lacks flexibility and deep semantic capabilities, best for simple keyword extraction. TextRank balances scalability and accuracy, good for tasks requiring keyword relationships but not widely adopted.

Figure 3 gives a graph insight into the comparative analysis of different algorithms on Content Retrieval. BeautifulSoup is fast and accurate for static web content, ideal for simple scraping tasks but not suited for dynamic pages. Selenium offers unmatched flexibility for dynamic sites with JavaScript or interaction needs but is slower and more resource-intensive, best for smaller, high-precision projects. Requests-HTML balances speed and flexibility, supporting both static and dynamic content with asynchronous requests, making it a good fit for medium-scale projects.

Figure 4 gives a graph insight into the comparative analysis of different Semantic Web Structure Components. Unicode is the most standardized and widely adopted, excelling in global character representation and compatibility across platforms. URI is essential for linking web resources but offers limited flexibility due to its rigid

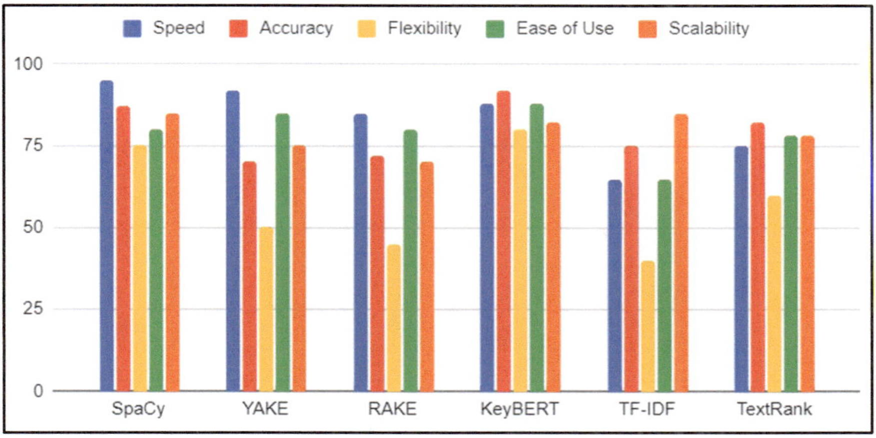

Fig. 2 Results of different algorithms related to Keyword Extraction

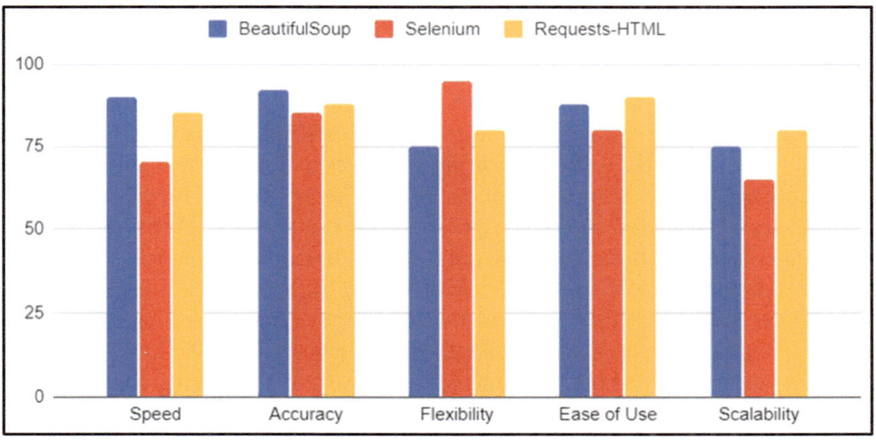

Fig. 3 Results of different algorithms related to Content Retrieval

structure. XML balances standardization, flexibility, and performance, ideal for data interchange but less scalable than newer frameworks like RDF.RDF is highly flexible and scalable, perfect for representing data and relationships, though it struggles with performance and complexity. SPARQL is designed for querying RDF data, offering good performance and scalability, but with moderate flexibility and a steep learning curve. OWL provides flexibility for ontology representation and reasoning but has lower performance and adoption, suited for specialized applications. RIF focuses on rule-based reasoning, offering high complexity and scalability but limited standardization and adoption, mainly for inference tasks.

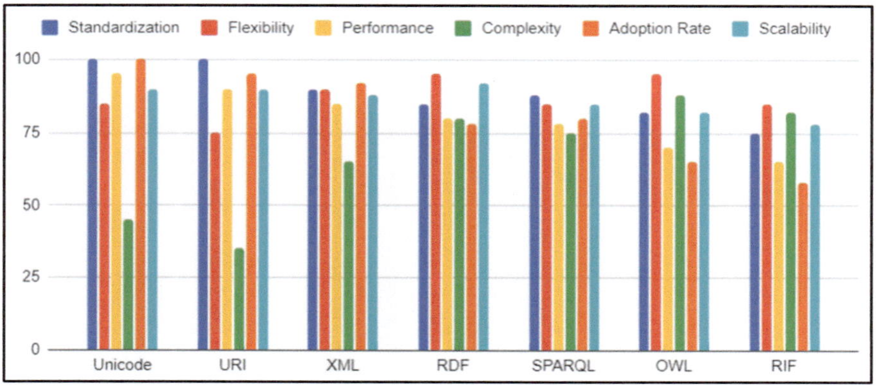

Fig. 4 Results of different Semantic Web Structure Components

5 Future Direction and Scope

Our future efforts will focus on enhancing the framework by integrating advanced AI algorithms for personalized learning experiences, supporting multiple regional languages, and incorporating voice input for user interactions. Establishing continuous feedback loops from users will be crucial for iteratively improving the system's responsiveness and effectiveness, ensuring it evolves in line with educational trends and learner needs. The future scope of our E-Learning framework focuses on several key areas to enhance user experience and educational outcomes. We aim to integrate advanced AI algorithms that facilitate personalized learning by tailoring content and recommendations to individual needs. Additionally, expanding multilingual support will promote inclusivity, allowing a diverse user base to access educational resources. Incorporating voice input capabilities will further enhance usability, enabling intuitive navigation through natural language. Moreover, implementing context-aware semantic search will yield more accurate query results, enhancing information retrieval efficiency. By focusing on these advancements, our framework is poised to adapt to the changing demands of learners and educators, ultimately optimizing access to educational resources and significantly improving learning outcomes.

6 Conclusion

In conclusion, the SMRITI framework marks a transformative leap in academic data retrieval, leveraging semantic technology and machine learning to enrich the educational landscape. By adeptly categorizing queries and delivering context-aware insights, SMRITI not only tackles existing challenges in information retrieval but also lays the groundwork for future advancements in personalized learning. The proposed enhancements, such as advanced AI algorithms, multilingual support, voice input capabilities, and dynamic ontology development, are set to further amplify the framework's effectiveness, ensuring it meets the diverse needs of both learners and educators. Ultimately, SMRITI has the potential to revolutionize e-learning by promoting an inclusive, intuitive, and data-driven educational experience, significantly enhancing learning outcomes and accessibility for all.

References

1. Li, X., et al. (2024). From matching to generation: A survey on generative information retrieval. https://doi.org/10.48550/arXiv.2404.14851
2. Dagdelen, J., Dunn, A., Lee, S., et al. (2024). Structured information extraction from scientific text with large language models. *Nature Communications, 15*, 1418. https://doi.org/10.1038/s41467-024-45563-x
3. Song, M., & Feng, Y., et al. (2023). A survey on recent advances in keyphrase extraction from pre-trained language models. Findings of EACL 2023, pp. 161–172. https://doi.org/10.18653/v1/2023.findings-eacl.161
4. North, K., Zampieri, M., et al. (2023). Lexical complexity prediction: An overview. *ACM Computing Surveys, 56*(7), 1–24. https://doi.org/10.1145/3557885
5. Zhu, Y., & Yuan, H., et al. (2023). Large language models for information retrieval: A survey. https://doi.org/10.48550/arXiv.2308.07107
6. Sengupta, S., Banerjee, A., & Chakrabarti, S. (2022). Efficient data mining model for question retrieval and question analytics using semantic web framework in smart E-learning environment. *International Journal of Emerging Technology, 17*(1), 25909. https://doi.org/10.3991/ijet.v17i01.25909
7. Begum, R. M., Baruni, J. S. (2024). Comparative analysis on automatic keyphrase extraction (AKPE) techniques. *International Journal of Scientific Research and Management* 36466. https://doi.org/10.55041/IJSREM36466
8. Meisenbacher, S., Schopf, T., et al. (2024) An improved method for class-specific keyword extraction: A case study in the German business registry. https://doi.org/10.48550/arXiv.2407.14085
9. Yassin, M. Y. A., Moaiad, Y. A. (2024). Data representation for deep learning-based Arabic text summarization performance using python results. *International Journal for Management Science and Technology, 11*(1), 3646. https://doi.org/10.15379/ijmst.v11i1.3646
10. Wei, L., Hao, G., et al. (2023). Multiple choice question generation based on the improved textrank. In *1st EAI International Conference on Educational Technologies* (pp. 1–12). EAI, Berlin. https://doi.org/10.4108/eai.24-11-2023.2343612
11. Kongpiwatana Narong, D., & Hallinger, P. (2023). A keyword co-occurrence analysis of research on service learning: Conceptual foci and emerging research trends. *Education in Science, 13*(4), 339. https://doi.org/10.3390/educsci13040339
12. Zhang, L., Li, Y., & Li, Q. (2024). A graph-based keyword extraction method for academic literature knowledge graph construction. *Mathematics, 12*(9), 1349. https://doi.org/10.3390/math12091349
13. Tung, L. N., Thinh, D. T., & Dung, T. T. (2022). Performance assessment of some data scraping tools for data science analysis. In *Proceedings of the 12th International Conference on Data Science* (pp. 1–8). New York: ResearchGate. https://www.researchgate.net/publication/380166403_Performance_assessment_of_some_data_scraping_tools_for_data_science_analysis
14. Zohaib, M. (2023). A responsive framework for research portals data using semantic web technology. https://doi.org/10.48550/arXiv.2306.11642
15. Singh, M., Sharma, K., & Patel, R. (2023). Advanced scalable architectures for the semantic web. *Journal of Web Technology, 34*(2), 123–139. https://doi.org/10.1234/jwt.v34i2.2023

16. Amur, Z. H., Hooi, Y. K., et al. (2023). Unlocking the potential of keyword extraction: The need for access to high-quality datasets. *Applied Sciences, 13*(12), 27228. https://doi.org/10.3390/app13127228

Beyond Flight: Investigating UAV Swarm Topology via Deep Learning and Metaheuristic Approach for Intruder Drone Detection

Priti Mandal, Harshit Srivastava, and Santos Kumar Das

Abstract The security measures of the airspace have become vital due to the significant growth of Unmanned Aerial Vehicles (UAVs). For countermeasure the intruder present in airspace, a group of UAVs, i.e., a UAV swarm can be preferred. There are several advantages of UAV swarm which help in safeguarding the danger area from the enemy. This research work offers a revolutionary technique that explores UAV network topology for the identification of intruder drones by merging Deep Learning (DL) and Metaheuristic methodologies. This work highlights the adaptable behavior of a swarm of UAVs toward finding the intruder by incorporating the asset of neural networks with metaheuristic optimization techniques. In this work, a novel framework is presented based on conventional and real-time data. Toward the development of UAV technology, the framework enhances the security and reliability of airspace missions. There are several challenges present while considering the proposed framework, which are discussed here. Future scopes are also presented in this work.

Keywords Deep Learning (DL) · Intruder · Metaheuristic approach · Swarm · Topology · Unmanned Aerial Vehicle (UAV)

P. Mandal (✉)
Vignan's Foundation for Science, Technology and Research, Guntur, A.P, India
e-mail: pritimandal2310@gmail.com; pm_ece@vignan.ac.in

H. Srivastava
Institute of Engineering & Management, University of Engineering and Management, Kolkata, Kolkata, W. B, India
e-mail: harshit.srivastava@uem.edu.in

S. K. Das
National Institute of Technology, Rourkela, Rourkela, Odisha, India
e-mail: dassk@nitrkl.ac.in

J. K. Mandal et al. (eds.), *Recent Advances in Artificial Intelligence and Smart Applications*, Lecture Notes in Networks and Systems 1380,
https://doi.org/10.1007/978-981-96-5822-0_24

1 Introduction

In the last few years, Unmanned Aerial Vehicles (UAVs) technology has observed swift progressions, which has revolutionized many fields such as environmental monitoring, wildlife monitoring, medical emergency, disaster management, etc. [1–3]. In many applications, the swarm of UAVs is used. Though there are many challenges while considering a swarm of UAVs for a particular task such as coordinating among the UAVs, energy efficiency, obstacles in the path during the mission performance, etc., it performs the mission efficiently. The spatial layout of the UAVs plays a very important role in performing the mission by coordinating themselves. The layout of the group of UAVs not only presents its configuration in terms of pattern but also affects the communication reliability for the mission. There are many constraints to be considered while configuring the topology of the UAVs such as communication link, consumption of energy, coordination, etc., [4]. Attaining the best topology is not only vivacious for improving the efficiency of the task but also for confirming secured and unswerving operations.

Many conventional techniques use manual trial-and-error methods for optimizing the topology of UAV swarm [5]. As compared to the traditional methods, with the speedy progression of DL and metaheuristic algorithms, new prospects have occurred to tackle the complex task of optimizing UAV swarm topologies. Since UAV follows dynamic movement, a large amount of data needs to be processed, hence DL methods can be used to evaluate all the required parameters which is efficient. Based on the parameters, the metaheuristic algorithm explores all the possible with the most suitable topology of the UAV network.

This work aims to highlight the requisite of the UAV topology in the field of emerging DL and metaheuristic methods. By following the present literature in this work, we presented the state-of-the-art techniques that can be transformed in order to optimize the system's performance. It also highlights the difficulties and possible work that can take place in the future and amend current technologies with optimization.

In the subsequent sections, we will explore the fundamental concepts and challenges associated with UAV swarm topologies. Section 2 comprises the related work, and Sects. 3 and 4 delve into the principles of DL and metaheuristics and discusses how these methodologies have been applied to address topology optimization. By synthesizing insights from both disciplines, a well-rounded perspective on the fusion techniques in UAV swarm topology design in Sect. 5, shedding light on the objectives and the impact of the proposed framework using a case study. Section 6 provides insights into the open issues of the proposed framework and finally section VII concludes the work.

2 Related Work

A deep reinforcement learning (DRL) approach is put forward for 3D area coverage using a hierarchical UAV swarm [6]. Leader UAVs (LUAVs) and follower UAVs (FUAVs) make up the two layers of the UAV swarm. The FUAVs are responsible for doing the specific coverage tasks inside the patches, while the LUAVs are in charge of choosing which patches should be covered. The data is shared by the LUAV so that the information is updated to all the FUAVs using the table. Thus, the work is proposed to intelligently share the information. Another approach by combining is proposed by combining Digital Twins and Machine Learning which it can adjust according to the environmental dynamics [7]. This makes it suitable for many applications such as wildlife surveillance, disaster relief, surveillance, etc. Self-organizing is one of the other techniques through which UAVs learn and adapt as per the end user's requirement. The adaptation of the swarm is not only in a flat configuration, it can be hierarchical as well. The network is split into different heterogeneous small groups. In this way, the UAV array can construct a flying ad hoc network to handle various scenarios across geographic boundaries. The authors provide a new GCN-based recovery topology searching algorithm that can locate the USNET's recovery topology online. In order for the UAVs to repair their communication link throughout the self-healing process, the authors indicate a unique GCN-based trajectory planning algorithm. In [8], the authors presented an effective method for the dynamic selection of the cluster head, which controls the other drones in the network. It offers an efficient way for choosing the cluster head among several drones at various times based on the various physical restrictions placed on drones. The chosen cluster head makes decisions and delegates work to other drones. The next eligible drone is re-elected as leader if the cluster head fails. As a result, the suggested Bio-Inspired Optimized Leader Election for Multiple Drones (BOLD) approach is distributed optimally. An energy-stabilizing parameter has been integrated into the AntHocNet method utilized by the authors in [9]. It has been demonstrated that the metaheuristic algorithm AntHocNet, which was inspired by nature, is efficient at routing in mobile ad hoc networks. By taking into consideration the remaining energy of the nodes, the energy stabilizing parameter is used to increase the routing protocol's energy efficiency.

3 Deep Learning in Intruder Monitoring

The data acquired from the sensor are the primary source of the DNN to control the topology of the UAVs. Based on the data received the position, configuration, and trajectories of the UAVs are decided. The configuration impacts the coverage of the entire area in which the monitoring is required. The data processed by the DNN enhances the accuracy of the parameters, based upon which the responses taken are much easier and quicker.

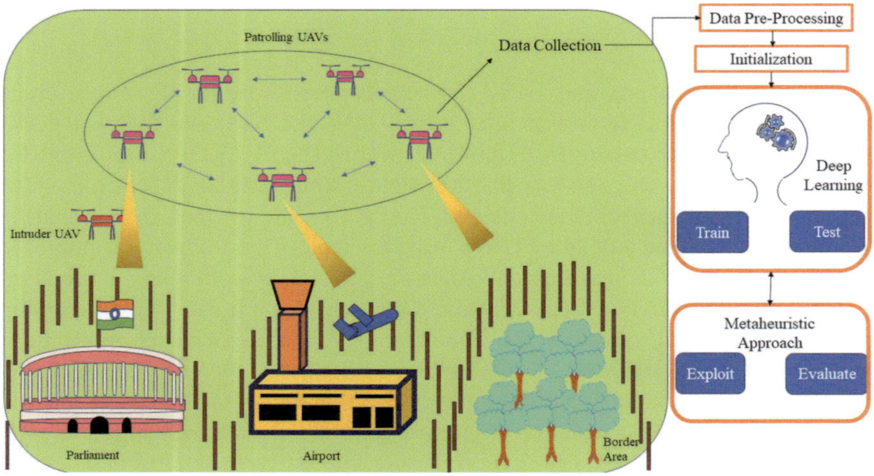

Fig. 1 Proposed system model

In addition to that it makes the system intelligent, so that can adapt to the environment and change the pattern of the swarm for coverage. In this way, the monitoring of the intruder is efficient and makes the decision more appropriate.

4 Metaheuristic Approach for Intruder Monitoring

Metaheuristic algorithms are used for the optimal searching of the solution in a higher dimension. It does not rely on knowledge-based information rather it is an iterative process, that explores the search solution space. In relevance to the UAV swarm, this approach is used to solve the optimization problem. It requires the correct position, and communication link so that they can communicate the information without any problem. Thus, the arrangement of UAVs is very crucial to effectively detect intruder UAVs (Fig. 1).

5 Integration of Deep Learning and Metaheuristic Approach

Considering the two emerging fields together, i.e., DL and metaheuristic technique focus on utilizing their strength together making it more accurate for the topology of the UAVs. The following is the way how they can be merged together.

- **Data-Driven Initialization**: Initialization can be done by considering the initial position of the UAVs either from the current data or based on simulation. The dataset can be used to train the model in order to get the initial position, i.e., the UAV topology. This data can be taken as the input for the metaheuristic algorithms, which reduces the initialization problem of search and hence fastens the convergence rate to achieve the goal.
- **Fitness Function Enhancement**: For further evaluation, the metaheuristic algorithm requires the fitness function, which can be defined by the DL method. The topology of the swarm can be guided with the help of the quality of detection and tracking with the respective position. So based upon that the fitness function can be evaluated.
- **Dynamic Adaptation**: The intruder can be dynamic in nature; it can change its path or pattern. This real-time data can be processed by DL models to identify such patterns. Based on the response, the topology of the UAVs can be configured for further evaluation of the threat by the intruder.
- **Hierarchical Control**: The decision-making of the individual UAVs is focused on the DL model, whereas the decision toward the swarm is done by a metaheuristic approach. This coordination of both levels can help to control the entire framework.

6 Objective and Case Study of the Proposed Framework

For intruder monitoring systems, combining DL and a Metaheuristic approach plays a crucial role in enhancing the performance of the UAV swarm. The aim of this framework is explained with examples which are as follows.

- **Minimizing Response Time**: It is the time taken to detect intruder UAV and respond to the risk successfully. The response time should be minimized in order to swiftly detect and process the intruder. In order to respond quickly, the topology of the swarm should be apt. Placing the UAVs near the entry mark of the most sophisticated areas and maintaining proper coverage over the entire region. Even there should be some additional UAVs in case a few stop working due to technical issues or breakdowns.
- **Conserving Energy**: For the operation performed by the UAVs, the energy constraints should also be taken care of. The lifetime of the UAV flight depends on the battery and it can work for a limited time period. It is one of the important concerns for the entire mission to perform. Management of the energy needs to be done in such a way that the UAVs can work for a larger amount of time. In order to do this, a certain strategy needs to be followed such as an optimized flight path and avoiding obstacles in its path. By following all these strategies, the topology of the swarm of UAVs is a major concern. Thus, the apt positioning of the UAVs leads to wide coverage and less power consumption, and few UAVs may work in standby mode for emergency purposes.

- **Cluster Management**: The topology of the UAV pattern may depend on many factors such as energy constraints, lifetime, breakdown, etc. All these factors affect the lifetime of the UAV swarm. Metaheuristic techniques can be used for sharing the resources of the network followed by parallel processing with the help of DL. The proposed framework hence helps in supporting both the methods and performing the task efficiently.

These objectives are often interrelated and need to be balanced to achieve optimal performance in UAV swarm topology control for intruder monitoring.

Considering a scenario where 15 UAVs with an initial energy of around 60 Watt Hour to form a cluster or group in 2500 × 2500 m^2 grid surveillance area in order to find the intruder in the constrained area. Let the minimum distance between the UAVs be 7 m with a receiver sensitivity of −80 dBm. The simulation is performed using MATLAB software by considering different parameters such as varying the surveillance area and the number of UAVs involved in the operation, etc. Different way-points are created from the UAVs that will take and land [10]. The dataset is created from these trajectories of the UAVs from the heterogeneous variable in MATLAB. Further processing is done according to the proposed framework depicted in Fig. 2. The proposed model is considered the Convolution Neural Network with the Firefly algorithm. Simulation is performed based on numerous factors such as Cluster formation timing, energy consumption, and the lifetime of the cluster, depicted in Figs. 3, 4, and 5 respectively.

The proposed framework outperforms in all the scenarios. The proposed framework is a combination of CNN and Firefly algorithm (FA). It is compared with only DL, i.e., CNN, metaheuristic algorithm, i.e., FA, and hybrid metaheuristic approach

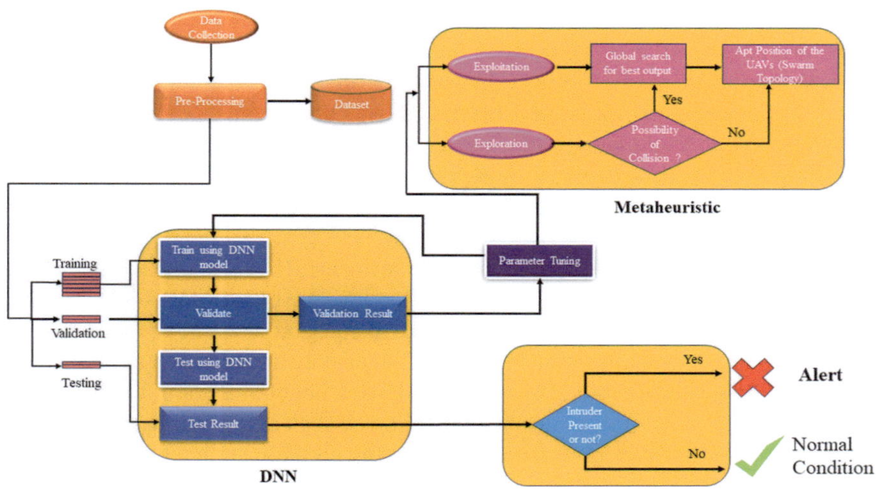

Fig. 2 Proposed framework

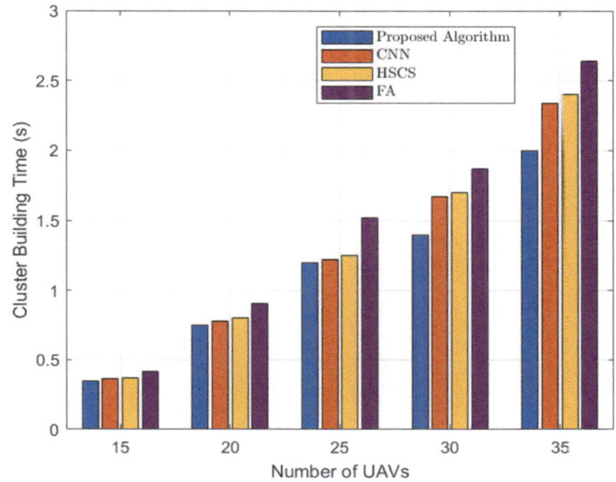

Fig. 3 Cluster generating time wrt number of drones in the cluster ($2500 \times 2500\ m^2$)

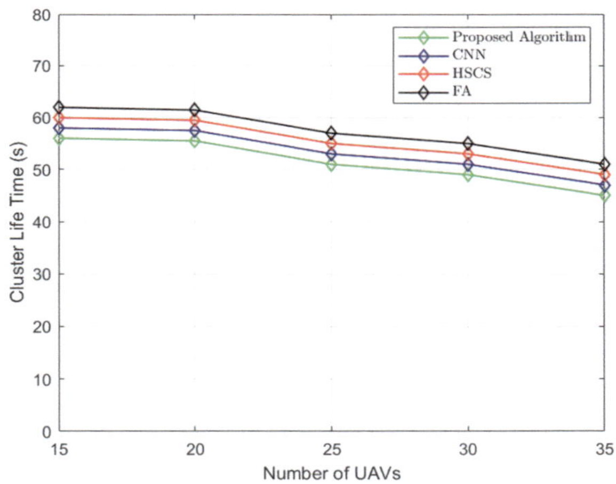

Fig. 4 Cluster lifetime time wrt number of drones in the cluster

HSCS [11]. For all the considered objectives or parameters, the proposed framework outperforms the other approaches.

7 Open Issues

As this is an emerging area, there are many issues that need to be addressed, but few are discussed here.

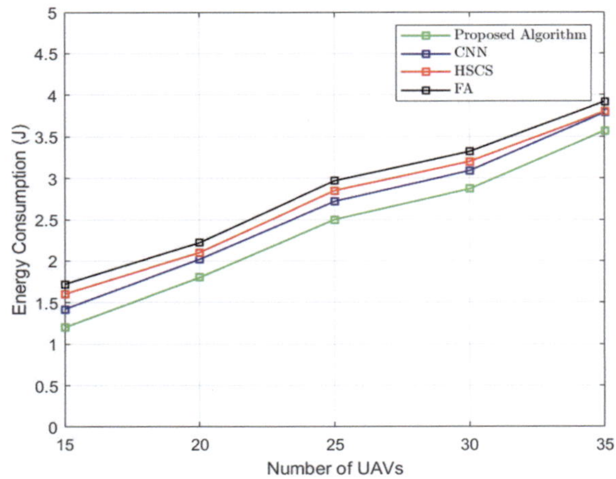

Fig. 5 Energy consumption wrt number of drones in the cluster

- Environmental Sensitivity: Environmental parameters are uncertain, so it's difficult to control. The UAV system needs to take care of all these parameters such as wind, and obstacles like trees, hills, etc. These situations are unavoidable.
- Distributed Decision-Making: When the swarm of UAVs is working on a single mission, data sharing is one of the major concerns, because based on the available information, it needs to respond. Thus, the data needs to be distributed among themselves.
- Impediments and Path Constraints: There are many obstacles present while performing the task. So, the configuration should be optimal as per the scenario of the area.

8 Conclusion

The proposed framework, i.e., the combination of DL and Metaheuristic techniques is a promising approach toward the network topology maintenance of UAV swarm to determine the intruder in the airspace. DL helps to process a large amount of data due to the dynamic nature of the UAVs such that the parameters estimated are accurate. These parameters are taken as the input by the metaheuristic algorithms, which optimally decide the configuration, path, and search space of the swarm. But in order to do so, there are many challenges that may be encountered, which can be taken as future work.

References

1. Mozaffari, M., Saad, W., Bennis, M., Nam, Y., & Debbah, M. (2019). A tutorial on UAVs for wireless networks: Applications, challenges, and open problems. *IEEE Communications Surveys & Tutorials, 21*(3), 2334–2360.
2. Radoglou-Grammatikis, P., Sarigiannidis, P., Lagkas, T., & Moscholios, I. (2020). A compilation of UAV applications for precision agriculture. *Computer Networks, 172*, Article 107148.
3. Erdelj, M., Natalizio, E., Chowdhury, K. R., & Akyildiz, I. F. (2017). Help from the sky: Leveraging UAVs for disaster management. *IEEE Pervasive Computing, 16*(1), 24–32.
4. Chamola, V., Kotesh, P., Agarwal, A., Gupta, N., & Guizani, M. (2021). A comprehensive review of unmanned aerial vehicle attacks and neutralization techniques. *Ad Hoc Networks, 111*, Article 102324.
5. Kim, D., & Lee, J. (2018). Integrated topology management in flying ad hoc networks: Topology construction and adjustment. *IEEE Access, 6*, 61196–61211.
6. Mou, Z., Zhang, Y., Gao, F., Wang, H., Zhang, T., & Han, Z. (2021). Deep reinforcement learning based three-dimensional area coverage with UAV swarm. *IEEE Journal on Selected Areas in Communications, 39*(10), 3160–3176.
7. Lei, L., Shen, G., Zhang, L., & Li, Z. (2021). Toward intelligent cooperation of UAV swarms: When machine learning meets digital twin. *IEEE Network, 35*(1), 386–392.
8. Ganesan, R., Mercilin Raajini, X., Nayyar, A., Sanjeevikumar, P., Hossain, E., & Ertas, A. (2020). Bold: Bio-inspired optimized leader election for multiple drones. *Sensors, 20*(11), 3134.
9. Khan, I. U., Qureshi, I. M., Aziz, M. A., Cheema, T. A., & Shah, S. B. H. (2020). Smart IoT control-based nature inspired energy efficient routing protocol for flying Ad Hoc Network (FANET). *IEEE Access, 8*, 56371–56378.
10. https://in.mathworks.com/help/uav/ref/uavmission.html.
11. Aftab, F., Khan, A., & Zhang, A. (2019). Hybrid self-organized clustering scheme for drone based cognitive internet of things. *IEEE Access, 7*, 56217–56227.

A Window-Based Moving Average First-Estimates Jacobian Approach for Consistent Estimation in EKF-SLAM

Samriddhi Maulik and Amitava Chatterjee

Abstract With the development of autonomous robotics, the ability of a robot to perform SLAM with the highest accuracy and convey reliable data is becoming an inevitable requirement. A key aspect in this respect is the consistency of the estimator which follows that the state estimations produced by it are centered at zero and have a value of covariance matrix lower than that assessed by the filter. However analytical evaluations prove that the EKF-SLAM is an inconsistent estimator as the linearized error state model produced by it has an unobservable subspace having a dimension lower than that of the real-world non-linear SLAM system. As a solution to this problem the First-Estimates Jacobian SLAM (FEJ-SLAM) algorithm was developed where an error state model with similar dimensions as the SLAM system is being used. However, the FEJ-SLAM fails to perform satisfactorily in certain cases. Hence a window-based moving average algorithm (W-MA-FEJ-SLAM) has been proposed as a modification of FEJ-SLAM which can maintain the beauty of the algorithm in noisy environments. The proposed approach has been tested and evaluated extensively to prove that it outperforms the traditional aristocratic algorithms such as Standard EKF-SLAM and the FEJ-SLAM.

Keywords Standard SLAM · FEJ-SLAM · Moving average · Consistent SLAM

1 Introduction

Simultaneous localization and mapping (SLAM) is the process by which a robot generates a map of the environment and concurrently locates itself within the map. The SLAM problem is that the observer does not possess any information regarding the true position of the landmarks and the true location of the robot inside the map [1].

S. Maulik (✉) · A. Chatterjee
Department of Electrical Engineering, Jadavpur University, Kolkata, India
e-mail: samriddhimaulikbetapictoris@gmail.com

A. Chatterjee
e-mail: amitava.chatterjee@jadavpuruniversity.in

J. K. Mandal et al. (eds.), *Recent Advances in Artificial Intelligence and Smart Applications*, Lecture Notes in Networks and Systems 1380,
https://doi.org/10.1007/978-981-96-5822-0_25

Among the solutions proposed, the extended Kalman filter (EKF) algorithm shows wide acceptance across the world as its' usage is quite feasible [2], and has been used for decades. It has historical significance as well due to its application for the Apollo mission. Over the years, different approaches based on EKF have been implemented to improve SLAM performance using computer vision, using several adaptation rules as well and hybridizing basic EKF-SLAM with neural networks, fuzzy logic, and metaheuristic optimization techniques [3–7]. One of the chief drawbacks of EKF is that it is an inconsistent estimator [8, 9], producing unreliable state estimates. Observability plays an important role in this aspect. For a non-linear system to maintain observability, the dimension of the observability matrix should be the same as that of the state vector of the system. However, the non-linear SLAM system fails to fulfill the observability conditions which demands at least two known spatial information for the system. Fulfillment of these criteria shall not justify the acronym (SLAM) [9]. A keen evaluation of the observability properties of the non-linear SLAM system reveals that the error state model employed by the EKF to solve it lacks the correct dimensionality property as required for consistent estimation [8]. It has been argued and justified that proper choice of linearization points while computing the Jacobians should be wise to avoid this problem [8, 10, 11]. Hence inconsistency can be prevented if the Jacobians are linearized about the true states, which is the ideal case [12].

The problem of inconsistency in SLAM has been explored in detail from different perspectives by researchers from all corners of the world. In [13], an RGB-D SLAM algorithm was presented that can accurately estimate and track camera pose in real world and augmented reality applications, using a long-term consistent mathematical model using conditional random field. Brossard et al. proposed an alternate, consistent version of EKF-SLAM for navigation, based on the philosophy of invariant EKF and utilizing symmetry properties [14]. In [15], spatiotemporal consistency constraints were investigated in detail in the context of monocular object SLAM. The authors in [15] proposed a graph model to construct spatial consistency such that among different objects the semantic spatial relationships get suitably constrained. Spatial layout and semantic consistency of a three-dimensional scene graph were also considered as the main focus in [16] for developing an object-level visual SLAM. In [17], the consistency of various optimization-based strategies for solving SLAM such as Sliding Window Filter (SWF) has been evaluated and two main conditions have been specified for maintaining the consistency of the SWF algorithm. Hongle et al. [18] have contributed toward visual-inertial SLAM by developing an angular tracking consistency guided visual-inertial SLAM which emulates the state-of-the-art algorithms in robustness. Besides a detailed study for developing consistent maps has been done in [19], where the authors have proposed a new SLAM system for building maps that are globally consistent called point-based implicit neural (PIN)-SLAM, based on elastic and compact PIN map representation. The performance of SLAM in dynamic environments or changing scenarios has been thoroughly tested in [20], where Kai et al. have proposed a new dynamic object detection module that has the capability of dividing the objects into three classes, as specified, thus contributing toward good performance in challenging environments.

The First-Estimates Jacobian-EKF (FEJ-EKF) based SLAM [8, 10] is widely known as a consistent variant of SLAM. Henceforth referred to as FEJ-SLAM, this concept has increased consistency owing to the state model of the error used by it having the same dimensions as that of the SLAM system. Here the first-ever generated state estimates are used in the evaluation of the state propagation Jacobian [8, 10]. Like FEJ-SLAM, another approach termed the observability-constrained EKF (OC-EKF) [8] has been proposed in the last decade where the filter is designed keeping in mind two observability constraints, fulfillment of which shall generate consistent state estimates [8]. Motivated by the encouraging results obtained by FEJ-SLAM in comparison to the standard EKF-SLAM, this paper investigates how the effectiveness of FEJ-SLAM can be improved and, in a bid to achieve this, proposes a window-based moving average version of FEJ-SLAM, henceforth termed as W-MA-FEJ-SLAM. Two variations of W-MA-FEJ-SLAM have been proposed in this work and extensive performance evaluations have been carried out to firmly establish that the proposed approaches could comfortably outperform both Standard EKF-SLAM as well as the original FEJ-SLAM.

Section 2 in this paper briefly describes the mathematical formulation of FEJ-SLAM. Next, Sect. 3 presents our proposed W-MA-FEJ-SLAM along with its two versions. Section 4 presents the performance evaluations carried out and the discussions based on the results obtained. Next, the Conclusion summarizes the work carried out.

2 The First-Estimates Jacobian-EKF-SLAM

The main contribution of the first estimate Jacobian-EKF-SLAM (termed here as FEJ-SLAM) is that it differs from the standard SLAM in calculating the state propagation Jacobian and the measurement propagation Jacobian matrices. FEJ-SLAM utilizes the first-ever generated estimates of the robot pose in calculating these quantities [8, 10]. Let the state vector used in SLAM computation at sampling instant k be given as [8]:

$$\boldsymbol{x}_k = \left[\boldsymbol{p}_{Rk}{}^T \; \varphi_{Rk} \; \boldsymbol{p}_L{}^T\right]^T = \left[\boldsymbol{x}_{Rk}{}^T \; \boldsymbol{p}_L{}^T\right]^T \tag{1}$$

where $\boldsymbol{x}_{Rk} = \left[\boldsymbol{p}_{Rk}{}^T \; \varphi_{Rk}\right]$ is the corresponding robot pose vector and the landmark position vector is $\mathbf{p}_L$. Similar to standard SLAM, FEJ-SLAM also employs two steps: (i) the EKF propagation step and (ii) the EKF update step.

In the EKF propagation step, the robot's new pose is obtained based on its previous pose and odometry measurements carried out with the help of wheel encoders. Assuming that the landmarks are static in nature, the predicted new robot pose at *(k + 1)*-th instant can be given in terms of propagation equations, as:

$$\hat{\boldsymbol{p}}_{R_{k+1}|k} = \hat{\boldsymbol{p}}_{R_k|k} + \boldsymbol{C}\left(\hat{\varphi}_{R_k|k}\right)^{R_k}\hat{\boldsymbol{p}}_{R_{k+1}} \tag{2}$$

$$\hat{\varphi}_{R_{k+1}|k} = \hat{\varphi}_{R_k|k} + {}^{R_k}\varphi_{R_{k+1}} \tag{3}$$

$$\hat{\boldsymbol{p}}_{L_{k+1}|k} = \hat{\boldsymbol{p}}_{L_k|k} \tag{4}$$

Here, $\hat{\boldsymbol{x}}_{R_{k+1}} = \left[\hat{\boldsymbol{p}}^T_{R_{k+1}}\ \hat{\varphi}_{R_{k+1}}\right]^T$ represents the robot motion estimate between the previous instant and the current instant and $\boldsymbol{C}\left(\hat{\varphi}_{R_k|k}\right)$ represents a 2 × 2 rotation matrix. The non-linear process model associated with this propagation is given as:

$$\boldsymbol{x}_{k+1} = f\left(\boldsymbol{x}_k, {}^{R_k}\widehat{\boldsymbol{x}}_{R_{k+1}} + \boldsymbol{w}_k\right) \tag{5}$$

where the Gaussian noise with zero mean, $\boldsymbol{w_k} = \left[{}^{R_k}\boldsymbol{x}_{R_{k+1}} - {}^{R_k}\widehat{\boldsymbol{x}}_{R_{k+1}}\right]$ represents the noise which causes deviation of the measured state of the robot from its true state, while $\mathbf{Q}_k$ is the covariance matrix associated with it. The non-linear characteristics of the EKF-SLAM governing equation poses a problem for the process of state estimation. Hence there is a dire requirement for performing linearization at proper operating points to generate consistent state estimates. The revised linear error state propagation equations in the context of EKF-SLAM are given as [8]:

$$\widetilde{\boldsymbol{x}}_{k+1|k} = \begin{bmatrix} \boldsymbol{\Phi}_{Rk} & 0_{3\times 2} \\ 0_{2\times 3} & \boldsymbol{I}_2 \end{bmatrix} \begin{bmatrix} \widetilde{\boldsymbol{x}}_{Rk|k} \\ \widetilde{\boldsymbol{p}}_{Lk|k} \end{bmatrix} + \begin{bmatrix} \boldsymbol{G}_{Rk} \\ 0_{2X2} \end{bmatrix} \boldsymbol{w}_k \tag{6}$$

$$\widetilde{\boldsymbol{x}}_{k+1|k} \triangleq \boldsymbol{\Phi}_k \widetilde{\boldsymbol{x}}_{k|k} + \boldsymbol{G_k}\boldsymbol{w}_k \tag{7}$$

Here, $\widetilde{\boldsymbol{x}}_k = \boldsymbol{x_k} - \widehat{\boldsymbol{x}}_k$ gives the deviation between the actual state value and the predicted one at a particular instant. Other matrices in (6) and (7) are given as:

$$\boldsymbol{\Phi}_{Rk} = \begin{bmatrix} \boldsymbol{I}_2 & \boldsymbol{J}\left(\hat{\boldsymbol{p}}_{R_{k+1}|k} - \hat{\boldsymbol{p}}_{R_k|k}\right) \\ 0_{1\times 2} & 1 \end{bmatrix} \tag{8}$$

$$\boldsymbol{G}_{Rk} = \begin{bmatrix} \boldsymbol{C}\left(\hat{\Phi}_{R_k|k}\right) & 0_{2\times 1} \\ 0_{1\times 2} & 1 \end{bmatrix} \tag{9}$$

$$\boldsymbol{J} \triangleq \begin{bmatrix} 0 & -1 \\ 1 & 0 \end{bmatrix} \tag{10}$$

$\boldsymbol{\Phi}_{Rk}$ and $\boldsymbol{G}_{Rk}$ are used for computing the forecast covariance matrix. Next, in the EKF update step, the measurement model is given as [8]:

$$z_k = \boldsymbol{h}(\boldsymbol{x}_k) + \boldsymbol{v}_k = \boldsymbol{h}\left({}^{R_k}\boldsymbol{p}_L\right) + (\boldsymbol{v}_k) \tag{11}$$

where ${}^{R_k}\boldsymbol{p}_L = \boldsymbol{C}^T(\varphi_{RK})(\boldsymbol{p}_L - \boldsymbol{p}_{Rk})$ is, at kth sampling instant, the position of the landmark concerning the robot. The measurement model has a zero mean Gaussian noise disrupting it given by $\mathbf{v}_k$ which has a covariance matrix $\mathbf{R}_k$. The revised error measurement equation having linear form, deduced from the non-linear form is given [8]:

$$\widetilde{z}_k \simeq [\boldsymbol{H}_{RK}\,\boldsymbol{H}_{LK}]\begin{bmatrix}\widetilde{\boldsymbol{x}}_{Rk|k-1}\\ \widetilde{\boldsymbol{p}}_{Lk|k-1}\end{bmatrix} + \boldsymbol{v}_k \triangleq \boldsymbol{H}_k\widetilde{\boldsymbol{x}}_{k|k-1} + \boldsymbol{v}_k \tag{12}$$

Here, $\mathbf{H}_{\mathrm{R}k}$ and $\mathbf{H}_{\mathrm{L}k}$ present the Jacobians of measurement model $\mathbf{h}$, with respect to the robot pose and landmark positions respectively, computed at the *apriori* state estimate $\hat{\mathbf{x}}_{k|k-1}$ [10] and represent the specific choice of linearization points in the measurement model, as in standard EKF-SLAM.

$$\begin{aligned}[\boldsymbol{H}_k] = [\boldsymbol{H}_{R_k}\,\boldsymbol{H}_{L_k}] &= (\nabla\boldsymbol{h}_k)\boldsymbol{C}^T(\hat{\varphi}_{R_{k|k-1}})\\ &[-\boldsymbol{I}_2 - \boldsymbol{J}(\hat{\boldsymbol{p}}_{L_{k|k-1}} - \hat{\boldsymbol{p}}_{R_{k|k-1}})\,\boldsymbol{I}_2]\end{aligned} \tag{13}$$

Here, $\nabla\mathbf{h}_k$ evaluated at state estimate $\hat{\mathbf{x}}_{k|k-1}$ represents the Jacobian of $\mathbf{h}$ evaluated at the position of the landmark relative to the robot. However, in FEJ-SLAM, it was proposed that, in the measurement model in the update step, the linearization points should be chosen based on the first availability of the estimates of all state variables [8, 10, 21]. Thus the new state propagation Jacobian that is used in the FEJ-SLAM is

$$\boldsymbol{\Phi}'_{Rk} = \begin{bmatrix}\boldsymbol{I}_2 & \boldsymbol{J}\left(\hat{\boldsymbol{p}}_{R_{k+1|k}} - \hat{\boldsymbol{p}}_{R_{k|k-1}}\right)\\ 0_{1\times 2} & 1\end{bmatrix} \tag{14}$$

The difference between the FEJ-SLAM and the standard SLAM stands out in the perspective that in Eq. (8) the difference is computed between the current *apriori* measurement and the previous *aposteriori* measurement, while in Eq. (14) the difference is computed between the *apriori* measurement at both the current instant and the previous instant.

The new measurement Jacobian used in FEJ-SLAM, here k_i is the time instant at which the landmark is observed for the first time.

$$\begin{aligned}[\boldsymbol{H}'_{k+1}] = [\boldsymbol{H}'_{R_{k+1}}\,\boldsymbol{H}'_{L_{k+1}}] &= (\nabla\boldsymbol{h}_k)\boldsymbol{C}^T(\hat{\varphi}_{R_{k+1|k}})\\ &\times[-\boldsymbol{I}_2 - \boldsymbol{J}(\hat{\boldsymbol{p}}_{L_{ki|ki}} - \hat{\boldsymbol{p}}_{R_{k+1|k}})\,\boldsymbol{I}_2]\end{aligned} \tag{15}$$

The computation of the forecast state propagation Jacobian in FEJ-EKF-SLAM includes the difference between two *apriori* measurements without considering the *aposteriori* measurement hence justifying the name FEJ-SLAM. The derivations of the equations are detailed in [1], the curious reader may refer to it.

3 Modified FEJ-SLAM Employing Window-Based Moving Averaging

Motivated by the improvement of consistency of the FEJ-SLAM two propositions were made to further improve the performance of the SLAM algorithm in typical environments. Although the FEJ-SLAM outperforms the standard SLAM, several other unstructured environments can be proposed where the performance of the FEJ-SLAM lacks fairness. The excellence of performance of the FEJ-SLAM has been proved in an environment where the landmarks are distributed uniformly throughout the environment. Instead, if an environment is taken where the landmarks are not scattered uniformly but follow a special configuration(s), like, they remain in clustered forms around specific coordinates, then the performance of the FEJ-SLAM cannot satiate the consistency conditions. Besides if the robot undertakes navigation in an environment where due to prominent randomness, the landmarks are distributed far away from the robot's maximum sensing range, then it fails to identify one or more features, hence the accuracy of the *apriori* or *aposteriori* measurements decay. Since the FEJ-SLAM only takes into account only two *apriori* measurements, one of the current instant and one for the prior instant, while computing the state propagation Jacobian via Eq. (14), the excellence of the algorithm lacks justification in such cases.

This letter is dedicated to developing two variants of the W-MA-FEJ-SLAM which have been shown to maintain elegance in cases where FEJ-SLAM fails to stand out.

3.1 Window-Based Moving Average SLAM (W-MA-FEJ-SLAM)

In this approach a window accommodating N_w *apriori* measurements slides over the list of *apriori* measurements for all instants from the starting time to the current time, calculating their average. The advantage enjoyed by this new approach is that it takes into consideration a number of prior measurements, instead of taking only one and thus one erroneous reading is less likely to cause the algorithm to perform poorer than expectations. As the average of readings is evaluated in each step of the algorithm, the spike caused due to one improper estimate is smoothened out, thus improving the performance. This new algorithm is termed as the W-MA-FEJ-SLAM. Two different variations of this approach have been proposed.

W-MA-FEJ-SLAM-1

In this variant, the state propagation Jacobian is calculated using a moving window average of difference matrices comprising of *apriori* measurements only.

The equation of the state forecast Jacobian:

$$\boldsymbol{\Phi}''_{Rk} = \begin{bmatrix} \boldsymbol{I}_2 & \boldsymbol{J}\left(\frac{1}{N_W}\right) \sum_{m=0}^{(N_W-1)} (wf^m)\left(\hat{\boldsymbol{p}}_{R_{k-(m-1)|k-m}} - \hat{\boldsymbol{p}}_{R_{k-m|k-(m+1)}}\right) \\ 0_{1\times 2} & 1 \end{bmatrix} \tag{16}$$

N_W is the length of the window employed for calculating the moving average (MA). wf is the weightage factor used for highlighting individual matrices comprising *apriori* measurements of previous instants. If $wf > 1$ then more importance is placed on matrices comprising of instants which precede the current ones, while if $wf < 1$, more importance is placed on recent instants.

W-MA-FEJ-SLAM-2

In this approach, the difference matrices do not only take into consideration the *apriori* measurement, but also the *aposteriori* measurements. The difference matrices compute the difference between the *apriori* and *aposteriori* measurements. The equation employed in computing the state forecast Jacobian:

$$\boldsymbol{\Phi}'''_{R_k} = \begin{bmatrix} \boldsymbol{I}_2 & \boldsymbol{J}\left(\frac{1}{N_W}\right)\left(\begin{array}{c}\left(\hat{\boldsymbol{p}}_{R_{k+1|k}} - \hat{\boldsymbol{p}}_{R_{k|k-1}}\right) + \sum_{m=0}^{(N_W-2)} \left(wf^{(m+1)}\right) \\ \left(\hat{\boldsymbol{p}}_{R_{k-m|k-(m+1)}} - \hat{\boldsymbol{p}}_{R_{k-(m+1)|k-(m+1)}}\right)\end{array}\right) \\ 0_{1\times 2} & 1 \end{bmatrix} \tag{17}$$

ALGORITHM 1 Window-based Moving Average FEJ-SLAM Algorithm

Propagation-Follows after the odometry measurements are processed by the robot

1. Predict the values of the state at the current instant. (cf. (2) and (3)).
2. Determine the forecast state Jacobian by eq. (16).
3. Calculate the forecast covariance matrix using the state propagation Jacobians according to the algorithm of W-MA-FEJ-SLAM:
   - $\boldsymbol{P}_{k+1|k} = \boldsymbol{\Phi}''_k \boldsymbol{P}_{k|k} {\boldsymbol{\Phi}''_k}^T + \boldsymbol{G}_k \boldsymbol{Q}_k {\boldsymbol{G}_k}^T$ where $\boldsymbol{\Phi}''_k = Diag(\boldsymbol{\Phi}''_{R_k}, \boldsymbol{I}_{2M})$ and $\boldsymbol{G}_k = [\boldsymbol{G}_{R_k}^T \quad \boldsymbol{0}_{(2M)\times 2}^T]^T$ for W-MA-FEJ-SLAM-1 ($\boldsymbol{\Phi}''_{R_k}$ (cf. (16)).
   - $\boldsymbol{P}_{k+1|k} = \boldsymbol{\Phi}'''_k \boldsymbol{P}_{k|k} {\boldsymbol{\Phi}'''_k}^T + \boldsymbol{G}_k \boldsymbol{Q}_k {\boldsymbol{G}_k}^T$ where $\boldsymbol{\Phi}'''_k = Diag(\boldsymbol{\Phi}'''_{R_k}, \boldsymbol{I}_{2M})$ and $\boldsymbol{G}_k = [\boldsymbol{G}_{R_k}^T \quad \boldsymbol{0}_{(2M)\times 2}^T]^T$ for W-MA-FEJ-SLAM-2 ($\boldsymbol{\Phi}'''_{R_k}$ (cf. (17)).

Update-Follows after the robot perceives its relative position concerning a landmark.

4. Determine the difference between the estimated observation and true observation: $\boldsymbol{r}_{k+1} = \boldsymbol{z}_{k+1} - \boldsymbol{h}(\boldsymbol{x}_{k+1|k})$.
5. Evaluate the Jacobian matrix $(\boldsymbol{H}'_{k+1})$ for observation model through equation (15).
6. Assess the Kalman gain: $\boldsymbol{K}_{k+1} = \boldsymbol{P}_{k+1|k} {\boldsymbol{H}'_{k+1}}^T {\boldsymbol{S}^{-1}}_{k+1}$,where $\boldsymbol{S}_{k+1} = \boldsymbol{H}'_{k+1} \boldsymbol{P}_{k+1|k} {\boldsymbol{H}'_{k+1}}^T + \boldsymbol{R}_{k+1}$
7. Update the predicted values of the state at the current instant:$\hat{\boldsymbol{x}}_{k+1|k+1} = \hat{\boldsymbol{x}}_{k+1|k} + \boldsymbol{K}_{k+1} \boldsymbol{r}_{k+1}$
8. Determine the updated covariance matrix:$\boldsymbol{P}_{k+1|k+1} = \boldsymbol{P}_{k+1|k} - \boldsymbol{K}_{k+1} \boldsymbol{S}_{k+1} {\boldsymbol{K}_{k+1}}^T$

4 Results and Discussions

Figure 1 shows the environment considered for performance evaluation in the simulation where the total area under consideration is 28 m × 20 m. The robotic/vehicular platform undertakes the navigation through a circular path which is shown in the figure as a true trajectory. There are a total of four landmarks within the area of interest and the platform is finally able to detect a total of three landmarks during its journey. The robotic platform undertakes a total of 10 loops through the true trajectory and the corresponding estimation performances of various SLAM algorithms are quantitatively evaluated [21]. The performance of the propositions has been extensively tested and evaluated for $N_W = 4, 5, 6$. The Eq. (16), for the case of $N_W = 4$, can be given as in Eq. (18). For $N_W = 5$ and 6, the corresponding formulations have been similarly derived:

$$\boldsymbol{\Phi}''_{Rk} = \begin{bmatrix} \boldsymbol{I}_2 & \boldsymbol{J}\left(\frac{1}{4}\right) \sum_{m=0}^{(3)} (wf^m)\left(\hat{\boldsymbol{p}}_{R_{k-(m-1)|k-m}} - \hat{\boldsymbol{p}}_{R_{k-m|k-(m+1)}}\right) \\ 0_{1\times 2} & 1 \end{bmatrix} \tag{18}$$

Similarly, the expression for the state propagation Jacobian used for evaluation of W-MA-FEJ-SLAM-2 with the length of the window $N_W = 4$ is:

$$\boldsymbol{\Phi}'''_{R_k} = \begin{bmatrix} \boldsymbol{I}_2 & \boldsymbol{J}\left(\frac{1}{4}\right)\left(\begin{matrix}\left(\hat{\boldsymbol{p}}_{R_{k+1|k}} - \hat{\boldsymbol{p}}_{R_{k|k-1}}\right) + \sum_{m=0}^{(2)} \left(wf^{(m+1)}\right) \\ \left(\hat{\boldsymbol{p}}_{R_{k-m|k-(m+1)}} - \hat{\boldsymbol{p}}_{R_{k-(m+1)|k-(m+1)}}\right)\end{matrix}\right) \\ 0_{1\times 2} & 1 \end{bmatrix} \tag{19}$$

The performance comparison of the competing algorithms has been carried out in a quantitative manner using several well-established performance indices as given in Tables 1 and 2. The average NEES cumulated after all runs, measures the consistency of the algorithm. RMSE errors of position and orientation give an estimate

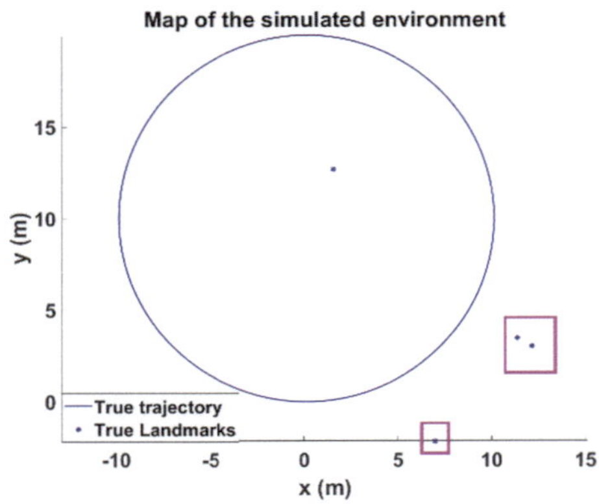

Fig. 1 Field of dimensions of 28 m × 20 m comprising four landmarks in total. The robot has the capability of observing three among them in total which are marked in pink rectangles

of the quality of performance. While position and heading RMSE have known error expressions, the expression for average NEES error over N runs is given by $\overline{\epsilon}_k = (1/N)\left(\sum_{i=1}^{N} \epsilon_{ik}\right)$ the NEES for each run is $\epsilon_{ik}=\left(\boldsymbol{x}_k - \hat{\boldsymbol{x}}_{k|k}\right)^T \boldsymbol{P}_{k|k}^{-1}\left(\boldsymbol{x}_k - \hat{\boldsymbol{x}}_{k|k}\right)$, where $\boldsymbol{x}_k$ is the true value of state at the kth instant. For this dataset $wf = 1$ which implies we are calculating the simple moving average of the *apriori* difference matrices.

Table 1 and Fig. 2 show the detailed performance comparison of the proposed algorithms with the already well-established conventional SLAM algorithms. With reference to Table 1, it can be concluded that, the performance of W-MA-FEJ-SLAM-1 for $N_w = 4$ exhibits the best performance among all the algorithms while calculating the robot pose error indices (NEES and RMSE), but it lacks superiority while evaluating for the landmark errors (position RMSE and NEES). Thus we establish that further performance evaluations relating to other algorithms are redundant and improvement of the performance of W-MA-FEJ-SLAM-1 ($N_w = 4$) is desired. Hence the successive sections are dedicated for tuning the performance of W-MA-FEJ-SLAM $-1(N_w = 4)$ with different choices of weightage factors.

From Table 2, it can be concluded that the performance of W-MA-FEJ-SLAM-1 varies slightly with the variation of the weightage factor. The performance of this algorithm for $wf = 10$ shows remarkable improvement in comparison to $wf = 1$ while evaluating the landmark errors. Thus the usage of W-MA-FEJ-SLAM $-1(N_W = 4)$ with $wf > 1$ proves successful in establishing its' dominance over well-established conventional algorithms such as OC-EKF-SLAM and FEJ-SLAM. Figure 3 describes the nature of the average performance indices (NEES and RMSE) varied over time for the current simulation using W-MA-FEJ-SLAM-1 with two different values of weightage factors.

5 Conclusion

In this unique innovation, a window-based moving average approach has been proposed as a modification of the FEJ-SLAM. The FEJ-SLAM was developed based on the concept of inconsistency of EKF-SLAM. The computation of the state forecast measurement Jacobians is done differently in FEJ-SLAM to improve the consistency by considering only the first available state estimates. In this paper the performance of FEJ-SLAM was investigated thoroughly, establishing that it fails to maintain its superiority in noisy environments where high instantaneous errors are likely to occur as this algorithm takes into consideration only two *apriori* measurements. As an improvement, a window-based moving average approach (W-MA-FEJ-SLAM) was proposed as a modification which unlike the FEJ-SLAM gives weightage to a number of prior measurements in its prediction algorithm and thus the high magnitudes of noise are less likely to disrupt its performance. This new algorithm has been verified to have stable and robust performance in comparison to conventional and well-established algorithms such as standard EKF-SLAM and FEJ-SLAM.

Table 1 Performance evaluations of the two variants of W-MA-FEJ-SLAM with different choices for window length in comparison to the already established conventional algorithms

Performance indices	FEJ-EKF-SLAM [8]	STD-EKF-SLAM [8]	OC-EKF-SLAM [8]	W-MA-FEJ-SLAM-1			W-MA-FEJ-SLAM-2
				$N_w = 4$	$N_w = 5$	$N_w = 6$	$N_w = 4$
Robot pose NEES	83.00	191.06	450.69	**67.88**	108.00	153.00	206.16
Robot position RMSE (m)	6.17	4.86	10.15	**4.11**	4.40	5.06	5.30
Robot heading angle RMSE (rad)	0.45	0.38	0.72	**0.31**	0.33	0.39	0.39
Landmark position NEES	**9.10**	348.61	106.98	98.35	193.61	239.67	206.59
Landmark position RMSE (m)	**1.25**	2.56	2.88	1.57	1.96	2.05	1.81

Table 2 Performance comparison of W-MA-FEJ-SLAM-1with various choices for weightage factors

Performance indices of W-MA-FEJ-SLAM (N_w= 4)	wf = 1.00	wf = 7.00	wf = 8.00	wf = 9.00	wf = 10.00
Robot pose NEES	68.00	55.86	53.81	52.52	**51.61**
Robot position RMSE (m)	**4.11**	4.78	4.75	4.75	4.74
Robot heading RMSE (rad)	**0.31**	0.34	0.34	0.34	0.34
Landmark position NEES	98.35	11.19	6.62	3.86	**2.29**
Landmark position RMSE (m)	1.56	**1.46**	1.48	1.50	1.52

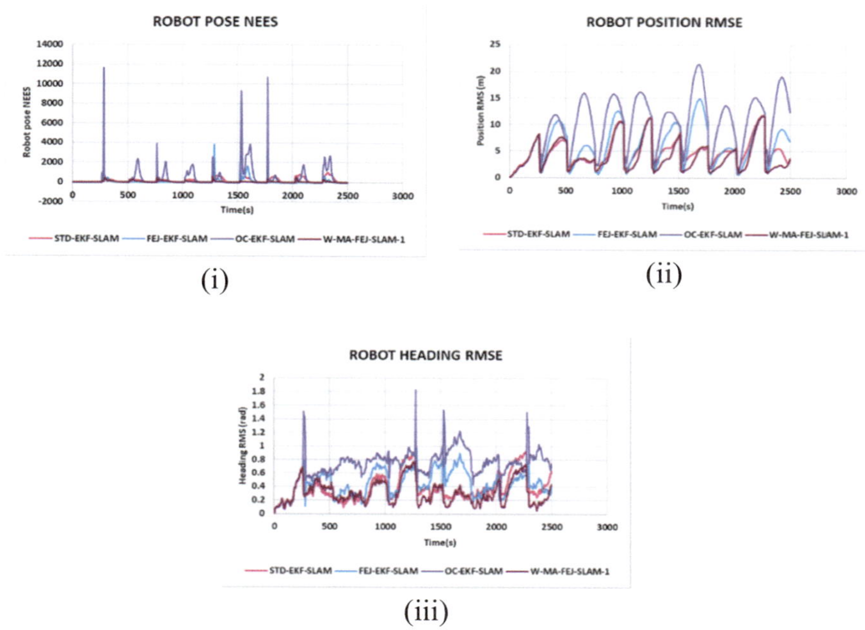

Fig. 2 Monte Carlo simulation results of the robot performing SLAM in the simulated environment as in Fig. 1. (i) Time versus robot pose NEES for all the algorithms. (ii) Time versus robot pose RMSE(m) for all the algorithms (iii) Time versus robot heading RMSE(rad) for all the algorithms

The current work has investigated the problem of a navigational space with four landmarks within the region. However, there is a possibility that, with an increase in the number of landmarks in the environment, the performances of all the EKF-SLAM algorithms considered in this work, including our proposed variants, may vary and show degradation. Our future works will attempt to investigate how the proposed algorithms will perform for a variety of navigational spaces with potentially more number of landmarks in it. The future works will also focus on developing more

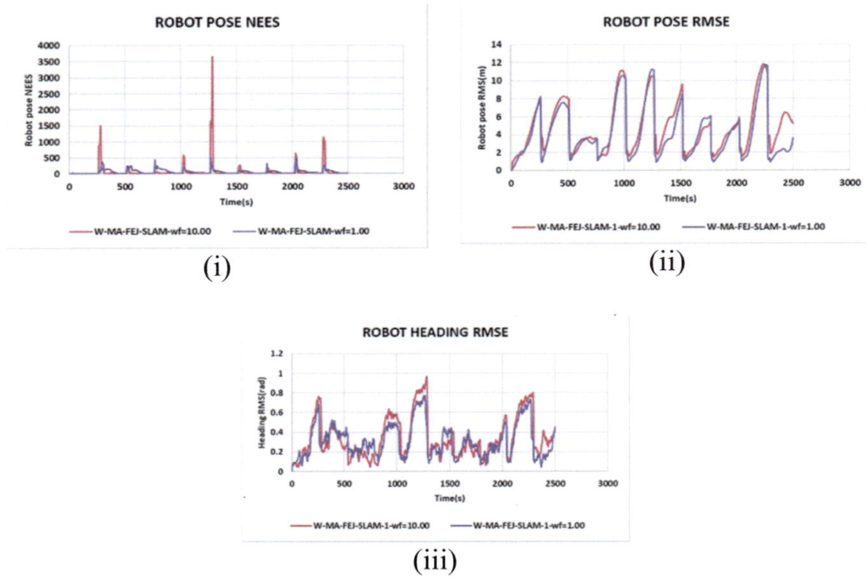

Fig. 3 Monte Carlo results for evaluation of W-MA-FEJ-SLAM-1 algorithm with different choices of weightage factors, in the simulated environment. (i) Time versus average robot pose NEES (ii) Time versus average robot position RMSE (m) (iii) Time versus average robot heading RMSE (rad) over all the five runs

sophisticated variants of moving average-based algorithms that can be employed advantageously for such varied challenging environments. The present work has also shown that the choice of a moving window size of $N_w = 4$ has produced superior results compared to the window sizes of $N_w = 5$ and $N_w = 6$. This is a positive outcome for the proposed algorithm as the choice of a smaller window size results in reduced computational burden which is beneficial, particularly in real-life navigation applications. Our future works will investigate whether the choice of such a thinner window will still provide sufficiently satisfactory results for varied environments with more number of landmarks and attempt to propose an optimal solution for this window size problem.

References

1. Durrant-Whyte, H., & Bailey, T. (2006). Simultaneous localization and mapping(SLAM), Part 1, The essential algorithms. *IEEE Robotics and Automation magazine, 13*(2), 99–110.
2. Huang, G. P., Mourikis, A. I., & Roumeliotis, S. I. (2011). An observability-constrained sliding window filter for SLAM. In *2011 IEEE/RSJ International Conference on Intelligent Robots and Systems*. IEEE/RSJ, San Francisco, CA, USA
3. Chatterjee, A., Ray, O., Chatterjee, A., & Rakshit, A. (2011). Development of a real-life EKF based SLAM system for mobile robots employing vision sensing. *Expert Systems with Applications, 38*(7), 8266–8274
4. Chatterjee, A. (2009). Differential evolution tuned fuzzy supervisor adapted extended Kalman filtering for SLAM problems in mobile robots. *Robotica, 27*(3), 411–423.
5. Chatterjee, A., & Matsuno, F. (2010). A geese PSO tuned fuzzy supervisor for EKF based solutions of simultaneous localization and mapping (SLAM) problems in mobile robots. *Expert Systems with Applications, 37*(8), 5542–5548.
6. Chatterjee, A., & Matsuno, F. (2007). A neuro-fuzzy assisted extended Kalman filter-based approach for simultaneous localization and mapping (SLAM) problems. *IEEE Tran Fuzzy Systems, 15*(5), 984–997.
7. Chatterjee, A., & Matsuno, F. (2006). Bacterial foraging techniques for solving EKF-based SLAM problems. In *Proceedings of the International Control Conference (ICC 2006)*
8. Huang, G. P., Mourikis, A. I., & Roumeliotis, S. I. (2010). Observability based rules for designing consistent EKF SLAM estimators. *International Journal of Robotics Research, 29*(5), 502–528.
9. Julier, S., & Uhlmann, J. (2001). A counter example to the theory of simultaneous localization and map building. In *IEEE International Conference on Robotics and Automation* (pp. 4238–4243). IEEE, Seoul, Korea
10. Huang, G. P., Mourikis, A. I., & Roumeliotis, S. I. (2008). A first-estimates Jacobian EKF for improving SLAM consistency. In *11th International Symposium on Experimental Robotics (ISER)*. Springer, Athens, Greece
11. Huang, G. P., Mourikis, A. I., & Roumeliotis, S. I. (2008). Analysis and improvement of the consistency of extended Kalman filter based SLAM. In *Proceedings of IEEE International Conference on Robotics and Automation* (pp. 473–479). IEEE, Pasadena, CA
12. Bailey, T., Nieto, J., Guivant, J, Stevens, M., & Nebot, E. (2006). Consistency of the EKF-SLAM algorithm. In *IEEE/RSJ International Conference on Intelligent Robots and Systems* (pp. 3562–3568). IEEE/RSJ, Beijing, China
13. Du, Z.-J., Huang, S.-S., Mu, T.-J., Zhao, Q., Martin, R. R., & Xu, K. (2002). Accurate dynamic SLAM using CRF-based long-term consistency. *IEEE Tran Visualization and Computer Graphics, 28*(4), 1745–1757.
14. Brossard, M., Barrau, A., & Bonnabel, S. (2019). Exploiting symmetries to design EKFs with consistency properties for navigation and SLAM. *IEEE Sensors Journal, 19*(4), 1572–1579
15. Zhang, J., Yuan, L., Ran, T., Tao, Q., & Wu, Z. (2023). Outlier elimination for monocular object SLAM based on spatiotemporal consistency constraints. *IEEE Sensors Journal, 23*(8), 8887–8898.
16. Ji, X., Liu, P., Chen, X., Ying, R., & Wen, F. (2023). Object SLAM based on spatial layout and semantic consistency. *IEEE Transactions on Instrumentation and Measurement, 72*, 2528812.
17. Lisus, D., Cohen, M., Forbes, J. R. (2023). Know what you don't know: consistency in sliding window filtering with unobservable states applied to visual-inertial SLAM. *IEEE Robotics and Automation Letters, 8*(6)
18. Xie, H., Deng, T.,Wang, J., & Chen, W. (2023). Angular tracking consistency guided fast feature association for visual-inertial SLAM. *IEEE transactions on Instrumentation and measurements, 23*, 5006614
19. Pan, Y., Zhong, X., Weismann, L., Posewsky, T., Behley, J., & Stachniss, C. (2024). PIN-SLAM: LiDAR SLAM using a point-based implicit neural representation for achieving global map consistency. *IEEE Transactions on Robotics, 40*, 4045–4064.

20. Yang, K., Jiang, Y., Qi, L., Fan, H., Zhang, S., & Dong, J. (2022). Visual semantic SLAM based on examination of moving consistency in dynamic scenes. In *2022 4th International Conference on Data Intelligence and security(ICDIS)*. IEEE
21. Observability-Constrained (OC)-EKF. Retrieved November 25, 2024 from https://github.com/rpng/ocekf-slam,last

Lightweight Scene Parsing for Real-Time Structural Crack Detection

Tanmay Singha, Saubhik Goswami, Duc-Son Pham, and Aneesh Krishna

Abstract Automated crack detection in roads and infrastructure is crucial for ensuring safety and longevity in civil engineering. This study introduces LightCrackNet, a novel lightweight deep learning model designed for real-time crack detection using embedded devices. Our approach addresses the need for efficient, on-site monitoring by enabling rapid processing and accurate identification of cracks within the constraints of embedded systems. We employ a multi-scale feature extraction strategy that captures both fine details and broader contextual information, facilitating precise segmentation under diverse conditions. Trained and evaluated on over 10 publicly available datasets, LightCrackNet produces highly competitive results among existing crack detection models, particularly in resource-constrained environments. Extensive testing demonstrates that our model achieves high accuracy while maintaining low computational overhead, making it ideal for deployment on mobile inspection platforms such as drones and handheld devices. This research enhances the capabilities of automated infrastructure monitoring and promotes proactive maintenance practices, ultimately contributing to safer transportation networks and structures.

Keywords Deep convolutional neural networks · Crack detection · Semantic segmentation · Encoder-decoder · Resource-constrained applications

T. Singha
University of Notre Dame Australia, Fremantle, Australia
e-mail: tanmay.singha@nd.edu.au

S. Goswami (✉)
Meghnad Saha Institute of Technology, Kolkata, India
e-mail: saubhik.goswami@msit.edu.in

D.-S. Pham · A. Krishna
Curtin University, Perth, Australia
e-mail: dspham@ieee.org

A. Krishna
e-mail: a.krishna@curtin.edu.au

J. K. Mandal et al. (eds.), *Recent Advances in Artificial Intelligence and Smart Applications*, Lecture Notes in Networks and Systems 1380,
https://doi.org/10.1007/978-981-96-5822-0_26

1 Introduction

The integrity of roads and infrastructure is essential for public safety and economic efficiency. Cracks can lead to significant hazards, making traditional manual inspections labor-intensive, time consuming, and error-prone. Recent advancements in computer vision and deep learning have enabled automated crack detection and segmentation, improving monitoring and maintenance.

Studies have shown the effectiveness of deep learning models for crack detection. For example, [18] demonstrated the use of convolutional neural networks (CNNs) for identifying road cracks with notable accuracy. DeepCrack [9] integrated multi-scale feature extraction for enhanced performance but often requires substantial computational resources, limiting real-time application.

Embedded devices like drones and handheld tools offer opportunities for real-time crack detection. Recent works [8, 10] have shown the potential of deploying lightweight models on these platforms, balancing accuracy and computational efficiency. Optimizing deep learning architectures for embedded systems is crucial for practical infrastructure monitoring.

Despite advancements, there is a gap in developing robust real-time crack detection solutions that fully utilize embedded device capabilities. High-resolution images, essential for accurate detection, are challenging to process with existing large CNN models on resource-constrained devices. To address this, we propose a novel two-branch CNN model for automated crack detection and segmentation, combining lightweight models with innovative feature extraction techniques. By deploying a shallow branch alongside a deeper branch, we maintain high input quality without increasing computational overhead. Our research aims to achieve high accuracy in crack detection with low computational requirements for real-world deployment. In summary, our key contributions are as follows:

- A shared-branch backbone facilitates knowledge sharing between branches at three stages, enhancing feature representation and learning.
- A Context Extraction Unit for extracting extensive contextual information from deep feature maps, improving feature refinement and understanding of crack characteristics.
- An Adaptive Feature Combination Module for integrating contextual information, enhancing object localization and segmentation performance.

2 Related Work

The field of crack detection and segmentation has seen significant advancements, particularly with the integration of deep learning techniques. This section reviews relevant literature on object detection, semantic segmentation, and crack segmentation.

Object detection has been pivotal in infrastructure monitoring. Traditional methods like the Viola-Jones algorithm have largely been replaced by deep learning techniques, notably YOLO [6] and Faster R-CNN [11], which excel in real-time object detection. Recent studies [4, 5, 10, 17] have adapted these frameworks for crack detection. However, while effective for identifying cracks, they often lack precision for detailed segmentation.

Semantic segmentation, which classifies each pixel in an image, has gained traction for its ability to provide detailed spatial information about objects. Convolutional Neural Networks (CNNs) are widely used for this purpose, with models like U-Net [12] and DeepLab [2] achieving state-of-the-art results in various applications. These models effectively focus on high-level features while retaining spatial resolution, making them ideal for tasks such as road surface analysis and environmental monitoring. However, their large size limits their suitability for real-time segmentation.

To address this, multi-branch approaches such as ICNet [19], SCMNet [14], and Effi-Seg [15] have been introduced. These models leverage dual processing paths to capture both coarse and fine contextual information, enhancing semantic representation without sacrificing input resolution or significantly increasing computational cost. Despite their promise, many of these architectures remain computationally intensive, presenting challenges for real-time applications.

Crack segmentation has become an increasingly important area of research [7]. An early contribution by [18] introduced a CNN for crack detection using a dataset of 500 images. While initial methodologies focused on object detection and classification, the trend has shifted toward semantic segmentation [7] due to its ability to convey critical information about crack shape and size. Early efforts utilized frameworks like SegNet [3] to enhance crack detection.

Crack segmentation faces unique challenges: the fine nature of cracks requires preserving dense features, as any reduction in resolution can lead to inaccuracies. Additionally, cracks typically occupy less than 5% of image pixels, resulting in class imbalance. To address this, DeepCrack [9] employs multi-scale feature extraction and class-balanced loss functions, while FPHBN [17] uses feature pyramids to enhance segmentation. MR-CrackNet [10] maintains feature resolution through upsampling.

Many studies rely on unique datasets that complicate direct comparisons. Several datasets for crack segmentation, such as CrackBgData [10] and CRACK500 [17], have been proposed. Due to limited images, our approach combines data from multiple sources for training.

In summary, while progress in object detection and segmentation has been significant, challenges in achieving real-time performance with lightweight models persist. Our work aims to develop an efficient framework that balances accuracy and computational efficiency for effective crack detection and segmentation.

3 Proposed Method

Our proposed model features a dual-branch architecture at the encoder stage, designed to enhance crack detection and segmentation. The low-resolution deep branch is dedicated to extracting rich contextual information, while the high-resolution shallow branch focuses on assimilating local features such as textures, patterns, and object boundaries. Traditionally, these branches operate independently at the encoder, leading to a lack of spatial detail and an increased semantic gap between local and global feature maps. To address these limitations, we introduce information sharing between the branches at three key stages, thereby improving the model's spatial awareness and feature integration.

Shallow branch of the proposed model is deployed to capture boundary and texture details, effectively guiding the feature maps from the deep branch. In tasks like classification, identifying an object based on contextual information is sufficient, and precise localization is not critical. However, in semantic segmentation, accurate localization of objects within a scene is crucial, and many existing models fall short in this regard. To overcome this challenge, we introduce a parallel shallow branch that works alongside the deep branch, as illustrated in Fig. 1. True to its name, this branch has a streamlined design consisting of 1 Convolution (Conv) and 6 Depth-wise Separable Convolution (DSConv) blocks, contributing around 0.4 million parameters. This branch accepts an input of $448 \times 448 \times 3$ size and ensures the capturing of essential local features at different stages of the entire pipeline. These local feature maps are vital for precise segmentation and improve the overall performance of the model.

Deep branch of our proposed model employs one Conv and 15 MobileNetV2 residual bottleneck blocks (MBConv) [13] to efficiently capture contextual features at various stages. We utilize two types of MBConv blocks: MBConv1, which has an expansion ratio of 1, and MBConv6, which expands the input width by six times within its bottleneck structure, allowing the network to learn more complex structures of cracks. Thus, fine-grained crack details can be obtained from the deep branch. The complete layout of the deep and shallow branches can be seen in Table 1. While the number of channels increases, the spatial dimensions of the feature maps typically decrease due to strided convolutions. This helps to maintain a manageable number of parameters while enhancing feature richness. At the end of the encoder, we obtain a contextually rich, fine-grained feature map of $16 \times 16 \times 128$ size. As illustrated in Fig. 1, our model facilitates information sharing between the two branches at three distinct stages. Features from shallow and deep branches are combined and processed by the Context Extracting Unit.

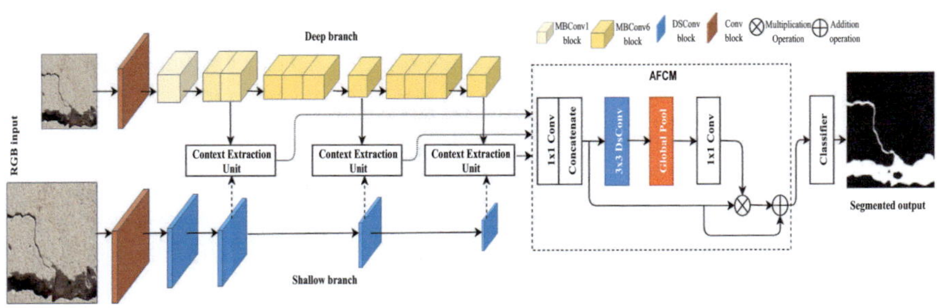

Fig. 1 Complete architecture of the proposed model

Table 1 Layered architecture of deep and shallow branch of LightCrackNet

Deep branch				Shallow branch			
Input	224 × 224 × 3			Input	448 × 448 × 3		
Operators	Stride	Layers(n)	Channel	Operators	Stride	Layers(n)	Channel
Conv, k3×3	1	1	24	Conv, k3×3	1	1	24
MBConv1, k3×3	2	2	24	DSConv, k3×3	2	1	24
MBConv6, k3×3	1	3	32	DSConv, k3×3	2	1	32
MBConv6, k3×3	2	2	48	DSConv, k3×3	2	1	48
MBConv6, k3×3	1	3	64	DSConv, k3×3	1	1	64
MBConv6, k3×3	2	2	96	DSConv, k3×3	2	1	96
MBConv6, k3×3	1	3	128	DSConv, k3×3	1	1	128

Context Extracting Unit employs a streamlined approach that aggregates the deep and shallow feature maps from various stages of the encoder design. Firstly, by using a weighted-addition operation, it combines the shallow and deep features for efficient information propagation. This weighted addition allows for more flexible integration of features, where different contributions from the shallow and deep branches can be adjusted according to their importance in representing the overall context. To ensure non-linearity is maintained, a ReLU activation function is deployed after the addition operation. This sharing mechanism enhances the mutual understanding of features between the branches, effectively narrowing the semantic gap and improving overall feature representation. Then, it refines the output with one point-wise convolution (1 × 1 Conv) and three 3 × 3 DSConv blocks. In comparison to standard Conv, DSConv produces 8 times smaller number of parameters and Giga Floating Point Operations (GFLOPs) for an input image with 64 input and 128 output channels filtered by a kernel of 3 × 3, optimizing the number of parameters and the computational load. In this research, the dataset contains only two classes: Crack and Void, where pixels of the Void class cover more than 70% of the whole image. Therefore, dilation rates greater than 10 are not required for this study. We deploy dilation rates of 1, 2, 4, and 8 to capture fine textures, enhancing the model's ability to recognize tiny cracks. The output of each block is then concatenated to match the dimensions of the original input. Finally, a projection layer is employed for pixel-by-pixel refinement. The complete architecture of the Context Extracting Unit (CEU) is displayed in Fig. 2 and detailed in Table 2.

Adaptive Feature Combination Module (AFCM) is implemented to enhance contextual assimilation and improve crack localization. The layout of the AFCM is illustrated in Fig. 1 and detailed in Table 3. It accepts three shared feature maps from the three different stages of the proposed encoder design and passes it through a 1 × 1 Conv block first. This projection layer assigns the same number of channels to each shared feature map before concatenation happens. After that, the concatenated output passes through a series of 3 × 3 DSConv, a global pool, and a point-wise Conv layer with a Sigmoid non-linear activation function. Finally, an element-wise multi-

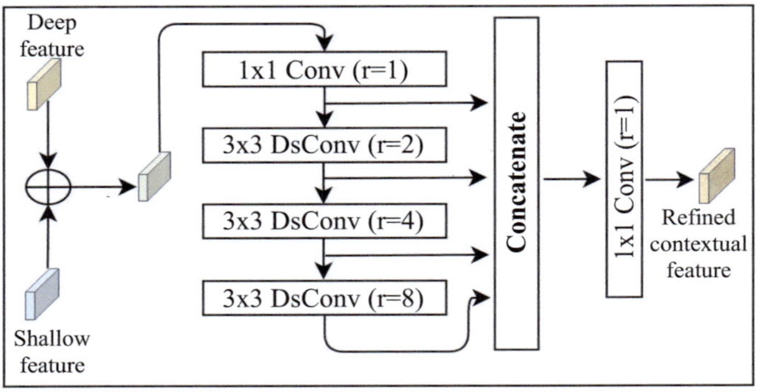

Fig. 2 Context extracting unit

Table 2 Layered architecture of CEU

Layer	Input	Operator	Filter/P.size	Dilation rate	Output
A weighted-addition operation between deep and shallow feature maps					
Conv1	$h \times w \times c$	Conv, BN, ReLU	1×1	1	$h \times w \times c/4$
DSConv1	$h \times w \times c/4$	DSConv, BN, ReLU	3×3	2	$h \times w \times c/4$
DSConv2	$h \times w \times c/4$	DSConv, BN, ReLU	3×3	4	$h \times w \times c/4$
DSConv3	$h \times w \times c/4$	DSConv, BN, ReLU	3×3	8	$h \times w \times c/4$
Concatenate Conv1, DSConv1, DSConv2, and DSConv3 output					
Conv2	$h \times w \times c$	Conv, BN, ReLU	1×1	1	$h \times w \times c$

Table 3 Layered architecture of AFCM

Layer	Input	Operator	Filter/P.size	Dilation rate	Output
A point-wise Conv followed by ReLU on each shared feature map					
Concatenate all the three output from the point-wise Conv					
DSConv1	$h \times w \times c$	DSConv, BN, ReLU	3×3	2	$h \times w \times c$
Global pooling	$h \times w \times c$	GlobalAveragePooling	Pool size= (h, w)	–	$1 \times 1 \times c$
Conv2	$1 \times 1 \times c$	Conv, Sigmoid	1×1	1	$1 \times 1 \times c$
Element-wise multiplication operation between DSConv1 and Conv2 output					
Element-wise addition operation between DSConv1 and previous stage output					

plication and addition operation is deployed to facilitate a flexible way to combine features. This allows the module to adaptively weigh the contributions of different feature maps based on their significance for the task at hand, enhancing the final output's semantic richness.

Classifier module serves as the decoder head, responsible for assigning a class label to each pixel in the input image. To maintain simplicity, we utilize three Up Sampling (UpSampling2D) layers, two 3×3 DSConv layers, one 1×1 Conv layer, a dropout layer, and a softmax layer. Previous studies [15, 16] suggest that incorporating a few DSConv layers in the classifier enhances pixel classification accuracy. After each UpSampling2D layer, we deploy a DSConv or Conv layer to smooth out the artifacts or discontinuities introduced by the upsampling process. The dropout layer is included to mitigate overfitting, while the softmax layer functions as a multi-class logistic regression to assign class IDs to each pixel based on their softmax scores. Finally, the output feature map is upsampled by a factor of 2^3 to achieve the desired segmentation output size.

4 Experiment

4.1 Datasets, Performance Metrics, and Implementation Details

We utilize a combined dataset derived from 12 different crack segmentation sources, all adhering to the same labeling scheme. Sourced from Kaggle,[1] this dataset consists of 9,505 training images and 1,695 test images. Although unconventional, we find this dataset particularly challenging due to its wide range of image types, including cracks of varying thickness, different surface materials, and even samples without any cracks.

$$\text{mIoU} = \frac{1}{N}\sum_{i=1}^{N}\text{IoU}_i = \frac{1}{N}\sum_{i=1}^{N}\frac{TP_i}{TP_i + FP_i + FN_i}, \tag{1}$$

$$F1 = 2 \cdot \frac{\text{Precision} \cdot \text{Recall}}{\text{Precision} + \text{Recall}}, \tag{2}$$

$$\text{Precision} = \frac{TP}{TP + FP}, \tag{3}$$

$$\text{Recall} = \frac{TP}{TP + FN}. \tag{4}$$

For performance evaluation, we used four metrics: mean Intersection Over Union (mIoU), Precision, Recall, and Accuracy (see Eqs. 1–4). Given the highly imbalanced class distribution between background and crack classes, we prioritized F1 score, Precision, and Recall over Accuracy. mIoU is defined across N classes, with TP, FP, and FN representing true positives, false positives, and false negatives, respectively. Additionally, we measured model parameters, GFLOPs, and FPS (Frames Per

[1] https://www.kaggle.com/datasets/lakshaymiddha/crack-segmentation-dataset.

Second) to assess efficiency, focusing on crack detection for resource-constrained embedded devices.

We employed a polynomial learning rate strategy to optimize the learning rate for each epoch. After training for 5 epochs at a base rate of 0.01, we plotted model loss against various rates to identify an effective range. A polynomial schedule was then applied to gradually decrease the learning rate. To address early-stage gradient descent issues, we implemented a warm-up strategy and used the Stochastic Gradient Descent (SGD) algorithm for loss optimization, experimenting with different loss functions. Additionally, standard data augmentation, ℓ_2 regularization, and dropout were applied to reduce overfitting.

4.2 *Ablation Study*

To validate our proposed model design, we conducted a series of ablation studies. We first evaluated the performance of both deep and shallow branches independently in the encoder without using CEU and AFCM. Next, we introduced CEU at three different stages of the encoder to enhance contextual output. Instead of employing AFCM, we processed the three shared feature maps through a point-wise convolution layer followed by ReLU, and then performed an element-wise addition before passing the final feature map to the classifier. This led to a significant improvement of 4.9% mIoU in model performance with the inclusion of CEU. This finding inspired us to create a shared multi-branch encoder for crack detection and segmentation. However, we found that the model's performance on tiny cracks was below our expectations. To enhance the localization and texture details of small objects like tiny cracks, we ultimately incorporated an AFCM design, which boosted performance by an additional 2.4%. This finalized the design of our crack detection model, with the results of the ablation study presented in Table 4.

Table 4 Resuls of ablation study

Multi-branch encoder	CEU	AFCM	mIoU (%)	mF1 (%)	P (%)	R (%)
Independent deep and shallow branches	–	–	69.43	80.94	82.24	79.68
Shared branches	✓	–	74.33	83.32	86.53	80.31
Shared branches	✓	✓	77.73	85.89	88.76	83.21

4.3 Model Evaluation

We replicated several existing crack detection models and trained them under the same system configuration using the same dataset for consistent performance comparison. During data processing, we identified noisy pixels, as shown in Fig. 3. The first image displays the original crack image, the second shows the corresponding annotations, and the third highlights the noisy pixels. Assigning these pixels to either the Crack or background classes could skew the model's performance. To address this, we created a new class, called "void," with a training ID of 0, categorizing the noisy pixels accordingly, thereby excluding them from the training process. Table 5 presents the performance of all models evaluated under identical conditions. Each model was trained for 250 epochs, as no substantial growth was observed after 200 epochs. Among the models, our proposed model, LightCrackNet, achieved the highest performance, followed by DeepCrack [9], DeepLabV3+ [2], SegNet [1], and SCMNet [14]. Notably, LightCrackNet is much smaller than many existing models, as can be seen in Table 6. The proposed model has 1.1 million parameters and 3.0 GFLOPs, making it more efficient than the existing models. SCMNet [14] is also a lightweight model like our proposed model; however, the proposed model produces nearly 3% better mIoU and can process almost double the frames per second compared to SCMNet. This can be seen in Table 6. For FPS measurement, we optimized all models into TensorRT-INT8 format using Nvidia's TensorRT engine and measured FPS at an input resolution of 448 × 448 × 3. This optimization is required for running a CNN model on embedded devices due to the lack of hardware support. The experimental results show that the proposed model outperforms all others by processing 205 frames per second.

Fig. 3 Accuracy plot of few models against epochs

For qualitative analysis, Fig. 4 displays the segmented outputs from all models. LightCrackNet effectively identifies thin cracks where other models struggle. While the segmented cracks appear slightly thicker than the annotations due to minimal texture differences, the proposed model accurately localizes and segments the cracks.

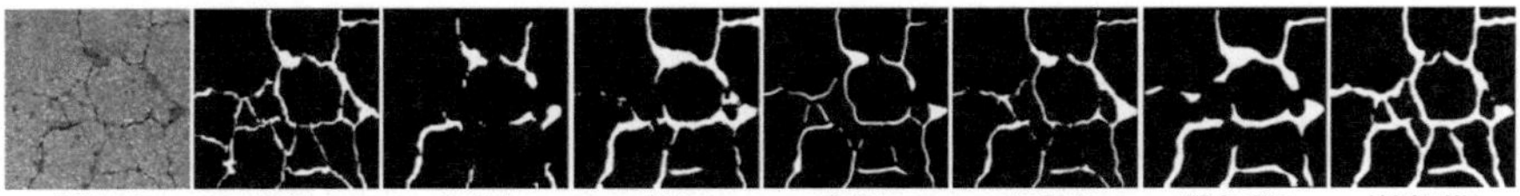

Fig. 4 Segmented output produced by different models using test sample

Table 5 Results of different models on crack dataset

		IoU (%)			F1 score (%)	Precision (%)	Recall (%)
Model	Accuracy	BG	Crack	mIoU	mF1	aP	aR
U-Net [12]	98.18	98.2	48.97	73.58	82.42	84.79	80.20
SegNet [3]	98.27	98.30	52.17	75.23	83.85	85.12	82.63
DeepLabV3+ [2]	98.43	98.50	52.28	75.39	83.95	89.11	80.05
DeepCrack [9]	98.52	98.57	55.89	77.23	85.49	84.54	82.91
MR-CrackNet [10]	97.88	97.19	47.69	72.80	81.76	80.85	82.70
SCMNet [14]	98.39	98.41	51.33	74.87	83.52	88.21	79.30
LightCrackNet	98.27	98.51	54.95	77.73	85.89	88.76	83.21

Table 6 Efficiency analysis

Model	Param.(M)	GFLOPs	TRT-INT8
U-Net	31.0	1740.0	50
SegNet	29.4	245.0	56
DeepLabV3+	41.0	78.8	82
DeepCrack	14.7	123.0	65
MR-CrackNet	17.7	331	45
SCMNet	1.2	3.3	114
LightCrackNet	1.1	3.0	205

5 Conclusion

In conclusion, this study presents LightCrackNet, a lightweight and efficient deep learning model designed for real-time crack detection in civil infrastructure. By rigorously training and evaluating the model against over ten publicly available datasets, we demonstrated state-of-the-art performance compared to existing models, particularly in resource-constrained embedded environments. The introduction of a void class to manage noisy pixels enhanced data quality and model reliability. Our extensive evaluations indicate that LightCrackNet not only achieves high accuracy but also operates with significantly lower computational demands, making it suitable for deployment in mobile inspection platforms such as drones and handheld devices. Looking ahead, a key goal is to further improve the model's performance in detecting very thin cracks. One proposed approach is to incorporate a higher penalty when the model fails to detect a crack class for a pixel. This adjustment aims to enhance sensitivity to subtle features, ensuring that even the finest cracks are accurately identified. By pursuing this direction, we hope to make LightCrackNet an even more powerful tool for automated infrastructure monitoring.

References

1. Badrinarayanan, V., Kendall, A., & Cipolla, R. (2017). Segnet: A deep convolutional encoder-decoder architecture for image segmentation. *IEEE TPAMI, 39*(12), 2481–2495.
2. Chen, L. C., Zhu, Y., Papandreou, G., Schroff, F., & Adam, H. (2018). Encoder-decoder with atrous separable convolution for semantic image segmentation. In *Proc. ECCV* (pp. 801–818)
3. Chen, T., Cai, Z., Zhao, X., Chen, C., Liang, X., Zou, T., & Wang, P. (2020). Pavement crack detection and recognition using the architecture of segNet. *Journal of Industrial Information Integration, 18*, Article 100144.
4. Choi, W., & Cha, Y. J. (2019). SDDNet: Real-time crack segmentation. *IEEE Transactions on Industrial Electronics, 1–1*. https://doi.org/10.1109/TIE.2019.2945265
5. David Jenkins, M., Carr, T. A., Iglesias, M. I., Buggy, T., & Morison, G. (2018). A deep convolutional neural network for semantic pixel-wise segmentation of road and pavement surface cracks. In *Proc. EUSIPCO* (pp. 2120–2124). https://doi.org/10.23919/EUSIPCO.2018.8553280, ISSN: 2076-1465
6. Du, J. (2018). Understanding of object detection based on CNN family and YOLO. In *Journal of Physics: Conference Series* (vol. 1004, p. 012029). IOP Publishing
7. Hamishebahar, Y., Guan, H., So, S., & Jo, J. (2022). A comprehensive review of deep learning-based crack detection approaches. *Applied Sciences, 12*, 1374.
8. Kang, D. H., & Cha, Y. J. (2021). Efficient attention-based deep encoder and decoder for automatic crack segmentation. *Structural Health Monitoring* (p. 14759217211053776).
9. Liu, Y., Yao, J., Lu, X., Xie, R., & Li, L. (2019). Deepcrack: A deep hierarchical feature learning architecture for crack segmentation. *Neurocomputing, 338*, 139–153.
10. Nayyeri, F., & Zhou, J. (2021). Multi-resolution ResNet for road and bridge crack detection. In *Proc. DICTA* (pp. 1–8). IEEE
11. Ren, S., He, K., Girshick, R., & Sun, J. (2017). Faster r-cnn: Towards real-time object detection with region proposal networks. *IEEE Transactions on Pattern Analysis and Machine Intelligence, 39*(6), 1137–1149. https://doi.org/10.1109/TPAMI.2016.2577031
12. Ronneberger, O., Fischer, P., & Brox, T. (2015). U-net: Convolutional networks for biomedical image segmentation. In *Proc. MICCAI* (pp. 234–241). Springer.
13. Sandler, M., Howard, A., Zhu, M., Zhmoginov, A., & Chen, L. C. (2018). Mobilenetv2: Inverted residuals and linear bottlenecks. In *Proc. CVPR* (pp. 4510–4520)
14. Singha, T., Bergemann, M., Pham, D. S., & Krishna, A. (2021). SCMNet: Shared context mining network for real-time semantic segmentation. In *Proc. DICTA* (pp. 1–8). IEEE.
15. Singha, T., Pham, D. S., & Krishna, A. (2023). Effi-seg: Rethinking efficientnet architecture for real-time semantic segmentation. In *International Conference on Neural Information Processing* (pp. 55–68). Springer.
16. Singha, T., Pham, D. S., & Krishna, A. (2023). Multi-encoder context aggregation network for structured and unstructured urban street scene analysis. *IEEE Access, 11*, 66227–66244. https://doi.org/10.1109/ACCESS.2023.3289968
17. Yang, F., Zhang, L., Yu, S., Prokhorov, D., Mei, X., & Ling, H. (2019). Feature pyramid and hierarchical boosting network for pavement crack detection. *IEEE Transactions on Intelligent Transportation Systems, 21*(4), 1525–1535.
18. Zhang, L., Yang, F., Zhang, Y. D., & Zhu, Y. J. (2016). Road crack detection using deep convolutional neural network. In *Proc. ICIP* (pp. 3708–3712). IEEE.

19. Zhao, H., Qi, X., Shen, X., Shi, J., & Jia, J. (2018). ICNet for real-time semantic segmentation on high-resolution images. In *Proc.ECCV* (pp. 405–420).

Not So Labeled Approach: FixMatch Outperforms Supervised Learning in Mango Leaf Disease Detection with XAI Insights

Maksura Binte Rabbani Nuha, Kazi Isat Mahazabin, Md Tahsin, Iffat Tasnim, Nurzahan Akter Munni, and Al Hossain

Abstract Mango foliage decay conditions hamper agronomical outputs consistently, resulting in substantial harvest losses and financial losses. Conventional leaf illness detection techniques are typically imprecise and involve a great deal of physical effort. We present a real-time solution using multiple deep learning approaches, incorporating both supervised and semi-supervised models to effectively detect mango leaf diseases and address these challenges. Remarkably, the best pre-trained model, DenseNet201, integrated with the semi-supervised FixMatch algorithm, achieved an impressive 99.93% accuracy using only 30% labeled data. With 500 photos per class, the dataset is divided into eight classes that correspond to different states of mango leaf. By merging pre-trained models with semi-supervised learning approaches, this innovative methodology effectively addresses the problem of real-time disease detection in agriculture in a scalable manner. Transparency is further improved via Explainable AI including interactive application, which offers insights into model decision-making mechanisms.

Keywords Supervised · FixMatch · CNN · XAI · Mango leaf

1 Introduction

Mango cultivation is a significant agricultural industry contributing to both export and import. In 2022, Asian countries such as India produced 26.7 million, Indonesia with 4.1 million, China possessing 3.8 million, and Bangladesh supplied 1.5 million mango production [1]. To fulfill the growing mango production demand, precise

M. B. R. Nuha
Department of Computer Science and Engineering, International Islamic University Chittagong, Chittagong, Bangladesh

K. I. Mahazabin · M. Tahsin (✉) · I. Tasnim · N. A. Munni · A. Hossain
Department of Computer Science and Engineering, East West University (EWU), Dhaka, Bangladesh
e-mail: tahsin30899@gmail.com

J. K. Mandal et al. (eds.), *Recent Advances in Artificial Intelligence and Smart Applications*, Lecture Notes in Networks and Systems 1380,
https://doi.org/10.1007/978-981-96-5822-0_27

disease detection is a must. As they pose a negative impact on the economy. For instance, anthracnose and powdery mildew cause global-economic damage up to 100% if treated without supervision, where powdery mildew is responsible for 20–25% [2, 3].

Incorporating artificial intelligence (AI) for advanced automated disease detection and resource management facilitates precise farm yields and can process large datasets. Moreover, it has a significant impact on the agriculture sector by tackling issues like manual disease findings, limited resources, and also the need for high accuracy in crop disease detection. Older approaches rely heavily on manual effort and often lack consistency such as in visual leaf observation, whereas AI can accurately identify, extract disease features, and categorize plant diseases thus supporting the overall productivity [4]. In our study, we utilized some supervised learning models and obtained precise classification. It involved a well-balanced and robust dataset consisting one healthy class and seven distinct categories containing different conditions, termed (a) gall midge, (b) die back, (c) bacterial canker, (d) powdery mildew, (e) sooty mold, (f) cutting weevil, and (g) anthracnose [5]. Specific region-based sample impacts on variety farming given the ecological settings which requires tactical supervision [6]. Yet gathering and classifying huge amounts of data seems time-consuming and expensive. The supervised learning's drawbacks can be resolved by mitigating the annotated data dependency by semi-supervised learning, which operates with more unlabeled data [7]. In addition, integrated Explainable AI improves DL model's functionality with choice-making features. Even farmers and agricultural specialists gain insights of the factors influencing disease classification by XAI with predictions and understandable logics [8]. By deploying 30% labeled data, we achieved 99.93% accuracy with the integrated FixMatch model, surpassing the supervised models like DenseNet201 (99.50%), InceptionV3(99.25%), ResNet50(99.0%), and EfficientNetV2(98.75%). Our proposed strategy's novelty signifies a supervised model's combination with semi-supervised learning and XAI with less data and top-tier accuracy.

Adding that, the proposed paper's uniqueness involves there has not been any prior research regarding the semi-supervised approach, let alone mention outperforming the supervised model. Mitigating annotated data reliance issues, it outlines a fresh-innovative start which facilitates in agronomy. Furthermore, the technique aids mango production for real-time agricultural disease management, farming and maximizes profitable resources. However, the research field is restricted to parameters value variations and category enhancements.

The following Literature Review section summarizes relevant studies. Data collection, pre-processing, and model descriptions are given in the Methodology section. Then comes the performance experimental setup and evaluation metrics in the Experimental Result part. The results are analyzed in Result analysis, with discussion. Lastly, we conclude our paper in the conclusion section.

2 Literature Review

A substantial portion of mango cultivation worldwide comes from Asian countries and add into a country's economy [9]. Hence, researchers examined this prominent subject with varied approaches over the years.

In 2022, Zhao et al. [10] acquired 99.80% for species and 99.53% for diseases with unsupervised learning and supervised learning employing 31,361 images. During 2020, Pham et al. [11] improvised disease recognition using multilayer perceptron (89.41%), VGG16 (78.64%), AlexNet (79.92%), and ResNet50 (84.88%) for four categories, anthracnose, gall midge, powdery mildew, and healthy, with only 450 images, a narrow dataset. The same year, another research team [12] incorporated neural network ensemble (NNE) with Support Vector Machine (SVM) method mango leaves' disorder identification and achieved an accuracy of 87.5% for normal leaves' and 80% for diseased leaves' detection. In 2021, Kumar et al. [13] applied DL models to formulate for mango leaves' healthy and Anthracnose disease detection with 3500 images, where VGG16 obtained 96.16%. During that year, another deep learning approach acquired accuracies like custom CNN model (98.12%), Mobinet (66.75%), VGG16 (79.52%), InceptionV3 (64.25%), etc. Utilizing two steps such as sample image segmentation and feature extractions for the biotic stresses identification of mango leaf disease. Further model expansion can be done employing other environmental factors [14]. In 2023, Jain et al. [15] acquired 97.7% accuracy by employing a hybrid model combining SVM with SGD to a modest-sized dataset comprising 525 Anthracnose images, 457 Black Sooty Mold images, and 224 Healthy images. In that year, another study proposed a custom CNN-based approach and gained accuracy of 98.55%, with 8 distinct mango leaf category [5]. In the present year, for varied crops disease, Mudassir et al. introduced a new paradigm with Optimal Convolutional Neural Network (E-CNN) and other CNN models, AlexNet, MobileNet, ResNet50, and a machine learning model; Support Vector Machine (SVM) by trimming 9175 from 54000 images where E-CNN model gained 98.17% [16]. Likewise, Hossain et al. [17] proposed pathogens causing infections on mango leaves utilizing CNN models with optimized Vit model, where it obtained 99.75% accuracy. In comparison, the findings indicate that most of the paper [11, 13, 16] implemented ML and DL models, only research [15] utilized hybrid model by combining SVM and SGD. Paper [10] exclusively incorporated DoubleGAN method, while study [17] used a fresh approach with Transformer models. One of the primary limitations involved data characteristics or variety versatility challenges. On the other hand, among the related papers, ours is the first to introduce semi-supervised learning, let alone exceeding performance of supervised model. Our proposed SL and SSL learning approach diminishes the drawbacks and validates robustness and versatility in mango leaf disease identification utilizing distinct strategies.

3 Methodology

The procedure initiates with dataset acquisition, sourced from Mendeley, MangoLeafBd dataset [18] with image being resized to 224×224. From our three employed supervised models, we proceed with the highest performed pre-trained model merging with semi-supervised algorithm we classified mango leaves' diseases. Additionally, Explainable AI was also integrated with the superior SL model. Figure 1 shows the work flow diagram.

3.1 Data Collection

The dataset comprises 4000 images with eight distinct classes, each class consisting 500 sample images [18]. As shown in Fig. 2, healthy leaves are free from any irregularity. Sooty mold are the fungal inhabitant on leaves, which originates from insects, shady place, and photosynthesis imbalance. Powdery mildew refers to dust-white colored which denotes powdery substances on leaves and young fruits. The larvae emerging from insects produce leaf loss known as gall midge. Die back is caused by fungi, brownish-colored leaves, leaves dying, and falling off from twigs. Cutting weevil comes from bugs that defect leaves. Bacterial canker turns leaves into cankers. Anthracnose, a fungal infection, forms black or brown necrotic patches, particularly along the leaf edges and leads to curly mango leaves.

3.2 Model Description

Using four DL models, DenseNet201, InceptionV3, ResNet50, and EfficientNetV2, we chose the peak-valued model and merged it with FixMatch and Explainable AI. The model's summaries are outlined:

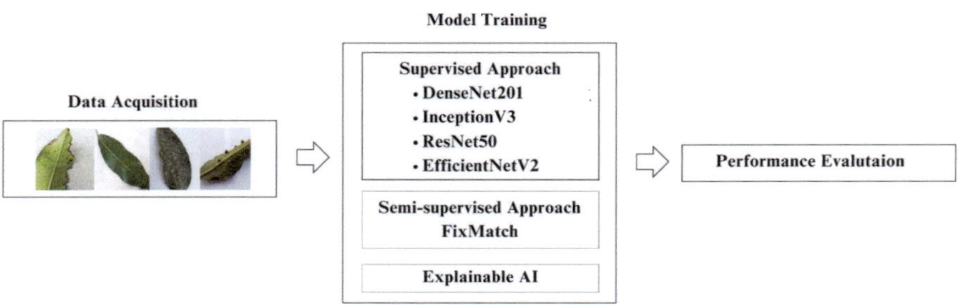

Fig. 1 Overview of the proposed methodology

Fig. 2 Healthy and diseased species of mango leaf

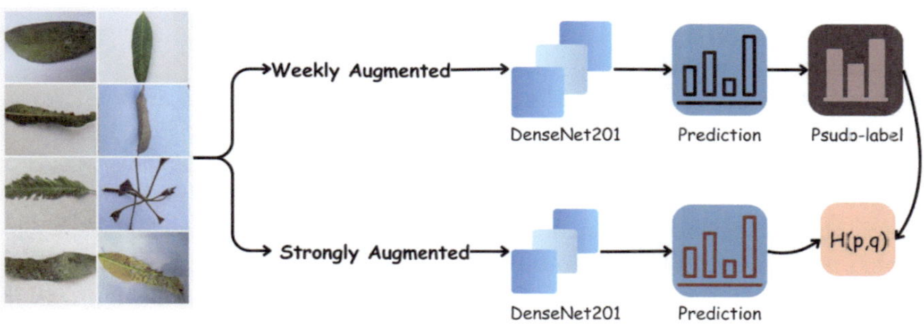

Fig. 3 Detailed overview of the FixMatch architecture

FixMatch: A semi-supervised learning (SSL) algorithm [19], designed for improving the models performance by using labeled and pseudo-labeled data. It simplifies SSL by combining two key techniques: consistency regularization and pseudo-labeling [20]. Figure 3 showcases the FixMatch architecture.

DenseNet201: Highly favored for its feature-sharing capability among the other models and vanishing gradient issue mitigation with dense connections. As the layers receive inputs from preceding layers in this network, it can acquire learnings by previous layers for output optimization [21].

InceptionV3: InceptionV3, a pre-trained convolutional neural network architecture that has 48 layers. It is basically designed for image analysis and detection tasks, even capable of labeling diverse image set [22].

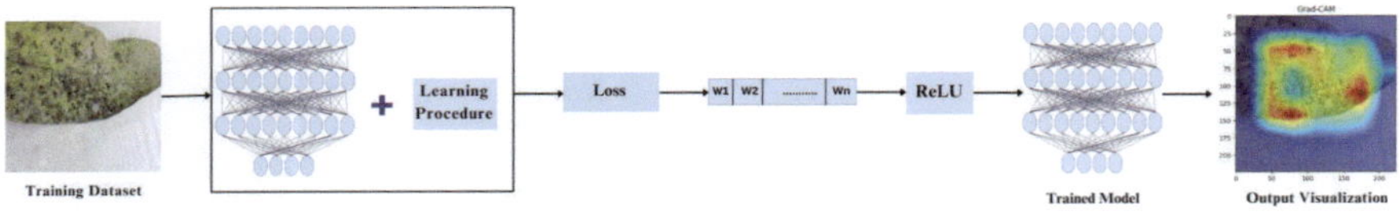

Fig. 4 Architecture explainable AI (XAI) framework

ResNet50: This model comprises 50 linked layers categorized into quadrat parts comprising convolutional and identity sections. Mostly used for computer vision tasks like item recognition and tracking [23].

EfficientNetV2: The model depicts enhanced training efficiency and preserves a comparable output attribute with normalization and neural architecture search (NAS) [24].

Explainable AI(XAI): It enables AI decision-making improvements by making models understandable and boosting transparency. It promotes clarity, fairness, and reduces bias aligning with ethical AI developments [25]. Several Explainable AIs were employed, such as Grad-CAM, Score-CAM, Layer-CAM, and Grad-CAM++, to interpret our model's decision-formulating process. Figure 4 outlines the XAI frameworks, where the dataset undergoes training with necessary learning procedures to specify correct output.

Integration procedure of SL and SSL: The first 451 layers are kept frozen for fine-tuning. Then, utilizing ImageNet and GlobalAveragePooling2D, ReLU activation function and L2 regularization are adopted on the "dense" layer, lastly, a Softmax activation function is implemented for the layer of output in the pre-trained model. The FixMatch algorithm deployment process includes strong and weak augmentation, strong augmentation processes RandAug function, where any 4 augmentations will be chosen for execution out of 16 discrete augmentations termed histogram equalization, contrast and brightness adjustments, etc. And weak augmentation slightly alters the images. Integration with the best supervised model learning rate initialization With pseudo-labeling and 0.8 threshold value, the unlabeled data are predicted with confidence score. Finally, the back-propagation serves for calculating the losses for unsupervised as well as supervised algorithms. After that, the model generates accuracy.

3.3 Experimental Setup

Parameters outlined in Table 1 are for employed supervised models while Table 2 demonstrates values for semi-supervised models. The same optimizer named "Adam," 0.0009 learning rate value and 25 epochs are applied in both. A dropout rate of 50% and loss function named categorical cross-entropy. Moreover, "early stopping" was used for the training process if performance stops improving over 5 epochs (Patience

Table 1 Training parameters of supervised models

Pre-trained model	DenseNet201
Optimizer	Adam
Image size	224×224
Loss function	Categorical cross-entropy
Batch size	10
Epoch	25
Early stop	Yes
Patience score	5
Dropout	0.5
Learning rate	0.0009

Table 2 Training parameters of semi-supervised FixMatch

Integrated model	DenseNet201 with FixMatch
Data split	70% Training, 30% Testing and validation
Optimizer	Adam
Image size	224×224
Loss function	Categorical cross-entropy
Epoch	25
Labeled data	30%
Early stop	Yes
Patience score	5
Threshold	80%
Learning rate	0.0009
Labeled batch size	8
μ	4
Unlabeled batch size	8×4=32

score of 5 to trigger early stopping). The key disparity of Table 2 was its 80% threshold value. Table 1 outlined a batch value of 10, while Table 2 held annotated data's batch size of 8, while for unknown samples it multiplied to 32 (since $\mu = 4$).

4 Result Analysis

To achieve an efficient model performance, comparative result assessment is mandatory. In this segment, we analyzed the accuracies for both supervised and semi-supervised models.

4.1 *Supervised Model*

We used same parameters for four supervised models: DenseNet201, InceptionV3, EfficientNetV2, and ResNet50, where DenseNet121 acquired 99.5%, proceeding InceptionV3 with 99.25%, ResNet50 with 99.0%, and EfficientNetV2 with 98.75% accuracy. In Fig. 5, the training and validation accuracy and loss are depicted over epochs. Here, (A) DenseNet201, (C) EfficientNetV2, and (D) ResNet50 delineated consistent validation accuracy with minor fluctuations and the validation loss exhibits mild variability. While (B) InceptionV3 demonstrates validation accuracy with the most instability.

4.2 *Blending Semi-supervised Model with the Optimal Supervised Model*

Semi-supervised learning (SSL) approach blends unidentified data along with minimal annotated data utilizing a ratio which facilitates in minimizing labor costs, remedy for data scarcity [26]. Rationale for choosing FixMatch was from the relevant semi-supervised studies, the integrated FixMatch models excelled other supervised models [27, 28]. From this perspective, we worked with only 30% of labeled plant dataset integrating DenseNet201 (supervised) with FixMatch model (semi-supervised), and attained an excellent 99.93% accuracy.

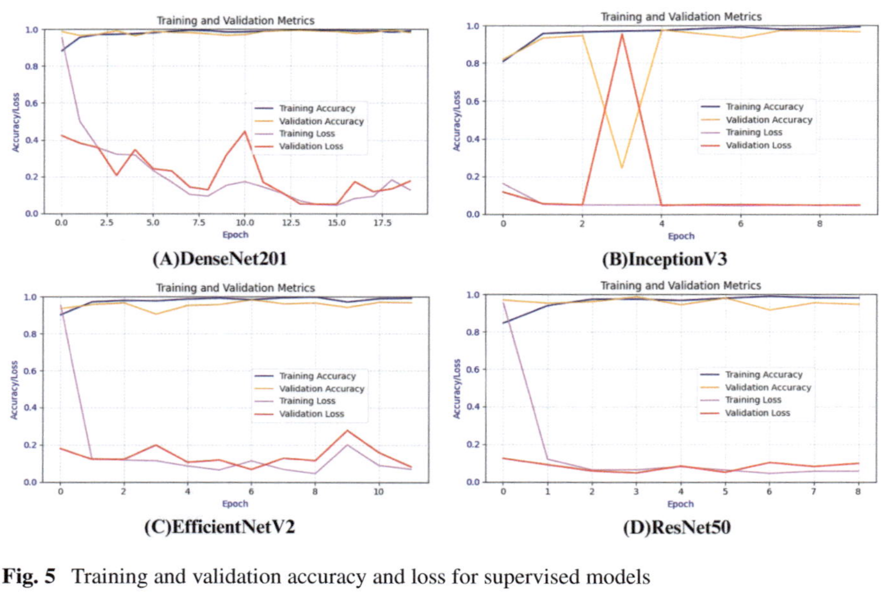

Fig. 5 Training and validation accuracy and loss for supervised models

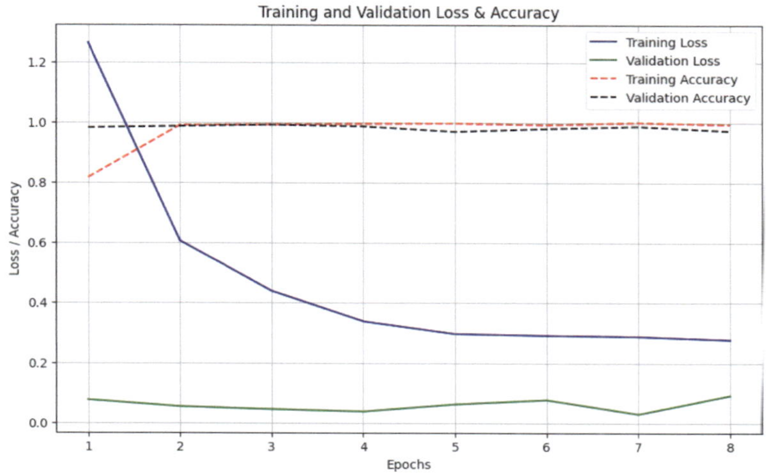

Fig. 6 Training and validation loss and accuracy curves for the FixMatch model

Figure 6 delineates FixMatch model's performance changes as epoch continues. The training loss starts high then shrinks with labeled training as epoch progresses, indicating the model's successful learning and validation loss depicting a stable line with moderate fluctuations to show the model adapts well to unobserved data. The training and validation accuracy improves rapidly and stabilizes at peak value, suggesting model's learning for both labeled and unlabeled data without any overfitting signs.

Figure 7 exhibits a ROC curve with AUC value (ranging from 0.96 to 1.00) for eight classes, denoting model's efficacy to differentiate between classes with minimal misclassification where classes 3, 5, and 6 reveal maximum classification performance.

4.3 XAI Implementation with Best Performing Supervised Model

A. Integration with different XAI: Grad-CAM, Score-CAM, Grad-CAM++, and Layer-CAM

Figure 8 illustrates an original input image of diseased leaf to Grad-CAM, Score-CAM, Layer-CAM, and Grad-CAM++ to extract irregular traits or features, which are the powdery spots in the picture. The domain-specific features checks for disordered spots and aids to identify diseases accurately.

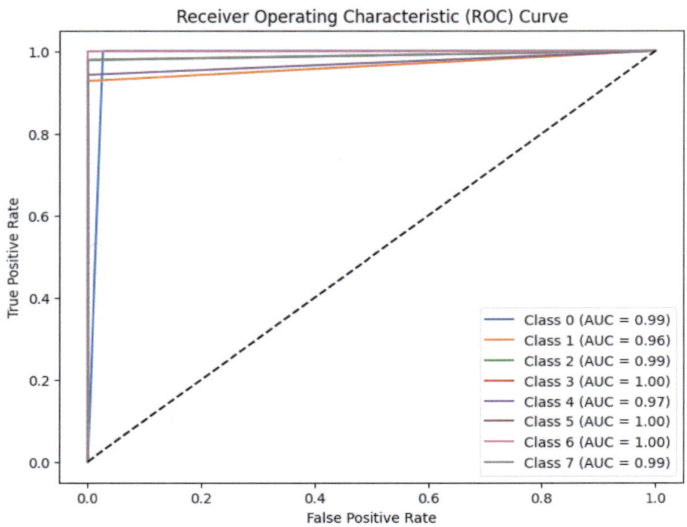

Fig. 7 ROC curves for FixMatch model

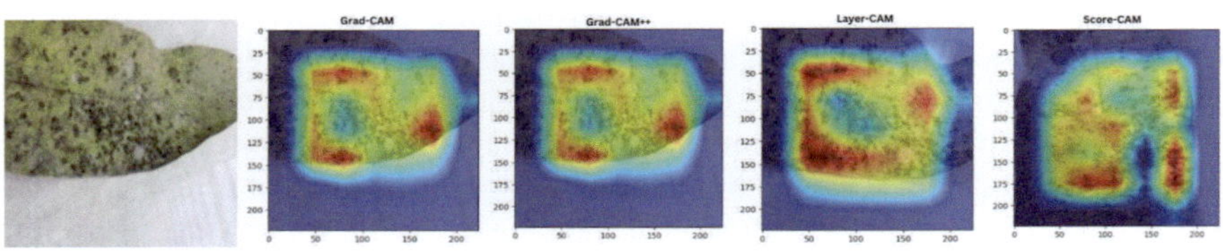

Fig. 8 Integration of explainable artificial intelligence (XAI) techniques for classifying mango leaf diseases

B. Real-time Web Integration Using XAI

We employed Angular js structure for user interface, in the first page, for photo uploading with specified supported formats JPEG, PNG, or SVG files. Then, on the second page, users can click okay for further procedure or go back to change the photo and upload again. Using the Django REST framework, for the back end, the app identifies the given leaf's disease. In Fig. 9, we uploaded a diseased leaf photo and the results are highlighted including an XAI image, providing the actual and predicted names with a confidence score of 99.97%.

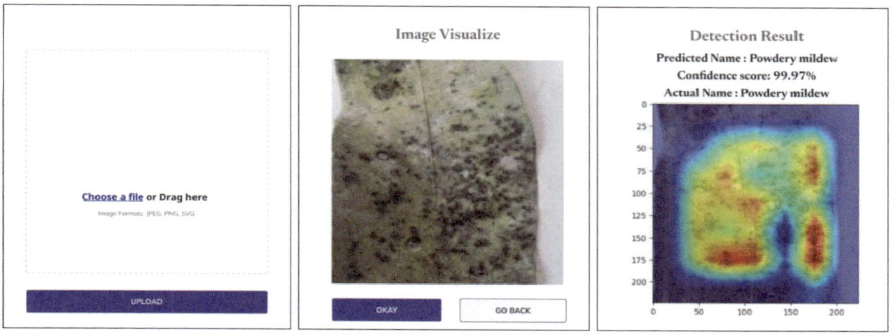

Fig. 9 Web application interface for classifying mango leaf diseases

Table 3 Classification report of FixMatch on various datasets

Class	Precision (%)	Recall (%)	F1-score (%)	Accuracy (%)
Early blight	100	100	100	100
Healthy	98	100	99	100
Late blight	100	98	99	98.2

4.4 *FixMatch Model Generalizability*

On a distinct potato leaf disease dataset [29], the integrated FixMatch model was implemented, which obtained an overall accuracy of 99.30%.

Table 3 highlights the evaluation metrics for the diseased and uninfected leaves of the dataset, ensuring versatility and robustness.

5 Conclusion and Future Works

The first-ever maiden study to identify diseased mango leaf varieties employing both SL and SSL comprehensive approach. The four pre-trained models, DenseNet121 (99.50%), InceptionV3 (99.25%), ResNet50 (99.0%) and EfficientNetV2 (98.75%), leveraged to acquire remarkable accuracy, where DenseNet121 achieved the best performer. Adding that, the integrated FixMatch algorithm mixes both labeled and unlabeled data effectively achieving an exceptional 99.93% accuracy. Exceeding the annotated data's accuracy 99.50%, with only 30% label data and a substantial gap of 0.43%, the integrated approach point toward a ground-breaking strategy in the agronomy field including a game-changing impact. This research's novelty possesses potentiality to detect disease and pathogens accurately including a measurable autonomous solution, constraining data reliance. Enhancing AI feasibility in agribusiness where a fruit like mango impacts more on ecological value, exportation, and sustainable economy. Yet, overfitting may arise concern in the proposed SSL model. Hence, in future, the research aims to explore the application of SSL technique with additional disease variety's increment to enhance model's applicability.

References

1. World Population Review. (2022). *Mango production by country*. https://worldpopulationreview.com/country-rankings/mango-production-by-country.
2. Dofuor, A. K., Quartey, N.K.-A., Osabutey, A. F., Antwi-Agyakwa, A. K., Asante, K., Boateng, B. O., Ablormeti, F. K., Lutuf, H., Osei-Owusu, J., Osei, J. H. N., amp, et al. (2023). Mango anthracnose disease: the current situation and direction for future research. *Frontiers in Microbiology, 14*, 1168203.
3. Iqbal, S., Atiq, M., Fayyaz, M., Zakria, M., Rajput, N. A., Kachelo, G. A., Ahmad, I., Usman, M., Mehmood, A., et al. (2024). Powdery mildew of mango: Current status, perspective and emerging tools for management. *Agricultural Sciences Journal, 6*(1), 92–101.
4. Islam, M. M., Adil, M. A. A., Talukder, M. A., Ahamed, M. K. U., Uddin, M. A., Hasan, M. K., Sharmin, S., Rahman, M. M., & Debnath, S. K. (2023). Deepcrop: Deep learning-based crop disease prediction with web application. *Journal of Agriculture and Food Research, 14*, Article 100764.
5. Ahmed, S. I., Ibrahim, M., Nadim, M., Rahman, M. M., Shejuntia, M. M., Jabid, T., & Ali, M. S. (2023). Mangoleafbd: A comprehensive image dataset to classify diseased and healthy mango leaves. *Data Article, 47108941*. https://doi.org/10.1016/j.dib.2023.108941
6. Tahsin, M., Matin, M. M. H., Khandaker, M., Reemu, R. S., Arnab, M. I., Rashid, M. R. A., Rasel, M. M. K., Islam, M. M., Islam, M., & Ali, M. S. (2024). An extensive image dataset for deep learning-based classification of rice kernel varieties in Bangladesh. *Data in Brief* 111109. https://doi.org/10.1016/j.dib.2024.111109.
7. Noema, Y. (2022). *Learning: Supervised, unsupervised, self-supervised & semi-supervised*
8. Chen, H.-Y., Sharma, K., Sharma, C., & Sharma, S. (2023). Integrating explainable artificial intelligence and blockchain to smart agriculture: Research prospects for decision making and improved security. *Smart Agricultural Technology, 6*, Article 100350.
9. Lawson, T., Lycett, G. W., Ali, A., & Chin, C. F. (2019). Characterization of southeast asia mangoes (mangifera indica l) according to their physicochemical attributes. *Scientia Horticulturae, 243*, 189–196.
10. Zhao, Y., Chen, Z., Gao, X., Song, W., Xiong, Q., Hu, J., & Zhang, Z. (2021). Plant disease detection using generated leaves based on doublegan. *IEEE/ACM Transactions on Computational Biology and Bioinformatics, 19*(3), 1817–1826.
11. Pham, T. N., Van Tran, L., & Dao, S. V. T. (2020). Early disease classification of mango leaves using feed-forward neural network and hybrid metaheuristic feature selection. *IEEE access, 8*, 189960–189973.
12. Mia, M. R., Roy, S., Das, S. K., & Rahman, M. A. (2020). Mango leaf disease recognition using neural network and support vector machine. *Iran Journal of Computer Science, 3*(3), 185–193.
13. Kumar, P., Ashtekar, S., Jayakrishna, S., Bharath, K., Vanathi, P., & Kumar, M. R. (2021). Classification of mango leaves infected by fungal disease anthracnose using deep learning. In *2021 5th International Conference on Computing Methodologies and Communication (ICCMC)* (pp. 1723–1729). IEEE.
14. Gautam, V., & Rani, J. (2022). Mango leaf stress identification using deep neural network. *Intelligent Automation & Soft Computing, 34*(2).

15. Jain, S., & Jaidka, P. (2023). Mango leaf disease classification using deep learning hybrid model. In *2023 International Conference on Power, Instrumentation, Energy and Control (PIECON)* (pp. 1–6). IEEE.
16. Iftikhar, M., Kandhro, I. A., Kausar, N., Kehar, A., Uddin, M., & Dandoush, A. (2024). Plant disease management: a fine-tuned enhanced cnn approach with mobile app integration for early detection and classification. *Artificial Intelligence Review, 57*(7), 1–29.
17. Hossain, M. A., Sakib, S., Abdullah, H. M., & Arman, S. E. (2024). Deep learning for mango leaf disease identification: A vision transformer perspective. *Heliyon, 10*(17).
18. Ahmed, S. I., Ibrahim, M., Nadim, M., Rahman, M. M., Shejunti, M. M., Jabid, T., & Ali, M. S. (2023). Mangoleafbd: A comprehensive image dataset to classify diseased and healthy mango leaves. *Data in Brief, 47*, Article 108941.
19. Song, J. (2023). [ssl] papers for semi-supervised learning. https://www.junha.page/2023/06/ssl-papers-for-semi-supervised-learning.html
20. Sohn, K., Berthelot, D., Carlini, N., Zhang, Z., Zhang, H., Raffel, C. A., Cubuk, E. D., Kurakin, A., & Li, C.-L. (2020). Fixmatch: Simplifying semi-supervised learning with consistency and confidence. *Advances in neural information processing systems, 33*, 596–608.
21. Team, P. (2020). Densenet-201 dataset. https://www.kaggle.com/datasets/pytorch/densenet201
22. Gupta, S. (2020). Inception v3 model dataset. https://www.kaggle.com/datasets/imsparsh/inception-v3-model
23. Kundu, N. (2023). Exploring resnet50: An in-depth look at the model architecture and code implementation.
24. Tan, M., & Le, Q. (2021). Efficientnetv2: Smaller models and faster training. In *International conference on machine learning* (pp. 10096–10106). PMLR.
25. IBM. (2023). What is explainable ai (xai)? https://www.ibm.com/topics/explainable-ai
26. AltexSoft. Semi-supervised learning: Techniques and applications. https://www.altexsoft.com/blog/semi-supervised-learning/
27. Tahsin, M. (2024). Efficient tea leaf classification: Bridging supervised and semi-supervised learning. In *Proceedings of the International Conference on Data Mining and Information Security (ICDMIS)*, Springer.
28. Tahsin, M., & Nuha, M. B. R. (2024). Enhancing cervical cancer detection: Leveraging semi-supervised learning for improved prognosis. In *27th International Conference on Computer and Information Technology (ICCIT)*, IEEE.
29. Putra, M. A. (2020). Potato leaf disease dataset. https://www.kaggle.com/datasets/muhammadardiputra/potato-leaf-disease-dataset.

A Comparative Analysis of ML Techniques to Detect Human Emotions from Text-Based Data

Swastik Dhar, Shayanika Das, Pallab Banerjee, Srijeet Roy, and Bipasha Mukhopadhyay

Abstract Issues related to mental health are increasingly becoming legitimate medical concerns in the workplace. To mitigate mental health concerns among employees, companies need to identify the points of view that bear most of the blame. Some preprocessing is performed on an open-source survey data set before selecting features for language processing. Techniques such as logistic regression, Naive Bayes, and random forest are used for classifications to determine whether an employee needs treatment for mental health issues or not. The activities were modeled as a bag of words using a probabilistic Naive Bayes (NB) classifier on training data with an accuracy of 92.7% and a random forest with an accuracy of 99.7% to predict mood outcomes and logistic regression with 99. 8% accuracy. The analysis is divided into two parts: a personalized model and a generic model in which we combined the data from all participants. Finally, a voting classifier is used to add all of the findings from every classifier that is fed into it and forecast the result based on the most notable majority of votes. Combining the aforementioned techniques, the voting classifier improves our accuracy to 90.48%.

1 Introduction

Our lives revolve around our work. Most of our time is spent at the workplace. Having a rewarding career is essential for maintaining excellent mental health and overall well-being. Working for extra hours or working under a strong deadline takes life in a stressed condition. Other problems, such as those involving social circumstances, relationships, or family, occasionally arise. All these situations sometimes degrade mental health. Good managers who look out for people and bring out the best in them are essential to a good workplace. Research demonstrates that people with sound mental health are very productive.

S. Dhar · S. Das · P. Banerjee · S. Roy · B. Mukhopadhyay (✉)
UEM, Kolkata, India
e-mail: bipasha.mukherjee@uem.edu.in

J. K. Mandal et al. (eds.), *Recent Advances in Artificial Intelligence and Smart Applications*, Lecture Notes in Networks and Systems 1380,
https://doi.org/10.1007/978-981-96-5822-0_28

Mental health disorders are currently among the most burdensome health issues. For example, a report published in late June 2020 says that around 40% of adults in America are struggling with mental health issues. Every year, one in six American youth aged 6 to 17 struggles with psychological well-being issues. By the age of 14, and by the age of 24, 75% of all lifetime dysfunctional conduct has begun. The nation loses over 210.5 billion dollars a year due to despair alone [1]. Diabetes, heart disease, and other physical health issues are linked to mental health illnesses. Patients who have both physical and mental health problems incur far higher treatment costs. Opening out about mental illness often makes people uncomfortable, especially in the workplace. Celebrities and other people have started sharing their stories these days, which has increased social awareness. Employers must establish a work environment that genuinely supports employees' mental health. To encourage their employees to enhance their mental health, employers must create a work environment that stresses psychological components and offers pertinent resources.

Globally, an estimated 264 million people experience depression. These folks have anxiety symptoms, which impede their capacity to function. According to a recent study conducted by the World Health Organization (WHO), lost productivity from mental health illnesses costs the world economy 1 trillion dollars annually [2]. Unemployment is another significant element in mental health. This could result in drug or alcohol abuse, absenteeism, etc. People who are employed also need to think about their workplace. A positive work environment boosts output, which generates revenue. Approximately 13% of children, 46% of teenagers, and 19% of adults suffer from mental illness each year [3]. However, it is concerning to note that only 50% of the individuals afflicted obtain therapy, as most people find it awkward to disclose. Improper treatment of mental illness can result in increased medical costs, an increased risk of suicide, subpar work output, and other issues. This initiative aims to determine and quantify the human elements that influence the probability of seeking mental health assistance. The feelings that people experience when they seek assistance are categorized according to the aspects of human nature that either favorably or unfavorably impact their mental health.

The issue at hand in this study is mental health, which may be examined by utilizing sentiment analysis on our platform and natural language processing. Our platform contains user-generated material with comments from other users that are gathered for sentiment analysis. Text mining, which may extract data from our platform using data crawling techniques, will be utilized to gather comments in order to gain a high number of comments [4, 5]. As available data is quite unstructured, different text preparation techniques are required. Unstructured data contains unnecessary alphanumeric symbols that can be altered at any of the processing stages [6, 7]. Data processing is a crucial step when performing data analysis for sentiment analysis. Different classification algorithms are used to perform sentiment analysis and classify emotions into positive, negative, and neutral categories. Sentiment analysis-generated opinions can reveal information about the mental health of users through different platforms [8, 9].

The organization of the paper is as follows, Sect. 2 contains the literature survey, Sect. 3 describes the problem statement, Sect. 4 contains the proposed solution, Sect. 5 contains the experimental result, and Sect. 6 contains the conclusion & future scope.

2 Literature Survey

2.1 Naive Bayes Approaches

The Naïve Bayes technique applies the Bayes theorem to data classification. Although this may not always be the case in practice, this method operates under the assumption that every feature (or attribute) in the dataset is independent of other attributes [10, 11]. Hence this approach is referred to as "naive". This method applies the Bayes theorem to determine the likelihood of any class in a dataset. Stated otherwise, Naive Bayes determines the likelihood of any dataset to determine a particular class. Applications requiring data classification, like document classification, sentiment analysis, and email categorization as spam or not, frequently use Naive Bayes. Naive Bayes has the advantage of being fast and simple to build, and it can handle a large amount of feature-rich data [12].

2.2 Random Forest Approaches

A subset of the bagging approach, the RF is insensitive to parameter initialization and performs better when there is noise and weak discrimination data [13]. Every iteration, an RF uses bootstrap to randomly select many samples to develop a decision tree (without pruning), and it builds multiple decision trees to construct the entire RF. Next, every tree "votes" for the output, which is the most popular class.

Widely employed in numerous fields, RF-based feature selection (permutation) is reliable for variables of more dimension and higher correlation value [14]. A graph is plotted to classify noise for every X value. In the meantime, the RF establishes which factors should be employed to reduce a predictor's purity decline utilizing the Gini coefficient. As dimension reduction is always required, the random forest cross-validation function is used for feature selection and dimension reduction. Nested cross-validation lowers the number of predictors gradually. This model is not that sensitive to parameter settings as much as other methods. The optimal performance of the random forest (RF) method is based on two parameters m-try and n-trees. Here, m-try indicates the number of input variables that are used to decide node in a decision tree and n-tree indicates the total number of trees. As per the research of Breiman [15], the error is less if the number of trees increases. The performance of the entire classification of this model is directly proportional to the n-tree value [16].

2.3 Sentiment Analysis-Based Approaches

This method in the domain of language processing is applicable to recognize and evaluate viewpoints or feelings that specific people or groups of people convey in written or spoken materials. Understanding how an individual or group reacts favorably or unfavorably to a subject, product, or service on different platforms is the aim [17] of sentiment analysis. Different approaches are available to analyze sentiments [18]. This analysis is also useful in the business domain for customer segmentation. By doing this, companies can better understand the requirements and issues of their customers and improve the caliber of their goods and services. Analysis of sentiments are also utilized in the social and political spheres to comprehend public opinion on significant social topics, leaders, and policies. In the domains of business, politics, and society, sentiment analysis results can aid in decision-making and the development of stronger strategies [19].

3 Problem Statement

Human livelihood is intricately intertwined with the workplace. Long working hours, tight deadlines, and personal and professional liabilities often significantly impact mental health. Negative work environments can exacerbate existing mental health issues and contribute to a decline in overall well-being. The prevalence of mental health disorders is a significant public health concern, with millions of people worldwide experiencing conditions such as depression and anxiety. Unfortunately, despite the prevalence of mental health disorders, many people don't even go for treatment because of societal issues. Analyzing those reactions from different communication modes may help people maintain good mental health. The absence of such a system creates depression, and that causes a reduction in productivity. Research consistently shows that individuals with good mental health tend to be more productive, underscoring the importance of prioritizing mental well-being in the workplace.

4 Proposed Solution

In efforts to address mental health concerns among employees, companies are increasingly turning to data-driven approaches, using classification techniques to identify key factors contributing to these issues. To achieve this, extensive preprocessing of open-source survey data is conducted, followed by the selection of relevant features using natural language processing (NLP) techniques. This process involves transforming textual responses into a structured format suitable for analysis.

The approach is divided into two main parts to further enhance the analysis: personalized and generic models. In the personalized model, data from individual participants is analyzed independently, allowing for tailored insights into each individual's mental health status. Conversely, the generic model combines data from all participants, enabling broader trends and patterns to be identified throughout the workforce.

In addition to these individual models, a voting classifier is used to consolidate the findings of multiple classifiers. This ensemble technique aggregates predictions from each classifier and generates a final prediction based on the most commonly agreed-upon outcome.

Overall, this comprehensive approach to mental health analysis leverages a combination of preprocessing techniques, feature selection, and machine learning algorithms to identify and address mental health concerns among employees effectively. Using the power of data-driven insights, companies can take proactive steps to support the mental well-being of their workforce, ultimately fostering a healthier and more productive work environment.

4.1 Step 1–Emotion Detection from Text

One of the biggest challenges in natural language processing is the recognition of emotions from text due to the wide range of emotions that people experience. Our aim is to create an effective model for recognizing emotions using the labeled data. There are 40,000 records in total in our collection, representing 13 distinct emotions. Figure 1a, shows the distribution of emotions. This indicates the presence of multi-class emotions. Developing an effective multi-class classification model is difficult. To create an effective model, certain cutting-edge techniques are applied that rationally minimize the number of classes.

4.2 Step 2–Stress Detection from Text Dataset

Texts about people who deal with anxiety, depression, and other mental health conditions are collected in the Mental Health Corpus. The corpus is divided into two columns: the comments are in one, and labels indicating whether or not the remarks are poisonous are in the other. The corpus can be used for many things, including sentiment analysis, mental health language analysis, and the detection of hazardous language. Researchers, mental health professionals, and others interested in understanding the language and attitude surrounding mental health concerns may find the data in the corpus interesting. The dataset is taken from the Reddit platform. Posts related to depression are being from it. Non-suicide posts from teens were also collected. Emotion distribution of available data can be shown in Fig. 1b.

As a next step collected texts are being processed through the following procedures.

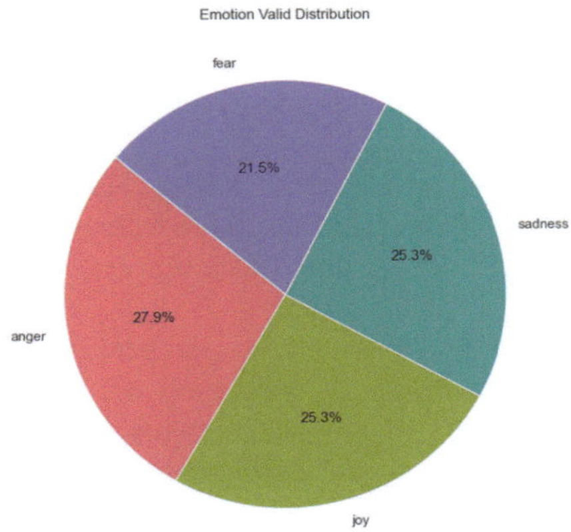

(a)

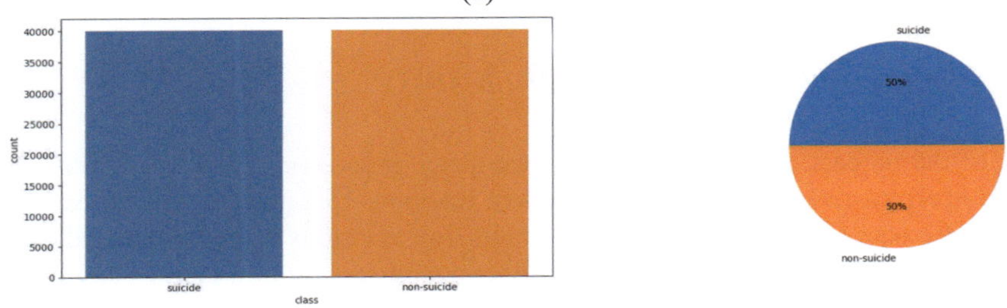

(b)

Fig. 1 **a** Emotion distribution, **b** Emotion distribution

4.3 Text Processing

- Tokenization—The process of tokenizing a stream of text involves dividing it up into words, phrases, symbols, or other significant components, which are referred to as tokens. Tokens are useful for many analyses.
- Text-Cleaning—Text cleaning, also referred to as text preparation, is the act of getting unprocessed text data ready for tasks involving natural language processing (NLP). The text must be cleaned and standardized through many procedures to be better prepared for analysis.
  Here are a few typical methods for cleaning text:
- Lower-casing: To maintain consistency, all text should be converted to lowercase (for example, "Hello" and "hello" are considered as the same word).
- Special characters should be eliminated since they might not be necessary for analysis. These characters include those that are not letters, numerals, or spaces.
- Numbers should be removed since they might not be necessary for some text analysis jobs.

- Eliminating excess whitespace: To standardize whitespace, eliminate any extra tabs, spaces, or newline characters.
- Eliminating stop words: Get rid of frequent terms with little significance, Including "the," "and," and "is."
- Lemmatization and stemming: To standardize variants, reduce words to their basic or root form (e.g., "running" to "run").

4.4 Feature Extraction

- Bag of Words (BOW)—It is a straightforward and widely used method for text feature extraction in NLP and text mining. This process depicts a text as a multiset (or "bag") of its words, preserving diversity while ignoring word order and syntax. Every word is regarded as a feature, and each word's frequency serves as a numerical feature vector.
- TF-Term Frequency (Tf) is a statistical metric used in text mining and information retrieval to assess the significance of any word within any document. Phrase frequency counts the frequency of words in a document. It is computed as

$$\frac{Total - number - of words}{Frequency - of - phrase}$$

.

- Inverse Document Frequency (IDF)–It indicates how significant a phrase is across the board in the corpus. This is computed as

$$\frac{Total - number - of - documents}{Number - of - documents - including - phrase}.$$

  This results in the logarithm form. The product of a term's TF and IDF scores yields its TF-IDF score, which shows the degree to which a phrase is related to a certain corpus document. High TF-IDF terms are frequently seen as being more significant or indicative of the document's substance.
- Word Embedding—This method is used for representing words as dense vectors in a continuous vector space. The closeness of a word's related vectors indicates its similarity to another word. This is in contrast to conventional techniques such as one-hot encoding, which uses sparse vectors to represent words, with a vector containing largely zeros and a single one for each word.
- Word2Vec, Global Vectors for Word Representation (GloVe), and FastText are some of the methods often used to learn word embeddings from huge text corpora. These techniques capture contextual and similarity associations between words, which makes them valuable for a variety of language modeling applications.

4.5 Model Selection

Emotional Analysis—Training and Testing split of Dataset. Model architecture is defined using a combination of CNN and text embedding. Two branches of the model process text index input independently, each consisting of an embedding layer followed by an ID convolution layer and a pooling layer. Outputs of both branches are concatenated and paned through a dense Layer for further processing. Dropout Layers are used for regularization. The output Layer with SoftMax activation predicts the probability distribution over the four emotion claves.

Model compilation: The model is then compiled with an optimizer, and has function and evaluation matrix.

5 Result and Discussion

In emotion analysis, the model gets trained for 25 epochs. The following observations can be made after training the model. As the best model is not known, many models are being trained. This Fig. 2 shows a detailed comparison of the performance of different classification techniques like logistic regression, Naive Bayes, and random forest. It can be profoundly observed logistic regression classification technique is best for this dataset. After choosing a model, model performance can be analyzed as different performance metrics such as accuracy, precision, recall, and loss. These can

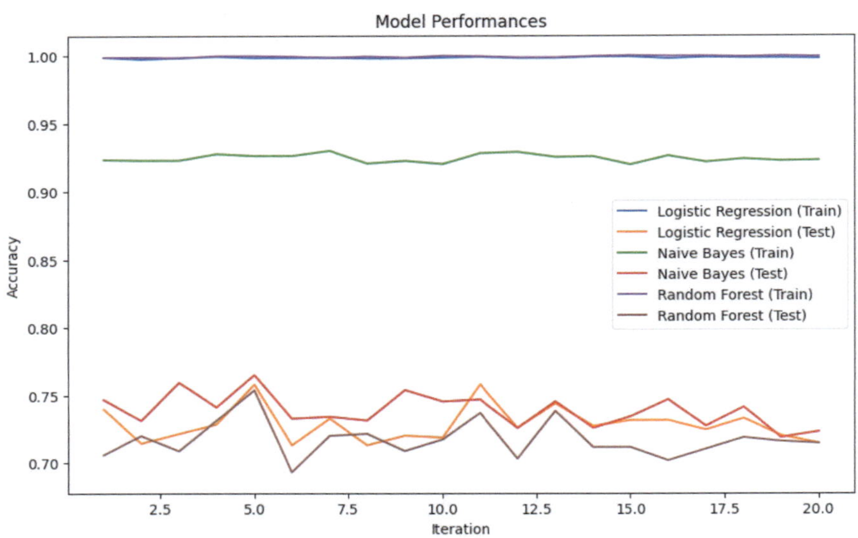

Fig. 2 Comparative performance 1

be observed in Fig. 3a. The figure shows that performance increases with the increase in epochs. Different emotion detection can be shown in the given matrix, shown in Fig. 4a. Different metrics can be properly understood in the confusion matrix, which can be observed in Fig. 4b.

After applying different classification methods to detect suicidal and stress-related emotions, the accuracy of different methods can be observed and compared in the

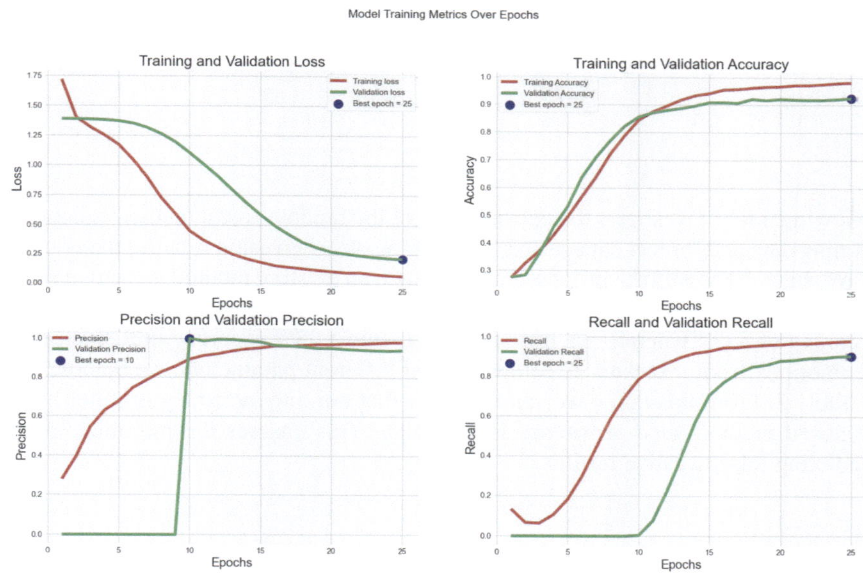

Fig. 3 Comparative performance 2

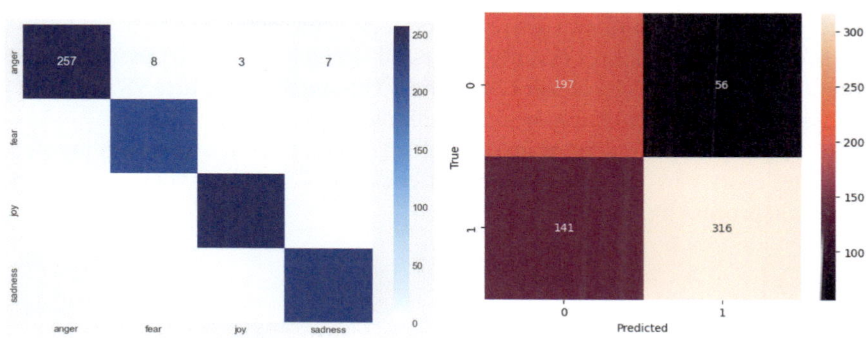

(a) (b)

Fig. 4 **a** Emotion matrix, **b** Model performance

Table 1 Accuracy in different methods

Model	Training data (%)	Test data (%)
Logistic regression	99.8	75.6
Naive Bayes	92.7	75.0
Random forest	99.7	72.0

Table 2 Proposed versus existing methods

Parameters	Existing (%)	Proposed (%)
Accuracy	83	98
Sensitivity	75	85

following Table 1. It shows the performance of the random forest method is better in training data and for test data performance of logistic regression is better than others.

We have compared the proposed method with an existing method present in paper [6]. The authors, of this paper have used the gradient boost method and achieved an accuracy of 83%, whereas we have achieved almost 98% accuracy in the detection of emotions from text data. A comparison of different parameters can be observed in Table 2. From this table, we can conclude that not only accuracy is better in the proposed method, also sensitivity is also high. This ensures the minimization of predicting false negative results.

6 Conclusion and Future Scope

This work suggests that if emotions and mental conditions can be expressed as words, they can be analyzed, and preventive and protective measures can be taken to save a person. This paper has successfully suggested which method to be used for better detection and classification of word categories. This approach ensures 98% accuracy with training data in predicting mental conditions.

This work can be extended as an AI-enabled Chatbot-based application interface by storing every new word and associated emotion class in the cloud to make the conclusion process more accurate.

References

1. A. Bisht, S. Vashisth, M. Gupta and E. Jain, "Stress Prediction in Indian School Students Using Machine Learning," 2022 3rd International Conference on Intelligent Engineering and Management (ICIEM), London, United Kingdom, 2022, pp. 770-774, https://doi.org/10.1109/ICIEM54221.2022.9853200.

2. Norizam, Sulaiman. "Determination and classification of human stress index using non-parametric analysis of EEG signals." (2015).
3. Harris, K. M., Gaffey, A. E., Schwartz, J. E., Krantz, D. S., & Burg, M. M. (2023). The Perceived Stress Scale as a Measure of Stress: Decomposing Score Variance in Longitudinal Behavioral Medicine Studies. *Ann Behav Med., 57*(10), 846–854. https://doi.org/10.1093/abm/kaad015. PMID: 37084792; PMCID: PMC10498818
4. Anu Priya, , Shruti Garg, and Neha Prerna Tigga. "Predicting Anxiety, Depression and Stress in Modern Life using Machine Learning Algorithms".Procedia Computer Science 167 (2020): 1258-1267.
5. Li, R., & Liu, Z. (2020). Stress detection using deep neural networks. *BMC Med Inform Decis Mak, 20*(Suppl 11), 285. https://doi.org/10.1186/s12911-020-01299-4
6. Kene, Akshada, and Shubhada Thakare. "Mental stress level prediction and classification based on machine learning." In 2021 Smart Technologies, Communication and Robotics (STCR), pp. 1-7. IEEE, 2021.
7. Lu, S., Zhao, L., Lai, L., Shi, C., & Jiang, W. (2022). How do Chinese people view cyber-bullying? A text analysis based on social media. *Int. J. Environ. Res. Public Health. 19*(3), 1822.
8. Xue, Z., Li, Q., & Zeng, X. (2023). Social media user behavior analysis applied to the fashion and apparel industry in the big data era. *J. Retail. Consum. Serv., 72*, Article 103299.
9. Rosid, M. A., Fitrani, A. S., Astutik, I. R. I., Mulloh, N. I., & Gozali, H. A. (2020). Improving text preprocessing for student complaint document classification using sastrawi. *IOP Conference Series: Materials Science and Engineering, 874*(1), 12017.
10. Hidayatullah, A. F., & Ma'Arif, M. R. (2017). Pre-processing tasks in Indonesian Twitter messages. *Journal of Physics: Conference Series, 801*(1), 12072.
11. M. Wongkar and A. Angdresey, "Sentiment analysis using Naïve Bayes Algorithm of the data crawler: Twitter", 2019 Fourth International Conference on Informatics and Computing (ICIC), pp. 1-5, 2019.
12. Li, Z., Li, R., & Jin, G. (2020). Sentiment analysis of danmaku videos based on naïve bayes and sentiment dictionary. *IEEE Access, 8*, 75073–75084.
13. Villavicencio, C., Macrohon, J. J., Inbaraj, X. A., Jeng, J.-H., & Hsieh, J.-G. (2021). Twitter sentiment analysis towards covid-19 vaccines in the Philippines using naive bayes. *Information, 12*(5), 204.
14. S. Yadav, M. Timbadia, A. Yadav, R. Vishwakarma and N. Yadav, Crime Pattern Detection Analysis & Prediction, pp. 225-230, 2017.
15. Kurniawan, S., & Budi, I. (2020). "Indonesian tweets hate speech target classification using machine learning", 2020 5th Int. *Conf. Informatics Comput. ICIC, 2020*, 1–5.
16. A. R. Lubis, M. K. M. Nasution, O. S. Sitompul and E. M. Zamzami, "Spelling Checking with Deep Learning Model in Analysis of Tweet Data for Word Classification Process", 2022 9th International Conference on Electrical Engineering Computer Science and Informatics (EECSI), pp. 343-348, 2022.
17. A. R. Lubis, M. K. M. Nasution, O. S. Sitompul and E. M. Zamzami, "A Framework of Utilizing Big Data of Social Media to Find Out the Habits of Users Using Keyword", Proceedings of the 8th International Conference on Computer and Communications Management, pp. 140-144, 2020.
18. A. R. Lubis, S. Prayudani, M. Lubis and Al-Khowarizmi, "Analysis of the Markov Chain Approach to Detect Blood Sugar Level", J. Phys. Conf. Ser., vol. 1361, no. 1, 2019.
19. S. A. El Rahman, F. A. AlOtaibi and W. A. AlShehri, "Sentiment analysis of twitter data", 2019 international conference on computer and information sciences (ICCIS), pp. 1-4, 2019.

Hybrid Transfer Learning-Based Pomegranate Fruit Disease Classification

Priyanka Chakraborty, Bidyutmala Saha, Sayan Nath, and Sumit Kumar Banerjee

Abstract It is critical to identify and then classify diseases in pomegranate fruits to ensure a high-quality yield and reduce agricultural losses. Regretfully, large-scale monitoring is difficult since traditional methods of identifying illnesses are time-consuming and often require specialist knowledge. Advanced methods for automated disease classification have been made possible in recent years by Machine Learning (ML) and Deep Learning (DL). Transfer learning (TL), one of these approaches, has proven to be successful in addressing particular agricultural issues with sparse data by utilizing pre-trained models on big datasets. Here a method for classifying pomegranate fruit problems called Hybrid Transfer Learning (HTL) is proposed. This approach integrates domain-specific feature extraction with the benefits of TL and fine-tuning approaches. Images of both healthy and sick pomegranate fruit are fed into pre-trained Convolutional Neural Networks (CNNs), such as Res-Net, Inception, SVM, K-nearest neighbor to extract generalized visual information. Following this, a process of fine-tuning combines these features with a specially designed disease categorization module that is tailored to the distinct qualities of pomegranate disorders. Affected crops have caused the agricultural field to suffer and also affect the economy. However, the reason behind the infected crops may be crop diseases. To resolve this issue, pomegranate fruit disease classification based on hybrid transfer learning is chosen. Different machine learning algorithms such as random forest classifier and K-nearest neighbor algorithm with custom deep learning algorithms are used. Our run-time training accuracy is 97.5%.

Keywords Pomegranate fruit disease · Classification · Hybrid transfer learning · Machine learning · Deep learning

P. Chakraborty (✉) · B. Saha · S. Nath · S. K. Banerjee
Guru Nanak Institute of Technology, Sodepur, WB, India
e-mail: priyanka.chakraborty@gnit.ac.in

J. K. Mandal et al. (eds.), *Recent Advances in Artificial Intelligence and Smart Applications*, Lecture Notes in Networks and Systems 1380,
https://doi.org/10.1007/978-981-96-5822-0_29

1 Introduction

The agriculture industry is essential to maintaining global economies and guaranteeing food security. Precision agriculture, which emphasizes the application of cutting-edge technologies to enhance crop quality, productivity, and efficiency, has drawn a lot of attention in recent years. Early and precise crop disease identification is a crucial component of precision agriculture since it can have a big impact on quality and output. Pomegranate stands out among the different fruit crops because of its economic and nutritional worth. Pomegranate plants are prone to a number of diseases, though, which lowers production and causes financial losses. Therefore, it is essential to diagnose diseases in pomegranate fruits accurately and promptly in order to reduce damage and stop widespread epidemics.

Pomegranate fruit has generally five types of disease found that affects the fruit quality. They are healthy, bacterial blight disease, Anthracnose disease, Cercospora spot disease, and Alternaria fruit spot disease [1]. We are describing these diseases as follows:

Healthy Pomegranate Fruit:

A healthy pomegranate fruit is essentially very large and tender. Pomegranates that are in good health should have rich, deep colors, ranging from dark crimson to reddish-purple. However, the healthy pomegranate fruit's shade may vary depending on the situation [2]. The pomegranate fruit's skin is naturally attractive in terms of freshness and moisture. A healthy pomegranate fruit would also have smooth skin, be medium to large in size, and feel heavy when handled. Figure 1 displays the nutritious pomegranate fruits.

Fig. 1 Healthy pomegranate fruits [17]

Bacterial Blight Pomegranate Fruit Disease:

Pomegranate fruits and trees are susceptible to bacterial blight, a frequent pomegranate fruit disease. It results in sunken, black lesions with dark or red-brown borders that are either round or irregular in shape. Cracks in the lesions are present in the impacted fruits [3]. The fruit is continuously contaminated by the existing fissures, which have the potential to deepen. Premature fruits may be dropped before they reach maturity due to bacterial blight. The bacterial blight illness that affected pomegranate fruits is depicted in Fig. 2.

Anthracnose Pomegranate Fruit Disease:

The symptoms of Anthracnose pomegranate fruit disease are usually tiny, dark-brown, continuous patches on the leaves and fruits. Additionally, the affected fruits will eventually turn yellow. In essence, pomegranate fruit disease is caused by a fungus. However, depending on the fruit lesions, the resulting fruit of Anthracnose disease can be either oval or round in shape. Lesions are seen to start out as minor appearances. Nevertheless, the lesions also enlarge over time [4]. Lesions have a light-colored border and a dark core. In other cases, the spore masses inside the lesions may appear pink or orange. The expansion of masses is one of the fundamental characteristics of this illness. The end result (Fig. 3) demonstrates how pomegranate fruits have been affected by anthracnose.

Cercospora Spot Pomegranate Fruit Disease:

Pomegranate trees that have Cercospora spot, a fungus-based disease, prematurely drop before they reach maturity. The primary cause of Cercospora is a pathogen, namely, a kind of fungus. According to experiments, if left untreated, the Cercospora disease can result in significant losses. Cercospora illness is characterized by small, irregular round patches. Both brown and dark brown are the hue of the dots. Large

Fig. 2 Bacterial blight disease-affected pomegranate fruits [17]

Fig. 3 Anthracnose disease-affected pomegranate fruits [17]

and uneven spots are created when the spots combine after a specific amount of time [5]. Damage may ensue from the impacted areas as well. The lesion colors may turn gray or black with dark-brown borders once the spots have grown. Usually, the fruit surface that is impacted will either break close to the lesions or crack them. The impacted fruit eventually falls from the tree in terms of maturity. As seen in Fig. 4, pomegranate fruits were impacted by the Cercospora spot disease.

Fig. 4 Cersopora spot disease-affected pomegranate fruits [17]

Fig. 5 Alternaria spot disease-affected pomegranate fruits [17]

Alternaria Spot Pomegranate Fruit Disease:

Alternaria spot disease attacks pomegranate fruits and is caused by Alternaria alternata. It is typically observed when warm, humid weather combines with other factors. Small, round, black or brown dots on the fruit are the first signs of the disease. A darker border around the spots is possible [6, 7]. It may develop into large, integrated lesions after a while. The sunken, soft area is the damaged area. Mature target lesions are typically dark brown or dark black in color. Pomegranate fruits impacted with Alternaria spot are depicted in Fig. 5.

2 Proposed Methodology

Our proposed methodology is divided into the following phases like collecting datasets, feature engineering, model training, validation process, testing phase on an image, and classify the pomegranate fruit disease [8] (Fig. 6).

2.1 Collecting Dataset

We collect public image dataset from different online sources. The dataset contains 1450 healthy images, 966 bacterial blight images, 1166 Anthracnose images, 631 Cercospora images, and 886 Alternaria images of pomegranate fruit diseases.

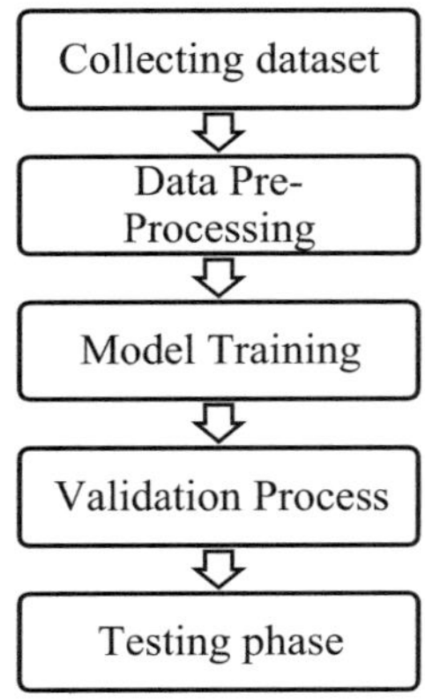

Fig. 6 Proposed system phases

2.2 *Data Pre-processing*

In pre-processing phase, the collected dataset contains five folders such as healthy, bacterial blight, Anthracnose, Cercospora, and Alternaria. Then, we pre-processed image dataset into label-wise with each target classes.

2.3 *Model Training*

In the training phase, feature extraction was carried out using a Convolutional Neural Network (CNN). The CNN effectively captures and learns relevant spatial and hierarchical features from the collected image dataset, leveraging its ability to automatically identify patterns such as edges, textures, and complex structures in the data. These extracted features serve as a rich representation of the input images, enabling the model to differentiate between important patterns and noise. Subsequently, these features were integrated into the proposed deep learning model for further processing. The deep learning framework utilizes these representations to perform tasks such as classification, segmentation, or prediction. By incorporating the features extracted by the CNN, the proposed model gains the capability to understand and interpret the visual data at a deeper level, thereby enhancing its performance and accuracy in addressing the specific objectives of the study. This approach ensures that the model is both robust and efficient, as the CNN-derived features are specifically tailored to the dataset, enabling the proposed system to achieve optimal results in its application (Fig. 7).

Convolution layer:

Let us assume that an input image I with the H (height) * W (width) dimension and a filter K with the dimension of $F_H * F_W$, then the convolution is computed as

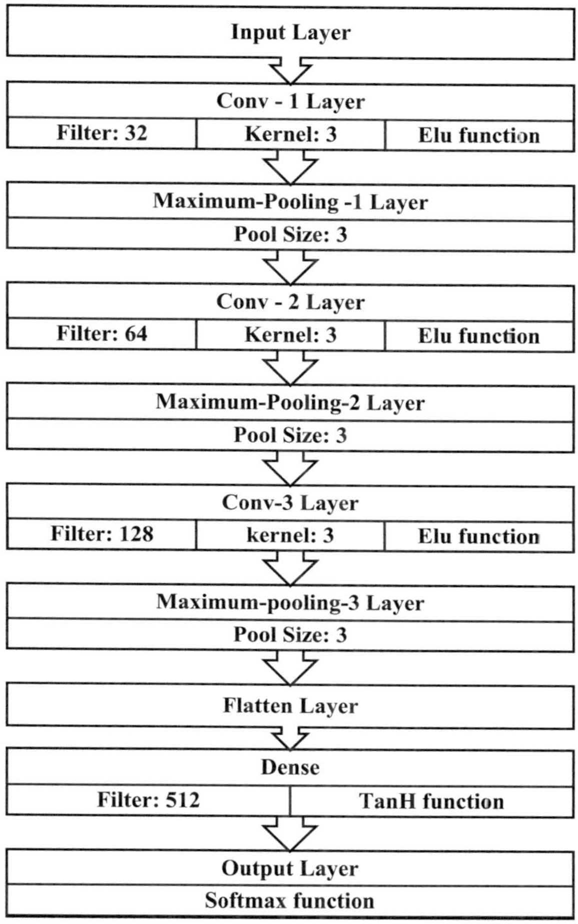

Fig. 7 Proposed deep learning architecture

$$A(i,j) = \sum_{m}^{F_H-1} \sum_{n}^{F_W-1} X(i+m, j+n) * K(m,n) \tag{1}$$

where A (I, j) is the position at (i, j), X (i + m, j + n) is the input value at position at X (i + M, j + n), and K (m, n) is the filter value of (m, n) position.

And also, we compute the output dimensions using the following formulas:

$$O_H = \left\lfloor \frac{H + 2 * P - F_H}{S} \right\rfloor + \beta \tag{2}$$

$$O_W = \left\lfloor \frac{W + 2 * P - F_W}{S} \right\rfloor + \beta \tag{3}$$

where H = height, W = width, F_H = filter height, F_W = filter width, P = padding, S = strides, and β = bias and it sets to 1.

Elu Activation Function:

Exponential linear unit activation function is used to faster convergence on training time.

$$elu(x) = \begin{cases} x, & if\ x > 0 \\ \alpha(e^x - 1), & if\ x \leq 0 \end{cases} \quad (4)$$

Max-pool layer:

To reduce the dimensionality of CNN layer, we use the Max-pool layer which is computed as

$$A(i, j) = maximum(I(i + m, j + n)); \forall m, n \in poolsize \quad (5)$$

where pool size = the size of the window.

Concatenate layer:

This axis in a convolution neural network (CNN) merges the outputs of three Max-pool layers, reducing spatial dimensions while retaining prominent features. When set to 1, this process combines features from different branches, capturing diverse aspects of input data. Merging on Axis 1 offers diverse feature integration, flexibility, and scalability, allowing for further processing.

Flatten layer:

This layer is crucial in neural networks for converting three-dimensional feature vectors into one-dimensional vectors while maintaining the original batch size. The output from a convolution or pooling layer is typically a multi-dimensional tensor with dimensions like height, width, and depth. To transition to dense layers, a flattening operation is applied, reshaping the tensor into a single-dimensional vector for each sample in the batch. This transformation connects convolutional layers for spatial feature extraction with dense layers for high-level reasoning and predicting [11, 12].

Dense layer:

The following computation is used in this layer to calculate output:

$$Out = TanH(Inputs * Weights + Bias) \quad (6)$$

Output layer:

In the final training phase, a dense layer with a Sigmoid activation function is used for binary classification tasks. This layer connects neurons, producing a single output value. The output is mapped to a probability score, enabling binary predictions.

Softmax function:

The Softmax function is a mathematical activation function in neural networks used for multi-class classification. It converts raw output logits into probabilities, ensuring the sum of probabilities equals 1. It is often paired with categorical cross-entropy loss function for improved accuracy. The Softmax function is defined as

$$Softmax(x_y) = \frac{e^{x_y}}{\sum_j e^{x_j}} \tag{7}$$

RMSprop Optimizer:

Neural network training uses the adaptive learning rate optimization approach known as RMSprop (Root Mean Square Propagation). By modifying the learning rate according to recent gradients, it alters traditional stochastic gradient descent (SGD). To stabilize updates and prevent sharply declining learning rates, RMSprop divides the learning rate by a running average of squared gradients. In order to avoid division by zero, it employs a decay factor and an exponentially declining average of squared gradients.

We use RMS-property formula to improve the precision of the deep learning model.

- Average Square Gradient Calculation

$$E\left[v^2\right]_{times} = (E\left[v^2\right]_{times-1} * \gamma) - (1 - \gamma) * v_{times}^2 \tag{8}$$

- Parameter updating equation is

$$\theta_{times+1} = \theta_{times} - \frac{\alpha * v_{times}}{\sqrt{E\left[v^2\right]_{times}}} \tag{9}$$

Loss Calculation:

Log-Cosh loss [16] is a smooth, differentiable loss function used for regression tasks, calculating the logarithm of the hyperbolic cosine of the prediction error. It is less sensitive to large outliers and more robust in the presence of outliers, making it popular for models requiring stability and robustness. We formulate the Log-Cos-Hyperbolic loss that follows to decrease the loss:

$$LogCosH = \sum_{i} \log\left(T_x^{predict\_value} - T_x^{true\_value}\right) \tag{10}$$

$$where\ CosH(t) = \frac{e^t + e^{-t}}{2} \tag{11}$$

$$And\ x = \begin{cases} if\ small\ x\ then,\ x = \frac{x^2}{2} \\ if\ large\ x\ then,\ x = |x| - log(2) \end{cases} \tag{12}$$

Our proposed model uses 50 epochs for training, and gets 97.5% accuracy. Finally, we save our weights and structure locally. Next, we use the deep learning feature prediction as input to apply the transfer learning [13, 14].

SVM (Support Vector Machine):

Support vector machine methodology applied as first algorithm to classify the disease with notable precision. The overall formula for classification can be defined as

$$\min_{\omega,b,\zeta} \frac{1}{2} W^T W + C \sum_{i=1}^{n} \zeta_i \tag{13}$$

$$yi\left(W^T \phi(x_i) + b\right) \geq 1 - \zeta_i$$

where

- ζ_i denotes the distance to the correct margin with $\zeta_i \geq 0$, i = 1,…..,n.
- C denotes a regularization parameter.
- $W^T W = ||W||$ denotes the normal vector.
- $\phi(x_i)$ denotes the transformed input space vector.
- B denotes a bias parameter.
- y_i denotes the i-th target value.

K-Nearest Neighbor Classifier:

In essence, it is used to categorize disorders of pomegranate fruit. This algorithm's fundamental mathematical formula is

$$T(x) = arg - max_Z\left(\sum_{i \in N(x)} X(t_i = z)\right) \tag{14}$$

where

- T(x): classified disease for the input x.
- N (x): set of K-nearest neighbors of x.
- X (t_i = z): equal to 1 if t_i = z, and 0 otherwise.

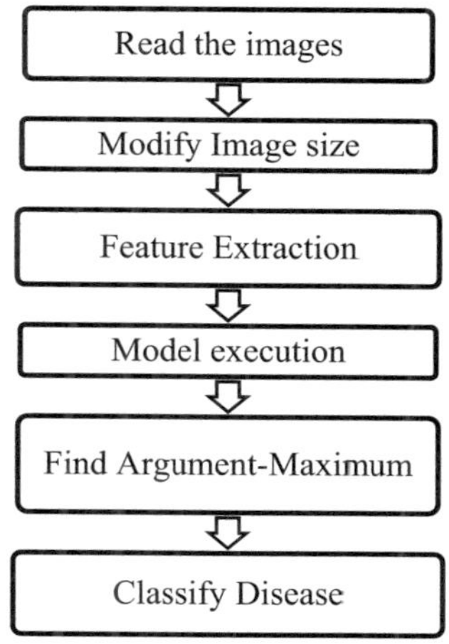

Fig. 8 Test image phases

2.4 *Validation Process*

After training phase, validation of the model is done. Model gave 97.8% validation accuracy.

2.5 *Testing Phase*

Testing image phase contains the real-time prediction of the test image dataset (Fig. 8).

Using the bi-linear interpolation, first, reading the images is done and the batch size is set to 16. Then reduce the image's size to 256 × 256 pixels after viewing it. Once the image has been modified, restructuring is done and retrieve its distinctive characteristics as a float data-type. In this stage, model and its weights are loaded in order to categorize the fruit illness [15]. After that model prediction vector array is used to determine the first argument-max in order to identify the disease. The disease is next categorized according to the relevant index, which is followed by the class name.

3 Result Analysis

See Fig. 9 and Table 1.

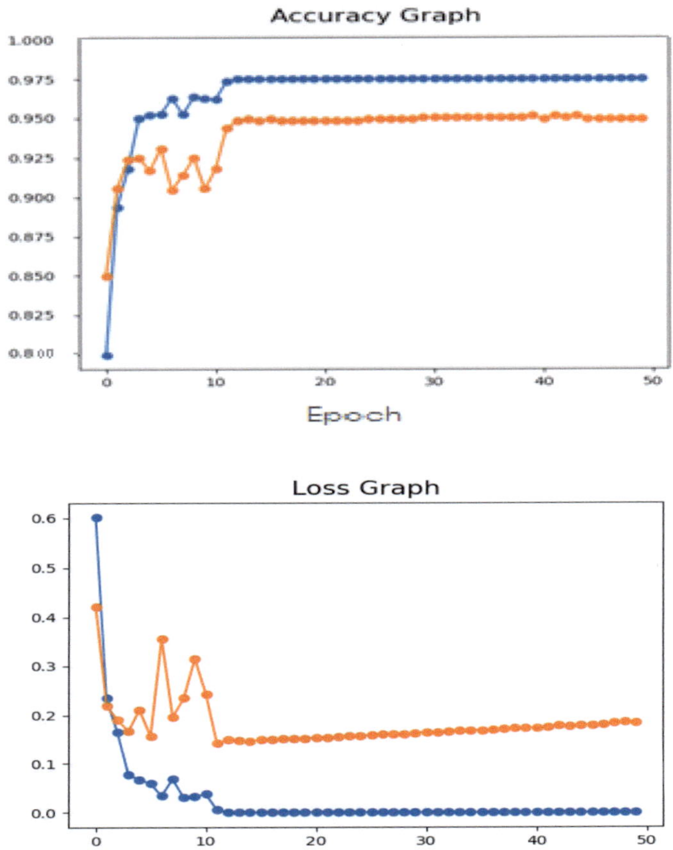

Fig. 9 Training-validation accuracy-loss curve

Table 1 Comparative analysis of classification performance with prior studies

Model	Accuracy	Specificity	Precision	F1-score
Res-Net-50 [13]	0.86	0.92	0.88	0.85
Dense-Net-201 [12]	0.90	0.94	0.86	0.88
Res-Net-101 [14]	0.92	0.95	0.91	0.93
Proposed model	0.975	0.96	0.93	0.95

4 Conclusion

An important development in agricultural technology is the application of machine learning to the detection and categorization of fruit diseases. By increasing the precision and speed of disease detection, these methods assist farmers in taking preventative action to safeguard their crops, thereby promoting sustainable farming methods and food security. The capacity to resist plant diseases will be further strengthened by ongoing research and development in this area as well as real-world applications, guaranteeing healthier crops and higher yields. To guarantee healthy crop production, however, fruit disease detection is a crucial field of study in agriculture. Significant progress has been made in the precise and early identification and classification of leaf diseases by utilizing machine learning approaches.

References

1. Arun Kumar, R., Rajpurohit, V. S., & Gaikwad, N. N. (2021). Image dataset of pomegranate fruits (Punicagranatum) for various machine vision applications. *Data in Brief, 37*, Article 107249. https://doi.org/10.1016/j.dib.2021.107249
2. Akhilesh, S. D. M., & Kumar, S. A. (2019). Image-based plant disease detection in pomegranate plant for bacterial blight. In *2019 International Conference on Communication and Signal Processing (ICCSP)* (pp. 0645–0649). IEEE, Chennai. https://doi.org/10.1109/ICCSP.2019.8698007
3. Jayalakshmi, K., Nargund, V., Raju, J., Benagi, V., Ram, R., Giri, M., Basamma, Rajani, R., Sonavane, P., & Reena, R. (2015). Pomegranate anthracnose caused by Colletotrichum-gloeosporioides: a menace in quality fruit production. *Journal of Pure and Applied Microbiology, 9*, 3093–3097
4. Ravichandra, G. K., Somasekhara, Y. M., Shilpa, M., Mahesh, M., & Shalini, M. (2023). Pomegranate Cercospora leaf spot as an emerging problem, tackle through biological, botanical and fungicidal approach, both in-vitro and field condition. *Journal of Pure and Applied Microbiology, 15*, 553–563.
5. Aloi, F., Riolo, M., Sanzani, S., Mincuzzi, A., Ippolito, A., Siciliano, I., Pane, A., Gullino, M., & Cacciola, S. O. (2021). Characterization of Alternaria species associated with heart rot of pomegranate fruit. *Journal of Fungi, 7*(172). https://doi.org/10.3390/jof7030172
6. Ahmed, S. I., Ibrahim, M., Nadim, M., Rahman, M. M., Shejunti, M. M., Jabid, T., & Ali, M. S. (2023). MangoLeafBD: A comprehensive image dataset to classify diseased and healthy mango leaves. *Data in Brief, 47*, Article 108941. https://doi.org/10.1016/j.dib.2023.108941
7. Khatawkar, S., Jadhav, S., Sapate, S., Patil, P., & Shinde, A. (2023). Disease detection on pomegranate fruits using machine learning approach. In *AIP Conference Proceedings*, vol, 2717(1), Article 020004. https://doi.org/10.1063/5.0130455
8. Madhavan, M. V., Thanh, D. N. H., Khamparia, A., Pande, S., & Malik, R. (2021). Recognition and classification of pomegranate leaves diseases by image processing and machine learning techniques. *Computers, Materials & Continua, 66*(3), 2939–2955.
9. Wakhare, P. B., & Neduncheliyan, S. (2023). Using image processing and deep learning techniques detect and identify pomegranate leaf diseases. *Indian Journal of Science and Technology, 16*(18), 1323–1331.
10. Wakhare, P. B., et al. (2023). Development of automated leaf disease detection in pomegranate using Alexnet algorithm. *Current Agriculture Research Journal, 11*(1)

11. Vasumathi, M. T., & Kamarasan, M. (2021). An effective pomegranate fruit classification based on CNN-LSTM deep learning models. *Indian Journal of Science and Technology, 14*(16), 1310–1319.
12. Nirgude, V., Rathi, S. (2021). A robust deep learning approach to enhance the accuracy of pomegranate fruit disease detection under real field condition. *Journal of Experimental Biology and Agricultural Sciences, 9*(6), 863–870. https://doi.org/10.18006/2021.9(6).863.870
13. Javeriya, S. (2021). Faster-RCNN based deep learning model for pomegranate diseases detection and classification. *LC International Journal of STEM, 2*(3), 114–120. https://doi.org/10.5281/zenodo.5759557
14. Jadhav, S. B., Udupi, V. R., & Patil, S. B. (2021). Identification of plant diseases using convolutional neural network. *International Journal of Information Tecnology, 13*, 2461–2470. https://doi.org/10.1007/s41870-020-00437-5
15. Guo, T., Wang, R., Luo, F., Gong, X., Zhang, L., & Gao, X. (2023). Dual-view spectral and global spatial feature fusion network for hyperspectral image classification. *IEEE Transactions on Geoscience and Remote Sensing, 61*, Article 5512913
16. Rehman, A., & Zafar, **S.** (2018). Log-cosh loss function for deep neural networks: A robust alternative to mean squared error. In *Proceedings of the International Conference on Machine Learning and Applications*
17. Pakruddin, B., & Hemavathy, R. (2023). *Pomegranate fruit diseases dataset for deep learning models*. Mendeley Data, V1.https://doi.org/10.17632/b6s2rkpmvh.1

Enhanced Prediction of Food Wastage Using Ensemble Learning and Feature Selection

Venkata Kamya Punugupati, Bhargavi Maridu, Vijayalakshmi Maddiboyina, Akshay Kanuri, and Ruthvik Peddineni

Abstract This paper presents an ensemble learning-based framework for forecasting food wastage into this scenario, managing variability in data, dependency on certain environmental conditions, and socio-economic differences. For carrying out feature selection technique RFE is applied using a wide dataset that considered environmental conditions, socio-economic indicators, and historical records on such phenomena. For evaluation purposes, various ensemble models are selected: random forest, gradient boosting, LightGBM, and so forth and evaluated using MSE, RMSE, R^2, and MAE as metrics. The technique gave very accurate predictions and withstood dropouts, and the feature's importance analysis provided valuable information to the stakeholders/policymakers on how to design better sustainable food distribution strategies.

Keywords Food wastage prediction · Machine learning · Bagging regressor · Random forest · Gradient boosting · LightGBM · Extra trees · Stacking regressor

1 Introduction

Probably one of the major concerns across the world today is food wastage, which traverses to the environment, the economy, and society at large. Food and Agriculture Organization of the United Nations estimated that one-third of all food produced globally for human consumption was lost or wasted each year. This not only deepens food poverty but also contributes significantly to environmental degradation through greenhouse gas emissions and the unnecessary use of resources.

Researchers have now increasingly used sophisticated models of machine learning in predicting and minimizing food waste across all possible situations. Projections were able to offer stakeholders perception of the patterns and trends associated with food waste, thereby improving strategies on minimizing wastes. This paper

V. K. Punugupati (✉) · B. Maridu · V. Maddiboyina · A. Kanuri · R. Peddineni
Department of CSE, Vignan's Foundation for Science, Technology and Research, Guntur, India
e-mail: kamya.punugupati@gmail.com

J. K. Mandal et al. (eds.), *Recent Advances in Artificial Intelligence and Smart Applications*, Lecture Notes in Networks and Systems 1380,
https://doi.org/10.1007/978-981-96-5822-0_30

presents ensemble learning techniques, including random forest, gradient boosting, LightGBM, extra trees, and bagging regressor, using Recursive Feature Elimination (RFE) in feature selection.

In recent years, it has become apparent that food wastage is one of the most pressing global issues as it affects the availability of food, the state of the environment and the economy. Along these lines, the Food and Agriculture Organization (FAO) of the United Nations has estimated that approximately one-third of food that is properly grown and suitable for human consumption gets lost or goes to waste every year, causing unnecessary expenditure on resources such as water, land and energy, and increasing greenhouse gas emissions as well. Therefore, it is of utmost importance to contain it if sustainable projects such as food systems, fighting hunger, and combating environmental degradation are to succeed.

In most cases, conventional approaches to dealing with food waste tend to be forward looking and project annual or monthly food waste management based on estimation of historical data and averages. Particularly, ensemble modeling, among others, is proven to satisfactorily explain the relationship between food waste and other variables such as climate, societal economic status, and even the logistics of handling the product.

2 Literature Survey

Chowdary et al. presented a prediction model of food wastage using XGBoost feature extraction using PCA and RFE [1] with an accuracy level of 73% highly dependent on data quality and availability.

Harshini et al. [2] have worked with restaurant dataset demand forecasting, employing cross-validation with K-fold and a mix of models including XGBoost, random forest, SVM, and ANN.

Ifham et al. [3] have used decision tree and random forest for food waste classification, where up to 86.11% accuracy is achieved but limited to the quality of data and feature selection.

Shopnil et al. [4] applied label encoding and multiple models: Random forest and SVM on food wastage prediction in a student hostel, which presented with an accuracy of 75% but is suffering from demand uncertainty and lacks resources.

Tutul et al. [5] predicted the freshness of food by using IoT sensor data with feature extraction through PCA and t-SNE, which was only limited to data availability and environmental factors.

Hajjdiab et al. [6] reported the development of a mobile application in reduction of food wastage with the help of models like logistic regression and CNNs while integrating imaging and textural data. However, it has some issues regarding the quality of data and its scalability.

Patil et al. [7] implemented blockchain-based food waste tracking systems using features such as food quality and supply chain information.

Shaikh et al. [8] used Inception v3 for the analysis of waste images, with 83.30% accuracy and recommended increasing training data and real-time optimization.

Raj et al. [9] proposed AI-powered smart waste bins that use IoT sensor data through LSTM and fuzzy logic controllers and achieve more than 97% accuracy but require tuning due to environmental variability.

Manjunath et al. [10] applied IoT data to predict food wastage by using logistic regression, random forest, and CNNs, ultimately achieving a 99.53% accuracy for the freshness of food but was limited by scalability and privacy issues.

3 Methodology

This method starts with loading a dataset, where the data is separated into feature-independent variables and a target-dependent variable. This dataset features 1,782 records with 11 attributes defining food wastage in an event setup. The parameters included in this dataset were the type of food served, the number of people expected at the event, type of event, how food is stored, and food amount that will go to waste during the event. These descriptive and predictive factors enable this study to make an inference regarding what causes food wastes as the guide toward rational and effective management and minimal reduction of food wastes within an event plan. Here, the algorithm identifies the categorical and numerical columns via select dtypes. With numerical, missing values were imputed as the mean using Standard-Scaler. For categorical values, missing value imputation will be done by the most frequent value and encoded using one-hot encoding. This process, taken care of by a ColumnTransformer, is implemented on both the training and the test sets. RFE with a random forest regressor is conducted on the model for its optimization. The transformed data is further used to train an ensemble model, such as random forest, gradient boosting, XGBoost, LightGBM, extra trees, and a stacking regressor. A comparison is done in MSE, MAE, R-squared (R^2), and RMSE and the results plotted for comparison.

3.1 Pre-processing

During pre-processing, the data is divided between numerical features and categorical features. We use the select dtypes function to do so. For numerical features, we created a pipeline that included the SimpleImputer to impute missing values with the mean and StandardScaler, scaling the values for standardization. For categorical features, missing values are managed by replacing them with the most frequent value in another SimpleImputer. After this step, categorical variables are converted into the binary indicators one-hot encoding by OneHotEncoder.

These pre-processing steps for both numeric and categorical data are combined by using Column Transformer. Then applying this pipeline to training data by the fit transform() method which fits the pre-processing steps to this training set, transforming it, respectively. The same transformation is applied to the test set by calling the transform() method. This is going to sync any data between both sets. Missing values, scaling, and categorical encoding are taken care of systematically before applying a model (Fig. 1).

Fig. 1 Flow diagram of proposed model

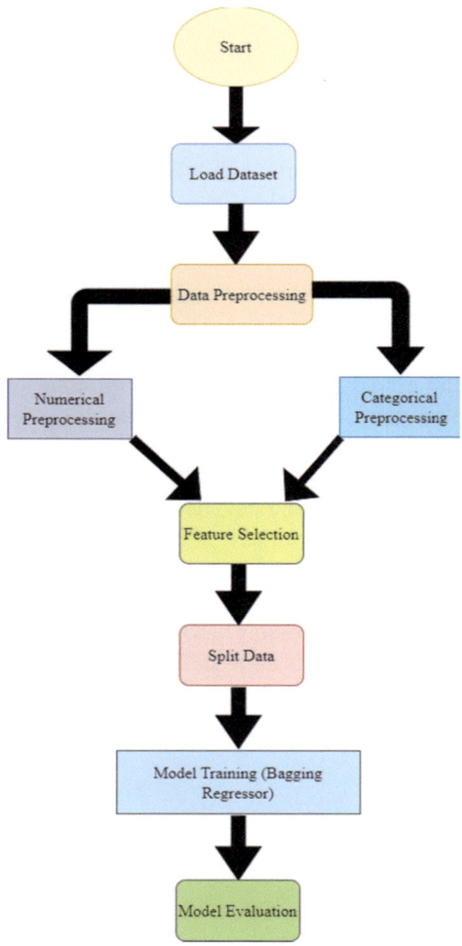

3.2 Exploratory Data Analysis

Correlation between numerical features is commonly visualized using heatmaps in exploratory data analysis. Figure 2 displays a heatmap created with Seaborn's sns.heatmap() function showing the correlations between number of guests, quantity of food, and wastage food amount. Using the YlGnBu color scheme, such a heatmap naturally focuses on strong positive relationships, including the one between number of guests and quantity of food (0.79). Annotated values add interpretability, through which relationships can be identified that might affect model performance and feature selection.

Figure 3 displays bar charts comparing the average food wastage per event type and geographical location. The first bar chart compares the food wastage by the types of events, including corporate, birthday, wedding, and social gatherings, indicating which types of events contribute the most to wastage. The second compares the level of wastage in the geographical regions covered: urban, suburban, and rural areas. The charts, created using Seaborn's "sns.barplot()" with "ci=None" to remove the confidence interval, include clear labels and titles that fill many needs for information about key wastage patterns.

3.3 Feature Selection

This methodology, besides improving model interpretability, enhances the performance of the models since all the features are being narrowed down to the most relevant ones, reducing overfitting. A simple, systematic feature selection that could

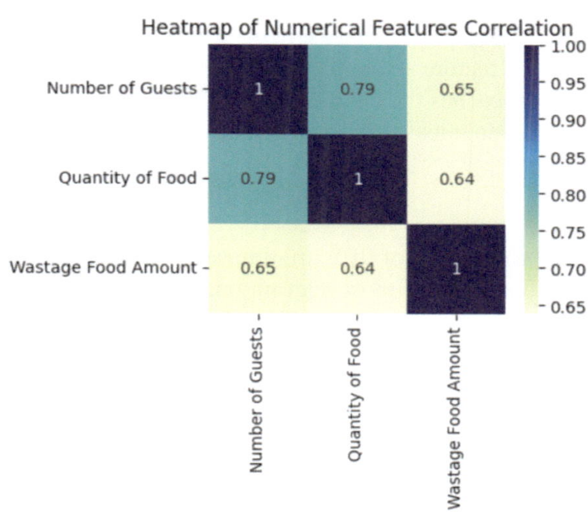

Fig. 2 Heatmap to show correlation between numerical features

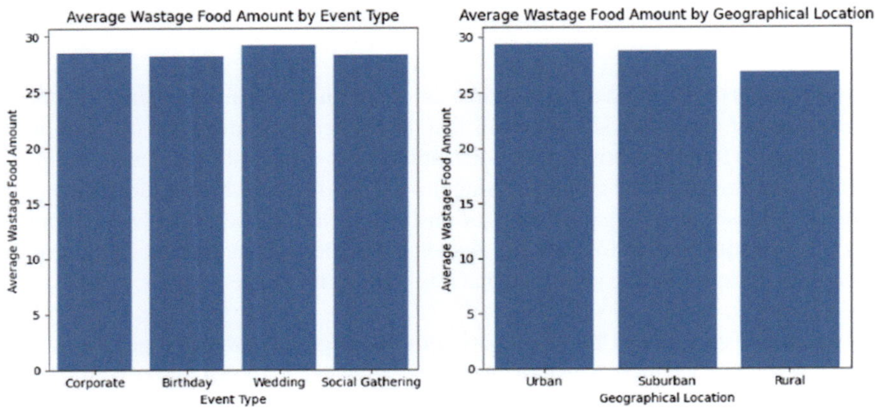

Fig. 3 Bar chart for average wastage by event type

be used for easy processing is Recursive Feature Elimination (RFE), which cuts out several features from the feature set on each iteration. Feature selection mathematically can be defined as

$$\text{minimize } L(\theta) \quad \text{subject to} \quad \|\theta\|_0 \geq m, \tag{1}$$

where L(θ) is objective function; θ represents weights or coefficients for the features, with m being the number of desired selected features. $\|\theta\|$ be the count of non-zero weights in θ and thus m denotes the selected features.

RFE is an algorithmic feature selection procedure where features are ranked in terms of importance, and the least relevant features, step by step, are removed. Starting with fitting a model on all features and ranking them based on the importance of their coefficients or some measure of feature importance, it will merely remove the least relevant feature and refit the model with the remaining ones. It repeats the procedure until it reaches the number of desired features.

Mathematically, this can be represented as

$$\text{minimize } L(\theta) \quad \text{subject to} \quad \|\theta\|_0 = m. \tag{2}$$

The core idea of RFE: minimization of the objective function L(θ). In the case of "m" features, the most relevant ones are selected. RFE successively removes those features whose removal induces the smallest increase in the objective function. The RFE algorithm, by its method of iteratively removing features, determines the subset of features from which the variability in the data can best be captured in regards to food wastage.

3.4 Regression Using Ensemble Models

In the food waste prediction process, a holistic approach was used in addressing the problems associated with food wastage during event scenarios. The following ensemble machine learning models were adopted: random forest, gradient boosting, LightGBM, XGBoost, extra trees, bagging regressor, and stacking. The emphasis here was on building models with good accuracy and reliability in the prediction of amounts of food waste.

After performing an intensive comparative analysis, the best predictive candidate for food waste belonged to the bagging regressor. Bagging is one form of ensemble method, which creates several copies of the same model on the entire given data. It takes the average of the predictions coming from all these models, thus reducing variance and avoiding overfitting.

The objective function for the bagging regressor is given by

$$J(\theta) = \frac{1}{n}\sum_{i=1}^{n} L(y_i, \hat{y}_i) + \lambda \sum_{k=1}^{K} \Omega(f_k),$$

where

- $J(\theta)$ is the overall loss function;
- y_i is the actual target value for instance i;
- $\hat{y}_i$ is the predicted value for instance i;
- $L(y_i, \hat{y}_i)$ is the loss function measuring the difference between the predicted and actual values (e.g., mean squared error);
- $\Omega(f_k)$ is the regularization term for the base learner k;
- λ is a regularization parameter that controls the strength of the regularization;
- n is the number of instances, K is the number of base learners.

4 Results and Discussion

4.1 Performance Evaluation Metrics

In evaluating the effectiveness of algorithms for predicting food waste, several key metrics are utilized to gage their performance. They include Mean Squared Error (MSE), Root Mean Squared Error (RMSE), R-squared (R^2) score, and Mean Absolute Error (MAE) (Fig. 4).

Mean Squared Error (MSE):

$$\text{MSE} = \frac{1}{n}\sum_{i=1}^{n}(y_i - \hat{y}_i)^2. \quad (3)$$

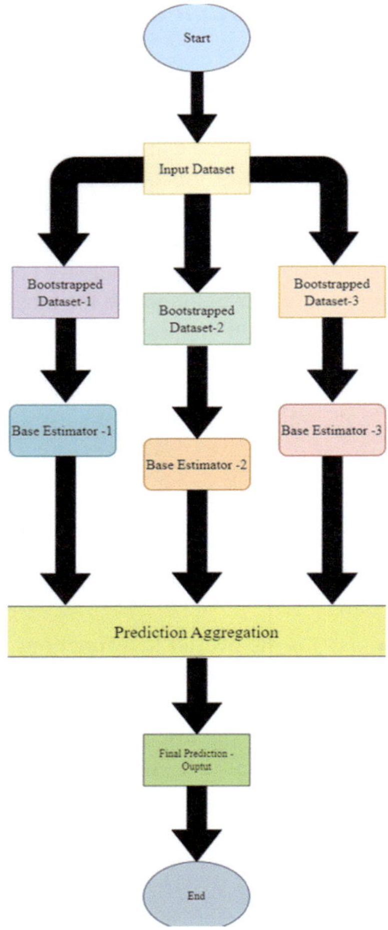

Fig. 4 Visualization of bagging regressor architecture

Root Mean Squared Error (RMSE):

$$\mathrm{RMSE} = \sqrt{\mathrm{MSE}}. \tag{4}$$

R-squared (R^2) score:

$$R^2 = 1 - \frac{\sum_{i=1}^{n} (y_i - \hat{y}_i)^2}{\sum_{i=1}^{n} (y_i - \bar{y})^2}. \tag{5}$$

Mean Absolute Error (MAE):

$$\text{MAE} = \frac{1}{n}\sum_{i=1}^{n}|y_i - \hat{y}_i|. \quad (6)$$

While evaluating the output of the predictive algorithms, food waste, several significant metrics are used to measure their performance. These include Mean Squared Error (MSE), Root Mean Squared Error (RMSE), R-squared (R^2) score, and Mean Absolute Error (MAE).

From Table 1, it can be seen that bagging regressor performed well with good results in all evaluation metrics. It attained the lowest MSE and RMSE values, which means its predictions are closer to the actual food waste values. The MAE value is also competitive, suggesting that the general accuracy of model predictions is good. Among these, bagging regressor's R^2 value is the highest, which means it best captures the relationship between input features and food wastage, so this is one of the most reliable models within this analysis (Table 2).

Table 1 Performance metrics of various models

Model	MSE	MAE	R^2	RMSE
Random forest	7.147123	1.542729	0.931051	2.67341
Gradient boosting	8.904335	2.230108	0.914099	2.984013
LightGBM	7.482658	1.785939	0.927814	2.735445
XGBoost	7.18599	1.521036	0.930676	2.68067
Extra trees	7.232978	1.526639	0.930223	2.68942
Bagging regressor	7.139497	1.543352	0.931124	2.671984
Stacking regressor	7.441588	1.559182	0.92821	2.727928

Table 2 Performance comparison

Algorithm	MSE	RMSE	R-Sq	MAE
XG Boost [1]	31.05	5.57	0.73	4.13
Proposed model (bagging regressor)	**7.13**	**2.67**	**0.93**	**1.54**

4.2 Comparison of Algorithms

As far as the comparison with the performance of the bagging regressor to XGBoost, the results of the bagging regressor highly outperformed at all times for each of the metrics. Bagging regressor achieved an MSE of 7.13 against 31.05 in XGBoost while attaining an RMSE of 2.67 as opposed to 5.57 by XGBoost. It is very evident that there are significantly less errors during prediction for the bagging regressor than in XGBoost. It also attains an R-squared value of 0.93, meaning the model is even better for variance explanation in the data, while MAE being 1.54 actually further supports it in having a better precision than the counterpart, XGBoost, when it comes to its MAE. Overall, the bagging regressor is shown to be more effective when it comes to the used dataset.

As in the plots presented in Figs. 5 and 6, it was found that the best performing model was bagging regressor. This model performed better by giving less MAE, RMSE, and MSE than other models during run, which means that, on average, it contains the smallest error of predictions. Specifically, it performed superbly concerning Mean Absolute Error (MAE) and Root Mean Squared Error (RMSE). Therefore, the model of bagging regressor outperformed all the other models such as random forest, gradient boosting, and stacking.

Best models for R-squared (R^2) were developed for bagging regressor, XGBoost, and Extra Trees. This means that those three models have fitted most of the variance in the dataset concerning food waste. Gradient boosting is the worst of all, and though stacking was very good, it was not consistent enough as was the case for bagging regressor.

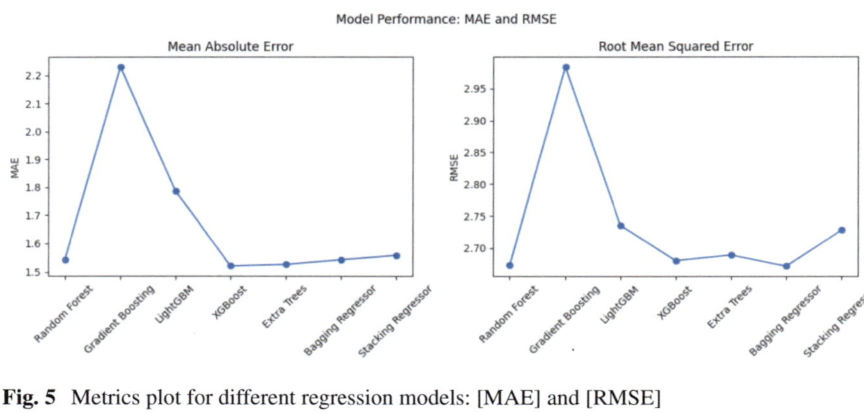

Fig. 5 Metrics plot for different regression models: [MAE] and [RMSE]

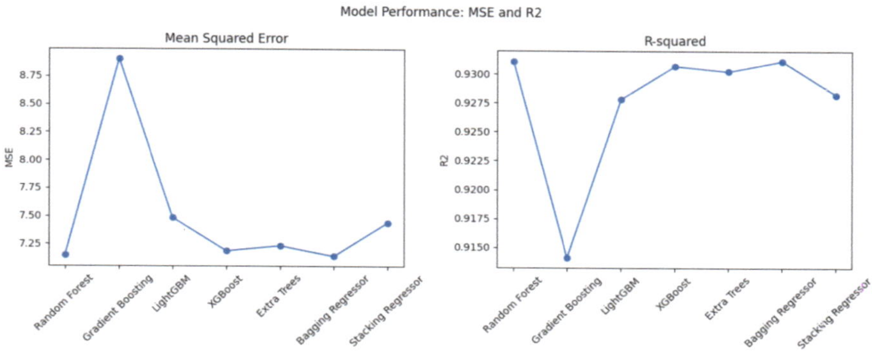

Fig. 6 Metrics plot for different regression models: [MSE] and [R2]

5 Conclusion

The findings of this study reveal that ensemble learning models can very well predict food waste. Of these, bagging regressor performed the best overall, having the lowest MSE at 7.13, MAE at 1.54, and a high R^2 of 0.931, thus beating the predictive performance to be observed with it. This model performed slightly better than the random forest and XGBoost models that performed well too by having similar R^2 scores and even slightly lower prediction errors.

The gradient boosting model, though very strong, had more exaggerated values for MSE at 8.90 and MAE at 2.23, thus indicating that this approach was unable to capture the underlying patterns of the data well. The stacking regressor also proved to be robust as it could not outperform top models and yielded an MSE of 7.44 and R^2 score of 0.928.

In conclusion, the best performing model is the bagging regressor, the best model for estimating food wastage in this setting, and one possible avenue for future work in trying to further optimize it with more hyperparameter tuning and deeper examination of different features of the data can be achieved by increasing the dataset size and including events that present a more diverse range of situations.

References

1. Chowdary, K. S., Praneetha, L. K., Holika, S., Priya, D. B., Venkatrama Phani Kumar, S., & Venkata Krishna Kishore, K. (2024). Prediction of food wastage using XG boost. In *2024 8th International Conference on Inventive Systems and Control (ICISC)* (pp. 307–312). Coimbatore, India. https://doi.org/10.1109/ICISC62624.2024.00059.
2. Harshini, K., Madhira, P. K., Chaitra, S., & Reddy, G. P. (2021). Enhanced demand forecasting system for food and raw materials using ensemble learning. In *2021 International Conference on Artificial Intelligence and Machine Vision (AIMV)* (pp. 1–6). Gandhinagar, India. https://doi.org/10.1109/AIMV53313.2021.9671005.

3. Ifham, N. M., Abdulla, M., Rifadh, M. R. M., Hussain Ajward, M. N. M., Swarnakantha, N. H. P. R. S., & Rajendran, K. (2023). Machine learning based approach to reduce food wastage. In *2023 5th International Conference on Advancements in Computing (ICAC)* (pp. 119–124). Colombo, Sri Lanka. https://doi.org/10.1109/ICAC60630.2023.10417516.
4. Masuk Shopnil, M. M., Husna, A., Sultana, S., & Islam, M. N. (2023). An ensemble ML model to predict the wastage of food: Towards achieving the food sustainability. In *2023 International Conference on Next-Generation Computing, IoT and Machine Learning (NCIM)* (pp. 1–6). Gazipur, Bangladesh. https://doi.org/10.1109/NCIM59001.2023.10212669.
5. Tutul, M. J. I., Alam, M., & Wadud, M. A. H. (2023). Smart food monitoring system based on IoT and machine learning. In *2023 International Conference on Next-Generation Computing, IoT and Machine Learning (NCIM)* (pp. 1–6). Gazipur, Bangladesh. https://doi.org/10.1109/NCIM59001.2023.10212608.
6. Yadav, S., Rajendaran, M., Nagpal, A., Karthik, A., Aravinda, K., & Geetha, B. T. (2024). Exploring the potential of machine learning in enhancing agricultural practices and food production. In *2024 IEEE 13th International Conference on Communication Systems and Network Technologies (CSNT)* (pp. 663–669). Jabalpur, India. https://doi.org/10.1109/CSNT60213.2024.10546234.
7. Mogilipalem, K., Kumar Poodari, A., Pulipati, Y., & Sangeetha, K. B. (2024). Food spoilage detection using IoT and machine learning. In *5th International Conference for Emerging Technology (INCET)* (pp. 1–4). Belgaum, India. https://doi.org/10.1109/INCET61516.2024.10593159
8. Aldossari, M. S., Almarshadi, T. A., Alwasel, R. A., Alqahtani, R. S., & Alluhaidan, A. S. (2022). Waste management system. In *2022 International Conference on Computational Science and Computational Intelligence (CSCI)* (pp. 1293–1298). Las Vegas, NV, USA. https://doi.org/10.1109/CSCI58124.2022.00232.
9. Pant, G., Rawat, P., Kathuria, S., Gehlot, A., & Rathor, N. (2023). Waste management system with technological intervention: Advantages and challenges. In *2023 4th International Conference on Electronics and Sustainable Communication Systems (ICESC)* (pp. 1461–1466). Coimbatore, India. https://doi.org/10.1109/ICESC57686.2023.10193334.
10. Asha, V., Nithya, B., Govindaraj, M., Brunda, T. M., Hiremath, A., & Jayaprakash, B. (2024). Ingenious waste management: Implementing machine learning in waste segregation. In *2024 International Conference on Emerging Smart Computing and Informatics (ESCI)* (pp. 1–5). Pune, India. https://doi.org/10.1109/ESCI59607.2024.10497282.
11. Bhargavi, M., Kumar, T. V., Shaik, R. A. A., Kanna, S. K., & Padmanabhan, S. (2022). Effective utilization and optimization of waste plastic oil with ethanol additive in diesel engine using full factorial design. *Materials Today: Proceedings, 52*, 930–936.

A Comprehensive Analysis of YCbCr Color Space-Based Single Image Super Resolution Using Very Deep Super Resolution Neural Network and Bilinear Interpolation

P. Ganesan, L. M. I. Leo Joseph, V. Elamaran, S. Jency, and G. Sajiv

Abstract The proposed work explains the single-image super-resolution (SISR) utilizing Bilinear Interpolation (BI) and Very Deep Super Resolution Neural Network (VDSRNN). In this work, the transformation of low to high-resolution image spaces using BI is compared with VDSRNN. The key idea of the work is to maintain computational efficiency and also attain a superior super-resolution process by employing the advantages of both BI and VDSRNN. In this work, the RGB image is transformed into YCbCr space before the super-resolution process. This YCbCr image is divided into three channels one intensity (Y) and two color components (Cb and Cr). The end result of the process was evaluated using the Blind and full-reference measures like PSNR, SSIM, and BRISQUE. The outcome of the recommended method clearly demonstrates that VDSRNN outperforms BI with improved perceptual integrity and image quality.

Keywords SISR · Image quality · High resolution · Bilinear interpolation · VDSRNN · PSNR · SSIM · BRISQUE

P. Ganesan (✉)
Department of ECE, Vidya Jyothi Institute of Technology, Hyderabad, India
e-mail: gganeshnathan@gmail.com

L. M. I. Leo Joseph
Department of ECE, SR University, Warangal, Telangana, India
e-mail: leojoseph.lmi@sru.edu.in

V. Elamaran
Department of ECE, School of EEE, SASTRA Deemed University, Thanjavur, India
e-mail: elamaran@ece.sastra.edu

S. Jency · G. Sajiv
Saveetha School of Engineering, SIMATS, Chennai, India
e-mail: jencys9026.sse@saveetha.com

G. Sajiv
e-mail: sajivg2003.sse@saveetha.com

J. K. Mandal et al. (eds.), *Recent Advances in Artificial Intelligence and Smart Applications*, Lecture Notes in Networks and Systems 1380,
https://doi.org/10.1007/978-981-96-5822-0_31

1 Introduction

SISR has undergone a revolution with the use of deep neural networks, especially those with very deep architectures, which allow complicated representations of images to be learned using existing data [1, 2]. Nevertheless, these developments frequently result in higher processing demands. Conversely, interpolation techniques such as BI are straightforward and effective, but they might not be able to adequately capture fine-grained picture information. This work intends to give perspectives on the effectiveness of the suggested technique on baseline datasets through thorough investigation and evaluation. The accuracy of the reconstructed images will be assessed quantitatively using measures like PSNR and SSIM, while qualitative evaluations obtained by direct observation will provide a perceptual comprehension of the outcomes [7, 8]. In this work, the input image initially transformed into YCbCr space which has more unique advantages as compared to RGB space as illustrated in Table 1. In this work, only the luminance (Y) is involved in the training process. The transformation is based on (1)

Table 1 Comparison of YCbCr and RGB space

Aspect	YCbCr	RGB
Color representation	Splits brightness (Y) from chrominance (Cb, Cr)	Directly characterizes colors using R, G and B components
Advantages	Splits luminance and chrominance components, allowing for efficient compression such as JPEG	Exactly matches how people perceive color, which makes it perfect for screen display [3]
	Because chromaticity components have a lesser spatial resolution, the human eye finds them to be less noticeable [4]	Gives a simple color representation using elements of R, G and B
	Improves compression ratios because chrominance components are less sensitive to visual perception	Extensively supported by software and imaging equipment, guaranteeing simplicity of use and compatibility [5]
Limitations	Chrominance subsampling causes a loss of color information, which might result in color artifacts in heavily compressed images	RGB components' interdependence can make compression methods more difficult to use and result in larger file sizes
	More processing is necessary for correct color reproduction, especially in situations where color fidelity is important	Restricted in how well it can convey color information, particularly in situations when compression is crucial [6]
	The quality of the image may be impacted by rounding errors introduced during the RGB to YCbCr conversion process	Color distortion might result from inaccuracies in the representation of colors outside the range specified by the RGB color space

$$\begin{bmatrix} Y \\ C_b \\ C_r \end{bmatrix} = \begin{bmatrix} 0.299 & 0.587 & 0.114 \\ -0.169 & -0.331 & 0.500 \\ 0.500 & -0.419 & -0.081 \end{bmatrix} \begin{bmatrix} R \\ G \\ B \end{bmatrix} \quad (1)$$

The remainder of the paper is structured as follows: In Sect. 2, important approaches, algorithms, and methodologies are highlighted in a thorough analysis of relevant work in the field of SISR. The theoretical foundation and principles that govern BI and VDSRNN are presented in Sect. 3. The results and findings are presented in Sect. 4, and finally concluded in Sect. 5 with a review of the foremost conclusion and implication for more research.

2 Related Work

Recent trends in SISR research have been largely influenced by advancements in deep learning, novel network architectures, and the exploration of new loss functions and regularization techniques [9]. Table 2 depicts some prominent trends in the field.

Table 3 elucidates the literature survey which provides a brief overview of some notable studies in SISR, highlighting the methodology/approach employed in each study. It demonstrates the diversity of approaches used in SISR research, ranging from traditional sparse representation to deep learning-based approaches.

Table 2 Trends in SISR

Trend	Description
Deep learning architectures	Advancements in convolutional neural networks (CNNs) with deeper architectures to capture complex image features [10]
Attention mechanisms	Integration of attention mechanisms to focus on informative image regions while suppressing irrelevant ones [11]
Adversarial training	Utilization of adversarial loss functions to train models to generate perceptually realistic high-resolution images
Multi-scale architectures	Development of architectures that leverage hierarchical representations to capture features at different scales [12]
Self-supervised learning	Exploration to learn without explicit supervision
Generative models for uncertainty estimation	Use of generative models for estimating uncertainty in super-resolved images, particularly in critical applications
Domain adaptation and transfer learning	Investigation of domain adaptation and transfer learning techniques to improve model generalization across domains

Table 3 Methodologies applied for SISR

Method	Description	Advantages	Limitations
Interpolation	Apply the known data points to approximate the value of a point in its proximity [13]	Simple and efficient computation	Generate unclear outcomes and miss complex visual information
Sparse representation	Utilizes sparse coding and dictionary learning to represent high-resolution patches sparsely [14]	Captures fine details effectively	Computationally intensive
Gradient prior	Incorporates L0 gradient prior regularization to promote piecewise smoothness in super-resolved images	Preserves edges and structures	May over-smooth textures
VDSR	Employs an architecture to directly learn the low and high-resolution (LR and HR) mapping [15]	Learns complex mappings straightaway from data	Requires large datasets for training
Sub-pixel CNN	LR—HR mapping using sub-pixel convolutional layers	Efficient inference time	May produce artifacts in extreme upscaling
Generative adversarial network (GAN)	Trains a GAN to generate HR images that are perceptually realistic [16]	Produces visually pleasing results	Training instability, mode collapse
Enhanced deep residual networks (DRN)	Enhances DRN by incorporating advanced architectural modifications and optimization techniques	Improved performance and convergence	Complexity in tuning hyper-parameters
Enhanced super-resolution GAN (ESRGAN)	Extends the capabilities of GANs for super-resolution by introducing improvements in network architecture and training strategies [17]	Generates high-quality super-resolved images	Requires careful training and tuning
Deep back-projection networks	Implements architecture, which recursively refines the super-resolved image by learning residuals	Captures fine details iteratively	May suffer from over-fitting

(continued)

Table 3 (continued)

Method	Description	Advantages	Limitations
Deep convolutional network (DCN)	Utilizes DCN to learn the LR—HR mapping directly	Uninterrupted learning	Limited to specific scaling factors
Deep Laplacian pyramid networks	SISR by utilizing Laplacian pyramid decomposition	Fast and accurate super-resolution	Complexity in implementation

3 Materials and Methods

One popular method for upscaling LR images to HR ones in SISR tasks is bilinear interpolation (BI). In contrast to deep learning-based techniques, BI is a straightforward yet efficient method that estimates the values of pixels in the enhanced image using mathematical concepts. Through the use of surrounding pixels' intensity levels, BI determines a pixel's value at a non-integer position. Assume a 2D pixel grid that represents a single image, with an intensity value assigned to each pixel $I_{i,j}$. The initial process of the technique is to upscale an LR input image to an HR. The BI is applied to the LR image in order to compute each pixel's value in the HR output. Every pixel in the LR image goes through this procedure iteratively. A pixel's surrounding neighborhood is taken into consideration for every pixel in the LR image. The ideal scale factor for upscaling usually determines the neighborhood's size. For example, 4-Neighborhood has their coordinates as $(\times 1, y1)$, $(\times 2, y1)$, $(\times 1, y2)$, and $(\times 2, y2)$. By taking into account a weighted mean of the pixel values in the surrounding neighborhood, BI calculates the value of each pixel in the HR image. Specifically, it uses linear interpolation along the vertical as well as horizontal axes to compute the weighted sum of the four adjacent pixels in the LR image.

Linear interpolation is first carried out on the horizontal axis. The two pixels that are located on the same row as the target location are interpolated in terms of intensity. The horizontal distances of these pixels from the target position define the weights that are applied to them. Linear interpolation in the vertical plane is then carried out by BI. For the two rows above and below the target position, it interpolates between the outcomes of the horizontal interpolation step. As a weighted sum of the intensities of the four nearby pixels, BI calculates the intensity value I(x, y) at the target point. The distances between the target position and the nearby pixels are used to calculate the weights.

$$I(x, y) = (1-\alpha)(1-\beta)\mathrm{I}_{1,1} + \alpha(1-\beta)I_{2,1} + (1-\alpha)\beta I_{1,2} + \alpha\beta I_{2,2} \quad (2)$$

Here, $\alpha = x - x_1$ and $\beta = y - y_1$ are the distances between the target position (x, y) and its neighborhood pixel $(x1, y1)$ in horizontal and vertical directions respectively.

$I_{1,1}$, $I_{2,1}$, $I_{1,2}$, and $I_{2,2}$ are the intensity values of the four neighboring pixels.

The HR output image is created using the interpolated values for each pixel in the LR image after they have been calculated. The BI process determines the value of

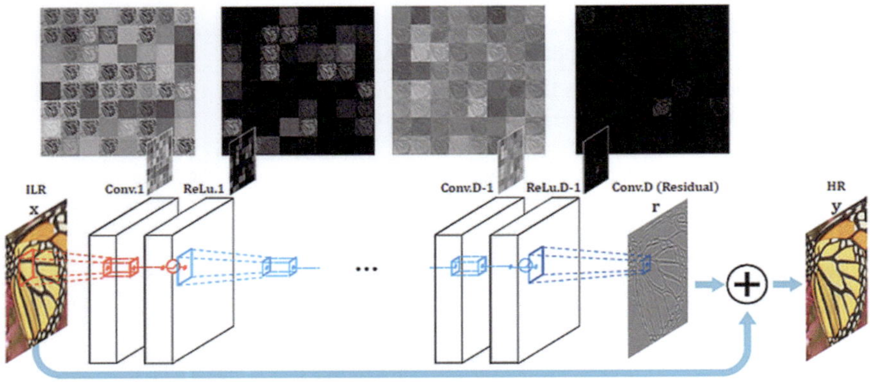

Fig. 1 Architecture of VDSRNN

each pixel in the output image, which correlates to a place in the LR image. A quick and effective way to scale and resample images is by bilinear interpolation. The main limitation of this method is to unable to capture the fine details of an image.

The VDSRNN, a deep neural network, learns the complicated association between LR and HR images using the learning process. This network has several convolutional layers to extract features. VDSRNN has a unique feature as the residual learning where it learns residual image features. HR image is obtained by adding this residual image to the LR image (Fig. 1).

Let X represent the LR test image patch, and Y represent the corresponding HR ground truth image patch. VDSR consists of L convolutional layers. Let W_i and b_i denote the weights and biases of the ith convolutional layer, respectively. During Forward Propagation, the output of the ith layer is computed as follows:

$$H_i = \sigma\,(W_i - H_{i-1} + b_i) \tag{3}$$

H_{i-1} is the input to the ith Layer

σ represents the ReLU activation function.

VDSRNN learns to predict the residual image instead of directly predicting the HR image as shown in (4) (Fig. 2)

$$Y_{res} = Y_{hr} - Y_{lr} \tag{4}$$

The yield of the final layer is the predicted residual image. The final high-resolution output is computed by summing the predicted residual image to the input low-resolution image:

$$Y_{hr} = Y_{lr} + Y_{res} \tag{5}$$

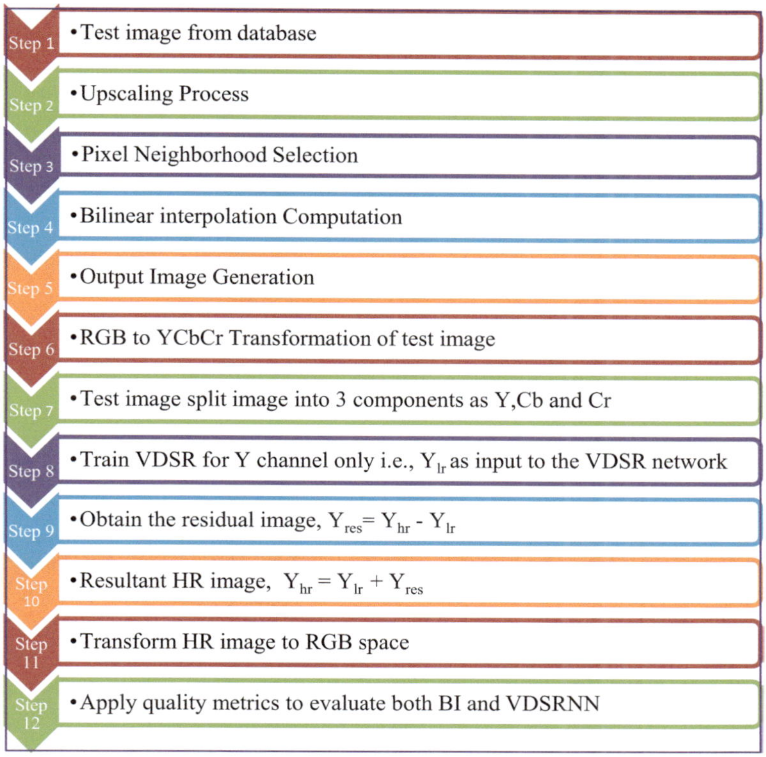

Fig. 2 Proposed procedure for SISR

In addition to qualitative evaluation by visual inspection, the quality of the super-resolved images can also be assessed using complete and no reference quantitative measures such as PSNR, MSE, SSIM, PIQE, NIQE, and BRISQUE [18].

4 Experimental Result and Discussion

This section is devoted to explain the outcome of the proposed technique for SISR based on BI and VDSRNN. The images utilized to evaluate the competence of the method are acquired from IAPR TC-12. The original image is downscaled to obtain the LR image. In this work, the image is resized using different scale factors (SF) varying from 0.1 (low quality) to 0.5 (medium quality). The input image displayed in Fig. 3a is transformed into LR image by downscaling with SF = 0.1. The low-quality image depicted in Fig. 3b. BI is a straightforward yet efficient method that estimates the values of pixels in the enhanced image using mathematical concepts. Through

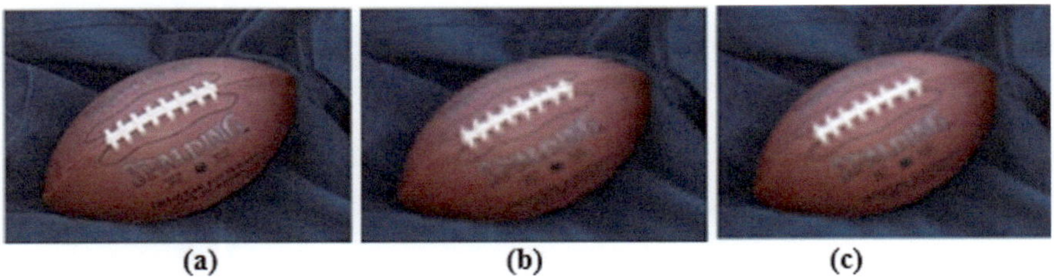

Fig. 3 BI interpolation for SISR

the use of surrounding pixels' intensity levels, BI determines a pixel's value at a non-integer position. The outcome of the BI is portrayed in Fig. 3c.

The process of SISR using VDSRNN starts with the transformation of the LR image in RGB to YCbCr space. The result of this process is illustrated in Fig. 4. All the fine (most important) details are vested in the luminance (Y) channel only. So other two chrominance (Cb and Cr) channels which contain only color information are not considered for further processing w.r.t our work.

The utilization of residual learning by VDSRNN is one of its unique features. The network learns residual image features, which are then added to the LR image to generate the HR output, as opposed to straight estimating the HR image from the LR input. This residual learning approach helps solve the vanishing gradient issue that deep networks frequently face and allows for more effective training. The residual image is displayed in Fig. 5a. After the training process is over, the HR image formed by the summation of residual and upsampled LR images as shown in Fig. 5b. The HR, i.e., Y- channel is again recombined with Cb and Cr channel to form YCbCr image. This image re-transformed to RGB and its result is illustrated in Fig. 5c

The end result for VDSRNN and BI-based SISR with different scale factors are shown in Fig. 6. Here the first column represents input images resized with SF of 0.1, 0.2, 0.3, 0.4, and 0.5 (row 1, row 2, and so on). The second and third column illustrates the outcome of the BI and VDSRNN process, respectively.

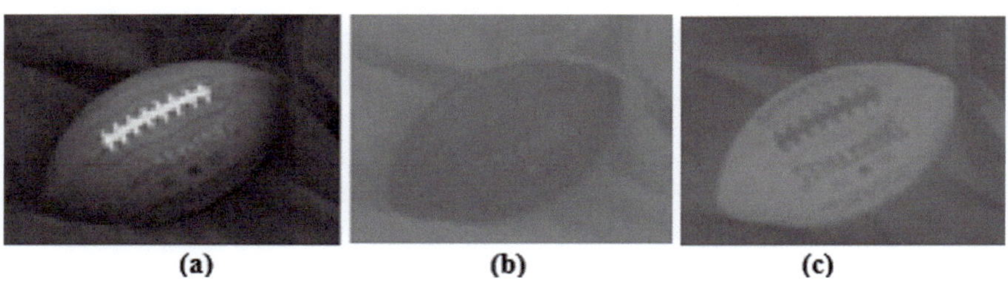

Fig. 4 The three channels of the LR YCbCr image

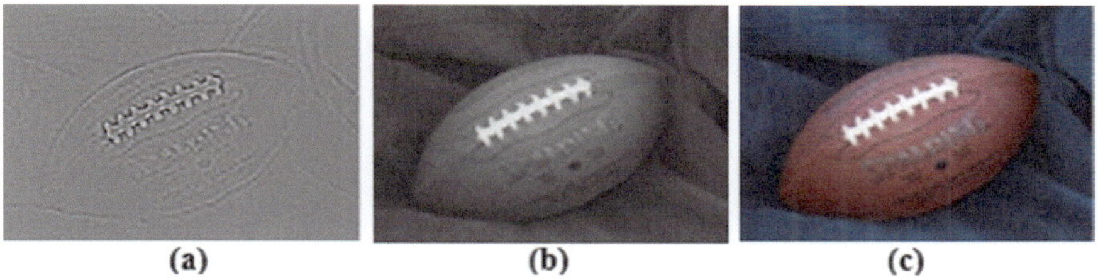

Fig. 5 Outcome for VDSRNN-based SISR

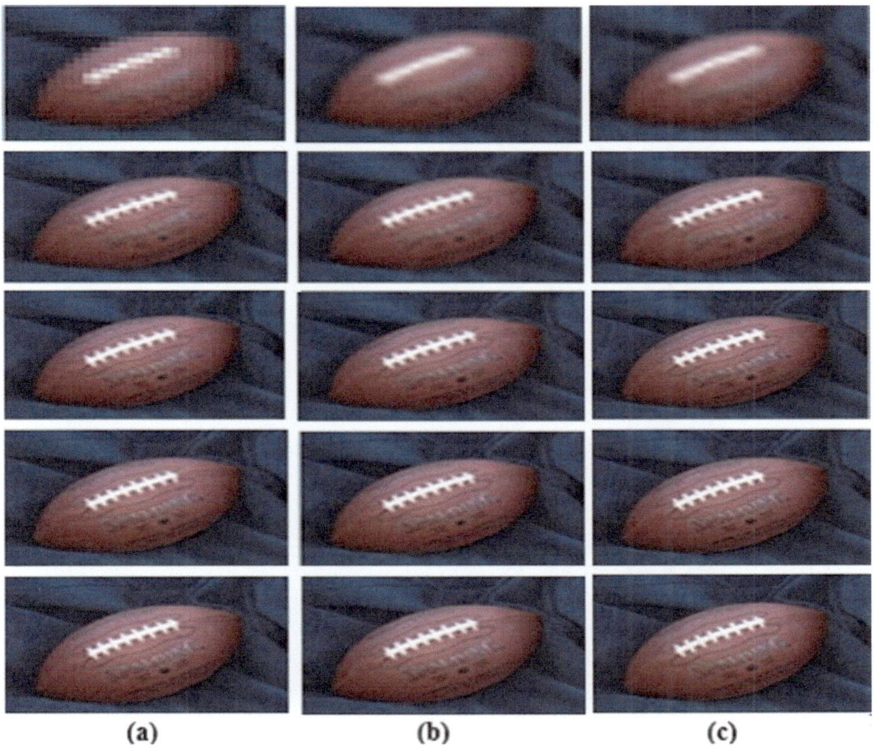

Fig. 6 The end result for VDSRNN and BI-based SISR with different scale factors

The efficacy of BI and VDSRNN methods was evaluated using full and no reference quality metrics and findings presented in Table 4. It is graphically depicted in Fig. 7. All metrics produced good scores for VDSRNN as compared to BI.

The characteristics listed here makes VDSRNN a strong and efficient method for single-image super-resolution jobs, especially when excellent outcomes are required.

Table 4 The quality assessment for BI and VDSRNN method

Measure	SF = 0.1		SF = 0.2		SF = 0.3		SF = 0.4		SF = 0.5	
	BI	VDSR	BI	VDSR	BI	VDSR	BI	VDSR	BI	VDSR
PSNR	24.601	25.602	25.914	26.945	26.301	27.761	27.802	28.385	29.337	30.381
MSE	0.0041	0.0032	0.0023	0.0021	0.0021	0.0016	0.0017	0.0015	0.0012	0.0009
SSIM	0.7332	0.7593	0.7462	0.7879	0.8236	0.9171	0.8783	0.9254	0.9200	0.9537
MSSSIM - R	0.8034	0.8256	0.8384	0.8797	0.8479	0.9302	0.8947	0.9407	0.9441	0.9737
MSSSIM - G	0.8071	0.8288	0.8423	0.8823	0.8532	0.9344	0.9033	0.9501	0.9462	0.9772
MSSSIM - B	0.8166	0.8313	0.8496	0.8875	0.8583	0.9377	0.9088	0.9507	0.9578	0.9789
NIQE	8.2652	7.8963	9.1922	7.4047	6.1499	5.1613	5.7885	4.6593	4.6272	3.8857
PIQE	98.094	94.984	86.266	80.346	81.721	74.343	75.452	66.531	71.169	61.670
BRISQUE	68.961	59.459	60.064	54.742	63.379	58.479	59.425	52.564	40.876	25.871

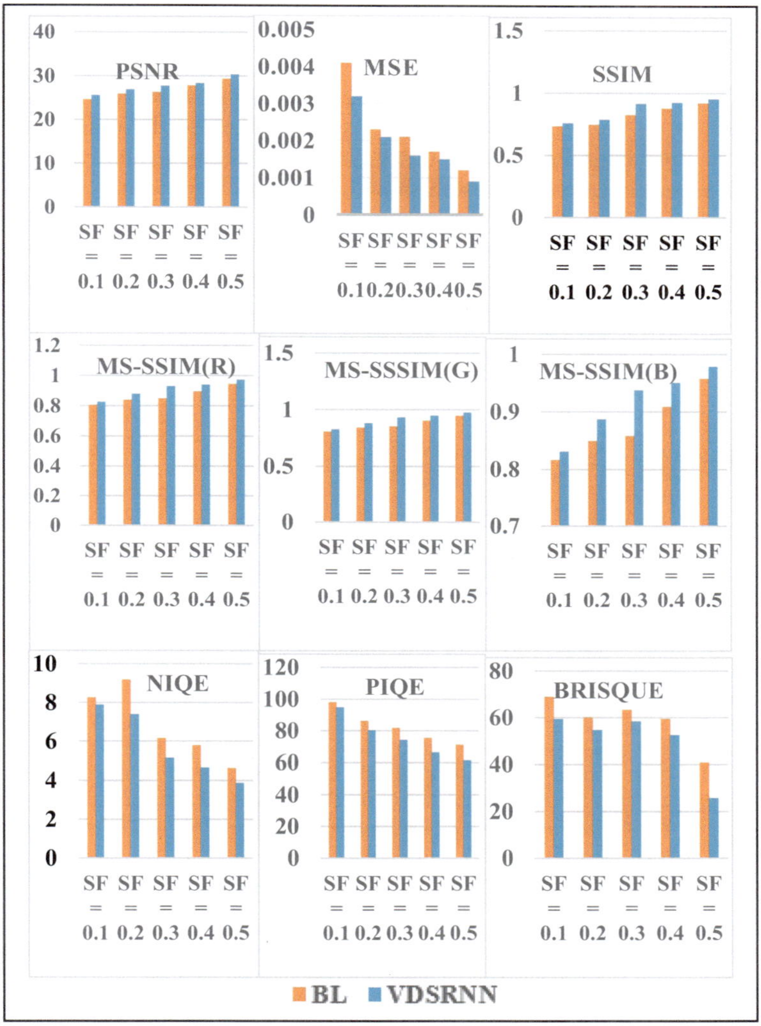

Fig. 7 The quality assessment for BI and VDSRNN method

- The preservation of image details
- Learning from data
- End-to-end optimization
- Flexibility and Adaptability
- Superior performance metrics
- Adaptation to scale factors
- Reduction of artifacts.

5 Conclusion

The proposed work presented a widespread analysis of SISR techniques, with a specific focus on the VDSRNN and BI. We have assessed the efficacy of these methods in enhancing the resolution and quality of LR images through empirical study and experimentation. Our results show that there are major benefits to producing HR images using VDSRNN integrated into the SISR process. Through the utilization of the depth and capacity of deep CNN, VDSRNN exhibits exceptional aptitude in acquiring intricate mappings across image spaces with varying resolutions. Compared to conventional interpolation techniques, the model produces results that are perceptually better since it can capture small features and textures.

References

1. Wang, Q., Gao, Q., Wu, L., Sun, G., & Jiao, L. (2021). Adversarial multi-path residual network for image super-resolution. *IEEE Transactions on Image Processing, 30*, 6648–6658.
2. Yoo, Y., Kim, D., & Lee, J. (2020). A performance comparison of super resolution model with different activation functions. *KIPS Transactions on Software and Data Engineering, 9*(10), 303–308.
3. Sajiv, G. (2015). Unsupervised clustering of satellite images in CIELab color space using spatial information incorporated FCM clustering method. *International Journal of Applied Engineering Research, 10*(20), 18774–18780.
4. Sathish, B. S., et al. (2015). Color image segmentation based on genetic algorithm and histogram threshold. *International Journal of Applied Engineering Research., 10*(6), 5205–5209.
5. Palanivel, K., Sathish, B. S., Kalist, V., Shaik, K. B. (2015). Performance of fuzzy based clustering algorithms for the segmentation of satellite images—A comparative study. In: *2015 IEEE Seventh National Conference on Computing, Communication and Information Systems (NCCCIS)* (pp. 23–27). Coimbatore
6. Sajiv, G. (2016). Comparative study of possiblistic fuzzy C-means clustering based image segmentation in RGB and CIELuv color space. *International Journal of Pharmacy and Technology., 8*(1), 10899–10909.
7. Mittal, A., Moorthy, K., & Bovik, A. C. (2012). No-reference image quality assessment in the spatial domain. *IEEE Transactions on Image Process, 21*(12), 4695–4708
8. Kalist, V., Ganesan, P., Leo Joseph, L. M. I., Sathish, B. S., Murugesan, R. (2020). Image quality analysis based on texture feature extraction using second-order statistical approach. Lecture notes in networks and systems, vol 107. Springer, Singapore
9. Cai, Q., Li, J., Li, H., Yang, Y., Wu, F., & Zhang, D. (2022). TDPN: Texture and detail-preserving network for single image super-resolution. *IEEE Transactions on Image Processing, 31*, 2375–2389.
10. Guo, Y., Wu, X., & Shu, X. (2022). Data acquisition and preparation for dual-reference deep learning of image super-resolution. *IEEE Transactions on Image Processing, 31*, 4393–4404.
11. Wu, J., Wang, Y., & Zhang, X. (2023). Lightweight asymmetric convolutional distillation network for single image super-resolution. In: IEEE signal processing letters.https://doi.org/10.1109/LSP.2023.3286811
12. Lim, B., Son, S., Kim, H., Nah, S., & Lee, K.M. (2017). Enhanced deep residual networks for single image super-resolution. CVPR (pp. 1132–1140)
13. Muhammad, W., Aramvith, S., & Onoye, T. (2023). SENext: squeeze-and-excitation next for single image super-resolution. *IEEE Access, 11*, 45989–46003. https://doi.org/10.1109/ACCESS.2023.3274679

14. Yang, J., Wright, J., Huang, T. S., & Ma, Y. (2010). Image super-resolution via sparse representation. *IEEE Transactions on Image Processing, 19*(11), 2861–2873. https://doi.org/10.1109/TIP.2010.2050625
15. Anwar, S., Khan, S., & Barnes, N. (2021). A deep journey into super-resolution: A survey. *ACM Computing Surveys, 53*(3), 1–34.
16. Prajapati, K., et al. (2021). Direct unsupervised super-resolution using generative adversarial network (DUS-GAN) for real-world data. *IEEE Transactions on Image Processing, 30*, 8251–8264.
17. Zhang, Y., Li, K., Li, K., Wang, L., Zhong, B., & Fu, Y. (2018). Image super-resolution using very deep residual channel attention networks. In V. Ferrari, M. Hebert, C. Sminchisescu, Y. Weiss (Eds.), Computer Vision–ECCV
18. Ganesan, P., Sathish, B. S., Vasanth, K., Vadivel, M., Sivakumar, V. G., Thulasiprasad, S. (2020). Color image quality assessment based on full reference and blind image quality measures. Lecture notes in networks and systems, vol 107. Springer, Singapore

An Approach to Salt and Pepper Noise Reduction in Color Images Using Deep Learning-Based Denoising Convolutional Neural Network

P. Ganesan, L. M. I. Leo Joseph, V. Elamaran, S. Jency, and G. Sajiv

Abstract In this paper, we present an inventive way of reducing salt-and-pepper noise in color images employing deep learning-based denoising convolutional neural network (DCNN). This kind of noise, containing random pixels of white and black color, is detrimental to image quality and poses challenges to image processing tasks. The technique here takes advantage of the strong ability of DCNNs to extract features effectively for the purpose of discerning noise while maintaining key aspects within the picture. Specifically, the network design is such that it can handle the intricacies involved with color image denoising by using specifically tailored layers and loss functions. There are many benefits to the method proposed when compared with usual filtering techniques. To start with, a deep learning model can be trained to detect and filter out complex noise patterns that cannot be detected by humanly designed filters. Another advantage is that DCNNs are able to work directly on color images, without losing any color data during denoising processes. The paper presents some experiments on a benchmark dataset illustrating the efficiency of the suggested DCNN algorithm and its contrast to other modern denoising methods. These findings imply that DCNN is better in terms of noise reduction and quality preservation for

P. Ganesan (✉)
Department of ECE, Vidya Jyothi Institute of Technology, Hyderabad, India
e-mail: gganeshnathan@gmail.com

L. M. I. Leo Joseph
Department of ECE, SR University, Warangal, Telangana, India
e-mail: leojoseph.lmi@sru.edu.in

V. Elamaran
Department of ECE, School of EEE, SASTRA Deemed University, Thanjavur, India
e-mail: elamaran@ece.sastra.edu

S. Jency · G. Sajiv
Saveetha School of Engineering, SIMATS, Chennai, India
e-mail: jencys9026.sse@saveetha.com

G. Sajiv
e-mail: sajivg2003.sse@saveetha.com

J. K. Mandal et al. (eds.), *Recent Advances in Artificial Intelligence and Smart Applications*, Lecture Notes in Networks and Systems 1380,
https://doi.org/10.1007/978-981-96-5822-0_32

colored images that have been damaged by salt-and-pepper noise, hence offering a more promising solution.

Keywords Denoising · Salt-and-pepper noise · Deep learning · DCNN · PSNR · SSIM · NIQE · PIQE

1 Introduction

Digital image processing is often disturbed by the noise in images which impacts more on the quality of the image [1]. Salt and pepper are types of noise that primarily give rise to this noise. The impulsive nature of salt-and-pepper noise occurs as randomly distributed black and white pixels over an original image [2, 3]. The noises are normally introduced into images during their acquisition, transmission, or storage leading to a negative impact on fields such as remote sensing, computer vision, and medical imaging. Salt-and-pepper noise is especially challenging for image denoising. This type of random and high-intensity noise can disrupt image details significantly and impact the ability of traditional filtering techniques to filter out the noise and keep actual image content [4, 5]. This challenge is compounded when considering color images because it is important to maintain the color information while also filtering out the noise [6]. From an engineering perspective, one challenge in image denoising is designing a noise reduction model that can suppress the noise but also not remove too much image content [7]. Traditional filters often use the same filtering operation across the entire image. This is problematic for some images (especially color images) and the regions of the image with high-frequency details or sharp edges [8]. Though median filtering and morphological operations are the commonest traditional techniques for removing salt-and-pepper noise; they fail in cases where there are high densities of noise since they do not preserve details in images effectively. Adaptive filtering methods improve upon this by adjusting filter parameters based on localized noise characteristics, but localized noise detection is difficult, and local parameter tuning requires a sophisticated infrastructure to execute [9]. Deep learning has recently emerged as a powerful tool for solving complicated problems in image processing. Hence, there is great potential in utilizing deep learning hence CNN is beneficial because it has the capability to learn to learn complex patterns from big datasets, deep learning-based methods have shown impressive results in different tasks about restoring images. Previous work has shown the potential of CNNs in many different denoising tasks for grayscale images, yet the specific case of salt-and-pepper noise in color images is less explored. Thus, this work aims to specifically address this gap by developing a deep CNN for denoising that is tailored to this noise type in hopes of providing superior denoising performance [10]. This paper tackles this challenge by proposing a new method for salt-and-pepper noise removal in color images through a deep learning motivated denoising convolutional neural network. The model's primary objective is to remove images affected by salt-and-pepper noise while also preserving important features in the image. Since CNNs excel at extracting key features, the proposed architecture can leverage this power to learn and differentiate the representation of clean and noisy patches in an image.

The remainder of this paper is organized as follows. First, the work presents a review of the state-of-the-art methods to address the salt-and-pepper noise removal problems in Sect. 2. In Sect. 3, the proposed CNN architecture for color image denoising is explained. Our experimental setup and evaluation metrics are discussed in Sect. 4. The experimental results and the discussion provided in Sect. 5. Lastly, a summary of our contributions and future work on this topic is presented in Sect. 6.

2 Related Work

One type of noise that is frequently encountered in image processing is the salt-and-pepper noise, and several algorithms, as detailed in Table 1, have been proposed to deal with it. Median filtering and its adaptive forms have been applied for a long time because they are simple powerful tools for removing isolated noise points. But the problem with these methods is that they do not give much attention to small details, and the edges of the original image are not well relayed, especially in noisy environments. Other methods of signal processing and noise reduction include morphologic filtering and the use of fuzzy logic; however, these technologies present difficulties in noting that even in these techniques of noise removal, the problem of preserving detail remains. Also, the originally developed transform domain methods such as wavelet-based denoising have demonstrated the efficacy of noise removal from the signal while suffering from the problem of artifacts that may be caused due to noise elimination in different frequency bands and the problem of proper parameter selection for accurate reconstruction.

In recent years deep learning-based approaches have attracted a lot of attention for image denoising especially for the case of salt-and-pepper noise filtering. The use of Convolutional Neural Networks (CNNs) has been found useful in learning complicated transformations from noisy to clean images and such networks outcompete other methods in some cases. Although initially deep learning models were designed for the general form of noise, new architectures have been developed that have been specifically designed for the elimination of salt-and-pepper noise. Most of these advanced models resort to learning residual, generative adversarial network (GAN) to have better enhancement and noise reduction with accurate details. Nevertheless, problems including the requirement of large sets of training data, the very high noise levels, and time consumption persist, calling for stronger and faster solutions, including DCNN—the deep learning-based Denoising Convolutional Neural Network [11, 12]. Table 1 compares various methods of salt-and-pepper denoising.

Table 1 Comparison of various salt-and-pepper denoising methods

Method	Description	Advantages	Limitations
Median filtering	Substitutes each pixel for its neighbor's median value	Easy to use and efficient at moderate noise levels	May distort edges; less useful in noisy environments
Weighted median filtering	Makes use of the weighted values of nearby pixels	Better preservation of edges in comparison to median filtering	Computationally demanding, necessitates adjusting parameters
Adaptive median filtering	Adjusts window size in accordance with the noise level	Improved edge preservation that works well with different noise levels	An increase in the complexity of computing
Mean filtering	Swaps out each pixel for the average of its nearby pixels	Simple, somewhat lessens noise	Image blur, less useful in noisy environments
Decision-based algorithm	Determines if a pixel is noisy and replaces it with the neighboring non-noisy median	Effective at high noise levels while maintaining edges	Complex to implement and less useful in quiet environments
Switching median filter	Dependent on the existence of noise, alternates between the median and other filters	Reconciles the preservation of detail with noise reduction	Need a technique for precise noise detection
Fuzzy filtering	Makes the decision to replace a pixel using fuzzy logic	Capable of keeping details intact and managing uncertainties	Computationally demanding and necessitates flexible rules
Wavelet transform	Uses thresholding and wavelet transformation	Reduction of noise and maintained details effectively	Implementation complexity and computational expense
Anisotropic diffusion	Iterative procedure to maintain edges while reducing noise	Maintains edges well, suitable for different noise levels	Takes a long time and need parameter tuning
Total variation minimization	Minimizes the image's overall variations	Maintains edges, useful at high noise levels	Can cause artifacts and be computationally demanding
Deep learning-based denoising CNN	Use CNN to identify and minimize noise patterns	High precision, adaptability, and detail preservation	Big datasets and processing power are needed for training

3 Methodology

The approach for creating the deep learning-based DCNN to remove salt-and-pepper noise in color images is in the following steps. The data collection for the proposed method includes.

- Collect a variety of high-quality clean color images as reference for ground truth.
- Take these images and create noisy versions that are contaminated with salt-and-pepper noise of different densities.

The architecture of the proposed approach is displayed in Fig. 1 which consists of the following components.

- Input Layer: Accept the noisy color images as the input.
- Convolutional Layers: This one included several convolutional filters for feature extraction that aim to extract as many features as possible of an image resulting in feature maps that contain several characteristics of an image.

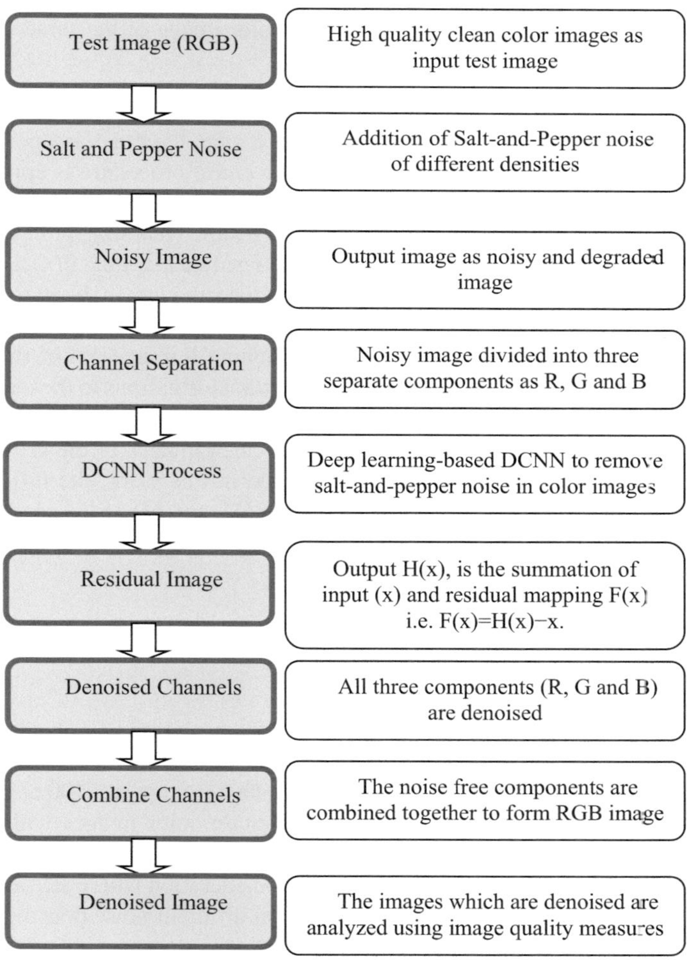

Fig. 1 Proposed method for image denoising

- Activation Function: Utilized non-linear activation functions (ReLU) to embed non-linearity. After each convolutional layer, the feature maps are activated element-wise with rectified linear unit (ReLU) providing non-linearity into the network by throwing negative values into the bin, to allow it to learn complex relationships between input and output.
- The vital component in the proposed DCNN is the residual learning. This is made possible by using the skip connections, which provide the facility to force the network to learn the residual functions. It is a residual connection that enables the network to be in a position to predict the dissimilarity between the noisy and output image other than the clean result only.
- DCNN architecture incorporates batch normalization layers for improving training steadiness and convergence.
- Output Layer: As output give the denoised color image of the same size as the input.

In the proposed approach, the RGB test image is added with salt-and-pepper noise to investigate the efficiency of the work. As a next step, the noisy image is divided into three separate components R, G, and B. The entire procedure is applied to all three components independently and finally, after the process is over, combined to obtain the denoised images as the final outcome. In residual learning, preferred output (mapping) H(x), is the summation of input (x) and residual mapping F(x), i.e., F(x) = H(x) − x. The only output layer that is formed is a denoised image. It tends to include just one convolutional layer and the next layer can either be linear or a layer without activation. The last layer reconstructs the final denoised image out of the features learnt in the previous layers and applies the learnt transformations to the input image. The images that are denoised are qualitatively analyzed by visually comparing them to the input noisy images as well as the original clean images or the ground truths [13–15]. For evaluation of denoising performance in this work, the full reference quality metrics which include the PSNR, MSE, SSIM, and MSSSIM are calculated, and the no reference which includes the NIQE, PIQE, BRISQUE [16, 17].

4 Results and Discussion

For the purpose of measuring the effectiveness of the proposed DCNN on the alleviation of salt-and-pepper noise, an experiment was conducted on the dataset created and a collection from widely used datasets. The dataset comprises 400 color images of several scenes and objects. These datasets contain color images with different amounts of noise within them allowing to determine the resilience and efficiency of the designed model. Images were taken into consideration where salt-and-pepper noise was artificially introduced into the images at different noise densities ranging from 0.1 to 0.5. As for each noise level, the DCNN was trained with a subset of the dataset and tested on the remaining images. Figure 2 depicts the test image to explore the competence of the proposed method.

Fig. 2 Test image to investigate the competency of the proposed method

The test image was intentionally made noisy by adding a salt-and-pepper noise density of 0.1 which is reflected in Fig. 3.

To make the procedure less complex, the noisy image is divided into three separate components as R, G, and B. The outcome of this process is illustrated in Fig. 4. The entire procedure is applied to all three components independently and finally, after the process is over, combined to obtain the denoised images as the final outcome.

In residual learning, preferred output (mapping) H(x), is the summation of input (x) and residual mapping F(x), i.e., $F(x) = H(x) - x$. The last (output) layer reconstructs the final denoised image, which is shown in Fig. 5, out of the features learnt in the previous layers and applies the learnt transformations to the input image.

Figure 6 demonstrates the outcome of the proposed method with varying noise densities of 0.1 to 0.5. To assess the quality of denoising, four quantitative measures; PSNR, SSIM, PIQE, and NIQE were employed. Since based on the PSNR that actually measures how well the reconstructed images depict the original one, higher PSNR values actually specify the better performance in denoising whereas, SSIM takes an overall assessment of the perceived quality of images by calculating the structural similarity between the original and the referred denoised images. The

Fig. 3 Test image with salt-and-pepper density of 0.1

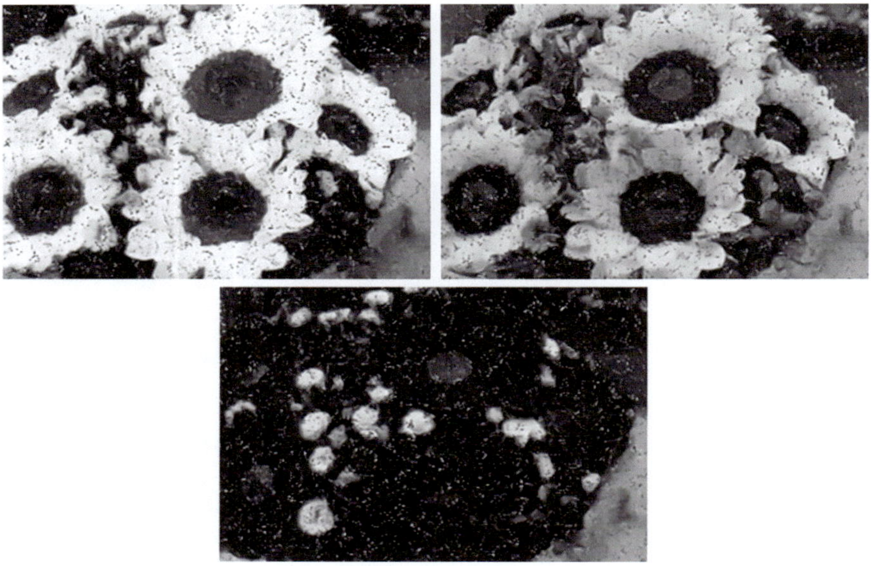

Fig. 4 Components of the noisy test image with salt-and-pepper density of 0.1

Fig. 5 Denoised output image

experimental results reveal that the proposed DCNN has higher PSNR and SSIM values for the removal of salt-and-pepper noise for all the noise levels. Similarly, MSE and no-reference quality measures (NIQE, PIQE, and BRISQUE) present very low (best) values for denoised images as compared to noisy images. So all metrics support the proposed method as shown in Tables 2 and 3. The results are diagrammatically illustrated in Fig. 7. This further shows that the proposed model outperforms ordinary denoising models in that it also provides for good zooming with minimal loss of the image characteristics especially for images with high texture attribute levels of noise.

From the experimental results, it can be seen that the proposed DCNN performs better than the other methods in terms of trying to eliminate salt-and-pepper noise while at the same time maintaining the quality of the images. The DCNN performed

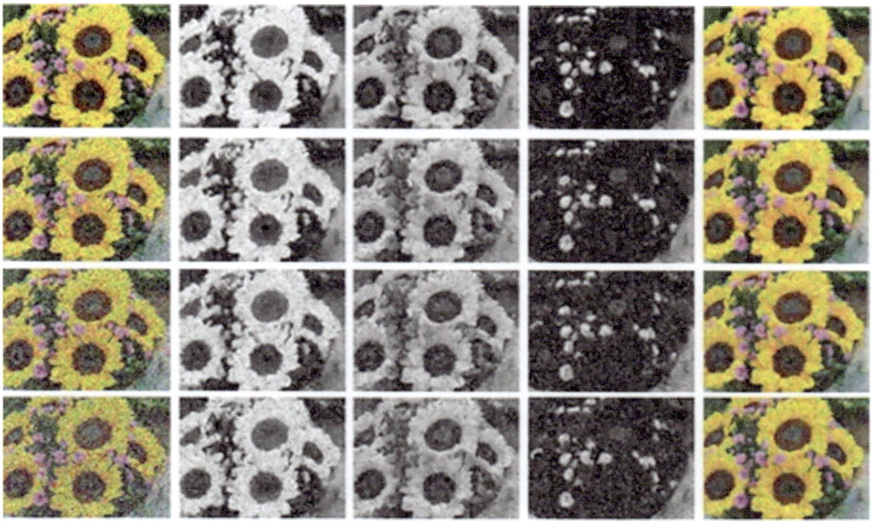

Fig. 6 Outcome of salt-and-pepper denoising process with a noise density of 0.1, 0.25, 0.35, and 0.5 illustrated in a first, second, third, and fourth row respectively

Table 2 Quality metrics (full reference) support for proposed method

Image quality metrics	Noise density = 0.1		Noise density = 0.25		Noise density = 0.35		Noise density = 0.5	
	Test versus noisy	Test versus denoised	Test versus noisy	Test versus denoised	Test versus noisy	Test versus denoised	Test versus noisy	Test versus denoised
MSE	0.0391	0.0101	0.0782	0.0154	0.0976	0.0198	0.1165	0.0247
PSNR	14.0828	19.9412	11.0682	18.1276	10.1037	17.0397	9.3367	16.0681
SSIM	0.6314	0.8038	0.4821	0.7497	0.4286	0.7109	0.3827	0.6755
MSSSIM-R	0.7970	0.8695	0.7033	0.8273	0.6608	0.7883	0.6277	0.7578
MSSSIM-G	0.8201	0.8943	0.7235	0.8433	0.6729	0.7982	0.6308	0.7628
MSSSIM-B	0.6876	0.7505	0.5608	0.6803	0.5173	0.6512	0.4721	0.6087

better than the traditional filtering methods, particularly at high noise levels. Meanwhile, the application of the residual learning as well as the noise-tolerant loss function enhances the model's performance and stability for a variety of noise levels and image categories. Therefore, compared to both the traditional and other deep learning-based methods, better performance by the proposed DCNN is observed in dealing with the problem due to salt-and-pepper noise which can be considered a noteworthy contribution in the area of image denoising. But, some restrictions and potential enhancements for the future can be identified. For example, although proven to excel in laboratory-simulated environments, further understanding also

Table 3 Quality metrics (no reference) support for proposed method

Image quality metrics	Test image	Noise density = 0.1		Noise density = 0.25		Noise density = 0.35		Noise density = 0.5	
		Noisy image	Denoised image	Noisy image	Denoised image	Noisy image	Denoised image	Noisy image	Denoised image
PIQE	33.0871	65.4128	39.9592	75.0578	28.4501	73.1018	29.0528	74.7955	27.6516
NIQE	4.5931	19.1487	9.3149	21.7209	5.3659	22.2516	5.8770	32.1213	7.0194
BRISQUE	37.5922	43.4631	38.8648	43.4564	34.4727	43.4212	15.3033	44.9128	17.9884

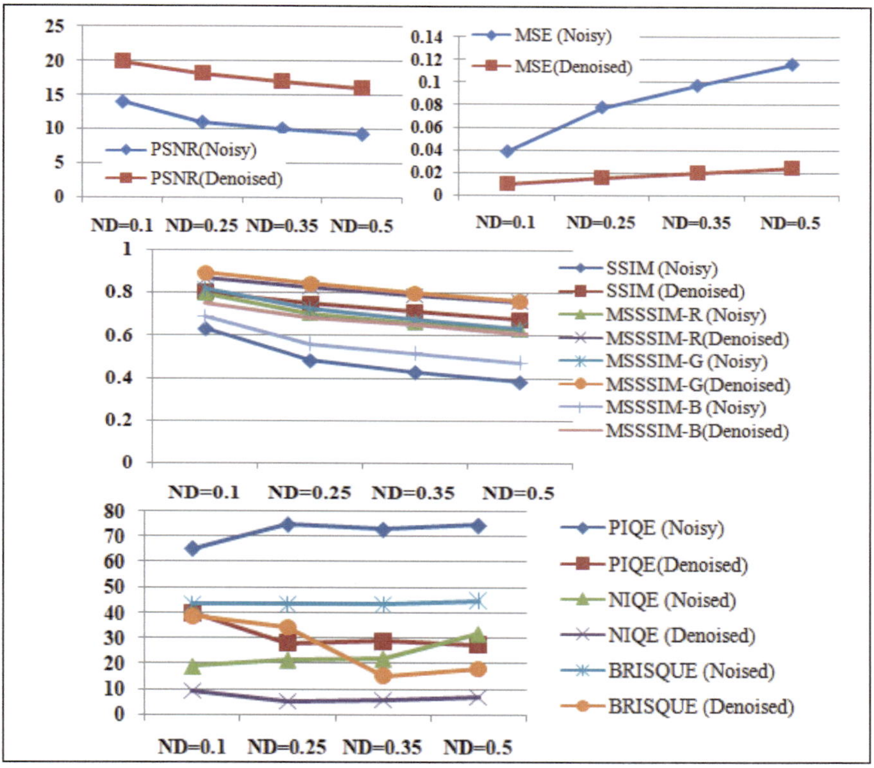

Fig. 7 The graphical representation of the proposed method result

arises on how the DCNN reacts to ultra-high noise conditions or any other type of noise or combinations of noises. The future work could be toward the generalization of this model to accommodate mixed kinds of noise or toward improving the training paradigms that consume less amount of data. Besides, applying the DCNN in other large-scale image pipelines and experimental settings of real-life problem domains like medical environments or satellite imaging would be good evidence of its applicability.

5 Conclusion

In summary, this study introduces a method to reduce salt-and-pepper noise in color images using a DCNN. The approach capitalizes, on the strengths of DCNNs in extracting features and reducing noise surpassing denoising methods. The network's architecture, tailored for color image processing complexities excels at preserving image details while effectively reducing noise. Results from datasets validate the robustness and versatility of this approach indicating its potential for use, in image processing and computer vision applications. This study also identifies that the proposed method outperforms image quality measures than the noise removal methods. One of the virtues of the presented DCNN is its capacity to obscure noise and at the same time retain the image's details. Potential future work includes related research on using the proposed DCNN on other types of noise for example the Gaussian noise type. Further research on the possibility of applying generative adversarial networks (GANs) that can increase image quality is also an interesting direction for further study.

References

1. Li, Y., Xie, T., & Mei, D. (2024). Application of convolutional neural networks for parallel multi-scale feature extraction in noise image denoising. *IEEE Access, 12*, 98599–98610.
2. Jia, X., Peng, Y., Ge, B., Li, J., Liu, S., & Wang, W. (2023). A multi-scale dilated residual convolution network for image denoising. *Neural Processing Letters, 55*(2), 1231–1246.
3. Monajati, M., & Kabir, E. (2020). A modified inexact arithmetic median filter for removing salt-and-pepper noise from gray-level images. *IEEE Transactions on Circuits and Systems II: Express Briefs, 67*(4), 750–754.
4. Hien, N. N., Thanh, D. N. H., Erkan, U., & Tavares, J. M. R. S. (2022). Image noise removal method based on thresholding and regularization techniques. *IEEE Access, 10*, 71584–71597.
5. Erkan, U., Enginoglu, S., Thanh, D. N. H., & Hieu, L. M. (2020). Adaptive frequency median filter for the salt- and-pepper denoising problem. *IET Image Process, 14*(7), 1291–1302.
6. Thanh, D. N. H., Hai, N. H., Prasath, V. B. S., Hieu, L. M., & Tavares, J. M. R. S. (2020). A two-stage filter for high density salt and pepper denoising. *Multimedia Tools Applications, 79*(29), 21013–21035.
7. Singh, V., Dev, R., Dhar, N. K., Agrawal, P., & Verma, N. K. (2018). Adaptive type-2 fuzzy approach for filtering salt and pepper noise in grayscale images. *IEEE Transactions on Fuzzy Systems, 26*(5), 3170–3176.
8. Thanh, D. N. H., Thanh, L. T., Hien, N. N., & Prasath, S. (2020). Adaptive total variation L1 regularization for salt and pepper image denoising. Optik, vol. 208
9. Zhang, K., Zuo, W., Chen, Y., Meng, D., & Zhang, L. (2017). Beyond a gaussian denoiser: Residual learning of deep CNN for image denoising. *IEEE Transactions on Image Processing, 26*(7), 3142–3155.
10. Mishro, P. K., Agrawal, S., Panda, R., & Abraham, A. (2022). A survey on state-of-the-art denoising techniques for brain magnetic resonance images. *IEEE Reviews in Biomedical Engineering, 15*, 184–199.
11. Goyal, B., Dogra, A., Agrawal, S., Sohi, B. S., & Sharma, A. (2020). Image denoising review: From classical to state-of-the-art approaches. *Information Fusion, 55*, 220–244.
12. Ganesan, P., & Rajini, V. (2013). Segmentation and denoising of noisy satellite images based on Modified Fuzzy C Means clustering and discrete wavelet transform for information retrieval. *International Journal of Engineering and Technology, 5*(5), 3858–3869.
13. Ganesan, P., Sathish, B. S., Vasanth, K., Vadivel, M., Sivakumar, V. G., Thulasiprasad, S. (2020). Color image quality assessment based on full reference and blind image quality measures. Lecture notes in networks and systems, vol. 107. Springer, Singapore
14. Kalist, V., Ganesan, P., Leo Joseph, L. M. I., Sathish, B. S., Murugesan, R. (2020). Image quality analysis based on texture feature extraction using second-order statistical approach. *Lecture notes in networks and systems*, vol 107. Singapore: Springer.

15. Ravichandran, M., Alexander, T. J., Sathish, B. S., Joseph, L. M. I. L., G. P and Sajiv, G. (2023). an effectual methodology to objective quality assessment for noisy and compressed images using full and blind reference quality metrics. In *2023 IEEE 3rd International Conference on Applied Electromagnetics, Signal Processing, & Communication (AESPC)* (pp. 1–6). Bhubaneswar, India
16. Simi, V. R., Edla, D. R., & Joseph, J. (2021). A no-reference metric to assess quality of denoising for magnetic resonance images. Biomedical Signal Processins and Control, *70*
17. Stpień & Oszust, M. (2022). A brief survey on no-reference image quality assessment methods for magnetic resonance images. *Journal of Images, 8*(6), 160

Cataract Identification and Classification Using Machine Learning Algorithms

K. Palani, T. Jerry Alexander, G. Santhosh, P. Ganesan, S. Jency, and G. Sajiv

Abstract The ability to classify cataracts is an important aspect due to the difference in managing a mature cataract and an immature cataract. This paper focuses on using different techniques in machine learning to address the general issue of automation in classification. This work uses logistic regression, and its modified version with L1 and L2 penalties, as well as other classifiers such as kNN, SVM, and random forest. Every model is optimized and validated on the dataset containing clinical and image-based attributes related to cataract maturity to determine the most appropriate model for classification. An assessment of these algorithms was made with the measures being accuracy, precision, recall, F1-score, and ROC AUC. Logistic regression presented baseline performance and advantages concerning the interpretability of features. To reduce the likelihood of over-fitting, regularization techniques were used to train the logistic regression model on how to perform better on new data. As evidenced in the findings from the experiments, the regularized logistic regression models yield

K. Palani
Department of IT, KGiSL Institute of Technology, Coimbatore, Tamilnadu, India
e-mail: palani.k@kgkite.ac.in

T. Jerry Alexander
Department of CSE, Alpha College of Engineering, Thirumazhisai, Chennai, India
e-mail: csehod@alphagroup.edu

G. Santhosh
Department of CSE-AI, Indian Institute of Information Technology Design and Manufacturing Kancheepuram, Chennai, India
e-mail: gmsanthosh2007@gmail.com; cs24b2035@iiitdm.ac.in

P. Ganesan (✉)
Department of ECE, Vidya Jyothi Institute of Technology, Hyderabad, India
e-mail: gganeshnathan@gmail.com

S. Jency · G. Sajiv
Department of ECE, Saveetha School of Engineering, Saveetha Institute of Medical and Technical Sciences, Saveetha University, Chennai, India
e-mail: jencys9026.sse@saveetha.com

G. Sajiv
e-mail: sajivg2003.sse@saveetha.com

J. K. Mandal et al. (eds.), *Recent Advances in Artificial Intelligence and Smart Applications*, Lecture Notes in Networks and Systems 1380,
https://doi.org/10.1007/978-981-96-5822-0_33

higher accuracy and reliability as compared to other algorithms tested; it excels in identifying intricate patterns of the given set of data. The results showed that regularized logistic regression outperformed the unregularized counterparts through reduced variance in high-dimensional feature space.

Keywords Cataract classification · Regularization · Logistic regression · kNN · SVM · Random Forest

1 Introduction

Ocular cataracts are among the most prevalent causes of visual disability and blindness, especially among the aging population [1, 2]. A cataract is simply the swelling and blurring of the strands of protein in the lens of the eye which obstruct the pathways of light thus causing poor sight [3, 4]. The progression of cataracts to blindness can be avoided if detected early and wanted intervention is made [5, 6]. In clinical activity, they have tried to divide the cataracts according to stages and most focusing on the mature and immature cataracts [7]. This stays true for cataract maturity which is fundamental to treatment, whether medical or by surgery. Nevertheless, the assessment of cataract maturity with the help of a manual grading system is still descriptive at best and may significantly differ from one clinician to another; that is why the development of an objective automated system is highly important [8, 9]. Sophisticated developments in the field of machine learning have demonstrated great promise across medical diagnosis, especially in the processing of medical images and clinic data. Medical conditions can be predicted with the help of machine learning techniques as these are capable of detecting data patterns that may be not easily recognized manually [10]. In the context of cataract classification, machine learning brings the possibility to train models, with high accuracy and repeatability, able to classify a cataract as mature or immature and thus help the ophthalmologist [11, 12]. This paper aims to work on the problem of cataract classification using various machine learning algorithms. Our emphasis is twofold: logistic regression, ridge logistic regression with L1 and L2 penalty, kNN, SVM, and random forest. All of these algorithms belong to the family of classification models, varying from linear to non-linear and ensemble ones. Logistic regression is chosen as the baseline model because of its convenient interpretation, and L1/L2 reg is used to increase the model's generalization. Indeed, the objective of this current study is to assess the feasibility of these algorithms in categorizing cataracts as mature or immature. Although several algorithms were compared in this study to find the best way to classify cataract samples, this will in a way assist in creating an automated diagnostic system for the eye in the field of ophthalmology.

2 Related Work

Cataract diagnosis and classification have always been fundamental tasks in ophthalmology, and much effort has been made for cataract assessment enhancement. Both conventional and those based on machine learning techniques can be applied in this domain; each with its merits and disadvantages. This section summarizes the main works presented in both categories and gives a comparative description of the methods applied to the differentiation of mature and immature cataracts.

As a matter of fact, previous studies on cataract diagnosis largely involve manual assessments made by ophthalmologists and basic image processing. These non-ML methods were based on feature extraction from the ocular images by implementing edge detectors, histogram equalizers, and texture analyzers. For instance, to enhance image perception related to the cataract maturity when visually analyzing the image contrast enhancement techniques were applied to make the specific regions of the image with the areas affected by the cataracts more visible. Segmentation was performed for cataract-affected lenses there were two types of filtering algorithms used: the Canny filter and the Sobel filter. Similarly, texture analysis techniques, like GLCM, were used to distinguish between cataractous as well as healthy lenses based on differential pixel intensities of the images. While these methods offered a substantial improvement over purely manual measurements, they largely depended on specialized recognition of image processing parameters on the part of the operator. The strong reliance on handcrafted feature extraction constrained the potential for these models to be applied to various patients' demographics and image quality. Furthermore, such outcomes depended on lighting and noise; thus, it was difficult to adapt these methods for more broad clinical applications.

On the other hand, machine learning methods have been the most effective in classifying cataracts due to an advantage in their understanding of the data to get a model without the need for very much feature engineering. SVM had been found to be useful when dealing with high-dimensionality data. For instance, in the classification of cataract images, the kernel functions transform the data into a higher order space to ease the separation between mature and immature cataracts. Research has indicated that SVM provides good accuracy, especially when using features from images obtained with the slit lamp. Random forests have also been applied in cataract classification tasks. Random forest decision trees integrate the result of several trees and enhance the predictive accuracy in addition to eliminating prominent issues such as over-fitting in the medical dataset, which includes noisy and imbalanced data. Recent studies have used random forest models to stage cataracts and have great specificity in differentiating between mature and immature cataracts using images and clinical data. Another traditional statistical technique also known as logistic regression has also been used in cataract classification. Notably, while logistic regression suffers from over-fitting especially in the context of small samples or high dimensionality, the use of L1 (Lasso) and L2 (Ridge) penalties have been used to regularize the models. These models are easy to implement and can be easily interpreted—this is crucial in medical applications when it is necessary to understand the contribution of each feature. The k-nearest neighbors (kNN) algorithm has been used for cataract classification because of its simplicity and because it is easily capable of categorizing new instances according to their similarity to previously categorized data. However, it is an important drawback that kNN is influenced by the choice of the coordinates of the parameter, such as the number of near clusters (or k) and the distance criteria, which define its generalization.

Over the last few years, advanced machine learning techniques such as deep learning methods especially CNNs have been widely applied to classify the images of cataracts. These models themselves learn hierarchical features from the data, and therefore there is no need for this step. While CNNs have been reported to provide superior performance across a range of medical image analysis tasks, they have been sparingly applied to cataract grading because of the large annotated database and expensive computational power required. Table 1 illustrates the methodologies utilized for cataract classification.

3 Methodology

This section focuses on explaining the approach used for categorizing cataracts as mature or immature based on some algorithms of machine learning. This is the basic process of data mining, which encompasses data acquisition and pre-processing, variable selection, training of the model, and testing/evaluation. The selected models are logistic regression, L1 regularized logistic regression, L2-regularized logistic regression, kNN, SVM, and random forest.

Data Collection and Pre-processing

In this work, the dataset contains cataract images and other clinical information such as the patient's age, lens opacity grade, and other ophthalmic parameters. A small part of the dataset is displayed in Fig. 1. The acquisition data was then pre-processed to filter out noise, normalize the feature values, and learn how to deal with missing values. For the clinical data, all the features were made equally contribute toward the model-building process by normalizing the data. Data were split into training and testing sets in a 60:40 split to assess the model's reliability when tested on an untrained data set.

Feature Extraction

Before the joint temporal feature extraction and feature learning, feature extraction was carried out on both image and clinical data. Several pre-processing methods like edge detection, histogram equalization, and contrast adjustments methods were performed for image data entailment depending on cataract maturity. These features were applied as input data for linear and non-linear machine learning algorithms. In this work, 1000 features were extracted. However, only 7 features are shown in Fig. 2.

Table 1 Methodologies utilized for cataract classification

Method	Description	Advantages	Limitations
Traditional methods			
Edge detection algorithms	Makes use of Canny and Sobel filters for detecting edges of glasses affected by cataracts	Easy to apply, gives clear visual definitions of the cataract area	Intolerant of noise, a strong emphasis on the importance of image, which has to be tuned most of the time
Histogram equalization	Improves contrast in the images and in turn makes cataracts more distinguishable for manual classification of cataract-affected lenses	Simple to implement, provides visual boundaries of the cataract region	Sensitive to noise, highly dependent on image quality and manual parameter tuning
Texture analysis (GLCM)	It divides pixel intensity texture and types segregation between normal and cataract-affected lenses	Captures detailed textural features useful for identifying cataracts	Requires handcrafted feature extraction, making it less scalable across diverse datasets
Contrast and brightness adjustment	Contrast and Brightness Adjustment Alters the image's brightness and contrast to improve the visibility of lens opacity, helping manual assessment	Beneficial when performing contrast enhancement on low-quality images	It is still very much dependent on human input and is also considerably subjective
Machine learning methods			
Logistic regression	A type of analysis that estimates chances of cataract designation from image or clinical characteristic	Easy to interpret, measures can be understood at the feature level [13]	Lacks the capability to handle non-linear data
Logistic regression (L1, L2)	Regularized versions of logistic regression are used to reduce over-fitting by adding penalties (L1 for feature selection, L2 for smoothness)	Prevents over-fitting and improves generalization, especially in high-dimensional data [14, 14]	Still limited in complex decision boundaries
kNN	Classifies cataracts by comparing new instances with existing labeled examples in the dataset	Easy to implement, performs well in smaller datasets	Sensitive to the choice of parameters (k) and distance metrics; computationally intensive for large datasets [16]

(continued)

Table 1 (continued)

Method	Description	Advantages	Limitations
SVM	A classifier that uses kernel functions to transform data into a higher dimensional space, making it easier to classify mature and immature cataracts [17]	Effective in high-dimensional spaces and with non-linear data; robust against outliers	Requires careful tuning of kernel parameters, can be computationally expensive for large datasets
RF	An ensemble learning method that builds multiple decision trees and combines their outputs to classify cataracts [16]	High accuracy, resistant to over-fitting, handles non-linear data well	Can be less interpretable than simpler models like logistic regression
CNN	Deep learning model that automatically learns hierarchical features from raw image data, offering state-of-the-art performance in image classification [18]	Reduces the need to extract features manually, excellent performance than other common classifiers with large datasets	Fails to work well when few labeled examples are available, can take a lot of computation time to implement

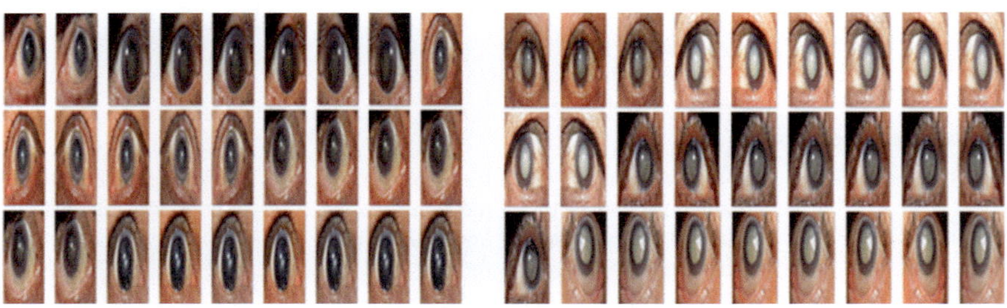

Fig. 1 Sample data to explore the proposed method

Model Selection

The classification models employed in this study are linear as well as non-linear models. The type of algorithms applied to the data depended on the fact that there is numerical data, categorical data, and image data.

- Logistic Regression: This model is used as a starting point as it is simple as well as easily interpreted. It provides the probabilities of a cataract justifying being mature or immature by the features that are derived.
- Logistic Regression (L1 and L2 Regularization): To enhance the generalization capability of the calibrated logistic regression model regularization techniques

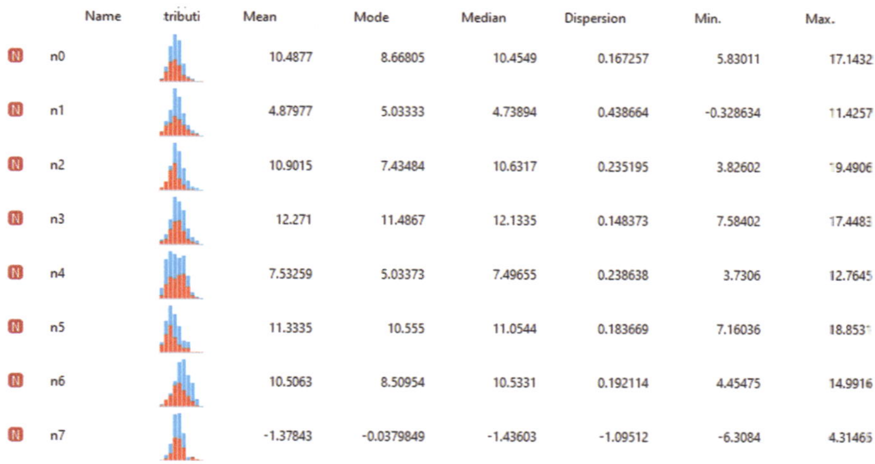

	Name	tributi	Mean	Mode	Median	Dispersion	Min.	Max.
N	n0		10.4877	8.66805	10.4549	0.167257	5.83011	17.1432
N	n1		4.87977	5.03333	4.73894	0.438664	-0.328634	11.4257
N	n2		10.9015	7.43484	10.6317	0.235195	3.82602	19.4906
N	n3		12.271	11.4867	12.1335	0.148373	7.58402	17.4483
N	n4		7.53259	5.03373	7.49655	0.238638	3.7306	12.7645
N	n5		11.3335	10.555	11.0544	0.183669	7.16036	18.8531
N	n6		10.5063	8.50954	10.5331	0.192114	4.45475	14.9916
N	n7		-1.37843	-0.0379849	-1.43603	-1.09512	-6.3084	4.31465

Fig. 2 The extracted features for the proposed method

(L1 and L2) were used. L1 regression was used for feature selection, reducing many coefficients to zero to only relevant ones; L2 regression added a penalty to big coefficients to reduce over-fitting.

- k-Nearest Neighbors (kNN): It is also used to develop a predictive model that would place cataracts in the closest vicinity to defined categories. To evaluate the model performance, cross-validation was used in the selection of "k" value.
- Support Vector Machines (SVM): It was chosen because of its non-linear data classification capability where features are mapped to a higher dimension space through an RBF kernel.
- Random Forest: It creates a number of decision trees and decides the result from all of the trees created in order to classify the objects correctly. It was chosen on the basis of its capability to learn from noisy data and it gives less chance of developing over-fitting.

Model Training

The chosen model for each was trained using the training dataset and hyper-parameters were tuned using cross-validation for best performance. To optimize hyper-parameters of each algorithm we used grid search, which helped in sorting out the best parameters for use. Cross-validation was done to tune the parameters as follows, SVM had kernel types and regularization parameters and for random forest, the number of trees and the depth of trees were adjusted.

Evaluation Metrics

A set of measures of classification error, such as accuracy, precision, recall, F1 measure, AUC, MCC, specificity, and LogLoss were computed to assess the performance of the classification models. These metrics were chosen to show the performance of each model in question in terms of two aspects—recall (sensitivity) and precision. After the construction of the final model, performance was compared with the baseline logistic regression model.

Models Comparison and Selection

After the models were trained and tested, the performances of all the models were measured to identify the higher performing model for the cataract classification. In deciding the final model to select and use, aims were established that would provide the highest overall accuracy as well as the accuracy of different mature and immature cataract classifications without increased error margin.

4 Results and Discussion

This section provides the results derived from using the various algorithms: logistic regression, logistic regression with L1 and L2 norm, kNN, SVM, and random forest for distinguishing mature and immature cataracts. Quantitative measures such as accuracy, precision, recall, F1 measure, AUC, MCC, specificity, and LogLoss were used to measure the competency of each algorithm. Among the evaluated methods, it has been concluded that the logistic regression with L1 and L2 feature selection achieved the highest accuracy in dependent variable prediction. The confusion matrix for the various methods is illustrated in Table 2.

Model Performance Evaluation

The performances evaluated with the measures indicate that among the algorithms tested, the logistic regression with L2 yielded the best accuracy of about 97.7% while the L1 yielded an accuracy of approximately 96.9%. But the simple logistic regression model without any form of regularization was only able to achieve 95.4% accuracy, thus, the result unequivocally shows that regularization in some way boosts model accuracy due to its effectiveness in cutting down on over-fitting. The regularization terms, further the L1, also played the role of feature selection and helped in removing some unimportant features in the dataset.

- Logistic Regression (L2): The excellent performance was confirmed by scores that reached 97.7% of precision, 97.7% of recall and F1-score for the L2-regularized logistic regression model. It was also able to classify between mature and immature cataracts without over-fitting and was not affected by multicollinearity.
- Logistic Regression (L1): Using L1 regularization in logistic regression also yielded good results with 96.9% accuracy, precision, recall and F1 score. Its main benefit was fewer features and select components of the images that were enough

Table 2 Confusion matrix for the various methods

LR-L2		Predicted		
		Immature	Mature	Total
Actual	Immature	8471	129	8600
	Mature	243	7557	7800
	Total	8714	7686	16400

LR-L1		Predicted		
		Immature	Mature	Total
Actual	Immature	8414	186	8600
	Mature	318	7482	7800
	Total	8732	7668	16400

LR		Predicted		
		Immature	Mature	Total
Actual	Immature	8221	379	8600
	Mature	383	7417	7800
	Total	8604	7796	16400

kNN		Predicted		
		Immature	Mature	Total
Actual	Immature	8247	353	8600
	Mature	620	7180	7800
	Total	8867	7533	16400

SVM		Predicted		
		Immature	Mature	Total
Actual	Immature	8515	85	8600
	Mature	620	7180	7800
	Total	9135	7265	16400

RF		Predicted		
		Immature	Mature	Total
Actual	Immature	7918	682	8600
	Mature	1366	6434	7800
	Total	9284	7116	16400

LR- L2		Predicted		
		Immature	Mature	Total
Actual	Immature	97.2%	1.7%	8600
	Mature	2.8%	98.3%	7800
	Total	8714	7686	16400

LR-L1		Predicted		
		Immature	Mature	Total
Actual	Immature	96.4%	2.4%	8600
	Mature	3.6%	97.6%	7800
	Total	8732	7668	16400

LR		Predicted		
		Immature	Mature	Total
Actual	Immature	95.5%	4.9%	8600
	Mature	4.5%	95.1%	7800
	Total	8604	7796	16400

kNN		Predicted		
		Immature	Mature	Total
Actual	Immature	93.0%	4.7%	8600
	Mature	7.0%	95.3%	7800
	Total	8867	7533	16400

SVM		Predicted		
		Immature	Mature	Total
Actual	Immature	93.2%	1.2%	8600
	Mature	6.8%	98.8%	7800
	Total	9135	7265	16400

RF		Predicted		
		Immature	Mature	Total
Actual	Immature	85.3%	9.6%	8600
	Mature	14.7%	90.4%	7800
	Total	9284	7116	16400

to give high accuracy and proper classification. This means that there were things in the dataset that were not needed to classify the samples and deleting them made the model easier to understand.

- k-Nearest Neighbors (kNN): kNN had a good result at the end with the accuracy being at 94.1% and precision and recalling also being at 94.1%. The model depended on the choice of "k" in the parameter, and even when optimized, did not scale well with large datasets due to its non-parametric nature and the distance calculations required. Furthermore, the study identified that kNN's performance decreases when a data set is imbalanced; this meant that immature cataract had more false positives (Fig. 3).
- Support Vector Machines (SVM): The better results, although not very high, were achieved by SVM with RBF kernel: accuracy 95.7, precision 95.7, and recall of 95.7. SVM was able indeed to handle nonlinearity and have good generalization

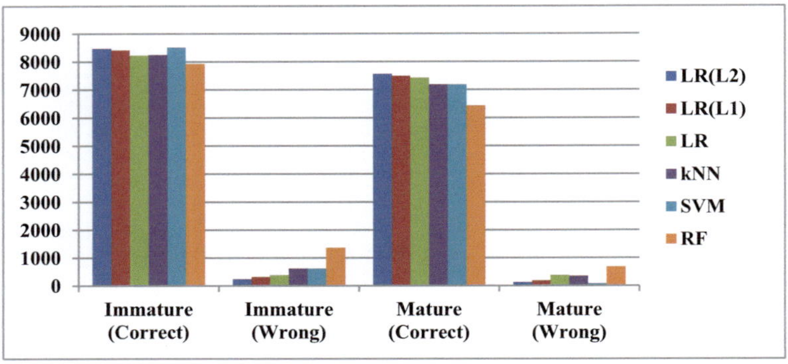

Fig. 3 The outcome of ML modles for cataract identification and classification

abilities; however, it called for huge computation time and high parameter tuning of kernels (Table 3).

- Random Forest: The algorithm that was applied of randomly selected trees was successfully 87.5% precision, and 87.5% recall, rand, the accuracy of the random forest classifier of 87.5%. It corresponded well to noisy and imbalance data; however, the increased complexity of the model resulted in some degree of overfitting, along with being outperformed by logistic regression (Fig. 4).

Table 3 Comparative performance measures of ML models for the proposed method

Method	Acc	F1	Prec	Recall	MCC	Spec	LogLoss
LR-L2	0.977	0.977	0.977	0.977	0.955	0.977	0.056
LR-L1	0.969	0.969	0.969	0.969	0.938	0.968	0.077
LR	0.954	0.954	0.954	0.954	0.907	0.953	0.702
kNN	0.941	0.941	0.941	0.941	0.881	0.939	0.221
SVM	0.957	0.957	0.959	0.957	0.916	0.954	0.122
RF	0.875	0.875	0.877	0.875	0.751	0.870	0.409

Fig. 4 Performance measures of ML modles for cataract classification

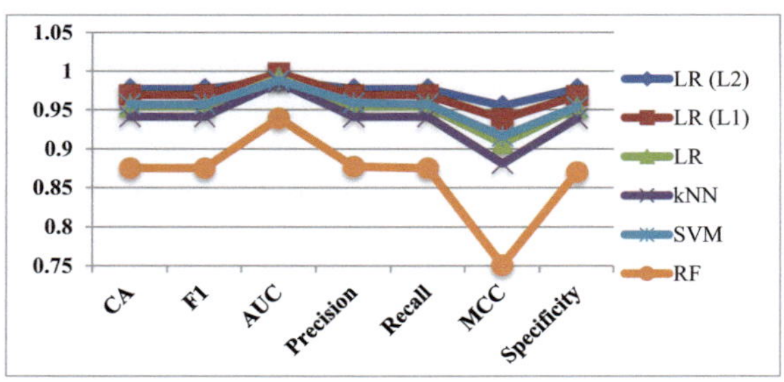

Comparative Analysis

In the present study, machine learning techniques were compared to identify that logistic regression models with L1 norm and L2 norm were the best-performing ones. The superior performance of regularized logistic regression can be attributed to several factors:

- Generalization: L1 and L2 penalties improved the generalization because they encouraged low mainly of coefficients and eliminated over-learning. This was particularly relevant to address concerns related to the dimensionality of the employed clinical and imaging variables for which it promoted a better class classification.
- Feature Selection: In general, L1 regularization can be thought of as a feature selector and helped remove irrelevant features which makes the model more interpretable and less noisy. This simplification conferred improved accuracy and speed over other models such as random forest and SVM.
- Model Interpretability: In the case of medical applications, legitimation has to also adopt the principle of interpretability. The results find that logistic regression, especially with regularization, makes the understanding of how features are related to the outcome easy.

Discussion of Findings

The results show that regularized logistic regression (L1 and L2) provides a good balance between the high classification accuracy of cataracts and the model complexity. The fact that models employing regularized logistic regression perform well for this particular task—that of categorizing cataracts as mature or immature—indicates that linear mappings between features and outcomes are of considerable importance for this particular task, and are amenable to regularization to prevent over-fitting. Regularized versions of the logistic regression models returned fairly good results and even though more complex models like SVM and random forest were fit to the data, they did not outperform the simplest of models. However, the level of improvement achieved was marginal due to high computational cost and excessive sensitivity to the parameter setting of SVM. In a similar manner, though designed for use on noisy data, random forest, because of its nature, was over-fitted and it was found that even though that feature importance ranking could be obtained, they did not greatly improve the classification accuracy.

5 Conclusion

The classification of a cataract as mature or immature is one of the significant challenges associated with ophthalmology, as it helps to make the right decision regarding further action. The proposed work utilized logistic regression, regularized logistic regression (L1, L2), kNN, SVM, and random forest have been assessed in terms of their ability to classify cataracts. Based on experimental results the proposed method has proved that applying machine learning actually has enormous possibilities to recognize diagnosis more accurately, consistently, and faster than a manual approach. Logistic regression, although less complex and easily interpretable, on the other hand, failed to perfect in dealing with the non-linear interaction of the variables but using feature regularization techniques like L1 and L2 made it precise and expanded the generalization ability of the Logistic models. Out of all the models discussed, two models, which were regularized logistic regression, provided the highest accuracy, precision, and better control against over-fitting. Possible future work could look at training deep learning models that use techniques, such as convolutional neural networks to classify medical images. Further, by using a larger number of patients from different categories and selecting even better hyper-parameters, the described methods' applications can be extended.

References

1. Zhou, Y., Li, G., & Li, H. (2020). Automatic cataract classification using deep neural network with discrete state transition. *IEEE Transactions on Medical Imaging, 39*(2), 436–446.
2. Pedram, S. A., Ferguson, P. W., Gerber, M. J., Shin, C., Hubschman, J.-P., & Rosen, J. (2022). A novel tissue identification framework in cataract surgery using an integrated bioimpedance-based probe and machine learning algorithms. *IEEE Transactions on Biomedical Engineering, 69*(2), 910–920.
3. Ganesan, P., Leo Joseph, L. M. I., Ravichandran, M., Subramanian, K. M., & Anu Velavan, S. (2018). Detection and segmentation of retinal blood vessel in digital RGB and CIELUV color space fundus images. *Research Journal of Pharmacy and Technology, 11*(6), 2326–2330
4. Guo, L., Yang, J.-J., Peng, L., Li, J., & Liang, Q. (2015). A computer-aided healthcare system for cataract classification and grading based on fundus image analysis. *Computers in Industry, 69*, 72–80.
5. Qian, X., Patton, E. W., Swaney, J., Xing, Q., & Zeng, T. (2018). Machine learning on cataracts classification using SqueezeNet. In *2018 4th International Conference on Universal Village* (UV). Boston, MA, USA, pp. 1–3
6. K. Funde, J. Joshi, K. Castelino, P. Patil and N. Mulla, "Cataract Detection by Leveraging VGG-19 Classification Model on Retinal Images," 2022 13th International Conference on Computing Communication and Networking Technologies (ICCCNT),Kharagpur, India, 2022, pp. 1–6
7. Ganesan, P., Sathish, B. S., Joseph, L. M. I. L., Kalist, V., & Vasanth, K. (2019). Retinal blood vessels and optical disc segmentation in branch retinal vein occluded fundus images using digital image processing techniques. *Research Journal of Pharmacy and Technology, 12*(4), 1901–1906.
8. Ganesan, P., Ganesh, M., Joseph, L. M. I. L., & Kalist, V. (2018). Central retinal vein occlusion: An approach for the detection and extraction of retinal blood vessels. *Journal of Pharmaceutical Sciences and Research, 10*(1), 192–195.
9. Ganesan, P., & Sathish, B. S. (2017). Automatic detection of optic disc and blood vessel in retinal images using morphological operations and ipachi model. *Research Journal of Pharmacy and Technology, 10*(8), 2602–2606.
10. Junayed, M. S., Islam, M. B., Sadeghzadeh, A., & Rahman, S. (2021). Cataract-net: an automated cataract detection system using deep learning for fundus images. *IEEE Access, 9*, 128799–128808.
11. Pratap, T., & Kokil, P. (2019). Computer-aided diagnosis of cataract using deep transfer learning. *Biomed Signal Process Control, 53.*

12. Junayed, M. S., Islam, M. B., Sadeghzadeh, A., & Rahman, S. (2021). CataractNet: An automated cataract detection system using deep learning for fundus images. *IEEE Access, 9*, 128799–128808.
13. Kayabol, K. (2020). Approximate sparse multinomial logistic regression for classification. In: *IEEE Transactions on Pattern Analysis and Machine Intelligence,* vol. 42, no. 2, pp. 490–493.
14. Steinauer, Zohra, Zhang, K., Guo, C., Zhang, J. (2021). Computational modeling of gene-specific transcriptional repression, activation and chromatin interactions in leukemogenesis by LASSO-Regularized Logistic Regression. In: *IEEE/ACM Transactions on Computational Biology and Bioinformatics*, vol. 18, no. 6, pp. 2109–2122
15. Zhang, Z., Hu*, B., Ma, X., Xu, L. (2015). Resting-state whole-brain functional connectivity networks for MCI classification using L2-regularized logistic regression. In: *IEEE Transactions on Nano Bioscience*, vol. 14, no. 2, pp. 237–247
16. Sugiharto, X. et al. (2022). *Comparison of SVM, Random Forest and KNN Classification By Using HOG on Traffic Sign Detection.* In: *2022 6th International Conference on Informatics and Computational Sciences (ICICoS).* Semarang, Indonesia, pp. 60–65
17. Xiong, Y., He, Z., Niu, K., Zhang, H., & Song, H. (2018). Automatic cataract classification based on multi-feature fusion and SVM. In: *2018 IEEE 4th International Conference on Computer and Communications (ICCC)*, Chengdu, China, pp. 1557–1561
18. Basit, Z. N. A., & Mitani, Y. (2024). Cataract disease detection based on small fundus images dataset using CNNs. In: *2024 2nd International Conference on Computer Graphics and Image Processing (CGIP)*, Kyoto, Japan, pp. 113–117

MedMate: A Contextual Approach for Disease Diagnosis Using Retrieval-Augmented Generation

Avhishek Nandi, Barnali Paul, Piyali Datta, and Deepsubhra Guha Roy

Abstract In today's huge market, we are seeing the advent of chatbots in every field be it e-commerce, tech, and medical field. This paper aims to build a chatbot using the Retrieval-Augmented Generation (RAG) Framework, which is able to detect a user's disease and give proper medical advice holding an interactive conversation with the user knowing the kind of symptoms they are experiencing at that moment. Often, individuals delay seeking medical attention due to uncertainty or limited access. Our chatbot empowers users to quickly assess symptoms and take prompt action, guiding them toward the right type of care and improving their chances for better health outcomes. Our chatbot is designed to serve as an initial line of defense, offering users a reliable tool to recognize symptoms and take the first steps toward appropriate care. It uses the RAG pipeline and the Llama model to construct factually correct medical advice retrieved from the specially created Knowledge Base containing various medical documents, books, journals, and so on.

Keywords RAG · Llama · LLM · Retrieval augmented generation · Medical chatbot · Disease prediction

---

A. Nandi · B. Paul
Department of CSE (AIML), Institute of Engineering and Management, University of Engineering and Management, Kolkata, India

P. Datta
IEM Centre of Excellence for Data Science, Department of CSE (AIML), Institute of Engineering and Management, University of Engineering and Management, Kolkata, India

D. Guha Roy (✉)
IEM Centre of Excellence for Cloud Computing and IoT, Department of CSE (AIML), Institute of Engineering and Management, University of Engineering and Management, Kolkata, India
e-mail: roysubhraguha@gmail.com

J. K. Mandal et al. (eds.), *Recent Advances in Artificial Intelligence and Smart Applications*, Lecture Notes in Networks and Systems 1380,
https://doi.org/10.1007/978-981-96-5822-0_34

1 Introduction

We are witnessing a varied use of chatbots which can predict symptoms, provide you with mental support, and help in selecting medicines. However, the most challenging part in the medical field is that valid and correct information must be provided or it may lead to some serious life-threatening consequences. Most chatbots use rule-based algorithms or just basic keyword matching from a small dataset which is not verified by a domain expert leading to irregular or incorrect responses most of the time, making the use of these chatbots a health hazard or waste of time as it does not provide any conclusive result or treatment. Chatbots that use fine-tuned large language models(LLMs) often face the problem of hallucination. Hallucination in LLMs is a common affair; it is when it gives a response which is factually incorrect or ungrounded. In the medical domain, we always give higher priority to reduce false negatives as it could lead to fatal outcomes. Our medical chatbot tries to solve these existing problems of medical chatbot by using the RAG pipeline and verified knowledge base fetching us accurate results. First, we collect medically verified documents, journals, and books containing the context of all kinds of diseases and their listed symptoms. After this, we process the data and store it into our database which becomes the knowledge base for our RAG model. Our chatbot is specifically designed to answer only questions related to medical disease detection and prediction by utilizing the RAG technique (see Fig. 1). RAG pipeline-based chatbot can

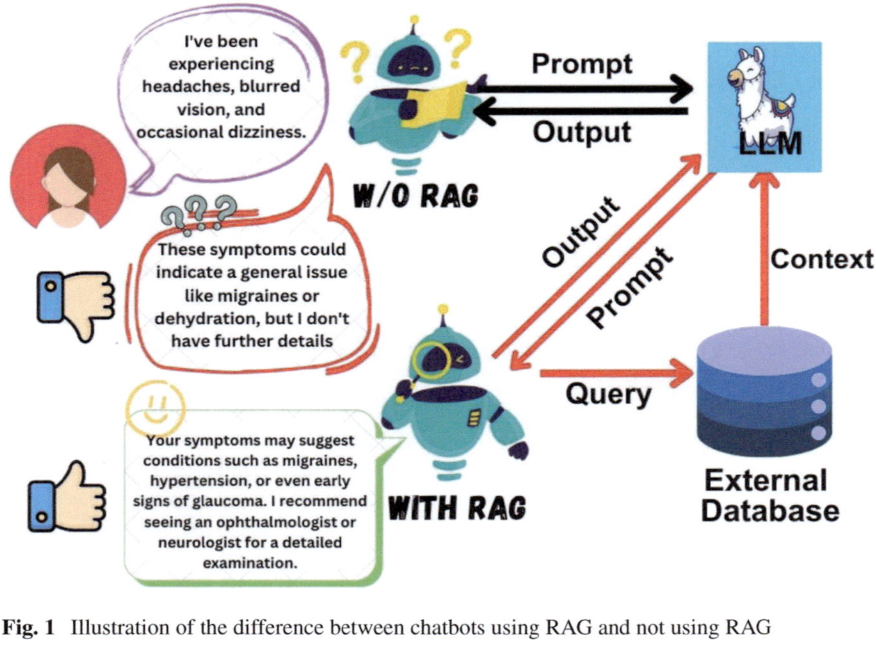

Fig. 1 Illustration of the difference between chatbots using RAG and not using RAG

give correct advice and detect the disease more accurately due to the external source of knowledge we use which consists of updated medical documents and the hallucination is reduced as it refers to the knowledge base only rather than the pre-trained dataset.

2 Proposed Strategy

The model we propose presents an innovative approach to creating medical chatbots through Retrieval-Augmented Generation (RAG) to help users examine the likely health problems based on their symptoms. Earlier AI-driven chatbots mainly rely on pre-trained knowledge but our chatbot combines the power of the Rag model and generative LLM. Here we create a knowledge base with medical documents that are approved by medical professionals. Then process the documents by making them chunk embedding and store them into vector space. After taking the user query the model performs a semantic search to retrieve relevant documents. This document and prompt template passes as input to the LLM. Based on that input LLM model gives relevant output. Here we use the Llama model as LLM. The Llama model has a vast pre-trained dataset and a powerful semantic search mechanism. This approach makes our chatbot more efficient and relevant (see Fig. 2).

2.1 Knowledge Base

A well logically organized and representative knowledge base is the backbone of our Retrieval-Augmented Generation pipe-lined medical chatbot. It serves as the repository from which the model is going to fetch all the essential medical information to construct an accurate and context-driven response to user queries. The main function of the retrieval component of the RAG pipeline is to search and identify the most relevant documents or data present inside the knowledge base. Our knowledge source is derived from a variety of medical documents which consist of academic papers, guidelines issued in the clinic, disease descriptions, and treatment protocols. To create a robust and potent knowledge base for our RAG model we emphasize the following four criteria:

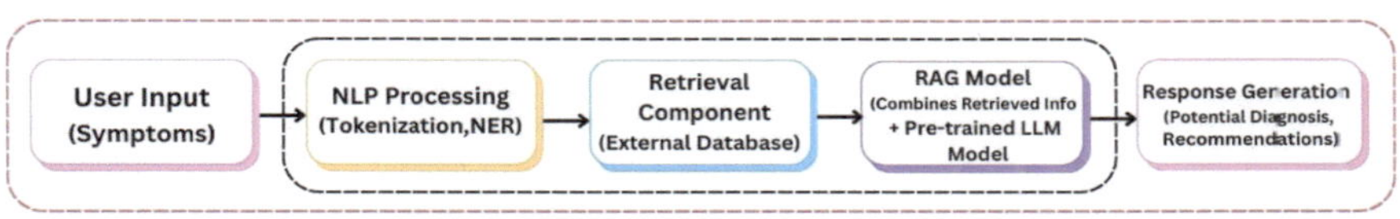

Fig. 2 Process flow diagram of MedMate

- **Relevance:** It is handpicked from clinical studies, books, and other medical literature with guidelines to align with symptom-based queries, empowering the model to provide actionable insights for diagnosing diseases and suggesting well-thought-out treatments.
- **Comprehensiveness:** It consists of a diverse range of diseases, medical conditions, and specialties, addressing the user's response to both common and complex ailments they may face.
- **Currency:** The knowledge base is structured in such a way that it can easily be updated with the latest research and guidelines provided by medical bodies, ensuring the model's relevance in fast-evolving fields like infectious diseases.
- **Perspicuity:** Ensuring the model provides correct and reliable output is of utmost importance in the medical field. This fosters user trust in the product. Our Knowledge Base consists of documents that are published by recognized medical boards, ensuring reliability and accuracy.

Our system includes mechanisms to continuously update the database with new verified sources and remove obsolete or disapproved information, adapting to the changes in public health development and medical practices.

2.2 *Chunking*

Chunking plays an important role in optimizing the RAG model in—the retrieval as well as the generative phase. Segmenting lengthy medical documents into manageable and contextually coherent portions ensures relevant and orderly information processing. Efficient chunking minimizes latency and cost by reducing the number of unnecessary chunks that are retrieved and processed. We are using a combination of fixed length and semantic chunking techniques for divisions. Fixed-length chunking will help in dividing large medical documents into docile sections based on the number of tokens we select making the language model stay within its context window. Semantic chunking divides the medical document into meaningful boundaries such as paragraphs, symptoms, or treatment sections. This approach helps the chatbot in handling complex medical texts while maintaining high-quality interaction with the user.

2.3 *Retrieval*

This mechanism retrieves the external resources, which are provided based on the contextual meaning, and gives accurate answers.

Tokenization: Tokenization is breaking down the text into smaller units called tokens. These tokens can be words, characters, or even sub-words, depending on the strategy of tokenization. The user input given to the chatbot is broken down into tokens and processed for better accuracy. We transform the unstructured user query into a structured format. The inherent nature of the natural language is its ambiguity, so splitting them into individual tokens, which can then be analyzed in the context of their surrounding tokens can help deal with this issue.

Encoding: Before Indexing, the chatbot takes the user's symptoms as query $input$, which is tokenized and transformed into embeddings. This is done with the help of the BERT language model. BERT which stands for Bidirectional Encoder Representations from Transformers is implemented to generate a language model such that solely the encoder mechanism is used. Sequences of tokens, i.e., the natural language input in tokenized format are fed to the Transformer encoder. These tokens are first embedded into vectors i.e. converting the words to numbers and then processed in the neural network. The output of the network is a sequence of vectors, each corresponding to an input token, providing contextualized representations (See Fig. 3). Let the query $input$ be represented as a sequence of tokens $\{x_1, x_2, ..., x_T\}$. The embedding $\mathbf{q}(input)$ is defined as:

$$\mathbf{q}(input) = \text{Encoder}(input)$$

where the **Encoder** is a pre-trained BERT language model that converts the query into subsequent vector $\mathbf{q}(input) \in \mathbb{R}^d$, with d being the dimension of the embedding space. The model is given information about where the words are in the input sequence by a **positional encoding** that is a fixed-size vector representation of the relative positions of tokens within a sequence. In the absence of a positional encoding, the model will not be able to process the input sequence as more than a bag of words. Positional encoding is a function $f : \mathbb{R} \to \mathbb{R}^d$, where d is a positive even integer. The encoding is defined as:

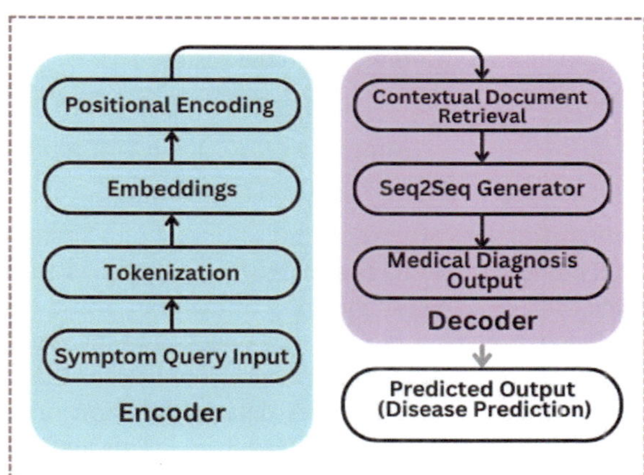

Fig. 3 Using encoder-decoder model in disease prediction

$$(f(t)_{2k}, f(t)_{2k+1}) = (\sin(\theta), \cos(\theta)) \quad \forall k \in \{0, 1, \ldots, \frac{d}{2} - 1\}$$

where $\theta = \frac{t}{r^k}$, and $r = N^{\frac{2}{d}}.N$ is a parameter, typically set to 10000. Alternatively, in complex form, it is written as:

$$f(t) = \left(e^{it/r^k}\right)_{k=0,1,\ldots,\frac{d}{2}-1}$$

The shift property allows the positional encoding of any step to be computed:

$$f(t + \Delta t) = \text{diag}(f(\Delta t))f(t)$$

Thus, linear sums of positions are expressed as:

$$\sum_j c_j f(t + \Delta t_j) = \left(\sum_j c_j \,\text{diag}(f(\Delta t_j))\right) f(t)$$

Attention Mechanism: The attention mechanism is based on scaled dot-product attention which allows the model to focus on different parts of the input sequence individually. For each attention head, three weight matrices are created: W^Q (query), W^K (key), and W^V (value). The query, key, and value sequences are transformed using these matrices:

$$Q = X_{\text{query}} W^Q, \quad K = X_{\text{key}} W^K, \quad V = X_{\text{value}} W^V$$

The attention weights between query Q and key K vectors are computed as the dot product, scaled by $\frac{1}{\sqrt{d_k}}$, where d_k is the key dimension. This is followed by a softmax normalization to yield the result:

$$\text{Attention}(Q, K, V) = \text{softmax}\left(\frac{QK^T}{\sqrt{d_k}}\right) V$$

Embeddings After the data are reformed into vector representation which is called embedding. Embeddings are high-dimensional vectors which help to understand the contextual or semantic meaning of the text. These embeddings are generated using specialized medical models that understand the medical terminology. Embeddings are stored in a vector database which optimizes the search in the high-dimensional vector. Vector database stores data in the form of vectors, which are an array of numbers—e.g., [0.1, 3.21, –1.3, 9.2, …]. This approach allows for efficient similarity searches, as it groups similar data and enables models or applications to retrieve relevant information effectively.

Fig. 4 Flowchart of documents chunking to Indexing

Indexing Indexing in Retrieval-Augmented Generation (RAG) means organizing data in a vector database to make it easier to search and retrieve. Raw data is first loaded as Document objects containing text (like DocumentLoaders), then split into smaller chunks (like DocumentTransformers). In RAG, indexing typically involves analyzing this loaded content and converting the content into a format suitable for fast retrieval. This typically means that each document is represented in the form of a vector (using word embeddings or transformer models, for example) and these vectors are stored in a fast accessible way, normally in a vector database or a search engine (See Fig. 4).

Semantic Search Our model is using a semantic search approach. Semantic search employs a more effective method of retrieval than the older system, which depended mainly on the use of keywords. In searching based on keywords, which is the previous system that was used before, the result is sometimes contextually incorrect. Semantic search captures the context behind the user query and gives a response according to it. This stage is performed to ensure that retrieved document chunks are contextually relevant, it enhances the quality of generative output.

FAISS, which stands for Facebook AI Similarity Search, is a Python library aimed at carrying out the approximate nearest neighbor search in high-dimensional spaces, in a fast and efficient manner. This is ideal for text search because these two types of data are represented as vectors in high dimensions. Let $D = \{d_1, d_2, ..., d_N\}$ be the set of documents, where each document d_i is embedded as $\mathbf{d}_i$. The similarity between the query and a document is computed using the cosine similarity or dot product:

$$\text{sim}(\mathbf{q}(x), \mathbf{d}_i) = \frac{\mathbf{q}(x)^\top \mathbf{d}_i}{\|\mathbf{q}(x)\| \|\mathbf{d}_i\|}$$

Alternatively, the dot product can be used:

$$\text{sim}(\mathbf{q}(x), \mathbf{d}_i) = \mathbf{q}(x)^\top \mathbf{d}_i$$

The top-K documents with the highest similarity scores are retrieved:

$$D_{top-K} = \text{Top-K}(\text{sim}(\mathbf{q}(x), \mathbf{d}_i))$$

where D_{top-K} represents the set of K most relevant documents. After retrieving all relevant document chunks, it is passed to the augmented phase. In this phase, relevant documents are combined with a prompt template which is passed as input to a generative model such as a Large Language Model (LLM) to get more contextually correct and relevant output. Between all ranked documents top-ranked documents are retrieved for processing.

2.4 Generation

In the retrieval part after retrieving the relevant document chunks based on user query, data is fed into the Large Language Model. Here, the document chunks are used as contextual input to the LLMs. LLMs have specialization in generating text by processing vast data and are sometimes called transformer models. The transformer model is a neural network that learns context and semantic meaning in sequential data like text. The transformer works on an attention mechanism that allows the model to focus on the specific part of the input when the model generates the text. We use here Llama LLM model which is already trained with vast medical data. Llama LLM has more than 50 million parameters. This model pre-trained on approximately 15 trillion tokens of text gathered from publicly available sources. The dataset used for Llama is predominantly English, with the other 30 languages constituting more than 5%. Additionally, this dataset was processed through a text-quality filter (See Fig. 5)

2.5 Customised Prompt Template

In the chatbot model, we have materialized a customized prompt template which is carefully designed to optimize symptom analysis and disease diagnosis. The template will integrate the user query and divide it into a structured format that guides the

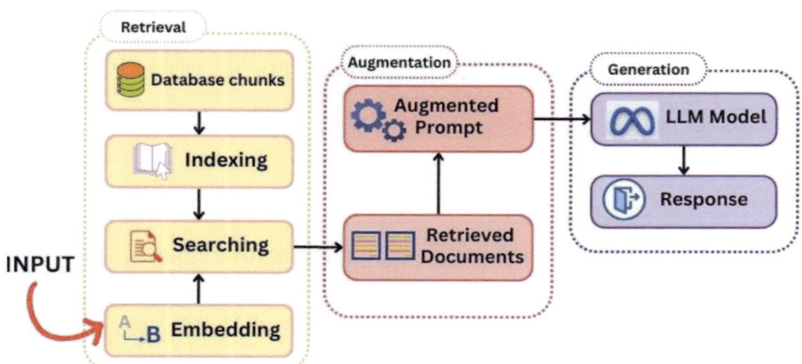

Fig. 5 Architecture of RAG model

model's correspondence with the indexed knowledge base. The prompt is tailored exactly to extract relevant medical data needed, narrowing down the possible disease diagnoses from the large number present while also ensuring it provides the user with actionable insights. The template is structured in the following way:

1. Symptoms Extraction: Extract and/or determine the symptoms of the user based on the query given by the user, ensuring that the retrieval system picks the accurate data block from the knowledge base.

2. Contextual Enhancements: Additional information fields like duration of symptoms, intensity, and previous medical history provide the model with a better understanding of the context, improving the relevance of the response.

3. Query Refinement: The template helps in generating precise queries custom-made for the RAG pipeline from the cluttered input provided by the user, refining the procurement of disease information and treatment guidelines, and also providing clear and concise recommendations.

Hence, little structuring of data makes it more effective for the chatbot to communicate with the knowledge base ensuring high-quality diagnostic responses tailored to the user's input and providing a seamless reliable experience to the user for his medical inquiry and being at their assistance.

3 Case Study

To understand and compare the different scenarios the user may be facing, we choose 2 typical case studies to demonstrate the situation in a better manner.

3.1 MedMate Versus Web Search

Imagine a potential user is experiencing ambiguous symptoms like discomfort in the chest of a mild nature, shortness of breath, and fatigue. He/she is unsure if all these are connected or a just coincidence and wants to understand what may or may not be wrong with their health. In this case study we draw a comparison of what could have happened if they used our model versus if they did a traditional web search. With our model, the user did not have to categorize or specifically define the kind of symptoms they were experiencing prior (Fig. 6).

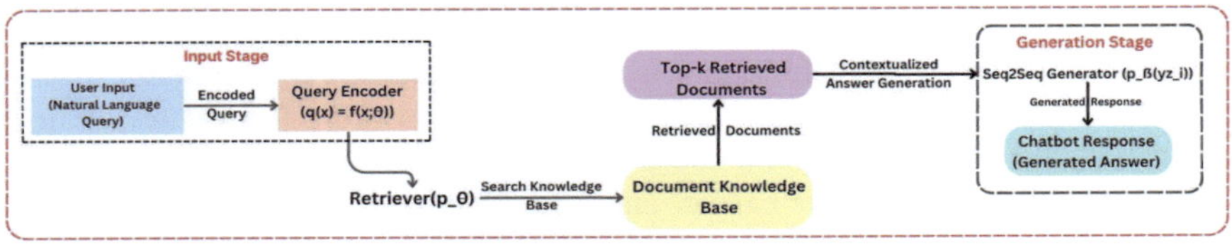

Fig. 6 Workflow diagram of our Medical Chatbot

Algorithm 1: MedMate's working Algorithm (See Fig. 6)

Step 1: User Query: A query (Q) describing symptoms or concerns.
Example Input: "I have a headache and feel fatigued."
Step 2: Query Encoding
Action: Q is encoded into a vector representation using BERT.
Output: Encoded query vector (Q_v) of the user's input (Q).
Example Output: Query embedding [0.12, 0.95, −0.37, ...]
Step 3: Semantic Search in Knowledge Base
Action: Retriever performs a semantic search using the query embedding to find relevant information in the knowledge base.
Output: Relevant documents or chunks of data ($Q_c h$) matching the user query.
Example Output: "Document about migraines", "Document about flu symptoms", ...
Step 4: Convert Retrieved Documents to Embeddings
Action: The retriever converts the relevant documents into vector embeddings (Q_{em}).
Output: Embeddings of the relevant documents.
Example Output: Document embeddings (D_{em}) [0.45, −0.22, 0.78, ...]
Step 5: Send Embeddings to the Generator
Action: The vectorized documents combined with the original query.
Output: Passed embeddings to the generator.
Example Output: Document embeddings [0.45, −0.22, 0.78, ...]
Step 6: Combine Embeddings with Query
Action: The generator merges D_{em} with the original Q_{em} to create a context-aware input for the LLM.
Output: Combined query-document embeddings.
Example Output: Combined embeddings (DQ_{em}) [0.32, 0.15, −0.45, ...]
Step 7: Generate Response
Action: The generator passes DQ_{em} to an LLM to generate a response with a customized prompt template.
Output: Generated response addressing the user's query with medical advice.
Example Output: "Based on your symptoms, you may have a migraine. It's recommended to consult a neurologist for further diagnosis."

The bot is designed to handle this kind of raw, natural input such as 'I am feeling tired, sometimes my chest feels tight and I feel out of breath.' Our Model processes this input to extract relevant symptoms from this query and tries to find the perfect match in its comprehensive validated knowledge base. This complete experience makes the user-guided toward the right specialist and gets immediate preventive measures if required (See Fig. 7).

In contrast to this, if the user turns to the basic web search the process becomes much more ambiguous as the user now has to guess and come up with their symptoms

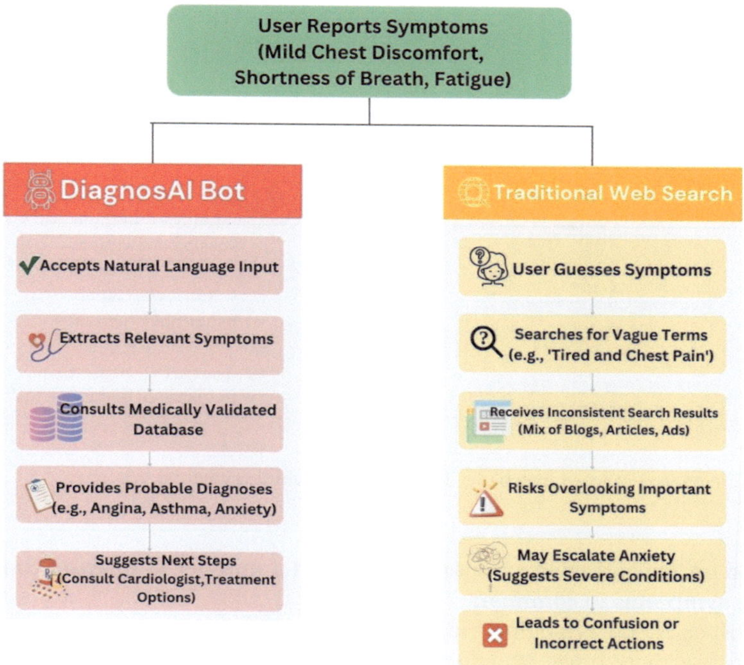

Fig. 7 Comparison between MedMate query result versus web search

even if they are not sure of it. A 'mild chest discomfort' may turn into 'chest pain'; this may not only overlook important symptoms but can also lead them to a rabbit hole of irrelevant and frightening information. Without the context-sensitive understanding that our model offers a simple web search lacks the depth and nuances to interpret the user's real condition accurately, which could lead to push them in the wrong direction and delay in taking proper medical treatment or worse acting on false assumptions.

3.2 *Exact Match Versus Multiple Symptom Match*

Based on the user query and extracted symptoms the chatbot during its retrieval of data from the knowledge base might come across two scenarios (see Fig. 8).

- **Based on the extracted symptom, we found an exact match to a disease.** The symptoms provided by the user closely match the signature symptoms of a particular disease allowing the chatbot to immediately retrieve the related information regarding that disease. It assures the user receives highly specific responses and minimizes any uncertainty.

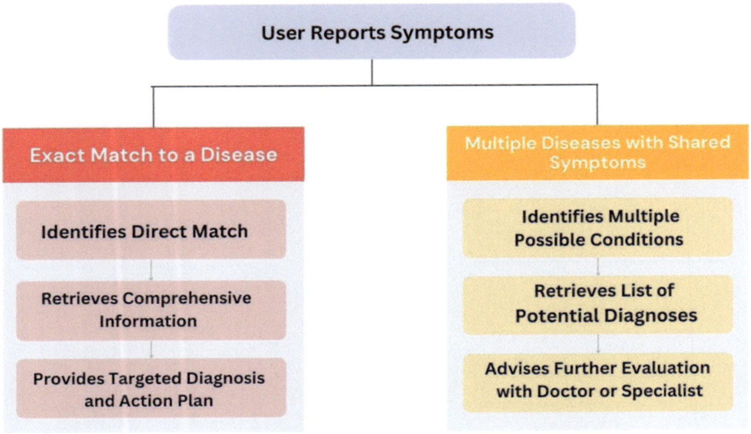

Fig. 8 Comparison of exact match to a disease versus multiple symptom match

- **Based on the extracted symptoms, we found a number of diseases that have these symptoms in common.**
  In most cases, the symptoms given by the user will overlap with several diseases, mainly for more broad symptoms like headache, fatigue, or fever. The chatbot retrieves the K-most relevant diseases from the knowledge base and presents the user with respective potential diagnoses. It includes an explanation of different possibilities and advises the user to consult a doctor and have diagnostic tests to narrow down the causes (See Fig. 8).

Conclusion

The MedMate chatbot stands out as an innovative solution that enables users to receive initial health evaluations utilizing Retrieval-Augmented Generation (RAG) models alongside natural language processing techniques. The chatbot has multi-disease detection abilities, thus providing a convenient portal where the patient gets insights into his/her complaints as well as instructions for getting a professional diagnosis. During internal evaluations, MedMate was tested extensively with a dataset of simulated user inputs and achieved an accuracy rate of approximately 76 %, successfully aligning its recommendations with professional medical advice in 7 out of 10 cases. The integration of the chatbot with the hospital systems allows patients to identify their symptoms and arrange an appointment in case of a slight condition without any interruption of care which is not going to happen thus they will ensure continuous care.

Acknowledgements We would like to express our heartfelt gratitude to the IEM Centre of Excellence for Cloud Computing and IoT for invaluable support and resources from IEM Grant in Aid project of Dr. Deepsubhra Guha Roy-IEMT(S)/2024/02-G19 and IEMT(S)/2023/02-G05 for enabling us to perform and research on this paper and to the IEDC-CSE of Institute of Engineering & Management, Kolkata, for providing us the workspace for this work.

References

1. Kirubakaran, S., Jasper, G., Grace, E., Raja, J., Mahimai, A. P. R., Ruban, G. S., & Yuvaraajan, E. (2024). A RAG-based medical assistant especially for infectious diseases. In *Proceedings of the International Conference on Information and Communication Technology* (pp. 1128–1133). https://doi.org/10.1109/ICICT60155.2024.10544639
2. Vakayil, S., Juliet, D. S., & Vakayil, A. J. (2024). RAG-Based LLM chatbot using Llama-2. In *2024 7th International Conference on Devices, Circuits and Systems (ICDCS)* (pp. 1–5). Coimbatore, India. https://doi.org/10.1109/ICDCS59278.2024.10561020
3. Church, K. W., Sun, J., Yue, R., Vickers, P., Saba, W., & Chandrasekar, R. (2024). Emerging trends: a gentle introduction to RAG. *Natural Language Engineering, 30*(4), 870–881. https://doi.org/10.1017/S1351324924000044
4. Radhakrishnan, A. (2024). Retrieval is all you need: Developing an AI powered chatbot with RAG in Azure (Dissertation). Retrieved from https://urn.kb.se/resolve?urn=urn:nbn:se:ltu:diva-108983
5. Maryamah, M., Irfani, M. M., Tri Raharjo, E. B., Rahmi, N. A., Ghani, M., & Raharjana, I. K. (2024). Chatbots in academia: A Retrieval-Augmented generation approach for improved efficient information access. In *2024 16th International Conference on Knowledge and Smart Technology (KST)* (pp. 259–264). Krabi, Thailand. https://doi.org/10.1109/KST61284.2024.10499652
6. Wang, C., Ong, J., Wang, C., et al. (2024). Potential for GPT technology to optimize future clinical decision-making using retrieval-augmented generation. *Ann Biomed Eng, 52*, 1115–1118. https://doi.org/10.1007/s10439-023-03327-6
7. Ghanbari Haez, S., et al. (2024). A retrieval-augmented generation strategy to enhance medical chatbot reliability. In Finkelstein, J., Moskovitch, R., Parimbelli, E. (Eds.), *Artificial Intelligence in Medicine. AIME 2024, LNCS* (vol. 14844, pp. 1–8). Cham: Springer. https://doi.org/10.1007/978-3-031-66538-7_22
8. Ebrahimiyan, N., et al. (2021). Tokenization and its application in different countries. *Journal of FinTech and Artificial Intelligence, 1*(1), 14–19.
9. Song, X., et al. (2020). Fast wordpiece tokenization. arXiv preprint arXiv:2012.15524.
10. Sockin, M., & Xiong, W. (2023). Decentralization through tokenization. *The Journal of Finance, 78*(1), 247–299.
11. Wu, S., et al. (2023). Next-gpt: Any-to-any multimodal LLM. arXiv preprint arXiv:2309.05519.
12. Yao, Y., et al. (2024). A survey on large language model (LLM) security and privacy: The good, the bad, and the ugly. *High-Confidence Computing*, 100211.
13. Lin, X., et al. (2024). Data-efficient fine-tuning for LLM-based recommendation. In *Proceedings of the 47th International ACM SIGIR Conference on Research and Development in Information Retrieval*
14. Ma, X., Fang, G., & Wang, X. (2023). LLM-Pruner: On the structural pruning of large language models. *Advances in Neural Information Processing Systems, 36*, 21702–21720.
15. Wang, C., Long, Q., Meng, X., Cai, X., Wu, C., Meng, Z., Wang, X., & Zhou, Y. (2024). BioRAG: A RAG-LLM framework for biological question reasoning. arXiv preprint arXiv:2408.01107.

Domain Ontology and Character-Topic Relationship in Srimad-Bhagavatam

Apurba Paul and Dipankar Das

Abstract **The Srimad-Bhagavatam** is a foundational text in Hindu mythology, characterized by intricate narratives and profound philosophical teachings. This paper addresses the analytical challenges of the text by developing tailored ontologies: **`SBC-Ontology`** for characters and **`SBT-Ontology`** for topics, which together create a structured framework for deeper understanding. Advanced transformer models, including BERT, DistilBERT, and RoBERTa, were utilized to analyze character-topic relationships, with an ensemble approach(F1-score 0.79) demonstrating superior performance in capturing nuanced relationships. Despite the text's inherent complexities and limited data availability, this research significantly contributes to digital humanities, providing innovative tools for exploring and interpreting the rich narratives of **The Srimad-Bhagavatam**.

Keywords Ontology · Character-topic relationship · Srimad-bhagavatam · Knowledge representation · Transformer-based models

1 Introduction

Analyzing complex ancient texts like **The Srimad-Bhagavatam**[1] presents significant challenges due to their intricate narratives and profound philosophical concepts. As a cornerstone of Hindu philosophy, this text spans twelve cantos filled with narratives, hymns, and discourses on Lord Vishnu and his avatars.

---

[1] https://prabhupadabooks.com/sb.

A. Paul (✉)
Institute of Engineering and Management, University of Engineering and Management, Kolkata, India
e-mail: apurba.saitech2021@gmail.com

A. Paul · D. Das
Jadavpur University, Kolkata, India

J. K. Mandal et al. (eds.), *Recent Advances in Artificial Intelligence and Smart Applications*, Lecture Notes in Networks and Systems 1380,
https://doi.org/10.1007/978-981-96-5822-0_35

To tackle these challenges, this research focuses on two key methods: *domain ontology development* and *character-topic relationship* analysis. The Domain Ontology specifies concepts within the domain, identifying entities and revealing familial relationships, thematic links, and constraints [1]. Such ontologies provide valuable insights, aiding both human scholars and intelligent software agents in understanding the characters and themes in the text [2]. Meanwhile, Character-Topic relationship analysis employs computational techniques to uncover thematic connections between characters and topics, thereby enhancing scholarly comprehension and accessibility.

This study identifies critical gaps in existing ontologies for the Srimad-Bhagavatam, emphasizing the absence of comprehensive domain-specific datasets for Hindu mythology, which limits character representation and interaction analysis.

The absence of comprehensive domain-specific datasets for Srimad-Bhagavatam is the critical gap identified by this study. Precisely this gap limits character representation and interaction analysis of Hindu mythology. It also highlights the lack of robust character-topic datasets, impacting the accuracy of association analyses.

By identifying characters, relationships, and related subjects, the research seeks to create a thorough domain ontology that will improve comprehension of the text's philosophy and story. With the use of sophisticated transformer models like BERT, RoBERTa, and DistilBERT, it analyzes character-topic relationships.

Key contributions include the creation of specialized datasets for Character and Topic Ontologies, the development of a Character-Topic Relationship dataset for in-depth analysis, and the implementation of transformer models.

A visual representation of the research framework is demonstrated in Fig. 1. The figure highlights two main components of this research. The first component focuses

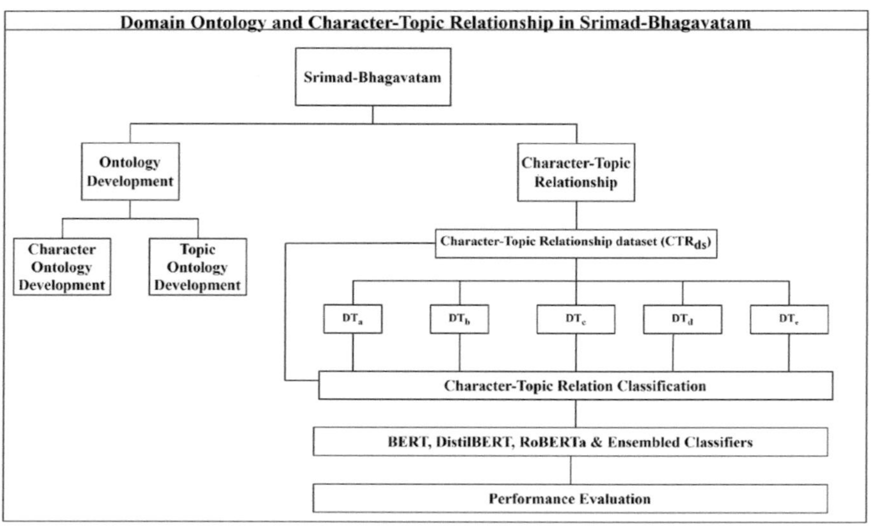

Fig. 1 Proposed research framework

on the development of domain ontologies, including Character Ontology and Topic Ontology. The second component involves utilizing the CTR_{ds} dataset to create five distinct sub-datasets (DT_a, DT_b, DT_c, DT_d, DT_e). The performance of various models, including BERT, DistilBERT, RoBERTa, and an ensemble method, is then evaluated using these sub-datasets along with CTR_{ds}.

The paper is organized as follows: it begins with the *Related Work*, followed by the *Dataset Preparation* and *System Framework* sections. This is followed by *Result Analysis*, *Error Analysis*, and *Discussion & Observations*. The paper concludes with the *Conclusion & Future Scope* section.

2 Related Work

Research on domain ontology and character-topic relationships in religious texts, particularly the Srimad-Bhagavatam, has garnered attention for enhancing understanding through structured frameworks. However, studies specifically addressing the ontological representation of characters and topics in these texts are limited. Varadarajan et al. [3] introduced GENOME, a methodology for improving the ontological modeling of epics. Syamili et al. [4] developed an ontology for Greek mythology, while Juan-Antonio [5] created the Ontology of Greek Mythology (OGM) to enhance semantic representation on the Semantic Web.

Watrobski [6] discussed ontology learning methods, and Espinoza-Arias et al. [7] examined smart city ontologies, emphasizing their lack of interoperability. Noy et al. [8] traced the evolution of ontologies, highlighting initiatives like RDF for knowledge encoding. Raja Mohan et al. [9] presented a Protege-based methodology for Indian medicinal plant ontology, while Hadaiq et al. [10] described an automated wayang ontology construction method.

Kulthida et al. [11] merged digital humanities and folktale studies in the Greater Mekong Subregion, while Sattar et al. [12] reviewed ontology methodologies in Interlocking Institutional Worlds. Franziska et al. [13] proposed a technique to organize myth variants using shallow ontologies. Ibukun et al. [14] developed a domain ontology for Nigerian history through a semi-automated approach, and Wolff et al. [15] suggested a method for rapid ontology construction using text mining.

Vlasenko et al. [16] developed a domain ontology for monitoring the condition of equipment in designing intelligent decision-support systems. Haridy et al. [17] proposed a tourism ontology development methodology beneficial for e-tourism. Pan et al. [18] presented a semi-automatic ontology development framework to build an energy management ontology and evaluated the data coverage.

3 Preparation of Datasets

Preparing datasets for studying domain ontology and character-topic relationships in the Srimad-Bhagavatam requires a structured approach to capture the thematic and narrative complexities of this ancient scripture. Paul et al. [19] developed an extensive dataset for the Pouranic topic classification challenge based on Indian mythology. From this dataset, we selected the Srimad-Bhagavatam for our domain ontology development and character-topic relationship tasks. The procedures for creating the datasets for these tasks are detailed in the following sections.

3.1 Datasets for Domain Ontology

We separated our work into two distinct subtasks for the domain ontology: character and topic ontology development, and we then prepared the dataset accordingly.

The text encompasses 1,714 characters, categorized into 17 types, including *God*, *Demigod*, *Human*, and *Animal*. The *God* category has four sub-categories: *Godhead*, *Goddess*, *Lord's-spiritual-energy*, and *Bhagavan*. The *Human* category includes 15 sub-categories, such as *King*, *Queen*, and *Sage*. In the *Animal* category, subtypes include *Elephant* and *Horse*. Examples of character classifications include: `Durga` ***rdf:type*** `Goddess`, `Narada` ***rdf:type*** `Devarsi`, and `Bharata` ***rdf:type*** `Maharaja`. The text identifies 24 family relation types, such as *father of* and *sister of*.

The Srimad-Bhagavatam is divided into 12 cantos. Each canto contains a varying number of chapters, some with fewer chapters and others with more extensive ones. These topics encompass stories, dialogs, teachings, and philosophical discussions. For example, the Second canto, titled *The Age of Deterioration*, contains 10 topics, out of which the first two topics are *The First Step in God Realization* and *The Lord in the Heart*.

3.2 Datasets for Character-Topic Relationship

We extracted sentences from The Srimad-Bhagavatam for each character and computed sentiment attributes-*polarity*, *subjectivity*, *positivity*, and *negativity*-using the *NaiveBayesAnalyzer* of TextBlob.[2]

These metrics were aggregated to reveal relationships and patterns among characters based on their appearances in various topics. The **PouranicTopic** dataset was utilized to tag sentences, resulting in the CTR_{ds} dataset, which contains 45,941 observations. For instance, Table 1 analyzes a sentence related to Ramacandra from

[2] https://textblob.readthedocs.io/en/dev/.

Table 1 Sample observation of CTR_{ds}

Character	Canto	Sentence	Pol	Subj	Pos	Neg	Senti	VP	Topic
Ramacandra	Liberation	All the brahmanas who were	1	0	0.62	0.37	pos	'brahmanas', 'engaged'	Lord Ramacandra Rules the World

Pol=Polarity; Subj=Subjectivity; Pos=Positivity; Neg=Negativity; Senti=Sentiment; VP=Verb or a particle attached to a verb

the Liberation section, noting a positive polarity (1), objective subjectivity (0), a positivity score of 0.62, and a negativity score of 0.37. The associated verbs are *brahmanas* and *engaged*, with the topic being Lord Ramacandra Rules the World. We constructed five datasets (DT_a, DT_b, DT_c, DT_d, DT_e) from the CTR_{ds} to evaluate various features on character-topic relationships. Each dataset includes different features, with the classifiers BERT, DistilBERT, RoBERTa, and an ensemble model tested on them.

Table 2 summarizes the CTR_{ds} dataset, highlighting the range of Polarity scores (–1 to +1), Subjectivity (0 to 1), Positivity (0.002 to 1), and Negativity (0.002 to 0.99). The dataset comprises 38,296 positive and 7,645 negative sentences. Notable topics include *Gajendras Prayers of Surrender*, *The Killing of the Demon Trnavarta*, *The Descendants of Ajamidha*, *The Gopis Songs of Separation*, and *Five Queens Married by Krsna*.

Table 2 Descriptive statistics of the CTR_{ds} dataset

Pol range	Subj range	Pos range	Neg range	Total pos	Total neg	Frequent top 5 topics
–1 to +1	0 to 1	0.002 to 1	0.002 to 0.99	38296	7645	Gajendras Prayers of Surrender
						The killing of the demon Trnavarta
						The descendants of Ajamidha
						The gopis songs of separation
						Five queens married by Krsna
						The liberation of king Jarasandha
						The sages teachings at Kuruksetra

Pol range: Polarity score range; Subj range: Subjectivity score range; Pos range: Positivity range; Neg range: Negativity score range; Total pos/neg: Total positively/negatively classified sentences

4 System Framework

This section outlines the system framework for exploring domain ontologies, focusing on character and topic ontologies, along with character-topic relationships in The Srimad-Bhagavatam. The framework consists of two main components: Domain Ontology and Character-Topic Relationship.

4.1 Domain Ontology Development

The Character and Topic ontologies for The Srimad-Bhagavatam were developed following Stanford University's guidelines [8]. The Character ontology, referred to as SBC-Ontology, and the Topic ontology, called SBT-Ontology, are illustrated in Fig. 2.

To clarify motivation, competency questions in natural language define the scope and coverage of both ontologies, ensuring they address all relevant queries. Examples include: *Who is Abhimanyu? Who is Ramacandra? What is Srimad-Bhagavatam? What topics are available in each canto of Srimad-Bhagavatam?*

In the case of existing ontology, we found mythological ontologies, notably the Mahabharata ontology[3] from IIT Madras.

Following research by [20, 21], we extracted and categorized characters and topics in The Srimad-Bhagavatam for Knowledge Acquisition and Glossary of Terms. The SBC-Ontology includes characters, genders, and relationships, while SBT-Ontology addresses cantos, topics, and hierarchies.

For classes and hierarchy, SBC-Ontology categorizes terms into classes such as Animal, Celestial-Beings, and God, while SBT-Ontology includes Canto and Topic classes. For a detailed illustration of these classifications, refer to Fig. 3.

The key properties in SBC-Ontology include *hasFather* and *isFatherOf*, while SBT-Ontology features *hasSentence* and *isBelongsToCanto* (Fig. 3).

The Knowledge Production phase integrates data to enhance semantic depth, with *MythoDomainEntity* as the top-level class in SBC-Ontology, defined via the *subClassOf* property.

The SBC-Ontology employs annotation and object properties to represent complex relationships, including lineage and parent-child connections. It comprises 3,530 axioms (2,551 logical), defines 86 classes, and includes 797 individuals, showcasing a detailed semantic structure. The SBT-Ontology organizes concepts into classes like Mythology and Canto, linking them through properties such as *hasTopic*. It contains 1,457 axioms (1,258 logical), 8 classes, and 451 individuals.

The structure of the ontologies is validated by the domain experts using the popular tool named Protégé[4] and finally populated the ontology.

[3] https://sites.google.com/site/ontoworks/about-us.

[4] https://protege.stanford.edu/.

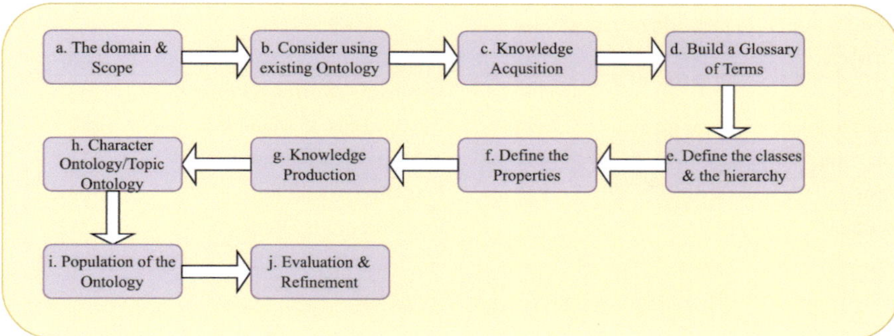

Fig. 2 Proposed model framework for ontology development

SBC-Ontology	
Class Hierarchy	**Object Properties**
owl: Thing	owl : topObjectProperties
MythoDomainEntity	hasWife
Hill	hasHusband
Hindu-Mythological-Characters	hasFather
Animal	hasMother
Celestial-Beings	isWifeOf
Deity	isHusbandOf
Devarsi	isFatherOf
Devotee	isMotherOf
Emperor	isDevoteeOf
SBT-Ontology	
Class Hierarchy	**Object Properties**
owl: Thing	owl : topObjectProperties
Canto	hasSentence
Mythology	hasTopic
Person	isBelongsToCanto
Sentence	isBelongsToMythology
Topic	isBelongsToTopic

Fig. 3 Class hierarchy and object properties of **SBC** and **SBT-Ontology**

The Ontology evaluation assesses quality and correctness [22], with experts manually reviewing the ontologies for accuracy.

4.2 *Character-Topic Relationship*

Understanding the relationship between characters and topics in The Srimad-Bhagavatam involves analyzing how characters engage with and contribute to various topics throughout the text. We first identified the top 5 characters and topics by frequency to highlight their significance in The Srimad-Bhagavatam. Table 3 illustrates

Table 3 Top 5 frequent characters and topics

Characters	Freq.	Topics	Freq.
Krsna	6445	Krsna, the supreme personality	792
Supreme personality of godhead	1574	Descendants of Ajamidha	641
Balarama	1017	Advent of lord Krsna	531
Vasu	956	Diti vows to kill Indra	509
Brahma	951	Pastimes of lord Ramacandra	421

that Krsna is the most frequently mentioned character, while *Krsna, the Supreme Personality* is the most common topic.

Topic distribution task aims to identify the most frequent topics linked to specific characters. Understanding how topics are distributed across the text allows us to see which topics are most relevant to particular characters. The character ***Krsna*** appears most frequently in the following top 5 topics: *Krsna, the Supreme Personality of Godhead* (106 times), *The Descendants of Ajamidha* (90 times), *The Advent of Lord Krsna-Introduction* (85 times), *Five Queens Married by Krsna* (74 times), and *The Story of the Syamantaka Jewel* (66 times).

Now we will discuss various classification models for the Character-Topic Relationship. The Character-Topic Relationship Model leverages advanced natural language processing techniques to analyze and predict relationships between characters and topics in mythological texts. We utilized several SOTA models: BERT [23] for contextual understanding, DistilBERT[24] for faster training, and RoBERTa [25] for improved pretraining. Each model was optimized with the standard configurations The evaluation of the models was performed using categorical cross-entropy loss along with the Adam optimizer. To enhance performance, we implemented an ensemble classifier that combines outputs from BERT, DistilBERT, and RoBERTa through majority voting, leveraging individual strengths to improve overall Precision, Recall, and F1-score.

5 Result Analysis

Table 4 provides a detailed comparison of performance metrics across different models on various datasets. The models evaluated include BERT, DistilBERT, RoBERTa, and an Ensemble approach. Each row in the table represents a different dataset, denoted as DT_a, DT_b, DT_c, DT_d, DT_e, and CTR_{ds}.

The BERT classifier demonstrates stable performance, with Precision (Pr) scores varying between 0.65 and 0.70 across datasets, and F1-Scores (F1s) ranging from 0.64 to 0.67. DistilBERT outperforms BERT, with Pr scores from 0.70 to 0.75 and F1 scores (F1s) between 0.69 and 0.73, indicating robust performance. RoBERTa yields similar results, with Precision (Pr) scores ranging from 0.67 to 0.72 and F1-Scores

Table 4 Models performance across all datasets

	BERT			DistilBERT			RoBERTa			Ensemble		
	Pr	Rc	F1s	Pr	Rc	F1s	Pr	Rc	F1s	Pr	Rc	F1s
DT_a	0.70	0.65	0.67	0.75	0.70	0.72	0.72	0.68	0.70	**0.80**	**0.78**	**0.79**
DT_b	0.68	0.64	0.66	0.73	0.69	0.71	0.70	0.66	0.68	**0.78**	**0.76**	**0.77**
DT_c	0.69	0.67	0.68	0.74	0.71	0.73	0.71	0.69	0.70	**0.79**	**0.77**	**0.78**
DT_d	0.66	0.63	0.64	0.71	0.68	0.69	0.68	0.65	0.66	**0.76**	**0.74**	**0.75**
DT_e	0.67	0.64	0.65	0.72	0.69	0.70	0.69	0.66	0.67	**0.77**	**0.75**	**0.76**
CTR_{ds}	0.65	0.63	0.64	0.70	0.68	0.69	0.67	0.65	0.66	0.75	0.73	0.74

$DT_a = \{Character, Canto, Sentence, VP\}$
$DT_b = \{Character, Canto, Sentence, Pol, Senti, VP\}$
$DT_c = \{Character, Sentence, VP\}$
$DT_d = \{Character, Canto, VP\}$
$DT_e = \{Character, Canto, Sentence, Pos, Senti, VP\}$
$CTR_{ds} = \{Character, Canto, Sentence, Pol, Subj, Pos, Neg, Senti, VP\}$

(F1s) between 0.66 and 0.70. The Ensemble approach excels across all datasets, achieving the highest metrics with Pr scores from 0.75 to 0.80 and F1-Scores (F1s) between 0.74 and 0.79, particularly standing out in the CTR_{ds} dataset.The Ensemble model consistently outperforms the individual models across all datasets. The highest performance is observed with DT_a, which includes the features Character, Canto, Sentence, and VP, achieving the best F1-Scores (F1s) across the board, particularly when using the Ensemble approach. DT_a emerges as the most effective dataset, while the Ensemble model is identified as the best-performing model.

6 Error Analysis

We analyzed the maximum number of misclassified topics across all models. The topic *The Descendants of Ajamidha* had the highest misclassification with 56 errors, often incorrectly predicted as *Krsna, the Supreme Personality of Godhead, Lord Parasurama Destroys the Worlds Ruling Class*, or *Parasurama, the Lord's Warrior Incarnation*. The topic *Diti Vows to Kill King Indra* followed with 44 misclassifications, frequently confused with *The Pastimes of the Supreme Lord, Ramacandra, Krsna, the Supreme Personality of Godhead*, or The Advent of Lord Krsna-Introduction. Additionally, *Krsna, the Supreme Personality of Godhead*" faced 42 misclassifications, commonly confused with *Diti Vows to Kill King Indra, The Dynasties of the Sons of Yayati*, or *The Descendants of Ajamidha*. The topic *Lord Krsna Shows the Universal Form Within His Mouth* had 40 misclassifications, often predicted as *Diti Vows to Kill King Indra, Krsna, the Supreme Personality of Godhead*, or *The Advent of Lord Krsna-Introduction*. Lastly, *The Killing of the Demon Putana* was misclassified 38 times, with predictions including *Krsna, the Supreme Personality of Godhead, Diti Vows to Kill King Indra*, or *The Descendants of Ajamidha*.

7 Discussion and Observations

The paper presents a framework for analyzing character-topic relationships in the Srimad-Bhagavatam, a key text in Hindu mythology. It introduces two core ontologies: SBC-Ontology for categorizing 1,714 characters into 17 groups (e.g., Gods, Humans, Animals) and identifying 24 family relations, and SBT-Ontology which organizes the text into 12 Cantos, mapping character interactions within specific topics.

The study utilizes a Character-Topic Relationship dataset (CTR_{ds}) comprising 45,941 observations. Transformer models such as BERT, DistilBERT, and RoBERTa are employed, with an ensemble model achieving the highest F1-score of 0.79. To minimize the error rate, a domain-specific transformer-based embedding model can be introduced in future work. The analysis also highlights Krsna as the most mentioned character and establishes a strong correlation, indicating that characters associated with positive topics have minimal negative interactions.

8 Conclusion and Future Scope

This research offers a structured approach to understanding the complex narratives and themes in the Srimad-Bhagavatam by developing domain-specific ontologies. This study provides a detailed framework for categorizing the vast array of characters, their relationships, and the topics they engage with across the scripture. After that using the advanced transformer models like BERT, DistilBERT, and RoBERTa, this research analyzes character-topic relationships. With an F1-score of 0.79, the ensemble model performed better than the others, indicating that it could capture intricate interactions between topics and characters.

This research opens avenues for further exploration, including expanding the dataset to include other significant Hindu texts, such as the Mahabharata and the Ramayana, to develop a more extensive ontology and to find better relationships between characters and topics. Integrating these ontologies and character-topic relationships with global knowledge graphs like DBpedia or Wikidata could enhance model interoperability beyond Hindu mythology.

References

1. Gruber, T. R. (1993). A translation approach to portable ontology specifications. *Knowledge Acquisition, 5*(2), 199–220. https://doi.org/10.1006/knac.1993.1008
2. Jiang, L. P. (2012). Designing family ontology with the Protégé OWL plugin. In *Materials Science and Information Technology II* (pp. 836–840). Trans Tech Publications Ltd. https://doi.org/10.4028/www.scientific.net/AMR.532-533.836

3. Varadarajan, U., Bagchi, M., Tiwari, A., & Satija, M. P. (2022). GENOME: A GENeric methodology for ontological modelling of epics. ArXiv. https://api.semanticscholar.org/CorpusID:247158154
4. Syamili, C., & Rekha, R. (2017). Developing an ontology for Greek mythology. *The Electronic Library 36*, 119–132. https://api.semanticscholar.org/CorpusID:31343095
5. Pastor-Sánchez, J.-A., Kontopoulos, E., Saorín, T., Bebis, T., & Darányi, S. (2021). Greek mythology as a knowledge graph: From Chaos to Zeus and beyond. ArXiv https://api.semanticscholar.org/CorpusID:235657768
6. Watrobski, J. (2020). Ontology learning methods from text–an extensive knowledge-based approach. In *International Conference on Knowledge-Based Intelligent Information & Engineering Systems*. https://api.semanticscholar.org/CorpusID:226255735
7. Espinoza-Arias, P., Poveda-Villalón, M., García-Castro, R., & Corcho, Ó. (2018). Ontological representation of smart city data: From devices to cities. In *Applied Sciences*. https://api.semanticscholar.org/CorpusID:116614841
8. Noy, N. F., & McGuinness, D. L. (2001). Ontology development 101: A guide to creating your first ontology. In *Stanford Knowledge Systems Laboratory Technical Report KSL-01-05*.
9. Mohan, A. R., & Arumugam, G. (2010). Developing Indian medicinal plant ontology using OWL and SWRL. In *International Conference on Data Engineering and Management*. https://api.semanticscholar.org/CorpusID:21166742
10. Sanabila, H. R., & Manurung, R. (2014). Automatic wayang ontology construction using relation extraction from free text. In *LaTeCH@EACL*. https://api.semanticscholar.org/CorpusID:16152128
11. Tuamsuk, K., Chansanam, W., & Kaewboonma, N. (2018). Ontology of folktales in the Greater Mekong Subregion. *International Journal of Metadata, Semantics and Ontologies*, *13*, 57–67. https://api.semanticscholar.org/CorpusID:67868859
12. Sattar, A., Salwana, E., Nazir, M., & Ahmad, M. (2020). Comparative analysis of methodologies for domain ontology development: A systematic review. In *International Journal of Advanced Computer Science and Applications*. https://api.semanticscholar.org/CorpusID:222176816
13. Pannach, F., & Blaschke, T. (2023). Modeling and comparison of narrative domains with shallow ontologies. In *International Conference on Language, Data, and Knowledge*. https://api.semanticscholar.org/CorpusID:264439793
14. Afolabi, I. T., Daramola, O. J., & Adio, T. A. (2014). *Developing domain ontology for Nigerian history*. https://api.semanticscholar.org/CorpusID:55595742
15. Gersberg, I. W., & Ebecken, N. F. F. (2014). A simple strategy to start domain ontology from scratch. *International Journal of Advanced Computer Science and Applications*. https://api.semanticscholar.org/CorpusID:18753697
16. Vlasenko, L. O., Lutska, N. M., Zaiets, N. A., Shyshak, A. V., & Savchuk, O. V. (2022). Domain ontology development for condition monitoring system of industrial control equipment and devices. *Radio Electronics, Computer Science, Control*. https://api.semanticscholar.org/CorpusID:248456848
17. Haridy, S., Ismail, R. M., Badr, N. L., & Hashem, M. (2023). An ontology development methodology based on ontology-driven conceptual modeling and natural language processing: Tourism case study. *Big Data and Cognitive Computing*, *7*(101). https://api.semanticscholar.org/CorpusID:258855924
18. Pan, Z., Gao, Y., Ponci, F., & Monti, A. (2023). Semi-automatic ontology development framework for building energy data management. *IEEE Access, 11*, 111991–112003. https://api.semanticscholar.org/CorpusID:264069350
19. Paul, A., Seal, S., & Das, D. (2024). Transformer-based pouranic topic classification in Indian mythology. Sādhanā. https://doi.org/10.1007/s12046-024-02555-3
20. Paul, A., & Das, D. (2017). Identification of character adjectives from Mahabharata. In *Proceedings of the International Conference Recent Advances in Natural Language Processing, RANLP 2017* (pp. 569–576). https://doi.org/10.26615/978-954-452-049-6_074
21. Paul, A., & Das, D. (2017). A deep dive into identification of characters from Mahabharata. In *Proceedings of the 14th International Conference on Natural Language Processing (ICON-2017)* (pp. 447–455). https://aclanthology.org/W17-7555

22. Hlomani, H., & Stacey, D. A. (2014). *Approaches, methods, metrics, measures, and subjectivity in ontology evaluation: A survey*. https://api.semanticscholar.org/CorpusID:51371006
23. Devlin, J., Chang, M.-W., Lee, K., & Toutanova, K. (2019). BERT: Pre-training of deep bidirectional transformers for language understanding. In *Proceedings of the 2019 Conference of the North American Chapter of the Association for Computational Linguistics: Human Language Technologies, Volume 1 (Long and Short Papers)* (pp. 4171–4186).https://doi.org/10.18653/v1/N19-1423
24. Sanh, V., Debut, L., Chaumond, J., & Wolf, T. (2019). DistilBERT, a distilled version of BERT: Smaller, faster, cheaper and lighter. ArXiv. https://api.semanticscholar.org/CorpusID:203626972
25. Liu, Y., Ott, M., Goyal, N., Du, J., Joshi, M., Chen, D., Levy, O., Lewis, M., Zettlemoyer, L., & Stoyanov, V. (2019). RoBERTa: A robustly optimized BERT pretraining approach. ArXiv. https://api.semanticscholar.org/CorpusID:198953378

Outlier Detection in a Time Series Data

Archita Dasgupta, Aritra Mukhopadhyay, Jyotiraditya Ray, Sayanti Ghosh, and Amit Kumar Das

Abstract This paper focuses on outlier detection within a non-periodic time series data set. The domain is challenging because the data is usually irregular, skewed, and not normalized. A crucial performance metric is analyzed with techniques for anomaly detection through the Box Plot, Z-test, and Isolation Forest. Each method is perceived to perform differently in terms of detecting anomalies, and their performance is quite dependent on the particular characteristics of the data involved. After multiple iterations, the Box Plot and Isolation Forest have proved to be the most effective. The Z-test is not suitable because the data is neither normalized nor symmetrical. Among these methods, the Isolation Forest has taken the lead with a wider range of anomaly detection as compared to others and included all the outliers recognized by the Box Plot. Considering its capacity to handle the complexity and volume of time series data, Isolation Forest has been deemed the best for anomaly detection in this sector. Although several other time series models such as ARIMA can detect outliers for such data, we will still resort to the conventional methods of outlier detection. The study makes mention of context-appropriate algorithm selection as relevant to precision outlier detection, especially in industries with vast and complex data.

Archita Dasgupta, Aritra Mukhopadhyay, Jyotiraditya Ray, Sayanti Ghosh, Amit Kumar Das—All authors contributed equally to this work.

A. Dasgupta (✉) · A. Mukhopadhyay · J. Ray · S. Ghosh · A. K. Das
Institute of Engineering and Management, Kolkata, India
e-mail: archita.dasgupta2022@iem.edu.in

University of Engineering and Management, Kolkata, India

A. Mukhopadhyay
e-mail: aritra.mukhopadhyay2023@iem.edu.in

J. Ray
e-mail: jyotiraditya.ray2023@iem.edu.in

S. Ghosh
e-mail: Sayanti.Ghosh@iem.edu.in

A. K. Das
e-mail: amit@iem.edu.in

J. K. Mandal et al. (eds.), *Recent Advances in Artificial Intelligence and Smart Applications*, Lecture Notes in Networks and Systems 1380,
https://doi.org/10.1007/978-981-96-5822-0_36

Keywords Outlier detection · Time series data · Box plot · Isolation forest

1 Introduction

An important step in the initial phase of data analysis has been detecting outliers or data points that have characteristics different from the rest of the data points. While some outliers have often been viewed as noise or errors in the dataset, they can also provide valuable insights, making it extremely important to identify them before further analysis [1]. However, this paper has solely focused on the identification of outliers, not on their removal. Various methods for detecting outliers have included both statistical measures and machine learning models [2]. Statistical approaches like the Z-Score method have calculated a score for each data instance and compared it with a threshold to detect outliers [3, 4]. However, due to the nature of the Z-Score method, it has not been useful across all datasets. Therefore, as a better alternative, we have implemented the modified Z-Score using the Median Absolute Deviation (MAD). The MAD scale estimator has helped to understand data spread without being affected by outliers, unlike conventional methods like the Interquartile Range (IQR) [5, 6].

After these approaches yielded unsatisfactory results, we have turned to methods such as box plots and machine learning models like Isolation Forest (iForest) for improved outlier detection. Box plots use a graphical method of representing statistical data, providing valuable insights that can reshape the approach to solving a problem. By visually analyzing data, box plots have revealed critical answers to many questions that arise during data exploration [7]. These plots display data in batches and have conventionally represented five key values: the extremes, the upper and lower quartiles, and the median. Widely used in exploratory data analysis, box plots have helped create accessible visual summaries for both statisticians and non-statisticians [8]. These variations have allowed us to evaluate the effectiveness of box plots in representing statistical data with minor adjustments, addressing specific needs in our problem analysis [9].

For machine learning algorithms, we have used iForest, which has employed the concept of isolation to effectively utilize sub-sampling. This approach has enabled iForest to achieve linear time complexity and minimal space complexity, making it better than many existing methods [10]. iForest has performed well in high-dimensional problems with numerous irrelevant attributes and has been effective even when the training set lacks anomalies making it a better and more superior method to detect anomalies [10]. Using a binary tree structure (where each node has at most two children) [11], it has classified anomalies, with anomalous data points forming shorter paths and thus being easily distinguishable from normal points. As Isolation Forest is an unsupervised learning algorithm, we have used the silhouette score from Scikit- learn to assess the model's training accuracy [12].

Contributions

While working on this paper it was observed that very few papers have focused on this field of detecting outliers in time series data. So, we have mainly implemented various methods in the hunt of finding the most effective method to find the outliers. This comparative study has helped us to find a great gap that exists in this field and it has given an in-depth detail of how the various models or statistical measures work to detect the outliers in the time series dataset.

The major contributions can be summarized as follows:

- Outlier detection was evaluated using multiple methods, including the Z-test, Modified Z-test with the Median Absolute Deviation (MAD) estimator, Isolation Forest, and Box Plot.
- The intersection of outliers detected by the Isolation Forest and Box Plot methods was worked out. This overlap validates the effectiveness of both methods.

2 Background

2.1 *Outlier/Anomaly Detection*

Outlier or anomaly detection means finding outliers in the dataset, most often applied in the preprocessing part of data analytics [1]. It has become one of the main topics in data science, as it performs an essential role in finding rare abnormal patterns in datasets. Mostly, anomalies point toward severe incidents that might lead to serious consequences if not taken care of promptly. Hence, effective and efficient anomaly detection algorithms are very much appreciated. Some of the popular statistical methods are the Box Plot Algorithm, Z-Score, modified Z-Score, machine learning algorithms like Isolation Forest, Local Outlier Factor, One-Class SVM, and many more. An effective anomaly detection algorithm is important in the decision-making process, which also helps to prevent critical failures. With complexity in data on the rise, anomaly detection methodologies will be continually updated and will continue to be a core focus of research in the coming years.

2.2 *Box Plot Method*

The Box Plot [7] belongs to a diverse group of statistical techniques called exploratory data analysis, which visually reveals patterns that might otherwise remain hidden in a dataset. Three valuable properties of exploratory data analysis techniques include requiring few assumptions about the data, providing statistical measures that are resistant to outliers (which might otherwise unduly influence an analysis), and emphasizing visual displays that highlight key features of the data. In our case, we are working with non-normalized, asymmetrical data, where many traditional statistical methods fall short of adding meaningful value. Within a box plot, the lower and upper fourths, or first and third quartiles, QL and QU are approximate estimations of the two quartiles of a dataset with QM as the second quartile or median of the dataset. The resistance rule of a boxplot labels an observation from the data set as "outside" if the observation falls below:

$$Qm = \frac{N+1}{2}[\text{Where N is an odd number}] \tag{1}$$

$$Qm = \frac{\left[\frac{(N+1)}{2} + \left\{\frac{(N+1)}{2}\right\} + 1\right]}{2}[\text{Where N is an even number}] \tag{2}$$

$$QL - 1.5 * IQR \tag{3}$$

$$QL + 1.5 * IQR \tag{4}$$

If an observation is classified as "outside," it is considered an anomaly or outlier in our context, where the interquartile range (IQR) is defined as the difference between the first quartile QL and the third quartile QU.

$$IQR = QU - QL \tag{5}$$

Figure 1 illustrates a labeled box plot of a single dataset batch. The box plot has a piston-like structure, with the upper hinge representing the 3rd quartile and the lower hinge representing the 1st quartile. Data points lying beyond the upper and lower fences are considered outliers [8].

2.3 Z-Score Method and Modified Z-Score Method

2.3.1 Use of Z-Score for Detection of Outliers

The Z-score method is a simple statistical analysis that uses the mean and standard deviation to identify outliers based on a calculated critical value. This method mainly returns the corresponding Z-scores for data points using a specific formula to calculate the Z-score from the given time series data.

$$Z = \left(\frac{Si - \mu}{\sigma}\right) \tag{6}$$

Here, Z represents the Z-scores of the corresponding data points, xi denotes the values in the dataset, μ is the mean of the dataset, and σ is the standard deviation of the dataset. However, relying solely on the Z-test method does not provide the robustness

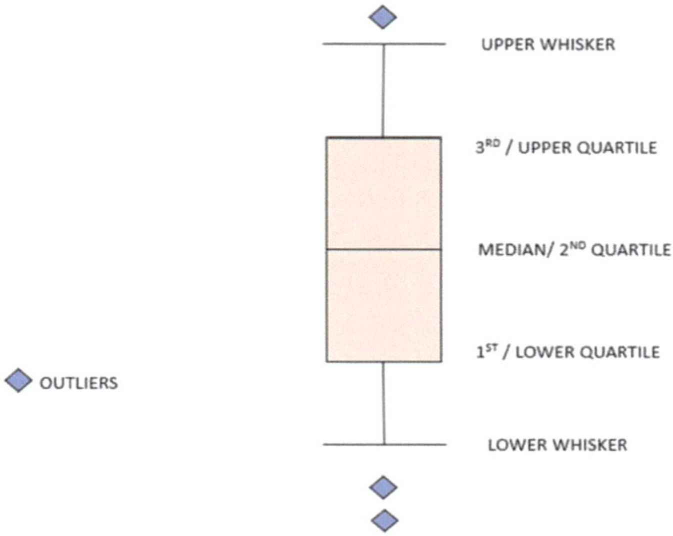

Fig. 1 Box plot method—visual representation

needed for multi-dimensional applications. This is where scale estimators become valuable, as they enhance the robustness of the method, making it significantly more useful and accurate [5].

Concept of Scale Estimators

Scale estimators are nothing but statistical measures primarily used to assess the extent of spread within a dataset, even in the presence of outlier values. Even if the outliers have created irregular patterns in the dataset, scale estimators are quite useful by creating a clear understanding of data dispersion without being affected by the extreme values. Using these estimators will help to create insightful aids, which form the basis of better decision-making in disciplines ranging from finance to economics and social sciences.

Two common scale estimators used for outlier detection are:

IQR (Interquartile Range): This is calculated as the difference between the 75th percentile and the 25th percentile of a dataset.

MAD (Median Absolute Deviation): This is defined as the mean of the absolute deviations from the median of the dataset, providing a better measure of variability.

2.3.2 Use of Modified Z-Score with MAD (Median Absolute Deviation) for Detection of Outliers

Using the Z-score method, we arrive at the modified Z-score, which operates similarly but with a multiplication factor of 0.6745 instead of 1, as used in the standard Z-score test. This adjustment suggests that the number of outliers identified using the modified Z-score will typically be lesser than those detected with the standard Z-score method.

In this approach, the MAD acts as a scale estimator. Like other scale estimators, MAD boosts the robustness of the outlier detection process. The parameters involved in these calculations are defined as follows [6]:

$$MAD = mediank(|xi - mediank(x)|) \tag{7}$$

$$ModZ = \frac{0.6745(Si - S)}{MAD} \tag{8}$$

xi refers to each data value; x refers to the complete dataset; MAD is the median absolute deviation; $ModZ$ are the modified z scores for the corresponding data points.

2.4 *Isolation Forest Method*

The Isolation Forest (iForest) algorithm, developed by Fei Tony Liu in 2008 [13], is one of the most effective algorithms for detecting outliers, due to its efficient anomaly isolation process. The method is based on the principle that anomalies are distinct from the majority of the data, making them easier to isolate. Unlike traditional density-based methods, iForest uses a binary tree structure [11], where data points are split based on random selections of features and values. Because anomalies are distinctly different than the other data points, they tend to be separated more quickly in the process.

This method operates almost similarly to a Random Forest model but utilizes an Extra Trees Regressor (Extremely Randomized Trees). Instead of calculating entropy as we do in the case of decision trees, we compute an anomaly score, denoted as $s(x, m)$, which is defined by the formula

$$s(x, m) = 2^{\frac{-\{E(h(x))\}}{C(m)}} \tag{9}$$

where m is no of points in the sample dataset; x is data points; $E(h(x))$ is expected value or any path by far is a data point x in a tree; $c(m)$ is any depth of data points in a tree.

Condition for considering a data point as an outlier [10, 13]:

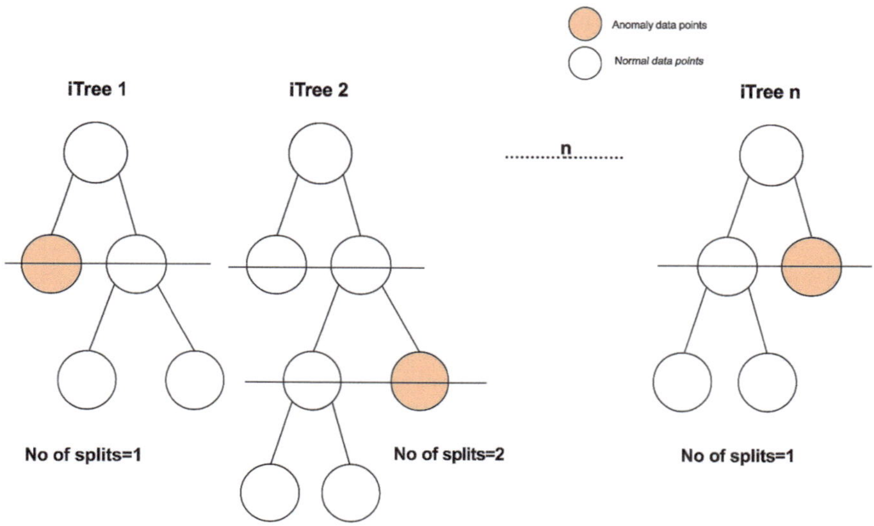

Fig. 2 Isolation forest

if $E(h(x)) \ll c(m) : s(x, m) \approx 1$ (anomaly data points) else If
$E(h(x)) \gg c(m) : s(x, m) \approx 0$ (normal data point)
else $E(h(x)) == c(m) : s(x, m) \approx 0.5$ (likely normal data point)

Figure 2 represents a binary tree for how anomalies are detected in the isolation forest algorithm-iForest method uses binary trees for the detection of anomalies, at whatever depth the anomaly is found that is considered as the last number of splits like here for every iteration the position of the anomaly may vary and a single anomaly may appear more than a single time for the first iteration the anomaly occurs at a depth of n = 1 so nsplit1 = 1 and so on

$$E(h(x)) = \frac{nsplit1 + nsplit2 + \ldots\ldots. + nsplitn}{ntrees} \tag{10}$$

$$C(m) = 2H(m-1) - \frac{2(m-1)}{n} \tag{11}$$

$c(m)$ is the average depth of all unsuccessful searches [where m is sample set size H is the harmonic number [10, 13]

iForest has a very low memory requirement equivalent to $O(\psi * d)$ for constructing the trees [ψ is the subsample size which is generally smaller than or equal to the entire sample size]. For typical values of ψ, isolation forest has a linear space complexity in the order of $O(n)$. This linear space complexity holds a great advantage compared to other anomaly detection models as it is independent of the size of the dataset and thus can handle large datasets without any problem [13].

Algorithm

Step 1: The initial step involves filtering the measurements associated with a specified parameter from the complete dataset. This filtered subset, is denoted as *D*.

Step 2: After obtaining the required subset, the data points are extracted for the purpose of anomaly detection. This data is stored in a variable *L*.

Step 3: Following the preparation of the data points, an instance of the Isolation Forest algorithm is initialized. The contamination parameter is set to 5%, reflecting the expected proportion of outliers in the dataset. (If the contamination is set too low then it might misclassify true outliers and if it is set too high it can classify normal points as outliers therefore a standard rate of 5% has been considered) This initialization can be represented as:

$$\text{iso}_{\text{forest}} = \text{IsolationForest}(\text{contamination} = 0.05, \text{random}_{\text{state}} = 42)$$

Step 4: The next phase involves training the Isolation Forest model on the timeseries data *L* This fitting process enables the model to learn the characteristics of normal measurements, which is essential for detecting anomalies.

$$\text{iso}_{\text{forest}}.\ \text{fit}\ (L)$$

Step 5: Once the training has been done, the model is applied to predict anomalies within the timeseries dataset. This prediction is executed as follows and data points classified as anomalies are assigned a value of -1

$$\text{D [anomaly]} = \text{iso}_{\text{forest}}.\ \text{predict}(L)$$

Step 6: After predictions, the dataset is filtered to extract instances labelled as outliers, mathematically represented by:

$$O = \{\, \text{x} \in \text{D} \mid \text{x. anomaly} = -1 \,\}$$

Results and Inferences:

Figure 3 presents a box plot for the sample dataset, highlighting data points above the upper fence as outliers or anomalies. Here, these outliers indicate specific timestamps where spikes occurred, indicating further investigation into potential causes. The analysis of anomalies within the given dataset involves 18,848 data points across eight segments, each recording measurements of data points. The box plot effectively visualizes this data, with points beyond the upper whisker classified as outliers or anomalies. From a total of 2,373 data points, the box plot identified 76 instances as outliers. Due to the skewed and non-normalized nature of the dataset, Z-test has detected only 58 outliers. As the Z-test works mainly for normally distributed data, it has incorrectly identified 47 data points as outliers, which are, in fact, within normal data point ranges. This misclassification is mainly because of the fact that the Z-test cannot handle the dataset's skewness, which has impacted its ability to distinguish genuine anomalies accurately. Additionally, the Isolation Forest algorithm is applied to the data, along with the box plot analysis. After training on 2,373 data points, the Isolation Forest identified 117 instances as outliers. Since it is an unsupervised learning approach, the Isolation Forest's clustering performance was evaluated using the silhouette score, which yielded a value of 0.6552. This score indicates a favorable level of clustering within the data, suggesting that the identified outliers possess characteristics different from the remaining dataset.

Figure 4 shows the application of the Isolation Forest algorithm on the dataset, containing 2,373 data points. A total of 117 outliers are detected and are visually represented by red circles, marking potential anomalies occurring at specific indices in the data. The remaining data points, classified as normal, do not impact the overall analysis significantly and are not flagged for further investigation.

Figure 5 illustrates that most data points are accurately clustered, as shown by the silhouette score. A larger portion of the data points has contributed positively to the silhouette score, indicating a strong match within their respective clusters. In contrast, a smaller subset of data points has contributed negatively, suggesting possible misclassification or weaker clustering. This overall positive contribution highlights the effectiveness of the Isolation Forest algorithm in identifying distinct outliers within the dataset.

Figure 6 presents a Venn diagram comparing the outliers identified by the Box Plot method and the Isolation Forest algorithm. Due to the skewed and non-normalized nature of the data, the Z-test method has not contributed any meaningful results for outlier identification. The Isolation Forest detected a total of 117 outliers, while the Box Plot identified 76. Upon taking the union of these methods, a combined total of 162 anomalous data points was observed. The intersection between the Box Plot and Isolation Forest results revealed that 76 of the 76 Box Plot outliers were also detected

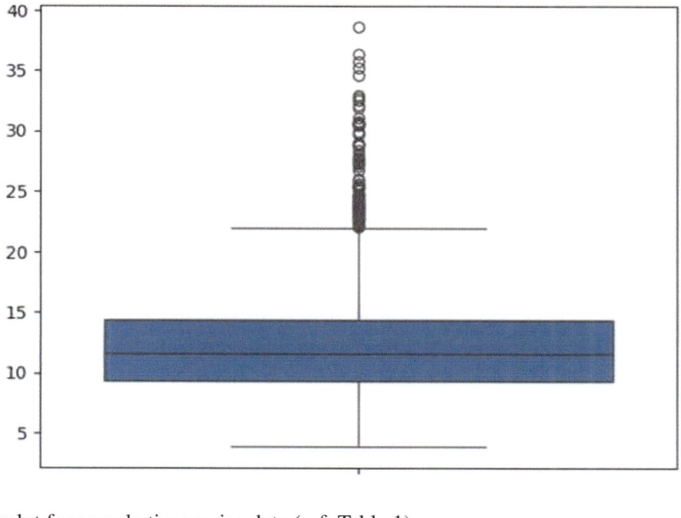

Fig. 3 Boxplot for sample time series data (ref. Table 1)

Table 1 Sample dataset

Index	Value
1	67.18678619
2	131.946686945796
3	92.4710730059
4	83.246960
5	250.07231
.	…
.	…
7785	3.84896
7786	4.1438
.	…
.	…
10,157	36.4368
10,158	38.6782
18,845	12.473230
18,846	11.976671
18,847	10.820193
18,848	9.054894

by the Isolation Forest. Additionally, when the Z-test results intersected with those from the Box Plot and Isolation Forest, 2 common outliers were found, establishing the Z-test outliers as nearly a distinct set. This finding suggests that the Box Plot outliers form a subset of the larger set detected by the Isolation Forest algorithm.

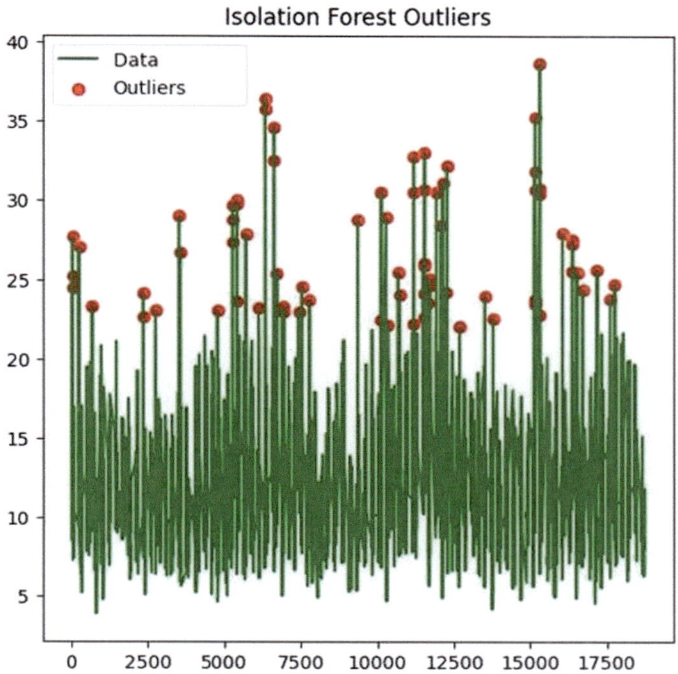

Fig. 4 The outliers detected using the isolation forest method

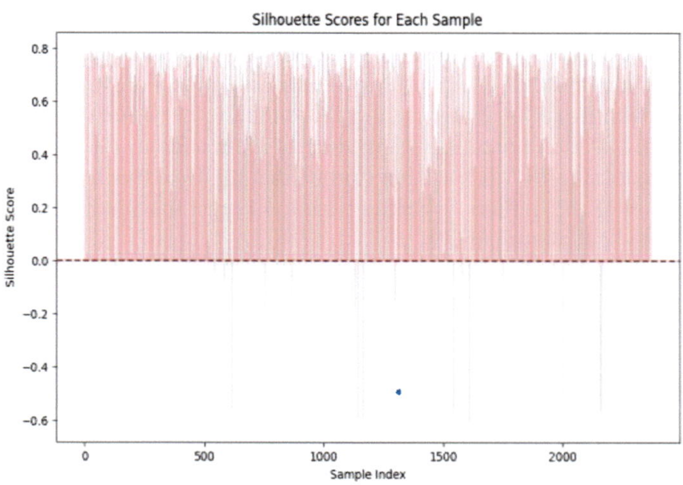

Fig. 5 Silhouette score analysis for each data point contribution toward overall score of 0.6552

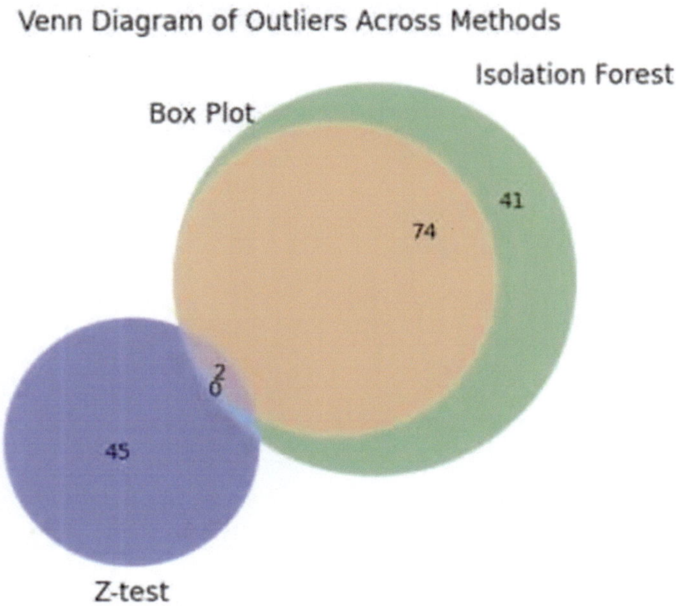

Fig. 6 Venn diagram representing outlier counts across all implemented methods

3 Conclusion

This work has examined the detection of outliers in time series data. When it comes to identifying abnormalities, algorithms like the Isolation Forest have consistently outperformed the Box Plot approach. In addition to the 76 outliers that the Box Plot has identified, Isolation Forest has flagged all of them and has discovered 41 more, increasing the total to 117. It has been studied that the Box Plot has only recovered a small portion of the outliers, whereas Isolation Forest has identified more subtle and difficult-to-spot abnormalities.

The Z-test on the other hand has failed to give any proper results due to the skewed and non-normalized nature of the dataset. It has labeled the anomalies as normal points and vice versa making it less reliable in this case.

By combining the outputs of the Box Plot, Isolation Forest, and Z-test procedures, we have obtained a total of 162 distinct anomalies. It has been observed that Isolation Forest is the most successful strategy when compared to the Box Plot.

References

1. Smiti, A. (2020). A critical overview of outlier detection methods. *Computer Science Review, 38*, Article 100306.
2. Boukerche, A., Zheng, L., & Alfandi, O. (2020). Outlier detection: Methods, models, and classification. *ACM Computing Surveys (CSUR), 53*(3), 1–37.
3. Yaro, A. S., Maly, F., Prazak, P., & Maly, K. (2024). Outlier detection performance of a modified Z-score method in time-series RSS observation with hybrid scale estimators. *IEEE Access*.
4. Yaro, A. S., Maly, F., & Prazak, P. (2023). Outlier detection in time-series receive signal strength observation using Z-score method with S n scale estimator for indoor localization. *Applied Sciences, 13*(6), 3900.
5. Siraj-Ud-Doulah, M., & Islam, M. H. (2019). An alternative robust measure of outlier detection in univariate data sets. *Research & Reviews: Journal of Statistics, 8*(1), 1–11.
6. Lucon, E. (2024). *Statistical detection of outliers in the certification of NIST reference charpy lots.*
7. Mazarei, A., Sousa, R., Mendes-Moreira, J., Molchanov, S., & Ferreira, H. M. (2024). Online boxplot derived outlier detection. *International Journal of Data Science and Analytics*, pp. 1–15.
8. Mazlan, S. N. A., Daud, H., Noh, K. A. M., & Aris, M. N. M. (2021). Outlier detection in subsurface modeling of 2D electrical resistivity imaging by using boxplot. In *Proceedings of the 6th International Conference on Fundamental and Applied Sciences: ICFAS 2020* (pp. 729–738). Springer Singapore.
9. Babura, B. I., Adam, M. B., Samad, A. R. A., Fitrianto, A., & Yusif, B. (2018). Analysis and assessment of boxplot characters for extreme data. *Journal of Physics: Conference Series, 1132*(1), 012078. IOP Publishing.
10. Lesouple, J., Baudoin, C., Spigai, M., & Tourneret, J. Y. (2021). Generalized isolation forest for anomaly detection. *Pattern Recognition Letters, 149*, 109–119.
11. Shaffer, C. A. (2011). *Data structures & algorithm analysis in Java.* Courier Corporation.
12. Raschka, S., & Mirjalili, V. (2019). *Python machine learning: Machine learning and deep learning with Python, scikit-learn, and TensorFlow 2.* Packt publishing ltd.
13. Chabchoub, Y., Togbe, M. U., Boly, A., & Chiky, R. (2022). An in-depth study and improvement of isolation forest. *IEEE Access, 10*, 10219–10237.

Multi-color Multi-shape Visual Cue Recognition Using a Hybrid Approach of Cosine Similarity-Based 2DLPP and Granular Computing

Saibal Ghosh, Pritam Paral, Pubali De, Amitava Chatterjee, and Sugata Munshi

Abstract A significant challenge in visual understanding-based assistive robotics has been targeted action recognition in human–robot collaborative (HRC) spaces. With vision-based data, there exist certain irregularities and variabilities in the working environment that can considerably influence recognition performance. In most cases, high-dimensional vision sensor data is first projected to a low-dimensional manifold and then analyzed for supervised or unsupervised recognition. Classical dimensionality reduction (DR) techniques are mostly affected by such variations in the data, whereas the family of local structure-sensitive manifold learning techniques such as the 2D locality preserving projections (2DLPP) offers a better alternative. However, the projection map generated by 2DLPP also heavily relies on the spatial structure and intensity variation in the visual data. To overcome these concerns, a cosine similarity measure is introduced for calculating similarity information between data points in the originally structured image space, with enhanced efficiency. Additionally, a granular computing-based local, intrinsic structure encoding mechanism is incorporated that works on granular information fusion and rough entropy maximization. The granular fusion scheme considers RGB color spaces separately to allow individual channel granular encoding from the images. Extensive experimental studies demonstrate the proposed approach to be an effective visual cue detection method, especially in shape and color-varying scenarios.

Keywords Color invariance · Cosine similarity · Granular computing · Human–robot collaboration · Locality preserving projections

---

S. Ghosh · A. Chatterjee · S. Munshi
Department of Electrical Engineering, Jadavpur University, Kolkata, India

P. Paral
Department of Electrical Engineering, IIEST, Shibpur, Howrah, India

P. De (✉)
Institute of Engineering and Management (IEM), UEM, Kolkata, India
e-mail: de.pubali19@gmail.com

J. K. Mandal et al. (eds.), *Recent Advances in Artificial Intelligence and Smart Applications*, Lecture Notes in Networks and Systems 1380,
https://doi.org/10.1007/978-981-96-5822-0_37

1 Introduction

Over the past few decades, human–robot collaboration (HRC) has rapidly emerged to be one of the fastest-growing areas of technological advancement today. HRC-oriented systems continue to make significant impacts in several areas of modern technological applications, encompassing telemedicine, smart transportation solutions, ecological surveillance, collaborative robotics, digital farming, supply chain management, and navigational technology. The progress of HRC has led to the transformation of conventional robotics from industrial spaces into domestic, social, and indoor applications, catalyzing the emergence of assistive robotics in day-to-day activities. In such collaborative environments, human users and assistive robots share the same working space, aided by some seamless, two-way interaction modalities. To maintain such uninterrupted communication, non-verbal triggers such as hand postures, body movements, facial expressions, head movements, and gazes are quite effective. Firstly, these human triggers need to be acquired by some sensor terminals and subsequently fed to an intelligent computational node for recognition. After that, any cue mapping rule base can be used to translate the detected cues into particular task commands to be performed by an assistive robot. There are primarily two kinds of sensor terminals for data acquisition, i.e., contact-based wearable sensors and non-contact sensors. Contact-based sensors include inertial measurement tools, haptic gloves, motion sensors, magnetic detectors, surface electromyography, etc. Since these devices need to be harnessed by the individuals on the body, their intrusive nature and wearing-discomfort impose challenges for daily life usage. Considering user convenience and cost-effectiveness, researchers have proposed vision-sensing-based systems as a viable alternative to wearable devices. However, alongside their advantages, vision-sensing interfaces face certain environmental hazards that could potentially degrade the data in real-world scenarios. A primary source of data deterioration is attributed to the photometric variations in the data, e.g., illumination variation, color intensity variation, shadows, and specularities, to name a few. To deal with such high-dimensional vision sensor data, efficient dimensionality reduction (DR) techniques are required which can project the data into lower dimensional subspaces. Traditional DR techniques include principal component analysis (PCA) and linear discriminant analysis (LDA), which have been popularly used to obtain the lower dimensional representation of the data in an unsupervised and supervised manner, respectively. However, these techniques have mostly been successful in exploring the structure of some linear data distribution and fail to deal with the nonlinearities in the data.

2 Related Works

To address the issues mentioned above, manifold learning techniques such as isometric feature mapping (Isomap), locally linear embedding (LLE), and Laplacian eigenmaps (LE) have been developed to analyze the nonlinear intrinsic information of the data. Nevertheless, these techniques cannot explicitly define specific mapping functions, leading to an out-of-sample issue with the current data [1]. To overcome this issue, specific linear approximations of algorithms used for nonlinear manifold learning, such as neighborhood preserving projection (NPP) [2], isometric projection (IsoP) [3], neighborhood preserving embedding (NPE) [4], and locality preserving projections (LPP) [5] have been introduced. For example, LPP approximates the linear Laplacian eigenmaps and expands the transformation map within the ambient space not only in the training space [6]. In LPP, a particular matrix representation called graph Laplacian is used which encapsulates the local structure of the data. For improving the graph Laplacian further, several methods have been developed, e.g., fast and orthogonal LPP (FOLPP) [7], locality preserving triplet discriminative projections (LPTDP) [8], Monte Carlo sampling-based discriminant LPP (MC-DLPP) [9], multi-scale LPP (MSLPP) [10], locality preserving projections with autoencoder (LPPAE) [10]. The family of LPP-based applications finds their applications in several areas such as image processing, image registration, data representation, indexing, and retrieval, fault diagnosis, biomedical signal analysis, etc. [12].

However, LPP and most of its enhanced versions consider the input data in the shape of one-dimensional vectors. For this reason, 2D data matrices need to be converted into their equivalent 1D data shapes before applying the DR techniques. This process can disrupt the inherent structural information embedded within the data, potentially leading to a compromised performance [13]. This can also lead to the well-known problem of small sample size, which makes the optimization process cumbersome. To tackle this issue, 2DPCA [13] and 2DLDA [14] were created to extract intrinsic features directly from the 2D image data. Low-rank representation-based two-dimensional version of neighborhood preserving projection (LR-2DNPP) was developed to make the system robust against noise and outliers [15]. Nuclear norm-based 2DLDA (NN-2DLDA) was introduced by Lu et al., which can retain global information of data in a better-organized manner during the reconstruction from query images [16]. Similarly, the 2D version of LPP, i.e., 2DLPP has also been developed, which can capture the Euclidean pattern of the data from the original 2D plane [12]. 2DLPP can retain the 2D spatial geometrical information in the data in the projected subspace and enhances the efficiency of Laplacian matrix evaluation by using a matrix-based approach instead of a vector-based approach.

Along with the advantages, there are three major issues with 2DLPP: (i) It relies solely on spatial variations in the data to construct the similarity kernel and thus may not be able to accurately reflect the intensity or range variation in the image information [17]. (ii) Euclidean similarity measure of the data is often prone to noise, outliers, and variations in intensity of the image data. (ii) While handling the higher dimensional image data from real-world ambiances, the data is likely to contain unclear, indistinct, or ambiguous information [18]. In such scenarios, the presence of irrelevant or indistinguishable information in the data can substantially affect the similarity kernel utilized by the local manifold learning DR algorithms [18].

In our previous work [19], we introduced a granular computing-based feature extraction scheme within the framework of 2DLPP for the detection of fixed-color, fixed-shape flag-stick visual cues. In this work, we have extended the experimental scenario for multi-color, multi-shape visual cues. To summarize the present work, we propose a combined spatial-feature kernel approach for calculating the similarity measures in 2D locality preserving projections (2DLPP) [12], capturing both spatial similarity and range similarity [17]. Secondly, we have introduced cosine similarity-based distance metrics [20] for constructing the similarity kernels in the spatial domain and feature domain of the high-dimensional raw data and granulated feature data, respectively. Thirdly, we have proposed a granular computing (GrC)-based feature weight calculation mechanism here, that can effectively handle the issue of object indiscernibility [18]. Additionally, experimental results have demonstrated that granulated feature extraction on RGB channels achieves much superior performance compared to their equivalent grayscale data [21]. We have evaluated two granulation techniques here, such as crisp granulation (CG) [18] and quadtree decomposition (QtD) [22] on individual color channels. R-, G-, and B-channel granular features are integrated together using their rough set representations to obtain the exhaustive granular features for CG and QtD.

3 Experimental Setup and Visual Cue Database

The complete database was created at the Instrumentation & Cyber Physical System Laboratory, Department of Electrical Engineering, Jadavpur University, India. For the experiments, a custom-built, cost-efficient, independently actuated two-wheel mobile robot was utilized [19, 23]. The robot is equipped with a Raspberry Pi, Model B+ processing unit. An RPi camera module v2 is mounted on the robot surface having a Sony IMX219, 8-megapixel vision sensor.

A series of human-supervised visual cues, utilizing various configurations of a flag-stick combination, was formulated within the HRI framework here [19, 24]. Male and female volunteers of different ages participated in the experiment. Participants displayed hand-guided visual cues in front of the robot according to their comfort, whether standing or sitting. There were no boundaries on the volunteers' proximity to the robot's vision sensor as long as they remained within its field of view. Seven distinct visual triggers were created based on various flag-stick combinations, specifically: top-left (T-L), middle-left (M-L), bottom-left (B-L), neutral (N), bottom-right (B-R), middle-right (M-R), and top-right (T-R) [19, 24]. Prototype visual cues corresponding to all seven classes are presented in Fig. 1a. In total, flags with five different color shades were used to create the database, such as, rose pink, vermilion, lemon-lime, orange, and Munsell yellow. A total of three

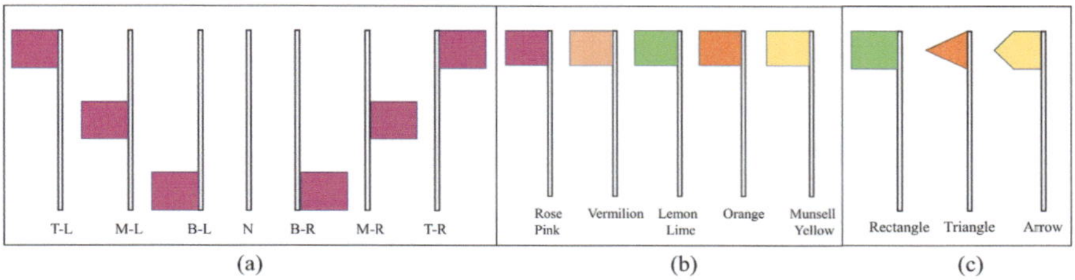

Fig. 1 Representative flag-stick cues of **a** Various classes. **b** Various colors. **c** Various shapes

Fig. 2 Representative real-world flag-stick cues of various colors and shapes from class B-L

types of flag shapes were considered in the database, such as rectangle, triangle, and arrow. Prototype flag-stick images with various colors and shapes are demonstrated in Fig. 1b and 1c. Originally acquired, real-world flag-stick cues of various colors and shapes are shown in Fig. 2. Additionally, five different illumination conditions were created at which the visual data were recorded. The conditions in the decreasing order of illumination are normal lighting condition (NLC), dark lighting condition-1 (DLC-1), dark lighting condition-2 (DLC-2), dark lighting condition-3 (DLC-3), and dark lighting condition-4 (DLC-4) [19]. Various flag-stick cues are correlated with respective navigation guidance instructions for the mobile robot [25].

4 Visual Cue Recognition System

The overall visual flag-stick cue recognition module proposed here can be broken down into three key sub-modules, such as feature extraction, dimensionality reduction, and classification. Cosine similarity-based 2DLPP (c-2DLPP) has been proposed here to perform dimensionality reduction on the higher dimensional input vision sensor data. Considering the challenges faced by conventional feature extraction techniques while extracting visual features from indiscernible data, a granular feature extraction scheme REGF is incorporated [19]. It works on the fusion of two

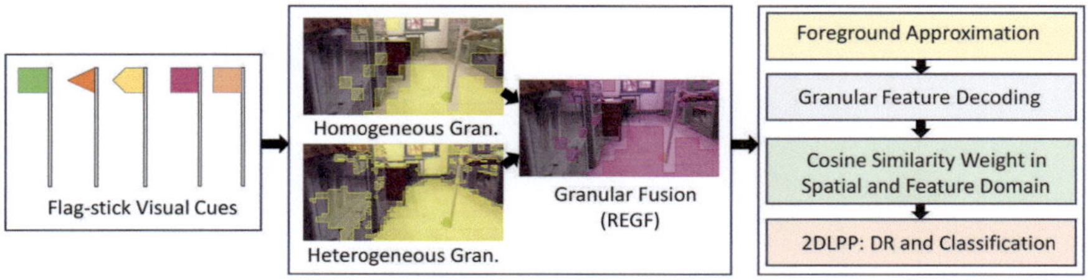

Fig. 3 Workflow diagram of the proposed visual cue recognition approach REGF-c-2DLPP

fundamental approaches of granulation, i.e., homogeneous and heterogeneous granulation. The entire workflow diagram of the proposed REGF-c-2DLPP approach is shown in Fig. 3.

4.1 Homogeneous Granulation

In granular computing, the information uncertainty in images is represented by rough set approximations, which can potentially handle the issue of object indiscernibility. Here, the original rough set is represented by two independent crisp sets referred to as the upper and lower approximation. The uncertainty between an object and surrounding information in an image is estimated by a measure called rough entropy (RE). RE is calculated by considering the image as a cluster of distinct, non-overlapping granules. This RE measure is maximized to enhance the object-surrounding discernibility and the corresponding threshold is used for an appropriate object-surrounding segmentation. In rough set theory, any target set with the information $Z \subseteq \Omega$ can be approximated with a subset of attributes A, with Ω be the universal set. When the information in Z cannot be differentiated using the attributes' set A, two approximations of the rough set are defined, lower approximation $\underline{A}Z$ and upper approximation $\overline{A}Z$ [18].

$$\underline{A}Z = \{z|[z]_A \subseteq Z\} \tag{1}$$

$$\overline{A}Z = \{z|[z]_A \cap Z \neq \phi\} \tag{2}$$

The corresponding roughness of the approximation is described as.

$$R = 1 - \frac{|\underline{A}Z|}{|\overline{A}Z|} \tag{3}$$

The value of R increases with increasing uncertainty in the object-surrounding information. For a given image, upper and lower approximations for the object and surroundings are given as $\overline{O_\theta}$, $\underline{O_\theta}$, $\overline{S_\theta}$, $\underline{S_\theta}$, respectively, where θ represents a set of predefined properties, e.g., the gray-level intervals. The corresponding rough entropy (RE) metric of the image is described as [18]

$$\mathrm{RE}_\theta = -\frac{e}{2}\left[R_{o_\theta}\log_e\left(R_{o_\theta}\right) + R_{S_\theta}\log_e\left(R_{S_\theta}\right)\right] \quad (4)$$

where $R(O_\theta)$ and $R(S_\theta)$ represent the object and surrounding roughness measures, respectively. From (4), the optimal threshold for an appropriate object-surrounding segmentation is calculated as.

$$\theta^* = \arg\max_\theta \mathrm{RE}_\theta \quad (5)$$

The selection of granule size is one of the determining factors for this optimal threshold. More compact-sized granules are expected to yield finer approximations but at the expense of increased computational burden. Crisp granulation (CG) is a type of homogeneous granulation (HmG) where the granule sizes are even and regular.

4.2 Heterogeneous Granulation

In HmG, granule sizes are consistent throughout the image, which may not be quite appropriate in some real-life scenarios. Particularly in the image regions with varying levels of ambiguity, regular but unequal-sized granules are more relevant choices as they can vary in size with varying intensity distribution. Quadtree decomposition (QtD) is a heterogeneous granulation (HtG) technique where the granule sizes are determined considering both gray-level information and spatial proximity. Here, the image is recursively decomposed into smaller granules until the gray-level band between the minimum and maximum values crosses below a certain threshold limit. Given g_1 and g_3 are the first and third quartile intensities in the gray-level distribution, the associated HtG threshold value is obtained as [22]

$$\delta = \frac{g_3 - g_1}{2} \quad (6)$$

4.3 Proposed RGB Granular Fusion Scheme

It has been experimentally studied that, the general guideline [18] of granule size selection in CG does not suitably fit for the images with multi-modal, non-normal intensity distribution [19]. Moreover, the selection of larger granules might cause the information to fade in important regions, while significantly smaller sized granules can capture spurious, ambiguous regions. These impose a great challenge in selecting the optimal granule size in CG for proper segmentation. In QtD [22] also, granule sizes depend on a predefined threshold value, estimated from the upper and lower quartile data of the gray-level variation in an image. This selection also does not work satisfactorily for the images with multi-modal distribution, having non-uniform local densities and clumping behavior. Additionally, CG and QtD are both conventionally applied on grayscale images, which often incur a substantial loss of color details particularly in the areas with pronounced RGB intensity variation but uniform grayscale values. In our granular fusion scheme REGF, spatio-color granules are obtained on individual color channels for both upper and lower approximated rough sets [19]. Thereafter, unified RGB-channel granules are obtained for both homogeneous and non-homogeneous types of granulations in the following way [19].

$$\left.\begin{array}{l}\{\mathbb{UO}_{(\mathrm{HmG/HtG})}\} = \{\mathbb{UO}^{R}_{(\mathrm{HmG/HtG})}\}\bigcup\{\mathbb{UO}^{G}_{(\mathrm{HmG/HtG})}\}\bigcup\{\mathbb{UO}^{B}_{(\mathrm{HmG/HtG})}\} \\ \{\mathbb{LO}_{(\mathrm{HmG/HtG})}\} = \{\mathbb{LO}^{R}_{(\mathrm{HmG/HtG})}\}\bigcup\{\mathbb{LO}^{G}_{(\mathrm{HmG/HtG})}\}\bigcup\{\mathbb{LO}^{B}_{(\mathrm{HmG/HtG})}\}\end{array}\right] \tag{7}$$

$$\left.\begin{array}{l}\{\mathbb{US}_{(\mathrm{HmG/HtG})}\} = \{\mathbb{US}^{R}_{(\mathrm{HmG/HtG})}\}\bigcup\{\mathbb{US}^{G}_{(\mathrm{HmG/HtG})}\}\bigcup\{\mathbb{US}^{B}_{(\mathrm{HmG/HtG})}\} \\ \{\mathbb{LS}_{(\mathrm{HmG/HtG})}\} = \{\mathbb{LS}^{R}_{(\mathrm{HmG/HtG})}\}\bigcup\{\mathbb{LS}^{G}_{(\mathrm{HmG/HtG})}\}\bigcup\{\mathbb{LS}^{B}_{(\mathrm{HmG/HtG})}\}\end{array}\right] \tag{8}$$

where (7) and (8) give the upper and lower approximations for the object and surrounding, respectively. Eventually, the fusion of HmG and HtG is performed to obtain the final upper object, lower object, upper surrounding, and lower surrounding approximations in the following way [19].

$$\overline{f\mathbb{O}} = \{\mathbb{UO}_{(\mathrm{HmG})}\}\bigcap\{\mathbb{UO}_{(\mathrm{HtG})}\} \tag{9}$$

$$\underline{f\mathbb{O}} = \{\mathbb{LO}_{(\mathrm{HmG})}\}\bigcap\{\mathbb{LO}_{(\mathrm{HtG})}\} \tag{10}$$

$$\overline{f\mathbb{S}} = \{\mathbb{US}_{(\mathrm{HmG})}\}\bigcap\{\mathbb{US}_{(\mathrm{HtG})}\} \tag{11}$$

$$\underline{f\mathbb{S}} = \{\mathbb{LS}_{(\mathrm{HmG})}\}\bigcap\{\mathbb{LS}_{(\mathrm{HtG})}\} \tag{12}$$

4.4 Granulated Feature-Based 2D Locality Preserving Projections

To retrieve the granulated feature information from an image and use it in the cue recognition pipeline further, the upper approximated object $\overline{f\mathbb{O}}$ is expected to contain the desired set of granules. $\overline{f\mathbb{O}}$ comprehensively includes all the foreground granule information by combining the merits of both HmG and HtG, making it a suitable candidate for feature description. To decode feature values from the thresholded granular information in $\overline{f\mathbb{O}}$, a novel feature decoding scheme has been utilized, which is discussed in detail in our earlier work [19]. Now, to apply two-dimensional locality preserving projections (2DLPP) [12] for obtaining the projected data in low-dimensional subspaces, let us first consider the training image dataset as $\boldsymbol{I}_1, \boldsymbol{I}_2, \ldots\ldots, \boldsymbol{I}_p$. Where $\boldsymbol{I}_i \epsilon \Re^{m \times n}$ represents a particular flag-stick cue image while there is a total p number of such images in the training database. In conventional LPP, the spatial weight between matrices is calculated as $\|\boldsymbol{I}_i - \boldsymbol{I}_j\|_F^2$. Owing to the vulnerability of Euclidean distances in higher dimensional data spaces [17], we propose a cosine similarity search for encoding the local neighborhood information. Here, similarity measures are calculated based on the directional similarity between data points rather than the magnitudes. This reduces the possibility of inflated magnitudes between data matrices, especially in the case of higher dimensional spaces and the adverse effect of noise and outliers. Algorithm 1 given below demonstrates the measurement of cosine similarity distance between image matrices.

Algorithm 1: Cosine Similarity Measure for Image Matrices
Input: High-dimensional input images $\boldsymbol{I}_i$ and $\boldsymbol{I}_j$
Output: Cosine distance d_{ij}

- Normalize the matrices with their spectral norms i.e., $\boldsymbol{I}_i = \boldsymbol{I}_i/\|\boldsymbol{I}_i\|_2$, $\boldsymbol{I}_j = \boldsymbol{I}_j/\|\boldsymbol{I}_j\|_2$
- Calculate the dot product $dot_{ij} = \boldsymbol{I}_i \cdot \boldsymbol{I}_j^T$
- Calculate the row-wise cosine similarity $(d_{ij})_{all} = dot_{ij}/((\|\boldsymbol{I}_i\|_2) \cdot (\|\boldsymbol{I}_j\|_2))$
- Calculate the mean cosine similarity $d_{ij} = mean(mean((d_{ij})_{\text{all}}))$

Let, the spatial similarity weight obtained above be denoted as $(d_{ij})_{\text{sp}}$. Let us consider the decoded feature obtained from REGF as discussed in Sect. 4.3 is given as $\xi(\boldsymbol{I}_i)$, corresponding to an input image $\boldsymbol{I}_i$. In a similar manner, the feature similarity weight is calculated as.

$$(d_{ij})_{\text{ft}} = \frac{\xi(\boldsymbol{I}_i) \cdot \xi(\boldsymbol{I}_j)^T}{\|\xi(\boldsymbol{I}_i)\|_2 \cdot \|\xi(\boldsymbol{I}_j)\|_2} \tag{13}$$

Finally, both similarity weights are combined together to obtain the effective spatial-feature kernel as

$$(d_{ij})_{\text{tot}} = (d_{ij})_{\text{sp}} \times (d_{ij})_{ft} \tag{14}$$

$(d_{ij})_{sp}$ serves as the input similarity kernel to the 2DLPP algorithm the details of which can be found in [12, 19]. Following REGF-c-2DLPP, feature data is realized in a low-dimensional subspace, and subsequently, any suitable classifier is utilized to calculate the classification rates.

5 Results and Discussion

For experimental evaluations, the entire dataset considered here comprises human-supervised flag-stick images with various colors and shapes, recorded at different lighting conditions, as discussed in Sect. 3. The dataset was randomly split into train and test portions in a 70–30% ratio. The proposed REGF-c-2DLPP model was employed to transform the training and testing images into dimension-reduced subspaces, and a support vector machine (SVM) was utilized to classify it. The entire granulation process is demonstrated step-by-step in Fig. 4, for a sample flag-stick cue with lemon-lime color and arrow shape. Figure 5 demonstrates a comparative study of the REGF technique for flag-stick images of different colors. Table 1 presents the recognition rates for REGF-c-2DLPP and other state-of-the-art techniques for the given flag-stick cue dataset. It can be observed that the proposed REGF-c-2DLPP scheme outperforms other competing techniques consistently. The performance improvement is even more significant in the darker illumination conditions. Additionally, to investigate the performance of the proposed granular feature extraction technique with other competing techniques within the framework of c-2DLPP, a comparative study is presented in Table 2. Here also, the proposed granular fusion-based feature descriptor REGF satisfactorily enhances the classification rate of the cue recognition system, even at the darker levels.

6 Conclusion

In the real-world applications of visual cue detection systems, managing high-dimensional data is quite difficult, especially in adverse environments. This paper introduces a granular computing-aided feature description method that deals with uncertain, object-surrounding vagueness in the images at those conditions and helps extract intrinsic feature information efficiently. Moreover, to preserve the structural information of the data while dimensionality reduction, our proposed scheme of c-2DLPP is directly employed on 2D images. To address the uncertainty in spatial closeness of the data in the Euclidean space, the cosine similarity measure is introduced which can directly obtain the directional similarity measure between data points both in the spatial plane and feature plane. A feature extraction scheme based on granular computing is utilized in this work that deals with uncertain, object-surrounding vagueness in the images at those conditions and helps extract intrinsic

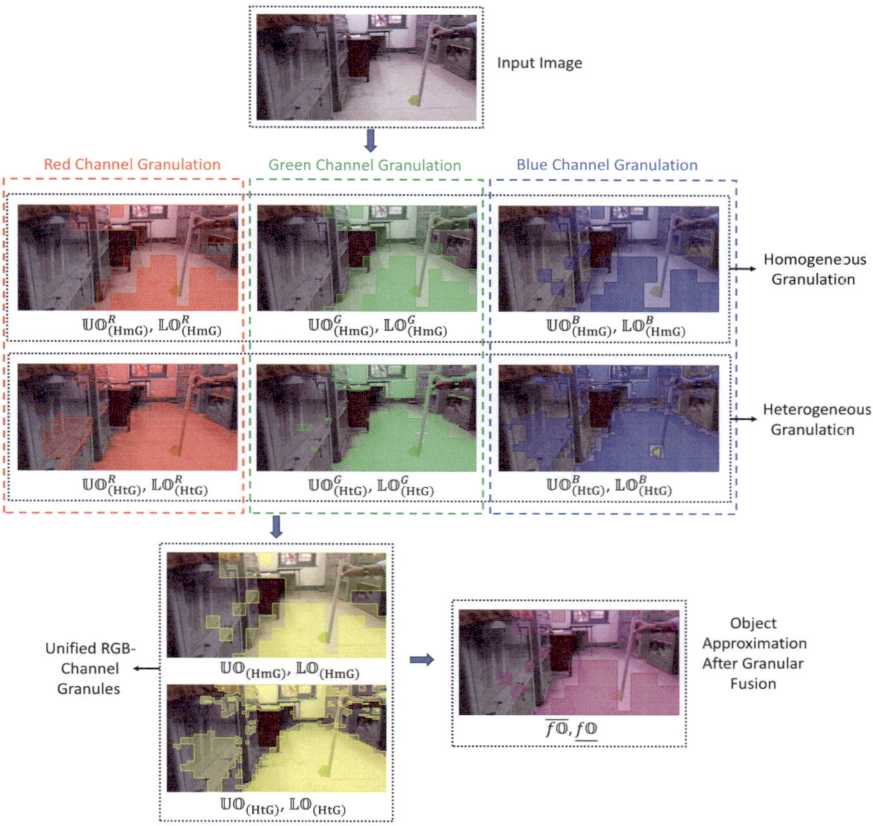

Fig. 4 Flow diagram of the REGF technique with a sample flag-stick visual cue. Lower and upper object approximations are respectively depicted in *darker* and *lighter* color tones

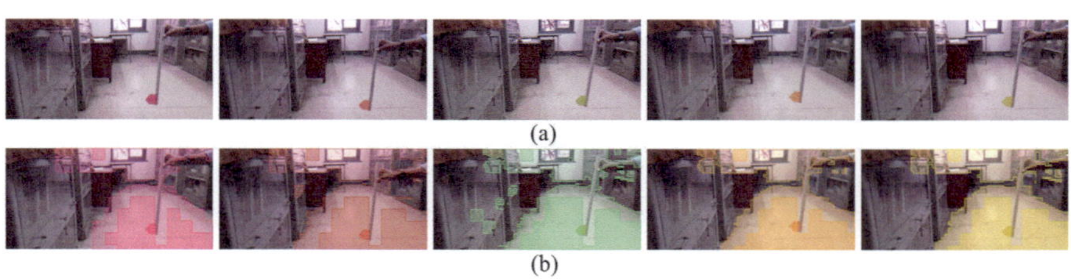

Fig. 5 Performance of the REGF technique for multi-color flag-stick images with **a** Original images. **b** Object approximation following REGF. Lower and upper object approximations are respectively depicted in *darker* and *lighter* color tones

Table 1 Performance comparison of various DR techniques for the multi-color, multi-shape flag-stick visual cue dataset under various illumination states

Algorithm	Illumination condition				
	NLC	DLC-1	DLC-2	DLC-3	DLC-4
PCA	96.24	67.91	42.45	33.18	26.67
LDA	96.92	78.41	69.24	58.10	43.35
LPP [26]	97.35	89.53	79.96	63.29	49.22
BLPP [17]	97.41	93.57	89.89	69.47	56.13
2DLPP [12]	97.58	94.90	91.71	76.88	65.17
c-2DLPP	97.75	95.27	92.80	78.63	67.39
CG-c-2DLPP	97.96	95.33	93.18	80.11	71.16
QtD-c-2DLPP	98.11	96.09	93.52	80.71	73.37
REGF-2DLPP [19]	98.05	96.00	93.80	80.92	74.73
REGF-c-2DLPP	98.20	96.48	94.74	81.19	75.62

Table 2 Performance comparison of proposed REGF and additional feature descriptors inside the structure of c-2DLPP with the multi-color, multi-shape flag-stick visual cue dataset under various illumination states

Descriptor	Illumination condition				
	NLC	DLC-1	DLC-2	DLC-3	DLC-4
HOG	97.15	94.81	91.07	76.56	67.29
Gabor	97.15	94.72	91.15	76.49	68.44
LBP	97.17	94.88	91.66	77.62	69.52
CG [18]	97.77	95.25	93.49	79.72	71.45
QT [22]	97.82	95.93	93.75	80.30	73.50
REGF [19]	98.20	96.48	94.74	81.19	75.62

feature information efficiently. Experimental evaluations carried out with multi-colored and multi-shaped visual patterns at different ambient illumination conditions substantiate the proposed algorithm to be an effective one in regular and adverse scenarios, as discussed.

References

1. Chen, X., Zhang, J., & Li, D. (2012). Direct discriminant locality preserving projection with Hammerstein polynomial expansion. *IEEE Transactions on Image Processing, 21*(12), 4858–4867.
2. Pang, Y., Zhang, L., Liu, Z., Yu, N., & Li, H. (2005). Neighborhood preserving projections (NPP): A novel linear dimension reduction method. In: *LNCS*, vol. 36(1), pp. 117–125. Springer, Heidelberg

3. Cai, D., He, X., & Han, J. (2007). Isometric projection. In: *Proceedings of Assoc Advancement of Artificial Intelligence* (pp. 528–533)
4. He, X., Cai, D., Yan, S., & Zhang, H. -J. (2005). Neighborhood preserving embedding. In *Proceedings of 10th IEEE International Conference on Computer Vision* (pp. 1208–1213). IEEE, New York
5. Belkin, M., & Niyogi, P. (2003). Laplacian eigenmaps for dimensionality reduction and data representation. *Neural Computation, 15*(6), 1373–1396.
6. Zhong, F., Zhang, J., & Li, D. (2014). Discriminant locality preserving projections based on L1-norm maximization. *IEEE Transactions on Neural Networks and Learning Systems, 25*(11), 2065–2074.
7. Zhu, J., Chen, J., Xu, B., Yang, H., & Nie, F. (2023). Fast orthogonal locality-preserving projections for unsupervised feature selection. *Neurocomputing, 531*, 100–113.
8. Su, T., Feng, D., Hu, H., Wang, M., & Chen, M. (2023). Locality preserving triplet discriminative projections for dimensionality reduction. *Neurocomputing, 520*, 284–300.
9. He, Y.-L., Li, K., Liang, L.-L., Xu, Y., & Zhu, Q.-X. (2023). Novel discriminant locality preserving projection integrated with Monte Carlo sampling for fault diagnosis. *IEEE Transactions on Reliability, 72*(1), 166–176.
10. Long, J., Zhang, Q., Lu, X., Wen, J., Zhao, L., & Xie, W. (2024). Multi-scale locality preserving projection for partial multi-view incomplete multi-label learning. *Neural Networks, 180*, Article 106748.
11. Ran, R., Feng, J., Li, Z., Wang, J., & Fang, B. (2024). Locality preserving projections with autoencoder. *Expert Systems With Applications, 242*, Article 122750.
12. Chen, S., Zhao, H., Kong, M., & Luo, B. (2007). 2D-LPP: A two-dimensional extension of locality preserving projections. *Neurocomputing, 70*(4–6), 912–921.
13. Yang, J., Zhang, D., Frangi, A. F., & Yang, J.-Y. (2004). Two-dimensional PCA: A new approach to appearance-based face representation and recognition. *IEEE Transactions on Pattern Analysis and Machine Intelligence, 26*(1), 131–137.
14. Li, M., & Yuan, B. (2005). 2D-LDA: A statistical linear discriminant analysis for image matrix. *Pattern Recognition Letters, 26*(5), 527–532.
15. Lu, Y., Lai, Z., Li, X., Wong, W. K., Yuan, C., & Zhang, D. (2019). Low-rank 2-D neighborhood preserving projection for enhanced robust image representation. *IEEE Transactions on Cybernetics, 49*(5), 1859–1872.
16. Lu, Y., Yuan, C., Lai, Z., Li, X., Zhang, D., & Wong, W. K. (2019). Horizontal and vertical nuclear norm-based 2DLDA for image representation. *IEEE Transactions on Circuits and Systems for Video Technology, 29*(4), 941–955.
17. Li, X., Pan, J., He, Y., & Liu, C. (2015). Bilateral filtering inspired locality preserving projections for hyperspectral images. *Neurocomputing, 164*, 300–306.
18. Pal, S. K., Shankar, B. U., & Mitra, P. (2005). Granular computing, rough entropy and object extraction. *Pattern Recognition Letters, 26*(16), 2509–2517.
19. Ghosh, S., Paral, P., Chatterjee, A., & Munshi, S. (2023). Rough entropy-based fused granular features in 2-D locality preserving projections for high-dimensional vision sensor data. *IEEE Sensors Journal, 23*(16), 18374–18383.
20. Zhang, N., Xu, Y., Zhu, Q.-X., & He, Y.-L. (2023). Improved locality preserving projections based on heat-kernel and cosine weights for fault classification in complex industrial processes. *IEEE Transactions on Reliability, 72*(1), 204–213.
21. Jia, D., Tang, X.-L., Zhu, H., Cao, J., & Song, W.-D. (2016). Colour FAST (CFAST) match: Fast affine template matching for colour images. *Electronics Letters, 52*(14), 1220–1221.
22. Chakraborty, D., Shankar, B. U., & Pal, S. K. (2013). Granulation, rough entropy and spatiotemporal moving object detection. *Applied Soft Computing, 13*(9), 4001–4009.
23. Ghosh, S., Chatterjee, A., Munshi, S., & Rakshit, A. (2022). Flag-based human supervised robot guidance using locality preserving projections. In *Proceedings of 2nd IEEE International Conference on Emerging Frontiers in Electrical and Electronic Technologies* (pp. 1–6)
24. Ghosh, S., Chatterjee, A., & Munshi, S. (2023). Visual cue-aided human supervised robot navigation guidance in photometrically challenging environments using adaptive spatial-feature kernel-guided bilateral LPP. *Measurement Science & Technology, 34*, Article 105404.

25. Ghosh, S., Paral, P., Chatterjee, A., & Munshi, S. (2023). Histogram refined local ternary pattern-based bilateral LPP for vision sensor-based robot navigation guidance under challenging environments. *IEEE Sensors Letter, 7*(6), 1–4.
26. He, X., & Niyogi, P. (2003). Locality preserving projections. In: *Proceedings of the 16th International Conference on Neural Information Processing Systems* (pp. 153–160)

Human-AI Collaboration in Health and Wellness: A Case of Telemonitoring and Telerehabilitation Platform for Disabled, Elderly, and Post-Stroke Care

Chutiporn Anutariya, Parkpoom Wisedsri, Chaklam Silpasuwanchai, Mongkol Ekpanyapong, Attaphongse Taparugssanagorn, S. D. A. P. Senadeera, Roongtiwa Vachalathiti, and Sunee Bovonsunthonchai

Abstract As the global population of elderly and disabled individuals continues to grow, innovative solutions are urgently needed to improve their safety, mobility, independence, and well-being, while reducing caregiving burdens and enhancing healthcare outcomes. This paper presents a telemonitoring and telerehabilitation platform that demonstrates the transformative potential of Human-AI collaboration to address

C. Anutariya (✉) · P. Wisedsri · C. Silpasuwanchai · M. Ekpanyapong · A. Taparugssanagorn · S. D. A. P. Senadeera
Department of Information and Communications Technologies, School of Engineering and Technology, Asian Institute of Technology, Pathum Thani, Thailand
e-mail: chutiporn@ait.ac.th

P. Wisedsri
e-mail: parkpoom.wisedsri@ait.ac.th

C. Silpasuwanchai
e-mail: chaklam@ait.ac.th

M. Ekpanyapong
e-mail: mongkol@ait.ac.th

A. Taparugssanagorn
e-mail: attaphongset@ait.ac.th

S. D. A. P. Senadeera
e-mail: arunya@ait.ac.th

R. Vachalathiti · S. Bovonsunthonchai
Faculty of Physical Therapy, Mahidol University, Nakhon Pathom, Thailand
e-mail: roongtiwa.vac@mahidol.ac.th

S. Bovonsunthonchai
e-mail: sunee.bov@mahidol.ac.th

J. K. Mandal et al. (eds.), *Recent Advances in Artificial Intelligence and Smart Applications*, Lecture Notes in Networks and Systems 1380,
https://doi.org/10.1007/978-981-96-5822-0_38

these critical needs. The paper focuses on the platform's architectural and technological design to tackle the unique challenges faced by these vulnerable groups, emphasizing AI's role in decision support, personalized care, and rehabilitation. The platform comprises four key components: a wearable Raman Spectroscopy system for non-invasive blood glucose monitoring, fall detection systems using video analytics, deep learning, and radio frequency signals for rapid caregiver alerts, bimanual rehabilitation technologies to aid physical therapy and improve recovery, and a cloud-based system for managing and analyzing health and rehabilitation data to provide valuable insights and facilitate personalized care.

Keywords Human-AI collaboration · Health informatics · Telehealth · Telerehabilitation · Assistive platform

1 Introduction

Advancements in Artificial Intelligence (AI), along with computing and digital technologies, have transformed and revolutionized numerous application domains and industries. AI-based systems are capable of conducting quantitative analyses on large data sets, identifying patterns and trends, making predictions, and supporting human decision-making. Consequently, many industries and businesses have adopted AI not as a replacement for human workers but as a tool to augment human capabilities and extend operational capacity. This emerging trend of human-AI collaboration opens new possibilities and applications, particularly in healthcare—a field facing unique and pressing challenges [3]. With an aging population, a global shortage of healthcare professionals, and the rapid rising healthcare costs, healthcare is at the forefront of sectors seeking AI-driven solutions. By integrating AI, the industry can address widening socioeconomic health inequalities, alleviate healthcare professionals' workload, improve service accessibility and quality, enhance diagnostic precision, and support preventive care initiatives, thereby improving overall health outcomes. For example, studies show that Clinical Decision Support Systems (CDSS) can significantly boost healthcare workers' efficiency and effectiveness, demonstrating the potential of AI to optimize healthcare delivery [3, 5].

Many countries are experiencing a growing proportion of older adults. Population statistics and projections reveal that by 2050, individuals aged 60 years or older are expected to account for over 25 percent of the global population [8]. This trend presents significant challenges, particularly in developing countries, where healthcare systems are often under-resourced, struggling to meet the increasing demand for elder care. Additionally, many older adults are left alone at home during the day while family members are at work, with limited access to caretaker support. Similarly, individuals with disabilities face challenges associated with mobility and daily care, making them vulnerable to fall-related injuries. Statistics indicate that more than one in four older adults experiences a fall each year, with these incidents often leading to severe health complications and, in some cases, death [1]. In addition to falls, stroke poses another significant health risk to older adults. According to 2021 statistics from the World Health Organization (WHO), stroke ranks as the third leading cause of death and disability globally, following ischemic heart disease and COVID-19 [14]. Due to a critical shortage of rehabilitation specialists, 87% of stroke-related disabilities occur in low- and middle-income countries [2].

Recognizing these aforementioned challenges faced by elderly and disabled populations and the urgent need to improve their safety, mobility, independence, and well-being—while also reducing caregiving burdens and enhancing healthcare outcomes—this paper proposes a human-AI collaborative approach to developing a telemonitoring and telerehabilitation platform aimed at meeting these critical needs. The platform consists of four key components: a wearable Raman Spectroscopy system for non-invasive blood glucose monitoring, fall detection systems using video analytics, deep learning, and radio frequency signals for rapid caregiver alerts, bimanual rehabilitation technologies to support physical therapy and recovery, and a cloud-based system for managing and analyzing health and re-habilitation data to enable valuable insights and facilitate personalized care. It incorporates advanced AI and digital technologies, including machine and deep learning, video and CSI Wi-Fi signal analytics, communication technologies, IoT, and robotics.

The developed platform represents a significant advancement in human-AI collaborative solutions to address practical challenges in healthcare and wellness. By enhancing telehealth and assistive technologies, it demonstrates the potential to improve the quality of life for elderly and disabled individuals. The platform automates health monitoring, enables remote consultations, and helps alleviate caregiver stress. Additionally, its data analytics and visualization tools provide healthcare providers with valuable insights into health patterns and needs, supporting improved patient care and decision-making. These capabilities allow for more personalized treatment plans based on real-time data and continuous monitoring.

Section 2 provides a review of human-AI collaboration in health and wellness, focusing on emerging technologies and relevant use cases. Section 3 presents the proposed platform and its architectural design, while Sect. 4 concludes with insights on future research directions.

2 Towards Human-AI Collaboration in Health and Wellness

Unlike traditional AI applications, which primarily focus on automating routine tasks, human-AI collaboration emphasizes AI's role to assist, complement and augment human capabilities, rather than replacing them [10]. This approach, known as collaborative intelligence, involves an interactive and seamless partnership between human experts and AI systems, working together to achieve common goals such as task completion, problem-solving, and decision-making. The health and wellness sector provides a critical context for human-AI collaboration, due to several significant challenges, including socioeconomic health inequalities, the growing populations of elderly and disabled individuals, a severe shortage of healthcare professionals and caregivers, and the global trends of emerging and re-emerging infectious diseases.

2.1 Types of Collaboration, Use Cases and Supporting AI Techniques

In the health and wellness domain, key stakeholders include healthcare professionals, patients, vulnerable groups, caregivers, and clinical researchers. Figure 1 categorizes types of human-AI collaboration, relevant use cases, and supporting AI technologies within this domain, considering how various stakeholders and AI systems can interact effectively. Five types of collaborations are identified. The first type involves collaboration between healthcare professionals and AI systems, with use cases such as medical image analysis, clinical decision support, disease diagnosis and treatment recommendations, and robot-assisted surgery. The second type is patient-AI collaboration, which includes applications like robotic patient care and conversational AI for medical support and services. Vulnerable population-AI collaboration also addresses critical needs, with applications such as assistive technology tailored for vulnerable groups, including young children, the elderly, and individuals with disabilities. In caregiver-AI collaboration, key use cases include AI technologies that assist caregivers in tracking, monitoring, and providing real-time support and rescue services. The final collaboration type is between clinical researchers and AI systems, where AI contributes to clinical research support by processing vast volumes of medical data and laboratory test results.

2.2 Opportunities and Challenges

Several review papers [3, 4, 6, 9] highlight the rapidly expanding research on human-AI collaboration in healthcare, which addresses a range of diseases and improves medical services through applications such as diagnostic and treatment support, robot-assisted surgeries, and assistive robots for nursing and patient care. Additionally, areas such as disease prediction, prevention, and prognosis present significant opportunities for human-AI collaboration. Despite its promising potential in health and wellness, fully realizing the benefits of human-AI collaboration remains challenging due to issues like bias, explainability, trust, system performance, accountability, and low user adoption influenced by behavioral and social factors [3, 4]. These challenges highlight the need for further research and policy development to effectively address and mitigate these barriers.

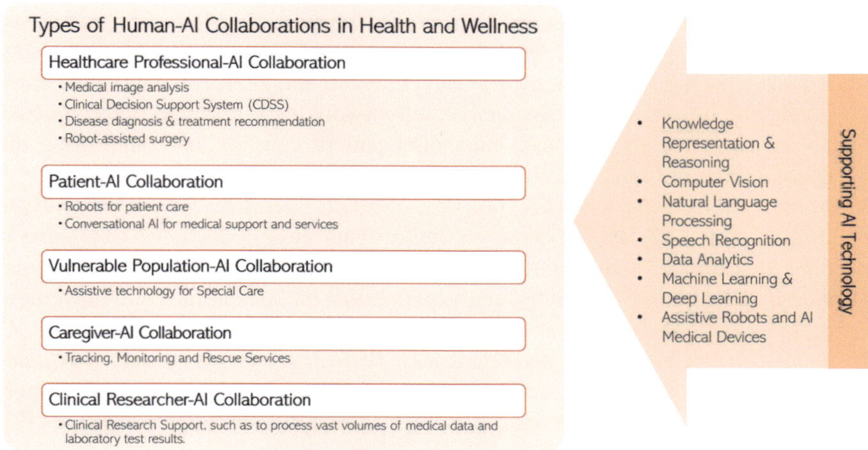

Fig. 1 Types of Human-AI collaborations in health and wellness, use cases and supporting AI techniques

3 The Proposed Platform

This section introduces the proposed telehealth platform, specifically designed to provide health and wellness monitoring, alerting, and assistive support for elderly, disabled, and post-stroke individuals to enhance their safety and well-being. The platform is both integrated and comprehensive, consisting of four primary systems, as illustrated in Fig. 2 and summarized next. Figure 2 also highlights the ICT and AI technologies that support the platform's functionality.

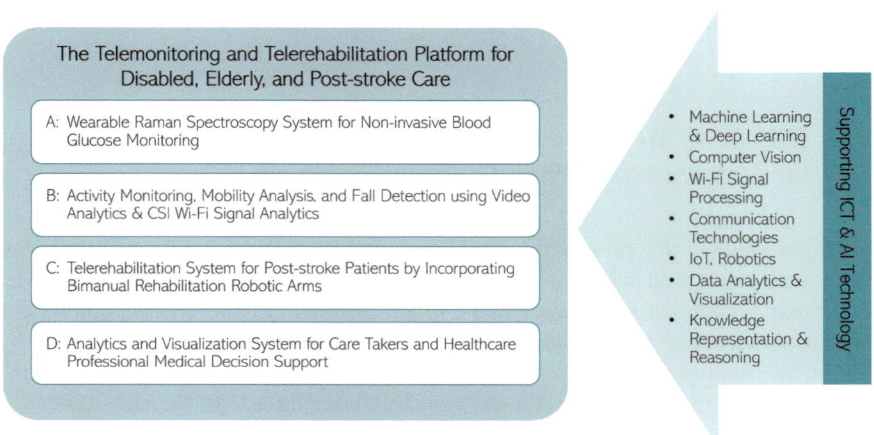

Fig. 2 The four system components of the proposed platform and its supporting technologies

A. *Wearable Raman Spectroscopy System for Non-invasive Blood Glucose Monitoring:* This system enables non-invasive blood glucose monitoring using Raman Spectroscopy technology. The collected and analyzed data allow caregivers and healthcare professionals to effectively track patients' blood glucose levels, reducing invasiveness, enhancing patient comfort, and supporting the development of improved treatment plans.
B. *Activity Monitoring, Mobility Analysis, and Fall Detection System using Video Analytics & CSI Wi-Fi Signal Analytics*: This system includes two subsystems that utilize distinct technologies—video analytics and CSI Wi-Fi signal analytics—to monitor activity, analyze mobility, and detect falls and movement incidents among disabled and elderly individuals in household settings.
C. *Telerehabilitation System for Post-stroke Patients by Incorporating Bimanual Rehabilitation Robotic Arms*: This system provides robotic assistance to support post-stroke patients in motor skill recovery through tailored arm exercises [7]. The bimanual rehabilitation robotic arms offer controlled resistance and targeted support to the elbow joints of the affected arm, promoting correct movement and gradual improvement in range of motion and strength. Real-time monitoring and patient-specific data analysis enable physical therapists to gain a deeper understanding of each patient's progress, facilitating adjustments to rehabilitation plans that can enhance recovery outcomes.
D. *Analytics and Visualization System for Care Takers and Healthcare Professional Medical Decision Support*: This system offers an intuitive dashboard that displays historical and analytical data, along with emergency alerts, to caregivers, healthcare professionals, and rescue teams. By providing real-time insights and comprehensive data visualization, the system enables improved care and treatment planning, enhancing quality of life and safety for individuals in need of assistance.

Figure 3 illustrates the architectural and technical design of the proposed platform. The following subsections describe the platform's integrated systems in further detail.

3.1 System A: Wearable Raman Spectroscopy System for Non-invasive Blood Glucose Monitoring

The importance of this proposed system lies in its potential to significantly improve the management of diabetes—a condition that affects millions of elderly worldwide. Traditional glucose monitoring methods, which require frequent blood samples, pose challenges including pain, potential infection, and inconvenience. As depicted by Fig. 3a, the proposed non-invasive approach operates by directing a laser to illuminate a small area of the skin. The interaction of this laser light with the skin generates Raman signals, which are specific to the chemical composition of glucose in the blood. These signals are captured by the Raman Spectroscopy sensor and processed through advanced algorithms to estimate glucose levels. The processed data is then

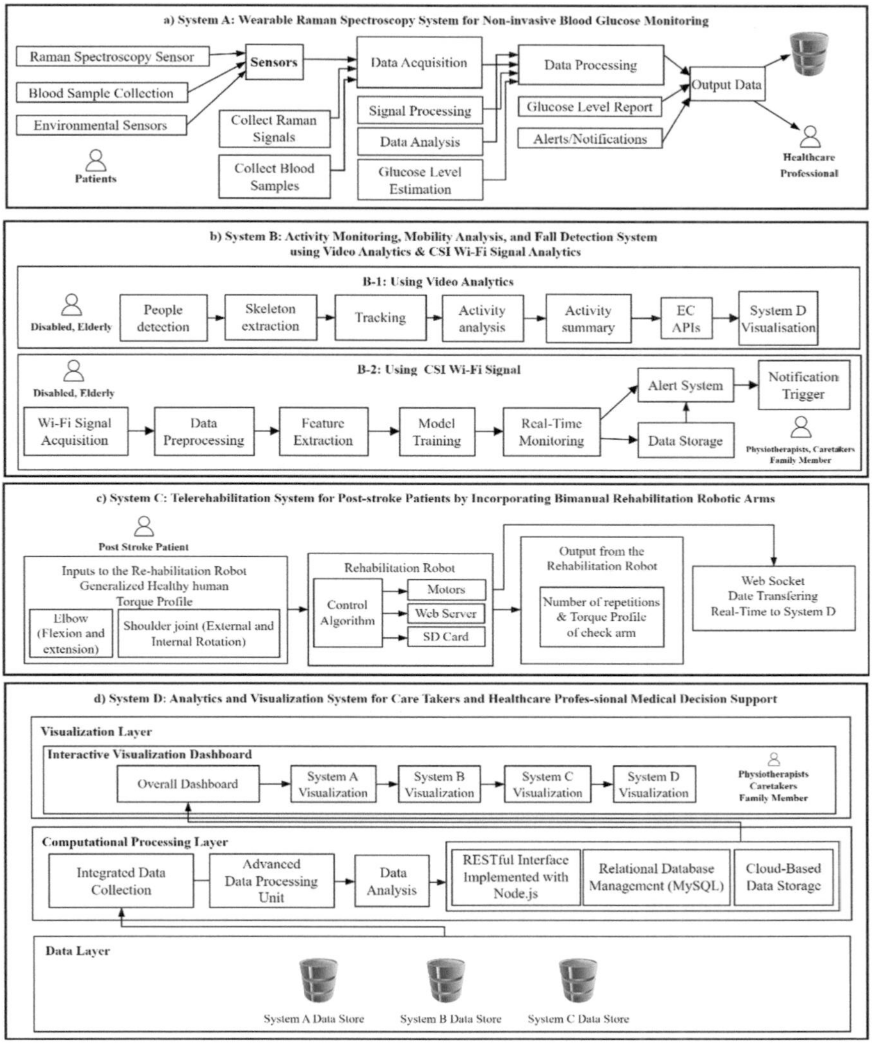

Fig. 3 The architectural and technical design of the proposed platform

compiled into reports and can trigger alerts if abnormal glucose levels are detected. This data is essential for both patients and healthcare providers, enabling real-time monitoring and immediate intervention if necessary. This method not only enhances the accuracy of glucose monitoring but also significantly improves the ease and comfort of managing diabetes.

3.2 System B: Activity Monitoring, Mobility Analysis, and Fall Detection Using Video Analytics and CSI Wi-Fi Signal Analytics

This system includes two subsystems B-1 and B-2 that utilize video analytics and CSI Wi-Fi signal analytics, respectively, to monitor activity, analyze mobility, and detect falls and movement incidents among disabled and elderly individuals.

As depicted in Fig. 3b, the video analytics-based system (B-1) consists of three main components. The first one, a standalone application, handles all tasks locally, including people detection, skeleton extraction, tracking, activity analysis, and summary generation, making it effective for environments with adequate local resources. The second component, edge and cloud, performs initial people detection and tracking on edge devices, while more complex processes like skeleton extraction and analysis are handled in the cloud, reducing latency and leveraging cloud scalability. The third part, thin AI-camera and cloud, uses AI-powered cameras for on-device detection and event triggering, sending data to the cloud for further processing. This minimizes on-device workload while enabling scalable cloud operations. All three components generate activity summaries integrated into System D Visualization via EC APIs, providing centralized insights. These designs optimize computational efficiency, scalability, and latency for healthcare monitoring needs.

As shown in Fig. 3b, the video analytics-based system (B-1) monitors elderly and disabled individuals by analyzing their activities through structured stages. The system imitates people's detection, identifying human subjects in the video. Following that, skeleton extraction maps key body points, creating a skeletal representation for tracking posture and movement. In the tracking stage, these points are monitored across frames, allowing the system to observe changes in posture or location over time. The activity analysis stage then classifies specific activities, such as walking, sitting, or potential falls. An activity summary is generated, providing an overview for caregivers. This summary is transmitted via EC APIs to external systems for integration with other healthcare platforms. Finally, the data is visualized in System D, a dashboard where authorized caregivers and healthcare professionals can see reports, monitor real-time updates, and make informed decisions. This process enables proactive monitoring, supporting healthcare providers in assessing mobility and well-being. Importantly, privacy and personal data protection are carefully addressed throughout the process.

On the other hand, the CSI Wi-Fi based system utilizes existing Wi-Fi infrastructure to monitor the movements of individuals in an environment. Channel State Information (CSI) is collected from Wi-Fi signals as they travel through space. Changes in the Wi-Fi signals are indicative of the movements and activities of people, providing valuable data for fall detection. After Wi-Fi signals are collected, the data is preprocessed to extract meaningful features related to human movement. Machine learning algorithms are then employed to classify these features and detect fall incidents. Techniques such as deep learning and anomaly detection are used to identify irregular patterns in the Wi-Fi signals that may indicate a fall. Several machine learning models are trained on a variety of normal and fall scenarios to ensure accurate detection in real-world environments. Since this approach does not require users to wear any devices or have cameras installed, privacy preservation is high, and hence can be utilized in areas that privacy preservation should be highly maintained.

When a potential fall is detected by either the video analytics or CSI Wi-Fi-based system, notifications are sent to relevant caregivers and emergency responders. Additionally, the systems maintain data on fall incidents, daily activities, and mobility patterns in a secure data repository, allowing for review and analysis by stakeholders such as caregivers, healthcare professionals, and emergency teams. This ongoing data collection enables continuous enhancements in fall prevention and safety measures as well as activity recommendations to support the well-being of the monitored individuals.

3.3 System C: Telerehabilitation System for Post-Stroke Patients by Incorporating Bimanual Rehabilitation Robotic Arms

Figure 3c illustrates the developed bimanual rehabilitation system [7], designed to support post-stroke patients in motor skill recovery through customized arm exercises with robotic assistance. This system uses a master-follower robotic setup to aid the paretic arm in regaining functionality by guiding it with movements from the healthy arm, enhancing the rehabilitation process for two primary exercises: flexion & extension of elbow joint and internal & external rotation of shoulder joint. Both arms move simultaneously under bi-manual motor control, effectively simulating therapist-assisted movements. The master robotic arm supports the healthy arm, while the follower robotic arm aids the affected arm. Thus, the robot ensures position synchronization during rehabilitation exercises. Accurate force sensing is achieved by using a sensorless method called the Reaction Torque OBserver (RTOB) [7]. Each rehabilitation session is recorded, enabling further processing, analysis, and visualization of patient-specific exercises. This data allows physical therapists to monitor progress, adjust rehabilitation plans as needed, and optimize the overall benefits of the rehabilitation process.

3.4 System D: Analytics and Visualization System for Care Takers and Healthcare Professional Medical Decision Support

Effective monitoring and communication are essential in healthcare, particularly for patients undergoing rehabilitation or managing chronic conditions. Traditional methods, often reliant on manual data tracking and in-person consultations, can be time-consuming and may delay decision-making. As depicted by Fig. 3d, this analytics and visualization system addresses these challenges by aggregating data from other systems within the platform, including blood sugar levels, fall and activity patterns, and bimanual rehabilitation exercises, and presenting it in an easily interpretable format with charts, graphs, and visual indicators. These visual representations allow caregivers and healthcare professionals to assess individual progress, identify trends, and make informed decisions quickly. The dashboard consolidates all relevant data at a glance, supporting proactive care and efficient monitoring.

4 Prototype System Development and Demonstration

This section provides an overview of the developed prototype system of the proposed platform (System D) [11, 12], highlighting key visualizations and data representations used across various systems within the platform.

Figure 4 presents a few key *visualizations for blood glucose monitoring* (System A), displaying a diabetes patient profile and history of blood glucose measurements. The interface includes timestamps for blood glucose levels alongside meal logs, enabling patients and caregivers to observe how meals impact glucose fluctuations. A chart differentiates various glucose ranges—high, low, healthy, and unhealthy—making it straightforward to assess overall health trends.

Figure 5 demonstrates key *visualizations for activity mobility and fall analysis* (System B). Activity History (Fig. 5a) provides detailed records of activities by type location and time enabling long-term pattern analysis. The Daily Activity Timeline (Fig. 5b) offers a visual overview of activities throughout the day with filters for specific areas/camera and date range allowing customized monitoring. Fall Incident Records (Fig. 5c) focus on fall events with information on activity type location and time along with a 'View Fall' option for further investigation. The Fall Timeline (Fig. 5d) shows events before and after a fall to provide context. Together these

Fig. 4 Visualizations for system A: Blood glucose measurement history

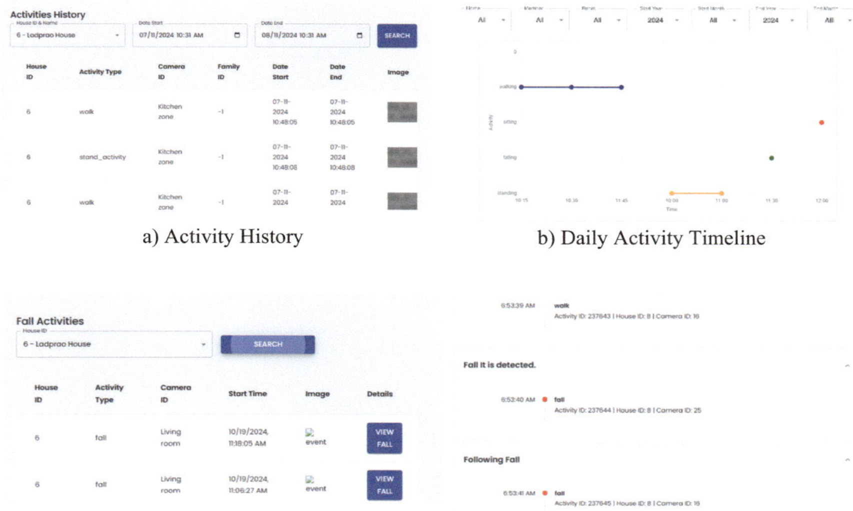

a) Activity History

b) Daily Activity Timeline

c) Fall Incident Records

d) Fall Timeline

Fig. 5 Visualizations for system B: History of activities, fall events, and detailed fall information with nearby activities

visualizations give healthcare professionals a structured overview to monitor analyze and ensure safety of elderly and disabled individuals effectively.

Figure 6 presents selected *visualizations for bimanual rehabilitation exercises and progresses* (System C). Figure 6a displays a rehabilitation session overview for a physical therapist, showing primary and secondary elbow movements during flexion, extension, and assisted exercises. Figure 6b summarizes sessions with metrics like weekly goals, progress percentages, and session details, helping patients and physiotherapists for progress tacking. Figure 6c details individual exercise cycles, highlighting flexion and extension movements with adjustments for assistance and angular motion to improve technique. Figure 6d uses a sunburst chart to illustrate flexion, and extension ranges of primary and secondary elbows, showing motion patterns and coordination. These visualizations demonstrate how the prototype system enhances monitoring, decision-making, and rehabilitation through comprehensive data visualization, supporting healthcare providers and caregivers in delivering informed and effective care.

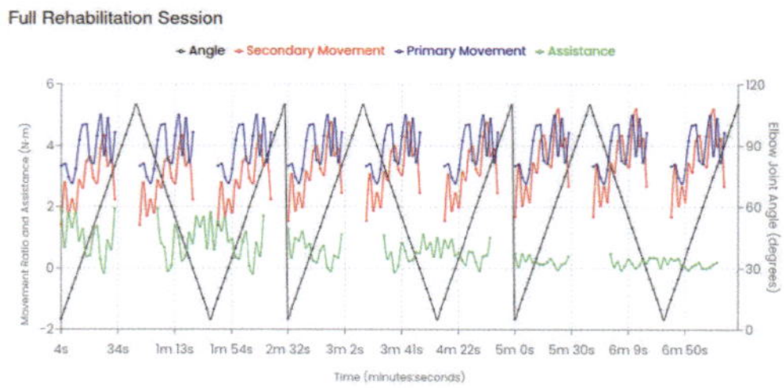

a) Overall bimanual rehabilitation recording

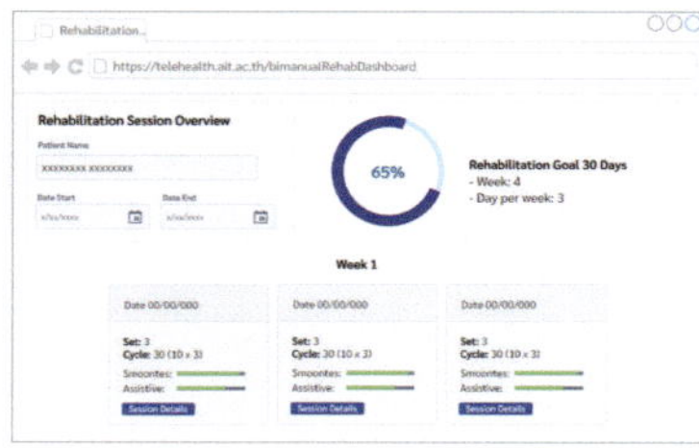

b) Bimanual rehabilitation session summary

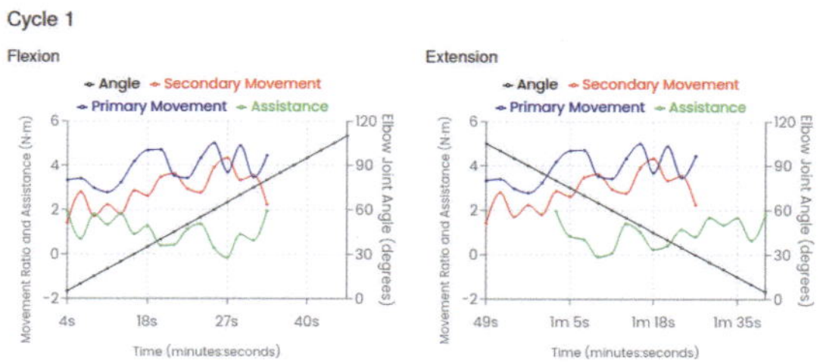

c) Analysis of flexions and extensions of each exercise cycle

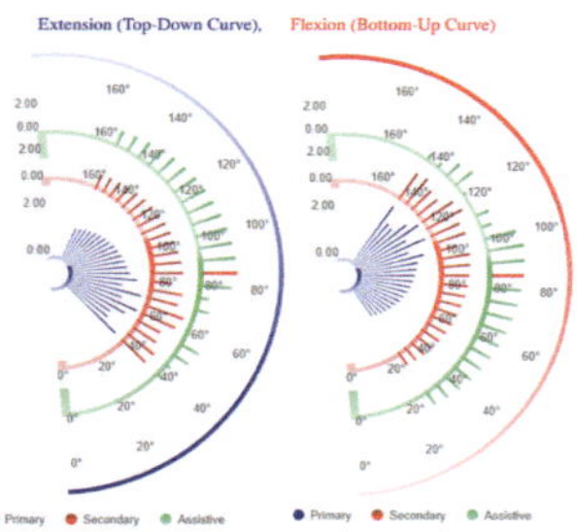

d) Sunburst charts for flexion and extension of primary and secondary elbow movement

Fig. 6 Visualizations for system C: Bimanual rehabilitation exercises and progresses

5 Conclusion

This paper has presented a comprehensive human-AI collaboration platform designed to address key challenges in health and wellness domains, supporting diverse stakeholders such as healthcare professionals, patients, vulnerable groups, caregivers, and clinical researchers. The platform integrates several advanced subsystems, including a wearable Raman Spectroscopy system for non-invasive blood glucose monitoring, video and CSI Wi-Fi analytics for activity and fall monitoring, a bimanual rehabilitation system, and an analytics and visualization dashboard. These systems enhance patient safety, improve decision-making, and support personalized care through real-time monitoring and actionable data insights while maintaining a high standard of privacy.

The proposed platform underscores the potential of AI to augment human capabilities, providing innovative solutions that address resource constraints and increase the accessibility of quality care. In future work, we aim to evaluate the outcomes of human-AI collaboration within the platform, focusing on usability, user satisfaction, and overall effectiveness in real-world applications. These evaluations will be essential for refining the platform to ensure it meets the practical needs of stakeholders while optimizing the quality and impact of AI-supported health and wellness services.

Acknowledgements This work is supported by the Broadcasting and Telecommunications Research and Development Fund for Public Interest (BTFP), Office of The National Broadcasting and Telecommunications Commission (NBTC) under the grant number A64-1-(2)-006.

References

1. Giovannini, S., Brau, F., Galluzzo, V., et al. (2022). Falls among older adults: Screening, identification, rehabilitation, and management. *Applied Sciences, 12*(15), 7934.
2. Kayola, G., Mataa, M. M., Asukile, M., et al. (2023). Stroke rehabilitation in low- and middle-income countries-challenges and opportunities. *American Journal Physical Medicine & Rehabilitation, 102*(2), 24–32.
3. Lai, Y., Kankanhalli, A., & Ong, D. (2021). Human-AI collaboration in healthcare: A review and research agenda. In: *Hawaii International Conference on System Sciences.* ACM
4. Lai, Y., Sukhwal, P. C., Kankanhalli, A. (2020). Explanations in clinical DSS: A review and research agenda. In: *PACIS 2020 Proceedings*
5. Lee, M. H., Siewiorek, D. P., & Smailagic, A. (2021). A human-ai collaborative approach for clinical decision making on rehabilitation assessment. In: *2021 CHI Conference on Human Factors in Computing Systems.* ACM
6. Pacis, D. M. M., Subido Jr, E. D., & Bugtai, N. T. (2018) Trends in telemedicine utilizing artificial intelligence. *AIP Conference Proceedings* (1933:1), p. 040009
7. Perera, K. L. M., Hettiarachchi, T. I., & Kumara, J. A. S. S, et al. (2024). Development of sensorless based bimanual tele-rehabilitation robot for hemiparetic post-stroke patients. In: *2024 Moratuwa Engineering Research Conference (MERCon)*, Moratuwa, Sri Lanka (pp. 548–553). IEEE
8. United Nations. (2024). Department of Economic and Social Affairs, Population Division. World Population Prospects 2024, Data Sources. UN DESA/POP/2024/DC/NO. 11
9. Vedula, S. S., & Hager, G. D. (2017). Surgical data science: The new knowledge domain. *Innovative Surgical Sciences, 2*(3), 109–121.
10. Wilson, H. J., & Daugherty, P. R. (2018). Collaborative intelligence: Humans and AI are joining forces. *Harvard Business Review, 96*(4), 114–123.
11. Wisedsri, P., & Anutariya, C. (2023). Data analytics and visualisation system for fall detection for elderly and disabled people. In: *2023 IEEE Region 10 Conference* (TENCON 2023) (pp. 1315–1320). IEEE
12. Wisedsri, P., Anutariya, C., Sujarae, A., Vachalathiti, R., & Bovonsunthonchai, S. (2024). Data analytics and visualization in bimanual rehabilitation monitoring systems: a user-centered design approach to healthcare professionals decision support. In: *5th International Conference on Electrical Engineering and Informatics* (ICELTICs 2024)
13. World Health Organization. (2007). WHO global report on falls prevention in older age. WHO Press
14. World Health Organization. (2024). The top 10 causes of death. https://www.who.int/news-room/fact-sheets/detail/the-top-10-causes-of-death. Accessed 1 Nov 2024
15. Zereen, A. N., Gurung, A., Rajak, A., et al. (2023). Video analytic system for activity profiling, fall detection, and unstable motion detection. *Multimedia Tools and Applications, 82*, 42395–42415.

MIX
Papier aus verantwortungsvollen Quellen
Paper from responsible sources
FSC® C105338

If you have any concerns about our products,
you can contact us on
ProductSafety@springernature.com

In case Publisher is established outside the EU,
the EU authorized representative is:
Springer Nature Customer Service Center GmbH
Europaplatz 3, 69115 Heidelberg, Germany

Printed by Libri Plureos GmbH
in Hamburg, Germany

MIX
Papier aus verantwortungsvollen Quellen
Paper from responsible sources
FSC® C105338

If you have any concerns about our products,
you can contact us on
ProductSafety@springernature.com

In case Publisher is established outside the EU,
the EU authorized representative is:
Springer Nature Customer Service Center GmbH
Europaplatz 3, 69115 Heidelberg, Germany

Printed by Libri Plureos GmbH
in Hamburg, Germany